Learning OpenCV

Gary Bradski and Adrian Kaehler

Beijing · Cambridge · Farnham · Köln · Sebastopol · Taipei · Tokyo

Learning OpenCV

by Gary Bradski and Adrian Kaehler

Published by O'Reilly Media, Inc., 1005 Gravenstein Highway North, Sebastopol, CA 95472.

O'Reilly books may be purchased for educational, business, or sales promotional use. Online editions are also available for most titles (*safari.oreilly.com*). For more information, contact our corporate/institutional sales department: (800) 998-9938 or *corporate@oreilly.com*.

Editor: Mike Loukides

Production Editor: Rachel Monaghan

Production Services: Newgen Publishing and Data Services

Cover Designer: Karen Montgomery

Interior Designer: David Futato

Illustrator: Robert Romano

Printing History:

September 2008:   First Edition.

 This book uses RepKover™, a durable and flexible lay-flat binding.

ISBN: 978-0-596-51613-0

[M]

Contents

Preface

This book provides a working guide to the Open Source Computer Vision Library (OpenCV) and also provides a general background to the field of computer vision sufficient to use OpenCV effectively.

Purpose

Computer vision is a rapidly growing field, partly as a result of both cheaper and more capable cameras, partly because of affordable processing power, and partly because vision algorithms are starting to mature. OpenCV itself has played a role in the growth of computer vision by enabling thousands of people to do more productive work in vision. With its focus on real-time vision, OpenCV helps students and professionals efficiently implement projects and jump-start research by providing them with a computer vision and machine learning infrastructure that was previously available only in a few mature research labs. The purpose of this text is to:

- Better document OpenCV—detail what function calling conventions really mean and how to use them correctly.

- Rapidly give the reader an intuitive understanding of how the vision algorithms work.

- Give the reader some sense of what algorithm to use and when to use it.

- Give the reader a boost in implementing computer vision and machine learning algorithms by providing many working coded examples to start from.

- Provide intuitions about how to fix some of the more advanced routines when something goes wrong.

Simply put, this is the text the authors wished we had in school and the coding reference book we wished we had at work.

This book documents a tool kit, OpenCV, that allows the reader to do interesting and fun things rapidly in computer vision. It gives an intuitive understanding as to how the algorithms work, which serves to guide the reader in designing and debugging vision applications and also to make the formal descriptions of computer vision and machine learning algorithms in other texts easier to comprehend and remember.

After all, it is easier to understand complex algorithms and their associated math when you start with an intuitive grasp of how those algorithms work.

Who This Book Is For

This book contains descriptions, working coded examples, and explanations of the computer vision tools contained in the OpenCV library. As such, it should be helpful to many different kinds of users.

Professionals
> For those practicing professionals who need to rapidly implement computer vision systems, the sample code provides a quick framework with which to start. Our descriptions of the intuitions behind the algorithms can quickly teach or remind the reader how they work.

Students
> As we said, this is the text we wish had back in school. The intuitive explanations, detailed documentation, and sample code will allow you to boot up faster in computer vision, work on more interesting class projects, and ultimately contribute new research to the field.

Teachers
> Computer vision is a fast-moving field. We've found it effective to have the students rapidly cover an accessible text while the instructor fills in formal exposition where needed and supplements with current papers or guest lecturers from experts. The students can meanwhile start class projects earlier and attempt more ambitious tasks.

Hobbyists
> Computer vision is fun, here's how to hack it.

We have a strong focus on giving readers enough intuition, documentation, and working code to enable rapid implementation of real-time vision applications.

What This Book Is Not

This book is not a formal text. We do go into mathematical detail at various points,* but it is all in the service of developing deeper intuitions behind the algorithms or to make clear the implications of any assumptions built into those algorithms. We have not attempted a formal mathematical exposition here and might even incur some wrath along the way from those who do write formal expositions.

This book is not for theoreticians because it has more of an "applied" nature. The book will certainly be of general help, but is not aimed at any of the specialized niches in computer vision (e.g., medical imaging or remote sensing analysis).

---

* Always with a warning to more casual users that they may skip such sections.

That said, it is the belief of the authors that having read the explanations here first, a student will not only learn the theory better but remember it longer. Therefore, this book would make a good adjunct text to a theoretical course and would be a great text for an introductory or project-centric course.

About the Programs in This Book

All the program examples in this book are based on OpenCV version 2.0. The code should definitely work under Linux or Windows and probably under OS-X, too. Source code for the examples in the book can be fetched from this book's website (*http://www.oreilly.com/catalog/9780596516130*). OpenCV can be loaded from its source forge site (*http://sourceforge.net/projects/opencvlibrary*).

OpenCV is under ongoing development, with official releases occurring once or twice a year. As a rule of thumb, you should obtain your code updates from the source forge CVS server (*http://sourceforge.net/cvs/?group_id=22870*).

Prerequisites

For the most part, readers need only know how to program in C and perhaps some C++. Many of the math sections are optional and are labeled as such. The mathematics involves simple algebra and basic matrix algebra, and it assumes some familiarity with solution methods to least-squares optimization problems as well as some basic knowledge of Gaussian distributions, Bayes' law, and derivatives of simple functions.

The math is in support of developing intuition for the algorithms. The reader may skip the *math* and the algorithm descriptions, using only the function definitions and code examples to get vision applications up and running.

How This Book Is Best Used

This text need not be read in order. It can serve as a kind of user manual: look up the function when you need it; read the function's description if you want the gist of how it works "under the hood". The intent of this book is more tutorial, however. It gives you a basic understanding of computer vision along with details of how and when to use selected algorithms.

This book was written to allow its use as an adjunct or as a primary textbook for an undergraduate or graduate course in computer vision. The basic strategy with this method is for students to read the book for a rapid overview and then supplement that reading with more formal sections in other textbooks and with papers in the field. There are exercises at the end of each chapter to help test the student's knowledge and to develop further intuitions.

You could approach this text in any of the following ways.

Grab Bag

Go through Chapters 1–3 in the first sitting, then just hit the appropriate chapters or sections as you need them. This book does not have to be read in sequence, except for Chapters 11 and 12 (Calibration and Stereo).

Good Progress

Read just two chapters a week until you've covered Chapters 1–12 in six weeks (Chapter 13 is a special case, as discussed shortly). Start on projects and start in detail on selected areas in the field, using additional texts and papers as appropriate.

The Sprint

Just cruise through the book as fast as your comprehension allows, covering Chapters 1–12. Then get started on projects and go into detail on selected areas in the field using additional texts and papers. This is probably the choice for professionals, but it might also suit a more advanced computer vision course.

Chapter 13 is a long chapter that gives a general background to machine learning in addition to details behind the machine learning algorithms implemented in OpenCV and how to use them. Of course, machine learning is integral to object recognition and a big part of computer vision, but it's a field worthy of its own book. Professionals should find this text a suitable launching point for further explorations of the literature—or for just getting down to business with the code in that part of the library. This chapter should probably be considered optional for a typical computer vision class.

This is how the authors like to teach computer vision: Sprint through the course content at a level where the students get the gist of how things work; then get students started on meaningful class projects while the instructor supplies depth and formal rigor in selected areas by drawing from other texts or papers in the field. This same method works for quarter, semester, or two-term classes. Students can get quickly up and running with a general understanding of their vision task and working code to match. As they begin more challenging and time-consuming projects, the instructor helps them develop and debug complex systems. For longer courses, the projects themselves can become instructional in terms of project management. Build up working systems first; refine them with more knowledge, detail, and research later. The goal in such courses is for each project to aim at being worthy of a conference publication and with a few project papers being published subsequent to further (postcourse) work.

Conventions Used in This Book

The following typographical conventions are used in this book:

Italic

Indicates new terms, URLs, email addresses, filenames, file extensions, path names, directories, and Unix utilities.

Constant width

Indicates commands, options, switches, variables, attributes, keys, functions, types, classes, namespaces, methods, modules, properties, parameters, values, objects, events, event handlers, XMLtags, HTMLtags, the contents of files, or the output from commands.

`Constant width bold`

Shows commands or other text that should be typed literally by the user. Also used for emphasis in code samples.

`Constant width italic`

Shows text that should be replaced with user-supplied values.

[...]

Indicates a reference to the bibliography.

Shows text that should be replaced with user-supplied values. his icon signifies a tip, suggestion, or general note.

This icon indicates a warning or caution.

Using Code Examples

OpenCV is free for commercial or research use, and we have the same policy on the code examples in the book. Use them at will for homework, for research, or for commercial products. We would very much appreciate referencing this book when you do, but it is not required. Other than how it helped with your homework projects (which is best kept a secret), we would like to hear how you are using computer vision for academic research, teaching courses, and in commercial products when you do use OpenCV to help you. Again, not required, but you are always invited to drop us a line.

Safari® Books Online

Safari When you see a Safari® Books Online icon on the cover of your favorite technology book, that means the book is available online through the O'Reilly Network Safari Bookshelf.

Safari offers a solution that's better than e-books. It's virtual library that lets you easily search thousands of top tech books, cut and paste code samples, download chapters, and find quick answers when you need the most accurate, current information. Try it for free at *http://safari.oreilly.com*.

We'd Like to Hear from You

Please address comments and questions concerning this book to the publisher:

O'Reilly Media, Inc.
1005 Gravenstein Highway North
Sebastopol, CA 95472

800-998-9938 (in the United States or Canada)
707-829-0515 (international or local)
707-829-0104 (fax)

We have a web page for this book, where we list examples and any plans for future editions. You can access this information at:

http://www.oreilly.com/catalog/9780596516130/

You can also send messages electronically. To be put on the mailing list or request a catalog, send an email to:

info@oreilly.com

To comment on the book, send an email to:

bookquestions@oreilly.com

For more information about our books, conferences, Resource Centers, and the O'Reilly Network, see our website at:

http://www.oreilly.com

Acknowledgments

A long-term open source effort sees many people come and go, each contributing in different ways. The list of contributors to this library is far too long to list here, but see the *.../opencv/docs/HTML/Contributors/doc_contributors.html* file that ships with OpenCV.

Thanks for Help on OpenCV

Intel is where the library was born and deserves great thanks for supporting this project the whole way through. Open source needs a champion and enough development support in the beginning to achieve critical mass. Intel gave it both. There are not many other companies where one could have started and maintained such a project through good times and bad. Along the way, OpenCV helped give rise to—and now takes (optional) advantage of—Intel's Integrated Performance Primitives, which are hand-tuned assembly language routines in vision, signal processing, speech, linear algebra, and more. Thus the lives of a great commercial product and an open source product are intertwined.

Mark Holler, a research manager at Intel, allowed OpenCV to get started by knowingly turning a blind eye to the inordinate amount of time being spent on an unofficial project back in the library's earliest days. As divine reward, he now grows wine up in Napa's Mt. Vieder area. Stuart Taylor in the Performance Libraries group at Intel enabled OpenCV by letting us "borrow" part of his Russian software team. Richard Wirt was key to its continued growth and survival. As the first author took on management responsibility at Intel, lab director Bob Liang let OpenCV thrive; when Justin Rattner became CTO, we were able to put OpenCV on a more firm foundation under Software Technology Lab—supported by software guru Shinn-Horng Lee and indirectly under his manager, Paul Wiley. Omid Moghadam helped advertise OpenCV in the early days. Mohammad Haghighat and Bill Butera were great as technical sounding boards. Nuriel Amir, Denver

Dash, John Mark Agosta, and Marzia Polito were of key assistance in launching the machine learning library. Rainer Lienhart, Jean-Yves Bouguet, Radek Grzeszczuk, and Ara Nefian were able technical contributors to OpenCV and great colleagues along the way; the first is now a professor, the second is now making use of OpenCV in some well-known Google projects, and the others are staffing research labs and start-ups. There were many other technical contributors too numerous to name.

On the software side, some individuals stand out for special mention, especially on the Russian software team. Chief among these is the Russian lead programmer Vadim Pisarevsky, who developed large parts of the library and also managed and nurtured the library through the lean times when boom had turned to bust; he, if anyone, is the true hero of the library. His technical insights have also been of great help during the writing of this book. Giving him managerial support and protection in the lean years was Valery Kuriakin, a man of great talent and intellect. Victor Eruhimov was there in the beginning and stayed through most of it. We thank Boris Chudinovich for all of the contour components.

Finally, very special thanks go to Willow Garage [WG], not only for its steady financial backing to OpenCV's future development but also for supporting one author (and providing the other with snacks and beverages) during the final period of writing this book.

Thanks for Help on the Book

While preparing this book, we had several key people contributing advice, reviews, and suggestions. Thanks to John Markoff, Technology Reporter at the *New York Times* for encouragement, key contacts, and general writing advice born of years in the trenches. To our reviewers, a special thanks go to Evgeniy Bart, physics postdoc at CalTech, who made many helpful comments on every chapter; Kjerstin Williams at Applied Minds, who did detailed proofs and verification until the end; John Hsu at Willow Garage, who went through all the example code; and Vadim Pisarevsky, who read each chapter in detail, proofed the function calls and the code, and also provided several coding examples. There were many other partial reviewers. Jean-Yves Bouguet at Google was of great help in discussions on the calibration and stereo chapters. Professor Andrew Ng at Stanford University provided useful early critiques of the machine learning chapter. There were numerous other reviewers for various chapters—our thanks to all of them. Of course, any errors result from our own ignorance or misunderstanding, not from the advice we received.

Finally, many thanks go to our editor, Michael Loukides, for his early support, numerous edits, and continued enthusiasm over the long haul.

Gary Adds . . .

With three young kids at home, my wife Sonya put in more work to enable this book than I did. Deep thanks and love—even OpenCV gives her recognition, as you can see in the face detection section example image. Further back, my technical beginnings started with the physics department at the University of Oregon followed by undergraduate years at

UC Berkeley. For graduate school, I'd like to thank my advisor Steve Grossberg and Gail Carpenter at the Center for Adaptive Systems, Boston University, where I first cut my academic teeth. Though they focus on mathematical modeling of the brain and I have ended up firmly on the engineering side of AI, I think the perspectives I developed there have made all the difference. Some of my former colleagues in graduate school are still close friends and gave advice, support, and even some editing of the book: thanks to Frank Guenther, Andrew Worth, Steve Lehar, Dan Cruthirds, Allen Gove, and Krishna Govindarajan.

I specially thank Stanford University, where I'm currently a consulting professor in the AI and Robotics lab. Having close contact with the best minds in the world definitely rubs off, and working with Sebastian Thrun and Mike Montemerlo to apply OpenCV on Stanley (the robot that won the $2M DARPA Grand Challenge) and with Andrew Ng on STAIR (one of the most advanced personal robots) was more technological fun than a person has a right to have. It's a department that is currently hitting on all cylinders and simply a great environment to be in. In addition to Sebastian Thrun and Andrew Ng there, I thank Daphne Koller for setting high scientific standards, and also for letting me hire away some key interns and students, as well as Kunle Olukotun and Christos Kozyrakis for many discussions and joint work. I also thank Oussama Khatib, whose work on control and manipulation has inspired my current interests in visually guided robotic manipulation. Horst Haussecker at Intel Research was a great colleague to have, and his own experience in writing a book helped inspire my effort.

Finally, thanks once again to Willow Garage for allowing me to pursue my lifelong robotic dreams in a great environment featuring world-class talent while also supporting my time on this book and supporting OpenCV itself.

Adrian Adds . . .

Coming from a background in theoretical physics, the arc that brought me through supercomputer design and numerical computing on to machine learning and computer vision has been a long one. Along the way, many individuals stand out as key contributors. I have had many wonderful teachers, some formal instructors and others informal guides. I should single out Professor David Dorfan of UC Santa Cruz and Hartmut Sadrozinski of SLAC for their encouragement in the beginning, and Norman Christ for teaching me the fine art of computing with the simple edict that "if you can not make the computer do it, you don't know what you are talking about". Special thanks go to James Guzzo, who let me spend time on this sort of thing at Intel—even though it was miles from what I was supposed to be doing—and who encouraged my participation in the Grand Challenge during those years. Finally, I want to thank Danny Hillis for creating the kind of place where all of this technology can make the leap to wizardry and for encouraging my work on the book while at Applied Minds.

I also would like to thank Stanford University for the extraordinary amount of support I have received from them over the years. From my work on the Grand Challenge team with Sebastian Thrun to the STAIR Robot with Andrew Ng, the Stanford AI Lab was always generous with office space, financial support, and most importantly ideas, enlightening conversation, and (when needed) simple instruction on so many aspects of vision, robotics, and machine learning. I have a deep gratitude to these people, who have contributed so significantly to my own growth and learning.

No acknowledgment or thanks would be meaningful without a special thanks to my lady Lyssa, who never once faltered in her encouragement of this project or in her willingness to accompany me on trips up and down the state to work with Gary on this book. My thanks and my love go to her.

CHAPTER 1

Overview

What Is OpenCV?

OpenCV [OpenCV] is an open source (see *http://opensource.org*) computer vision library available from *http://SourceForge.net/projects/opencvlibrary*. The library is written in C and C++ and runs under Linux, Windows and Mac OS X. There is active development on interfaces for Python, Ruby, Matlab, and other languages.

OpenCV was designed for computational efficiency and with a strong focus on real-time applications. OpenCV is written in optimized C and can take advantage of multicore processors. If you desire further automatic optimization on Intel architectures [Intel], you can buy Intel's Integrated Performance Primitives (IPP) libraries [IPP], which consist of low-level optimized routines in many different algorithmic areas. OpenCV automatically uses the appropriate IPP library at runtime if that library is installed.

One of OpenCV's goals is to provide a simple-to-use computer vision infrastructure that helps people build fairly sophisticated vision applications quickly. The OpenCV library contains over 500 functions that span many areas in vision, including factory product inspection, medical imaging, security, user interface, camera calibration, stereo vision, and robotics. Because computer vision and machine learning often go hand-in-hand, OpenCV also contains a full, general-purpose Machine Learning Library (MLL). This sublibrary is focused on statistical pattern recognition and clustering. The MLL is highly useful for the vision tasks that are at the core of OpenCV's mission, but it is general enough to be used for any machine learning problem.

Who Uses OpenCV?

Most computer scientists and practical programmers are aware of some facet of the role that computer vision plays. But few people are aware of all the ways in which computer vision is used. For example, most people are somewhat aware of its use in surveillance, and many also know that it is increasingly being used for images and video on the Web. A few have seen some use of computer vision in game interfaces. Yet few people realize that most aerial and street-map images (such as in Google's Street View) make heavy use of camera calibration and image stitching techniques. Some are aware of niche applications in safety monitoring, unmanned flying vehicles, or biomedical analysis. But few are aware how pervasive machine vision has become in manufacturing: virtually everything that is mass-produced has been automatically inspected at some point using computer vision.

The open source license for OpenCV has been structured such that you can build a commercial product using all or part of OpenCV. You are under no obligation to open-source your product or to return improvements to the public domain, though we hope you will. In part because of these liberal licensing terms, there is a large user community that includes people from major companies (IBM, Microsoft, Intel, SONY, Siemens, and Google, to name only a few) and research centers (such as Stanford, MIT, CMU, Cambridge, and INRIA). There is a Yahoo groups forum where users can post questions and discussion at *http://groups.yahoo.com/group/OpenCV*; it has about 20,000 members. OpenCV is popular around the world, with large user communities in China, Japan, Russia, Europe, and Israel.

Since its alpha release in January 1999, OpenCV has been used in many applications, products, and research efforts. These applications include stitching images together in satellite and web maps, image scan alignment, medical image noise reduction, object analysis, security and intrusion detection systems, automatic monitoring and safety systems, manufacturing inspection systems, camera calibration, military applications, and unmanned aerial, ground, and underwater vehicles. It has even been used in sound and music recognition, where vision recognition techniques are applied to sound spectrogram images. OpenCV was a key part of the vision system in the robot from Stanford, "Stanley", which won the $2M DARPA Grand Challenge desert robot race [Thrun06].

What Is Computer Vision?

Computer vision* is the transformation of data from a still or video camera into either a decision or a new representation. All such transformations are done for achieving some particular goal. The input data may include some contextual information such as "the camera is mounted in a car" or "laser range finder indicates an object is 1 meter away". The decision might be "there is a person in this scene" or "there are 14 tumor cells on this slide". A new representation might mean turning a color image into a grayscale image or removing camera motion from an image sequence.

Because we are such visual creatures, it is easy to be fooled into thinking that computer vision tasks are easy. How hard can it be to find, say, a car when you are staring at it in an image? Your initial intuitions can be quite misleading. The human brain divides the vision signal into many channels that stream different kinds of information into your brain. Your brain has an attention system that identifies, in a task-dependent

---

* Computer vision is a vast field. This book will give you a basic grounding in the field, but we also recommend texts by Trucco [Trucco98] for a simple introduction, Forsyth [Forsyth03] as a comprehensive reference, and Hartley [Hartley06] and Faugeras [Faugeras93] for how 3D vision really works.

way, important parts of an image to examine while suppressing examination of other areas. There is massive feedback in the visual stream that is, as yet, little understood. There are widespread associative inputs from muscle control sensors and all of the other senses that allow the brain to draw on cross-associations made from years of living in the world. The feedback loops in the brain go back to all stages of processing including the hardware sensors themselves (the eyes), which mechanically control lighting via the iris and tune the reception on the surface of the retina.

In a machine vision system, however, a computer receives a grid of numbers from the camera or from disk, and that's it. For the most part, there's no built-in pattern recognition, no automatic control of focus and aperture, no cross-associations with years of experience. For the most part, vision systems are still fairly naïve. Figure 1-1 shows a picture of an automobile. In that picture we see a side mirror on the driver's side of the car. What the *computer* "sees" is just a grid of numbers. Any given number within that grid has a rather large noise component and so by itself gives us little information, but this grid of numbers is all the computer "sees". Our task then becomes to turn this noisy grid of numbers into the perception: "side mirror". Figure 1-2 gives some more insight into why computer vision is so hard.

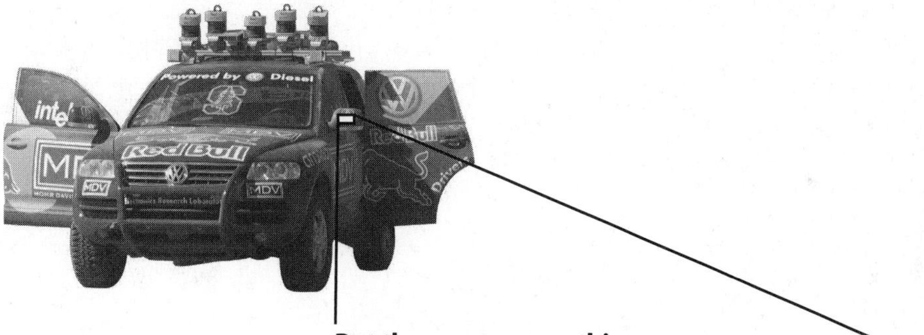

But the camera sees this:

194	210	201	212	199	213	215	195	178	158	182	209
180	189	190	221	209	205	191	167	147	115	129	163
114	126	140	188	176	165	152	140	170	106	78	88
87	103	115	154	143	142	149	153	173	101	57	57
102	112	106	131	122	138	152	147	128	84	58	66
94	95	79	104	105	124	129	113	107	87	69	67
68	71	69	98	89	92	98	95	89	88	76	67
41	56	68	99	63	45	60	82	58	76	74	65
20	41	69	75	56	41	51	73	55	70	63	44
50	50	57	69	75	75	73	74	53	68	59	37
72	59	53	66	84	92	84	74	57	72	63	42
67	61	58	65	75	78	76	73	59	75	69	50

Figure 1-1. To a computer, the car's side mirror is just a grid of numbers

In fact, the problem, as we have posed it thus far, is worse than hard; it is formally impossible to solve. Given a two-dimensional (2D) view of a 3D world, there is no unique way to reconstruct the 3D signal. Formally, such an ill-posed problem has no unique or definitive solution. The same 2D image could represent any of an infinite combination of 3D scenes, even if the data were perfect. However, as already mentioned, the data is

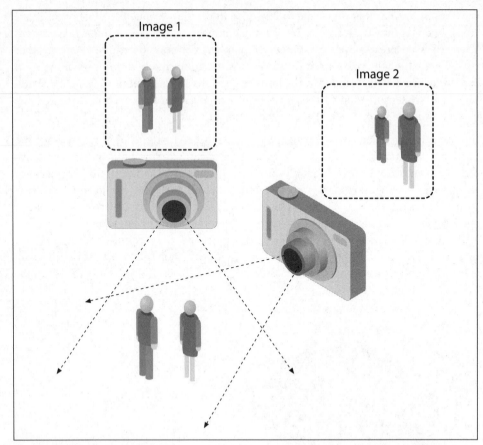

Figure 1-2. The ill-posed nature of vision: the 2D appearance of objects can change radically with viewpoint

corrupted by noise and distortions. Such corruption stems from variations in the world (weather, lighting, reflections, movements), imperfections in the lens and mechanical setup, finite integration time on the sensor (motion blur), electrical noise in the sensor or other electronics, and compression artifacts after image capture. Given these daunting challenges, how can we make any progress?

In the design of a practical system, additional contextual knowledge can often be used to work around the limitations imposed on us by visual sensors. Consider the example of a mobile robot that must find and pick up staplers in a building. The robot might use the facts that a desk is an object found inside offices and that staplers are mostly found on desks. This gives an implicit size reference; staplers must be able to fit on desks. It also helps to eliminate falsely "recognizing" staplers in impossible places (e.g., on the ceiling or a window). The robot can safely ignore a 200-foot advertising blimp shaped like a stapler because the blimp lacks the prerequisite wood-grained background of a desk. In contrast, with tasks such as image retrieval, all stapler images in a database may be of real staplers and so large sizes and other unusual configurations may have been implicitly precluded by the assumptions of those who took the photographs. That is, the photographer probably took pictures only of real, normal-sized staplers. People also tend to center objects when taking pictures and tend to put them in characteristic orientations. Thus, there is often quite a bit of unintentional implicit information within photos taken by people.

Contextual information can also be modeled explicitly with machine learning techniques. Hidden variables such as size, orientation to gravity, and so on can then be correlated with their values in a labeled training set. Alternatively, one may attempt to measure hidden bias variables by using additional sensors. The use of a laser range finder to measure depth allows us to accurately measure the size of an object.

The next problem facing computer vision is noise. We typically deal with noise by using statistical methods. For example, it may be impossible to detect an edge in an image merely by comparing a point to its immediate neighbors. But if we look at the statistics over a local region, edge detection becomes much easier. A real edge should appear as a string of such immediate neighbor responses over a local region, each of whose orientation is consistent with its neighbors. It is also possible to compensate for noise by taking statistics over time. Still other techniques account for noise or distortions by building explicit models learned directly from the available data. For example, because lens distortions are well understood, one need only learn the parameters for a simple polynomial model in order to describe—and thus correct almost completely—such distortions.

The actions or decisions that computer vision attempts to make based on camera data are performed in the context of a specific purpose or task. We may want to remove noise or damage from an image so that our security system will issue an alert if someone tries to climb a fence or because we need a monitoring system that counts how many people cross through an area in an amusement park. Vision software for robots that wander through office buildings will employ different strategies than vision software for stationary security cameras because the two systems have significantly different contexts and objectives. As a general rule: the more constrained a computer vision context is, the more we can rely on those constraints to simplify the problem and the more reliable our final solution will be.

OpenCV is aimed at providing the basic tools needed to solve computer vision problems. In some cases, high-level functionalities in the library will be sufficient to solve the more complex problems in computer vision. Even when this is not the case, the basic components in the library are complete enough to enable creation of a complete solution of your own to almost any computer vision problem. In the latter case, there are several tried-and-true methods of using the library; all of them start with solving the problem using as many available library components as possible. Typically, after you've developed this first-draft solution, you can see where the solution has weaknesses and then fix those weaknesses using your own code and cleverness (better known as "solve the problem you actually have, not the one you imagine"). You can then use your draft solution as a benchmark to assess the improvements you have made. From that point, whatever weaknesses remain can be tackled by exploiting the context of the larger system in which your problem solution is embedded.

The Origin of OpenCV

OpenCV grew out of an Intel Research initiative to advance CPU-intensive applications. Toward this end, Intel launched many projects including real-time ray tracing and 3D display walls. One of the authors working for Intel at that time was visiting universities and noticed that some top university groups, such as the MIT Media Lab, had well-developed and internally open computer vision infrastructures—code that was passed from student to student and that gave each new student a valuable head start in developing his or her own vision application. Instead of reinventing the basic functions from scratch, a new student could begin by building on top of what came before.

Thus, OpenCV was conceived as a way to make computer vision infrastructure universally available. With the aid of Intel's Performance Library Team,* OpenCV started with a core of implemented code and algorithmic specifications being sent to members of Intel's Russian library team. This is the "where" of OpenCV: it started in Intel's research lab with collaboration from the Software Performance Libraries group together with implementation and optimization expertise in Russia.

Chief among the Russian team members was Vadim Pisarevsky, who managed, coded, and optimized much of OpenCV and who is still at the center of much of the OpenCV effort. Along with him, Victor Eruhimov helped develop the early infrastructure, and Valery Kuriakin managed the Russian lab and greatly supported the effort. There were several goals for OpenCV at the outset:

- Advance vision research by providing not only open but also optimized code for basic vision infrastructure. No more reinventing the wheel.

- Disseminate vision knowledge by providing a common infrastructure that developers could build on, so that code would be more readily readable and transferable.

- Advance vision-based commercial applications by making portable, performance-optimized code available for free—with a license that did not require commercial applications to be open or free themselves.

Those goals constitute the "why" of OpenCV. Enabling computer vision applications would increase the need for fast processors. Driving upgrades to faster processors would generate more income for Intel than selling some extra software. Perhaps that is why this open and free code arose from a hardware vendor rather than a software company. In some sense, there is more room to be innovative at software within a hardware company.

In any open source effort, it's important to reach a critical mass at which the project becomes self-sustaining. There have now been approximately two million downloads

---

* Shinn Lee was of key help.

of OpenCV, and this number is growing by an average of 26,000 downloads a month. The user group now approaches 20,000 members. OpenCV receives many user contributions, and central development has largely moved outside of Intel.* OpenCV's past timeline is shown in Figure 1-3. Along the way, OpenCV was affected by the dot-com boom and bust and also by numerous changes of management and direction. During these fluctuations, there were times when OpenCV had no one at Intel working on it at all. However, with the advent of multicore processors and the many new applications of computer vision, OpenCV's value began to rise. Today, OpenCV is an active area of development at several institutions, so expect to see many updates in multicamera calibration, depth perception, methods for mixing vision with laser range finders, and better pattern recognition as well as a lot of support for robotic vision needs. For more information on the future of OpenCV, see Chapter 14.

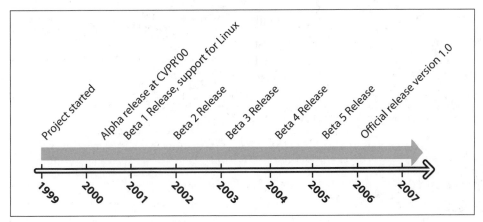

Figure 1-3. OpenCV timeline

Speeding Up OpenCV with IPP

Because OpenCV was "housed" within the Intel Performance Primitives team and several primary developers remain on friendly terms with that team, OpenCV exploits the hand-tuned, highly optimized code in IPP to speed itself up. The improvement in speed from using IPP can be substantial. Figure 1-4 compares two other vision libraries, LTI [LTI] and VXL [VXL], against OpenCV and OpenCV using IPP. Note that performance was a key goal of OpenCV; the library needed the ability to run vision code in real time.

OpenCV is written in performance-optimized C and C++ code. It does *not* depend in any way on IPP. If IPP is present, however, OpenCV will automatically take advantage of IPP by loading IPP's dynamic link libraries to further enhance its speed.

---

* As of this writing, Willow Garage [WG] (*www.willowgarage.com*), a robotics research institute and incubator, is actively supporting general OpenCV maintenance and new development in the area of robotics applications.

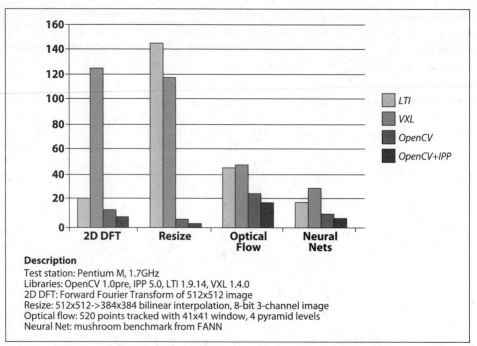

Description
Test station: Pentium M, 1.7GHz
Libraries: OpenCV 1.0pre, IPP 5.0, LTI 1.9.14, VXL 1.4.0
2D DFT: Forward Fourier Transform of 512x512 image
Resize: 512x512->384x384 bilinear interpolation, 8-bit 3-channel image
Optical flow: 520 points tracked with 41x41 window, 4 pyramid levels
Neural Net: mushroom benchmark from FANN

Figure 1-4. Two other vision libraries (LTI and VXL) compared with OpenCV (without and with IPP) on four different performance benchmarks: the four bars for each benchmark indicate scores proportional to run time for each of the given libraries; in all cases, OpenCV outperforms the other libraries and OpenCV with IPP outperforms OpenCV without IPP

Who Owns OpenCV?

Although Intel started OpenCV, the library is and always was intended to promote commercial and research use. It is therefore open and free, and the code itself may be used or embedded (in whole or in part) in other applications, whether commercial or research. It does not force your application code to be open or free. It does not require that you return improvements back to the library—but we hope that you will.

Downloading and Installing OpenCV

The main OpenCV site is on SourceForge at *http://SourceForge.net/projects/opencvlibrary* and the OpenCV Wiki [OpenCV Wiki] page is at *http://opencvlibrary.SourceForge.net*. For Linux, the source distribution is the file *opencv-1.0.0.tar.gz*; for Windows, you want *OpenCV_1.0.exe*. However, the most up-to-date version is always on the CVS server at SourceForge.

Install

Once you download the libraries, you must install them. For detailed installation instructions on Linux or Mac OS, see the text file named *INSTALL* directly under the

.../*opencv/* directory; this file also describes how to build and run the OpenCV test-ing routines. *INSTALL* lists the additional programs you'll need in order to become an OpenCV developer, such as *autoconf, automake, libtool,* and *swig.*

Windows

Get the executable installation from SourceForge and run it. It will install OpenCV, reg-ister DirectShow filters, and perform various post-installation procedures. You are now ready to start using OpenCV. You can always go to the .../*opencv/_make* directory and open *opencv.sln* with MSVC++ or MSVC.NET 2005, or you can open *opencv.dsw* with lower ver-sions of MSVC++ and build debug versions or rebuild release versions of the library.*

To add the commercial IPP performance optimizations to Windows, obtain and in-stall IPP from the Intel site (*http://www.intel.com/software/products/ipp/index.htm*); use version 5.1 or later. Make sure the appropriate binary folder (e.g., *c:/program files/intel/ipp/5.1/ia32/bin*) is in the system path. IPP should now be automatically detected by OpenCV and loaded at runtime (more on this in Chapter 3).

Linux

Prebuilt binaries for Linux are not included with the Linux version of OpenCV owing to the large variety of versions of GCC and GLIBC in different distributions (SuSE, Debian, Ubuntu, etc.). If your distribution doesn't offer OpenCV, you'll have to build it from sources as detailed in the .../*opencv/INSTALL* file.

To build the libraries and demos, you'll need GTK+ 2.x or higher, including headers. You'll also need *pkgconfig, libpng, zlib, libjpeg, libtiff,* and *libjasper* with development files. You'll need Python 2.3, 2.4, or 2.5 with headers installed (developer package). You will also need *libavcodec* and the other *libav** libraries (including headers) from *ffmpeg* 0.4.9-pre1 or later (*svn checkout svn://svn.mplayerhq.hu/ffmpeg/trunk ffmpeg*).

Download ffmpeg from *http://ffmpeg.mplayerhq.hu/download.html.*[†] The ffmpeg pro-gram has a lesser general public license (LGPL). To use it with non-GPL software (such as OpenCV), build and use a shared ffmpg library:

```
$> ./configure --enable-shared
$> make
$> sudo make install
```

You will end up with: */usr/local/lib/libavcodec.so.**, */usr/local/lib/libavformat.so.**, */usr/local/lib/libavutil.so.**, and include files under various */usr/local/include/libav**.

To build OpenCV once it is downloaded:[‡]

---

* It is important to know that, although the Windows distribution contains binary libraries for release builds, it does not contain the debug builds of these libraries. It is therefore likely that, before developing with OpenCV, you will want to open the solution file and build these libraries for yourself.

† You can check out ffmpeg by: *svn checkout svn://svn.mplayerhq.hu/ffmpeg/trunk ffmpeg.*

‡ To build OpenCV using Red Hat Package Managers (RPMs), use `rpmbuild -ta OpenCV-x.y.z.tar.gz` (for RPM 4.x or later), or `rpm -ta OpenCV-x.y.z.tar.gz` (for earlier versions of RPM), where *OpenCV-x.y.z.tar .gz* should be put in */usr/src/redhat/SOURCES/* or a similar directory. Then install OpenCV using `rpm -i OpenCV-x.y.z.*.rpm`.

```
$> ./configure
$> make
$> sudo make install
$> sudo ldconfig
```

After installation is complete, the default installation path is */usr/local/lib/* and */usr/local/include/opencv/*. Hence you need to add */usr/local/lib/* to */etc/ld.so.conf* (and run ldconfig afterwards) *or* add it to the LD_LIBRARY_PATH environment variable; then you are done.

To add the commercial IPP performance optimizations to Linux, install IPP as described previously. Let's assume it was installed in */opt/intel/ipp/5.1/ia32/*. Add *<your install_path>/bin/* and *<your install_path>/bin/linux32* LD_LIBRARY_PATH in your initialization script (*.bashrc* or similar):

```
LD_LIBRARY_PATH=/opt/intel/ipp/5.1/ia32/bin:/opt/intel/ipp/5.1
/ia32/bin/linux32:$LD_LIBRARY_PATH
export LD_LIBRARY_PATH
```

Alternatively, you can add *<your install_path>/bin* and *<your install_path>/bin/linux32*, one per line, to */etc/ld.so.conf* and then run *ldconfig* as root (or use sudo).

That's it. Now OpenCV should be able to locate IPP shared libraries and make use of them on Linux. See *.../opencv/INSTALL* for more details.

MacOS X

As of this writing, full functionality on MacOS X is a priority but there are still some limitations (e.g., writing AVIs); these limitations are described in *.../opencv/INSTALL*.

The requirements and building instructions are similar to the Linux case, with the following exceptions:

- By default, Carbon is used instead of GTK+.

- By default, QuickTime is used instead of ffmpeg.

- pkg-config is optional (it is used explicitly only in the *samples/c/build_all.sh* script).

- RPM and ldconfig are not supported by default. Use configure+make+sudo make install to build and install OpenCV, update LD_LIBRARY_PATH (unless ./configure --prefix=/usr is used).

For full functionality, you should install *libpng*, *libtiff*, *libjpeg* and *libjasper* from *darwinports* and/or *fink* and make them available to *./configure* (see ./configure --help). For the most current information, see the OpenCV Wiki at *http://opencvlibrary .SourceForge.net/* and the Mac-specific page *http://opencvlibrary.SourceForge.net/ Mac_OS_X_OpenCV_Port*.

Getting the Latest OpenCV via CVS

OpenCV is under active development, and bugs are often fixed rapidly when bug reports contain accurate descriptions and code that demonstrates the bug. However, official OpenCV releases occur only once or twice a year. If you are seriously developing a project or product, you will probably want code fixes and updates as soon as they become available. To do this, you will need to access OpenCV's Concurrent Versions System (CVS) on SourceForge.

This isn't the place for a tutorial in CVS usage. If you've worked with other open source projects then you're probably familiar with it already. If you haven't, check out *Essential CVS* by Jennifer Vesperman (O'Reilly). A command-line CVS client ships with Linux, OS X, and most UNIX-like systems. For Windows users, we recommend TortoiseCVS (*http://www.tortoisecvs.org/*), which integrates nicely with Windows Explorer.

On Windows, if you want the latest OpenCV from the CVS repository then you'll need to access the CVSROOT directory:

```
:pserver:anonymous@opencvlibrary.cvs.sourceforge.net:2401/cvsroot/opencvlibrary
```

On Linux, you can just use the following two commands:

```
cvs -d:pserver:anonymous@opencvlibrary.cvs.sourceforge.net:/cvsroot/opencvlibrary
login
```

When asked for password, hit return. Then use:

```
cvs -z3 -d:pserver:anonymous@opencvlibrary.cvs.sourceforge.net:/cvsroot/opencvlibrary
co -P opencv
```

More OpenCV Documentation

The primary documentation for OpenCV is the HTML documentation that ships with the source code. In addition to this, the OpenCV Wiki and the older HTML documentation are available on the Web.

Documentation Available in HTML

OpenCV ships with html-based user documentation in the *.../opencv/docs* subdirectory. Load the *index.htm* file, which contains the following links.

CXCORE
> Contains data structures, matrix algebra, data transforms, object persistence, memory management, error handling, and dynamic loading of code as well as drawing, text and basic math.

CV
> Contains image processing, image structure analysis, motion and tracking, pattern recognition, and camera calibration.

Machine Learning (ML)
> Contains many clustering, classification and data analysis functions.

HighGUI
> Contains user interface GUI and image/video storage and recall.

CVCAM
Camera interface.

Haartraining
How to train the boosted cascade object detector. This is in the *.../opencv/apps/ HaarTraining/doc/haartraining.htm* file.

The *.../opencv/docs* directory also contains *IPLMAN.pdf*, which was the original manual for OpenCV. It is now defunct and should be used with caution, but it does include detailed descriptions of algorithms and of what image types may be used with a particular algorithm. Of course, the first stop for such image and algorithm details is the book you are reading now.

Documentation via the Wiki

OpenCV's documentation Wiki is more up-to-date than the html pages that ship with OpenCV and it also features additional content as well. The Wiki is located at *http:// opencvlibrary.SourceForge.net*. It includes information on:

- Instructions on compiling OpenCV using Eclipse IDE
- Face recognition with OpenCV
- Video surveillance library
- Tutorials
- Camera compatibility
- Links to the Chinese and the Korean user groups

Another Wiki, located at *http://opencvlibrary.SourceForge.net/CvAux*, is the only documentation of the auxiliary functions discussed in "OpenCV Structure and Content" (next section). CvAux includes the following functional areas:

- Stereo correspondence
- View point morphing of cameras
- 3D tracking in stereo
- Eigen object (PCA) functions for object recognition
- Embedded hidden Markov models (HMMs)

This Wiki has been translated into Chinese at *http://www.opencv.org.cn/index.php/ %E9%A6%96%E9%A1%B5*.

Regardless of your documentation source, it is often hard to know:

- Which image type (floating, integer, byte; 1–3 channels) works with which function
- Which functions work in place
- Details of how to call the more complex functions (e.g., contours)

- Details about running many of the examples in the .../*opencv/samples/c/* directory

- *What* to do, not just how

- How to set parameters of certain functions

One aim of this book is to address these problems.

OpenCV Structure and Content

OpenCV is broadly structured into five main components, four of which are shown in Figure 1-5. The CV component contains the basic image processing and higher-level computer vision algorithms; ML is the machine learning library, which includes many statistical classifiers and clustering tools. HighGUI contains I/O routines and functions for storing and loading video and images, and CXCore contains the basic data structures and content.

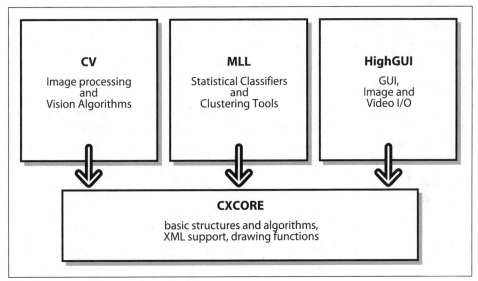

Figure 1-5. The basic structure of OpenCV

Figure 1-5 does not include CvAux, which contains both defunct areas (embedded HMM face recognition) and experimental algorithms (background/foreground segmentation). CvAux is not particularly well documented in the Wiki and is not documented at all in the .../*opencv/docs* subdirectory. CvAux covers:

- Eigen objects, a computationally efficient recognition technique that is, in essence, a template matching procedure

- 1D and 2D hidden Markov models, a statistical recognition technique solved by dynamic programming

- Embedded HMMs (the observations of a parent HMM are themselves HMMs)

- Gesture recognition from stereo vision support

- Extensions to Delaunay triangulation, sequences, and so forth

- Stereo vision

- Shape matching with region contours

- Texture descriptors

- Eye and mouth tracking

- 3D tracking

- Finding skeletons (central lines) of objects in a scene

- Warping intermediate views between two camera views

- Background-foreground segmentation

- Video surveillance (see Wiki FAQ for more documentation)

- Camera calibration C++ classes (the C functions and engine are in CV)

Some of these features may migrate to CV in the future; others probably never will.

Portability

OpenCV was designed to be portable. It was originally written to compile across Borland C++, MSVC++, and the Intel compilers. This meant that the C and C++ code had to be fairly standard in order to make cross-platform support easier. Figure 1-6 shows the platforms on which OpenCV is known to run. Support for 32-bit Intel architecture (IA32) on Windows is the most mature, followed by Linux on the same architecture. Mac OS X portability became a priority only after Apple started using Intel processors. (The OS X port isn't as mature as the Windows or Linux versions, but this is changing rapidly.) These are followed by 64-bit support on extended memory (EM64T) and the 64-bit Intel architecture (IA64). The least mature portability is on Sun hardware and other operating systems.

If an architecture or OS doesn't appear in Figure 1-6, this doesn't mean there are no OpenCV ports to it. OpenCV has been ported to almost every commercial system, from PowerPC Macs to robotic dogs. OpenCV runs well on AMD's line of processors, and even the further optimizations available in IPP will take advantage of multimedia extensions (MMX) in AMD processors that incorporate this technology.

	IA32	EM64T	IA64	Other (PPC, Sparc)
Windows	✔ (w. IPP; MSVC6, .NET2005+OMP, ICC, GCC, BCC)	✔ (w. IPP; MSVC6+PSDK.NET2005+OMP, PSDK)	± (w. IPP; PSDK, some tests fail)	N/A
Linux	✔ (w. IPP; GCC, BCC)	✔ (w. IPP; GCC, BCC)	✔ (GCC, ICC)	✘
MacOSX	✔ (w. IPP, GCC, native APIs)	? (not tested)	N/A	✔ (iMac G5, GCC, native APIs)
Others (BSD, Solaris…)	✘	✘	✘	Reported to build on UltraSparc Solaris

Figure 1-6. OpenCV portability guide for release 1.0: operating systems are shown on the left; computer architecture types across top

Exercises

1. Download and install the latest release of OpenCV. Compile it in debug and release mode.

2. Download and build the latest CVS update of OpenCV.

3. Describe at least three ambiguous aspects of converting 3D inputs into a 2D representation. How would you overcome these ambiguities?

CHAPTER 2

Introduction to OpenCV

Getting Started

After installing the OpenCV library, our first task is, naturally, to get started and make something interesting happen. In order to do this, we will need to set up the programming environment.

In Visual Studio, it is necessary to create a project and to configure the setup so that (a) the libraries *highgui.lib*, *cxcore.lib*, *ml.lib*, and *cv.lib* are linked* and (b) the preprocessor will search the OpenCV *...*/*opencv/*/include* directories for header files. These "include" directories will typically be named something like *C:/program files/opencv/cv/include,*[†] *...*/*opencv/cxcore/include*, *...*/*opencv/ml/include*, and *...*/*opencv/otherlibs/highgui*. Once you've done this, you can create a new C file and start your first program.

 Certain key header files can make your life much easier. Many useful macros are in the header files *...*/*opencv/cxcore/include/cxtypes.h* and *cxmisc.h*. These can do things like initialize structures and arrays in one line, sort lists, and so on. The most important headers for compiling are *...*/*cv/include/cv.h* and *...*/*cxcore/include/cxcore.h* for computer vision, *...*/*otherlibs/highgui/highgui.h* for I/O, and *...*/*ml/include/ml.h* for machine learning.

First Program—Display a Picture

OpenCV provides utilities for reading from a wide array of image file types as well as from video and cameras. These utilities are part of a toolkit called HighGUI, which is included in the OpenCV package. We will use some of these utilities to create a simple program that opens an image and displays it on the screen. See Example 2-1.

---

* For debug builds, you should link to the libraries *highguid.lib*, *cxcored.lib*, *mld.lib*, and *cvd.lib*.

† *C:/program files/* is the default installation of the OpenCV directory on Windows, although you can choose to install it elsewhere. To avoid confusion, from here on we'll use "*...*/*opencv/*" to mean the path to the opencv directory on your system.

Example 2-1. A simple OpenCV program that loads an image from disk and displays it on the screen

```
#include "highgui.h"

int main( int argc, char** argv ) {
    IplImage* img = cvLoadImage( argv[1] );
    cvNamedWindow( "Example1", CV_WINDOW_AUTOSIZE );
    cvShowImage( "Example1", img );
    cvWaitKey(0);
    cvReleaseImage( &img );
    cvDestroyWindow( "Example1" );
}
```

When compiled and run from the command line with a single argument, this program loads an image into memory and displays it on the screen. It then waits until the user presses a key, at which time it closes the window and exits. Let's go through the program line by line and take a moment to understand what each command is doing.

```
    IplImage* img = cvLoadImage( argv[1] );
```

This line loads the image.* The function cvLoadImage() is a high-level routine that determines the file format to be loaded based on the file name; it also automatically allocates the memory needed for the image data structure. Note that cvLoadImage() can read a wide variety of image formats, including BMP, DIB, JPEG, JPE, PNG, PBM, PGM, PPM, SR, RAS, and TIFF. A pointer to an allocated image data structure is then returned. This structure, called IplImage, is the OpenCV construct with which you will deal the most. OpenCV uses this structure to handle all kinds of images: single-channel, multichannel, integer-valued, floating-point-valued, et cetera. We use the pointer that cvLoadImage() returns to manipulate the image and the image data.

```
    cvNamedWindow( "Example1", CV_WINDOW_AUTOSIZE );
```

Another high-level function, cvNamedWindow(), opens a window on the screen that can contain and display an image. This function, provided by the HighGUI library, also assigns a name to the window (in this case, "Example1"). Future HighGUI calls that interact with this window will refer to it by this name.

The second argument to cvNamedWindow() defines window properties. It may be set either to 0 (the default value) or to CV_WINDOW_AUTOSIZE. In the former case, the size of the window will be the same regardless of the image size, and the image will be scaled to fit within the window. In the latter case, the window will expand or contract automatically when an image is loaded so as to accommodate the image's true size.

```
    cvShowImage( "Example1", img );
```

Whenever we have an image in the form of an IplImage* pointer, we can display it in an existing window with cvShowImage(). The cvShowImage() function requires that a named window already exist (created by cvNamedWindow()). On the call to cvShowImage(), the

---

* A proper program would check for the existence of argv[1] and, in its absence, deliver an instructional error message for the user. We will abbreviate such necessities in this book and assume that the reader is cultured enough to understand the importance of error-handling code.

window will be redrawn with the appropriate image in it, and the window will resize itself as appropriate if it was created using the CV_WINDOW_AUTOSIZE flag.

```
cvWaitKey(0);
```

The cvWaitKey() function asks the program to stop and wait for a keystroke. If a positive argument is given, the program will wait for that number of milliseconds and then continue even if nothing is pressed. If the argument is set to 0 or to a negative number, the program will wait indefinitely for a keypress.

```
cvReleaseImage( &img );
```

Once we are through with an image, we can free the allocated memory. OpenCV expects a pointer to the IplImage* pointer for this operation. After the call is completed, the pointer img will be set to NULL.

```
cvDestroyWindow( "Example1" );
```

Finally, we can destroy the window itself. The function cvDestroyWindow() will close the window and de-allocate any associated memory usage (including the window's internal image buffer, which is holding a copy of the pixel information from *img). For a simple program, you don't really have to call cvDestroyWindow() or cvReleaseImage() because all the resources and windows of the application are closed automatically by the operating system upon exit, but it's a good habit anyway.

Now that we have this simple program we can toy around with it in various ways, but we don't want to get ahead of ourselves. Our next task will be to construct a very simple—almost as simple as this one—program to read in and display an AVI video file. After that, we will start to tinker a little more.

Second Program—AVI Video

Playing a video with OpenCV is almost as easy as displaying a single picture. The only new issue we face is that we need some kind of loop to read each frame in sequence; we may also need some way to get out of that loop if the movie is too boring. See Example 2-2.

Example 2-2. A simple OpenCV program for playing a video file from disk

```
#include "highgui.h"

int main( int argc, char** argv ) {
    cvNamedWindow( "Example2", CV_WINDOW_AUTOSIZE );
    CvCapture* capture = cvCreateFileCapture( argv[1] );
    IplImage* frame;
    while(1) {
        frame = cvQueryFrame( capture );
        if( !frame ) break;
        cvShowImage( "Example2", frame );
        char c = cvWaitKey(33);
        if( c == 27 ) break;
    }
    cvReleaseCapture( &capture );
    cvDestroyWindow( "Example2" );
}
```

Here we begin the function main() with the usual creation of a named window, in this case "Example2". Things get a little more interesting after that.

```
CvCapture* capture = cvCreateFileCapture( argv[1] );
```

The function cvCreateFileCapture() takes as its argument the name of the AVI file to be loaded and then returns a pointer to a CvCapture structure. This structure contains all of the information about the AVI file being read, including state information. When created in this way, the CvCapture structure is initialized to the beginning of the AVI.

```
frame = cvQueryFrame( capture );
```

Once inside of the while(1) loop, we begin reading from the AVI file. cvQueryFrame() takes as its argument a pointer to a CvCapture structure. It then grabs the next video frame into memory (memory that is actually part of the CvCapture structure). A pointer is returned to that frame. Unlike cvLoadImage, which actually allocates memory for the image, cvQueryFrame uses memory already allocated in the CvCapture structure. Thus it will not be necessary (or wise) to call cvReleaseImage() for this "frame" pointer. Instead, the frame image memory will be freed when the CvCapture structure is released.

```
c = cvWaitKey(33);
if( c == 27 ) break;
```

Once we have displayed the frame, we then wait for 33 ms.* If the user hits a key, then c will be set to the ASCII value of that key; if not, then it will be set to –1. If the user hits the Esc key (ASCII 27), then we will exit the read loop. Otherwise, 33 ms will pass and we will just execute the loop again.

It is worth noting that, in this simple example, we are not explicitly controlling the speed of the video in any intelligent way. We are relying solely on the timer in cvWaitKey() to pace the loading of frames. In a more sophisticated application it would be wise to read the actual frame rate from the CvCapture structure (from the AVI) and behave accordingly!

```
cvReleaseCapture( &capture );
```

When we have exited the read loop—because there was no more video data or because the user hit the Esc key—we can free the memory associated with the CvCapture structure. This will also close any open file handles to the AVI file.

Moving Around

OK, that was great. Now it's time to tinker around, enhance our toy programs, and explore a little more of the available functionality. The first thing we might notice about the AVI player of Example 2-2 is that it has no way to move around quickly within the video. Our next task will be to add a slider bar, which will give us this ability.

---

* You can wait any amount of time you like. In this case, we are simply assuming that it is correct to play the video at 30 frames per second and allow user input to interrupt between each frame (thus we pause for input 33 ms between each frame). In practice, it is better to check the CvCapture structure returned by cvCaptureFromCamera() in order to determine the actual frame rate (more on this in Chapter 4).

The HighGUI toolkit provides a number of simple instruments for working with images and video beyond the simple display functions we have just demonstrated. One especially useful mechanism is the slider, which enables us to jump easily from one part of a video to another. To create a slider, we call cvCreateTrackbar() and indicate which window we would like the trackbar to appear in. In order to obtain the desired functionality, we need only supply a callback that will perform the relocation. Example 2-3 gives the details.

Example 2-3. Program to add a trackbar slider to the basic viewer window: when the slider is moved, the function onTrackbarSlide() is called and then passed to the slider's new value

```
#include "cv.h"
#include "highgui.h"

int        g_slider_position = 0;
CvCapture* g_capture         = NULL;

void onTrackbarSlide(int pos) {
    cvSetCaptureProperty(
        g_capture,
        CV_CAP_PROP_POS_FRAMES,
        pos
    );
}

int main( int argc, char** argv ) {
    cvNamedWindow( "Example3", CV_WINDOW_AUTOSIZE );
    g_capture = cvCreateFileCapture( argv[1] );
    int frames = (int) cvGetCaptureProperty(
        g_capture,
        CV_CAP_PROP_FRAME_COUNT
    );
    if( frames!= 0 ) {
      cvCreateTrackbar(
          "Position",
          "Example3",
          &g_slider_position,
          frames,
          onTrackbarSlide
      );
    }
    IplImage* frame;
    // While loop (as in Example 2) capture & show video
    ...
    // Release memory and destroy window
    ...
    return(0);
}
```

In essence, then, the strategy is to add a global variable to represent the slider position and then add a callback that updates this variable and relocates the read position in the video. One call creates the slider and attaches the callback, and we are off and running.* Let's look at the details.

```
int g_slider_position = 0;
CvCapture* g_capture   = NULL;
```

First we define a global variable for the slider position. The callback will need access to the capture object, so we promote that to a global variable. Because we are nice people and like our code to be readable and easy to understand, we adopt the convention of adding a leading g_ to any global variable.

```
void onTrackbarSlide(int pos) {
    cvSetCaptureProperty(
    g_capture,
    CV_CAP_PROP_POS_FRAMES,
    pos
);
```

Now we define a callback routine to be used when the user pokes the slider. This routine will be passed to a 32-bit integer, which will be the slider position.

The call to cvSetCaptureProperty() is one we will see often in the future, along with its counterpart cvGetCaptureProperty(). These routines allow us to configure (or query in the latter case) various properties of the CvCapture object. In this case we pass the argument CV_CAP_PROP_POS_FRAMES, which indicates that we would like to set the read position in units of frames. (We can use AVI_RATIO instead of FRAMES if we want to set the position as a fraction of the overall video length). Finally, we pass in the new value of the position. Because HighGUI is highly civilized, it will automatically handle such issues as the possibility that the frame we have requested is not a key-frame; it will start at the previous key-frame and fast forward up to the requested frame without us having to fuss with such details.

```
int frames = (int) cvGetCaptureProperty(
    g_capture,
    CV_CAP_PROP_FRAME_COUNT
);
```

As promised, we use cvGetCaptureProperty()when we want to query some data from the CvCapture structure. In this case, we want to find out how many frames are in the video so that we can calibrate the slider (in the next step).

```
if( frames!= 0 ) {
  cvCreateTrackbar(
      "Position",
      "Example3",
      &g_slider_position,
      frames,
      onTrackbarSlide
  );
}
```

---

* This code does not update the slider position as the video plays; we leave that as an exercise for the reader. Also note that some mpeg encodings do not allow you to move backward in the video.

The last detail is to create the trackbar itself. The function cvCreateTrackbar() allows us to give the trackbar a label* (in this case Position) and to specify a window to put the trackbar in. We then provide a variable that will be bound to the trackbar, the maximum value of the trackbar, and a callback (or NULL if we don't want one) for when the slider is moved. Observe that we do not create the trackbar if cvGetCaptureProperty() returned a zero frame count. This is because sometimes, depending on how the video was encoded, the total number of frames will not be available. In this case we will just play the movie without providing a trackbar.

It is worth noting that the slider created by HighGUI is not as full-featured as some sliders out there. Of course, there's no reason you can't use your favorite windowing toolkit instead of HighGUI, but the HighGUI tools are quick to implement and get us off the ground in a hurry.

Finally, we did not include the extra tidbit of code needed to make the slider move as the video plays. This is left as an exercise for the reader.

A Simple Transformation

Great, so now you can use OpenCV to create your own video player, which will not be much different from countless video players out there already. But we are interested in computer vision, and we want to do some of that. Many basic vision tasks involve the application of filters to a video stream. We will modify the program we already have to do a simple operation on every frame of the video as it plays.

One particularly simple operation is the smoothing of an image, which effectively reduces the information content of the image by convolving it with a Gaussian or other similar kernel function. OpenCV makes such convolutions exceptionally easy to do. We can start by creating a new window called "Example4-out", where we can display the results of the processing. Then, after we have called cvShowImage() to display the newly captured frame in the input window, we can compute and display the smoothed image in the output window. See Example 2-4.

Example 2-4. Loading and then smoothing an image before it is displayed on the screen

```
#include "cv.h"
#include "highgui.h"

void example2_4( IplImage* image )

    // Create some windows to show the input
    // and output images in.
    //
    cvNamedWindow( "Example4-in" );
```

---

* Because HighGUI is a lightweight and easy-to-use toolkit, cvCreateTrackbar() does not distinguish between the name of the trackbar and the label that actually appears on the screen next to the trackbar. You may already have noticed that cvNamedWindow() likewise does not distinguish between the name of the window and the label that appears on the window in the GUI.

Example 2-4. Loading and then smoothing an image before it is displayed on the screen (continued)

```
cvNamedWindow( "Example4-out" );

// Create a window to show our input image
//
cvShowImage( "Example4-in", image );

// Create an image to hold the smoothed output
//
IplImage* out = cvCreateImage(
    cvGetSize(image),
    IPL_DEPTH_8U,
    3
);

// Do the smoothing
//
cvSmooth( image, out, CV_GAUSSIAN, 3, 3 );

// Show the smoothed image in the output window
//
cvShowImage( "Example4-out", out );

// Be tidy
//
cvReleaseImage( &out );

// Wait for the user to hit a key, then clean up the windows
//
cvWaitKey( 0 );
cvDestroyWindow( "Example4-in" );
cvDestroyWindow( "Example4-out" );

}
```

The first call to cvShowImage() is no different than in our previous example. In the next call, we allocate another image structure. Previously we relied on cvCreateFileCapture() to allocate the new frame for us. In fact, that routine actually allocated only one frame and then wrote over that data each time a capture call was made (so it actually returned the same pointer every time we called it). In this case, however, we want to allocate our own image structure to which we can write our smoothed image. The first argument is a CvSize structure, which we can conveniently create by calling cvGetSize(image); this gives us the size of the existing structure image. The second argument tells us what kind of data type is used for each channel on each pixel, and the last argument indicates the number of channels. So this image is three channels (with 8 bits per channel) and is the same size as image.

The smoothing operation is itself just a single call to the OpenCV library: we specify the input image, the output image, the smoothing method, and the parameters for the smooth. In this case we are requesting a Gaussian smooth over a 3 × 3 area centered on each pixel. It is actually allowed for the output to be the same as the input image, and this would work more efficiently in our current application, but we avoided doing this because it gave us a chance to introduce cvCreateImage()!

Now we can show the image in our new second window and then free it: cvReleaseImage() takes a pointer to the IplImage* pointer and then de-allocates all of the memory associated with that image.

A Not-So-Simple Transformation

That was pretty good, and we are learning to do more interesting things. In Example 2-4 we chose to allocate a new IplImage structure, and into this new structure we wrote the output of a single transformation. As mentioned, we could have applied the transformation in such a way that the output overwrites the original, but this is not always a good idea. In particular, some operators do not produce images with the same size, depth, and number of channels as the input image. Typically, we want to perform a *sequence* of operations on some initial image and so produce a chain of transformed images.

In such cases, it is often useful to introduce simple wrapper functions that both allocate the output image and perform the transformation we are interested in. Consider, for example, the reduction of an image by a factor of 2 [Rosenfeld80]. In OpenCV this is accomplished by the function cvPyrDown(), which performs a Gaussian smooth and then removes every other line from an image. This is useful in a wide variety of important vision algorithms. We can implement the simple function described in Example 2-5.

Example 2-5. Using cvPyrDown() to create a new image that is half the width and height of the input image

```
IplImage* doPyrDown(
  IplImage* in,
  int        filter = IPL_GAUSSIAN_5x5
) {

    // Best to make sure input image is divisible by two.
    //
    assert( in->width%2 == 0 && in->height%2 == 0 );

    IplImage* out = cvCreateImage(
        cvSize( in->width/2, in->height/2 ),
        in->depth,
        in->nChannels
    );
    cvPyrDown( in, out );
    return( out );
};
```

Notice that we allocate the new image by reading the needed parameters from the old image. In OpenCV, all of the important data types are implemented as structures and passed around as structure pointers. There is no such thing as private data in OpenCV!

Let's now look at a similar but slightly more involved example involving the *Canny edge detector* [Canny86] (see Example 2-6). In this case, the edge detector generates an image that is the full size of the input image but needs only a single channel image to write to.

Example 2-6. The Canny edge detector writes its output to a single channel (grayscale) image

```
IplImage* doCanny(
    IplImage*   in,
    double      lowThresh,
    double      highThresh,
    double      aperture
) {
    If(in->nChannels != 1)
        return(0); //Canny only handles gray scale images

    IplImage* out = cvCreateImage(
        cvSize( cvGetSize( in ),
        IPL_DEPTH_8U,
        1
    );
    cvCanny( in, out, lowThresh, highThresh, aperture );
    return( out );
};
```

This allows us to string together various operators quite easily. For example, if we wanted to shrink the image twice and then look for lines that were present in the twice-reduced image, we could proceed as in Example 2-7.

Example 2-7. Combining the pyramid down operator (twice) and the Canny subroutine in a simple image pipeline

```
IplImage* img1 = doPyrDown( in, IPL_GAUSSIAN_5x5 );
IplImage* img2 = doPyrDown( img1, IPL_GAUSSIAN_5x5 );
IplImage* img3 = doCanny( img2, 10, 100, 3 );

// do whatever with 'img3'
//
...
cvReleaseImage( &img1 );
cvReleaseImage( &img2 );
cvReleaseImage( &img3 );
```

It is important to observe that nesting the calls to various stages of our filtering pipeline is not a good idea, because then we would have no way to free the images that we are allocating along the way. If we are too lazy to do this cleanup, we could opt to include the following line in each of the wrappers:

```
cvReleaseImage( &in );
```

This "self-cleaning" mechanism would be very tidy, but it would have the following disadvantage: if we actually did want to do something with one of the intermediate images, we would have no access to it. In order to solve that problem, the preceding code could be simplified as described in Example 2-8.

Example 2-8. Simplifying the image pipeline of Example 2-7 by making the individual stages release their intermediate memory allocations

```
IplImage* out;
out = doPyrDown( in, IPL_GAUSSIAN_5x5 );
out = doPyrDown( out, IPL_GAUSSIAN_5x5 );
out = doCanny( out, 10, 100, 3 );

// do whatever with 'out'
//
…
cvReleaseImage ( &out );
```

One final word of warning on the self-cleaning filter pipeline: in OpenCV we must always be certain that an image (or other structure) being de-allocated is one that was, in fact, explicitly allocated previously. Consider the case of the IplImage* pointer returned by cvCreateFileCapture(). Here the pointer points to a structure allocated as part of the CvCapture structure, and the target structure is allocated only once when the CvCapture is initialized and an AVI is loaded. De-allocating this structure with a call to cvReleaseImage() would result in some nasty surprises. The moral of this story is that, although it's important to take care of garbage collection in OpenCV, we should only clean up the garbage that we have created.

Input from a Camera

Vision can mean many things in the world of computers. In some cases we are analyzing still frames loaded from elsewhere. In other cases we are analyzing video that is being read from disk. In still other cases, we want to work with real-time data streaming in from some kind of camera device.

OpenCV—more specifically, the HighGUI portion of the OpenCV library—provides us with an easy way to handle this situation. The method is analogous to how we read AVIs. Instead of calling cvCreateFileCapture(), we call cvCreateCameraCapture(). The latter routine does not take a file name but rather a camera ID number as its argument. Of course, this is important only when multiple cameras are available. The default value is –1, which means "just pick one"; naturally, this works quite well when there is only one camera to pick (see Chapter 4 for more details).

The cvCreateCameraCapture() function returns the same CvCapture* pointer, which we can hereafter use exactly as we did with the frames grabbed from a video stream. Of course, a lot of work is going on behind the scenes to make a sequence of camera images look like a video, but we are insulated from all of that. We can simply grab images from the camera whenever we are ready for them and proceed as if we did not know the difference. For development reasons, most applications that are intended to operate in real time will have a video-in mode as well, and the universality of the CvCapture structure makes this particularly easy to implement. See Example 2-9.

Example 2-9. After the capture structure is initialized, it no longer matters whether the image is from a camera or a file

```
CvCapture* capture;

if( argc==1 ) {
    capture = cvCreateCameraCapture(0);
} else {
    capture = cvCreateFileCapture( argv[1] );
}
assert( capture != NULL );

// Rest of program proceeds totally ignorant
…
```

As you can see, this arrangement is quite ideal.

Writing to an AVI File

In many applications we will want to record streaming input or even disparate captured images to an output video stream, and OpenCV provides a straightforward method for doing this. Just as we are able to create a capture device that allows us to grab frames one at a time from a video stream, we are able to create a writer device that allows us to place frames one by one into a video file. The routine that allows us to do this is cvCreateVideoWriter().

Once this call has been made, we may successively call cvWriteFrame(), once for each frame, and finally cvReleaseVideoWriter() when we are done. Example 2-10 describes a simple program that opens a video file, reads the contents, converts them to a log-polar format (something like what your eye actually sees, as described in Chapter 6), and writes out the log-polar image to a new video file.

Example 2-10. A complete program to read in a color video and write out the same video in grayscale

```
// Convert a video to grayscale
 // argv[1]: input video file
 // argv[2]: name of new output file
 //
#include "cv.h"
#include "highgui.h"
main( int argc, char* argv[] ) {
    CvCapture* capture = 0;
    capture = cvCreateFileCapture( argv[1] );
    if(!capture){
        return -1;
    }
    IplImage *bgr_frame=cvQueryFrame(capture);//Init the video read
    double fps = cvGetCaptureProperty (
        capture,
        CV_CAP_PROP_FPS
    );
```

Example 2-10. A complete program to read in a color video and write out the same video in grayscale (continued)

```
    CvSize size = cvSize(
        (int)cvGetCaptureProperty( capture, CV_CAP_PROP_FRAME_WIDTH),
        (int)cvGetCaptureProperty( capture, CV_CAP_PROP_FRAME_HEIGHT)
    );
    CvVideoWriter *writer = cvCreateVideoWriter(
        argv[2],
        CV_FOURCC('M','J','P','G'),
        fps,
        size
    );
    IplImage* logpolar_frame = cvCreateImage(
        size,
        IPL_DEPTH_8U,
        3
    );
    while( (bgr_frame=cvQueryFrame(capture)) != NULL ) {
        cvLogPolar( bgr_frame, logpolar_frame,
                    cvPoint2D32f(bgr_frame->width/2,
                    bgr_frame->height/2),
                    40,
                    CV_INTER_LINEAR+CV_WARP_FILL_OUTLIERS );
        cvWriteFrame( writer, logpolar_frame );
    }
    cvReleaseVideoWriter( &writer );
    cvReleaseImage( &logpolar_frame );
    cvReleaseCapture( &capture );
    return(0);
}
```

Looking over this program reveals mostly familiar elements. We open one video; start reading with cvQueryFrame(), which is necessary to read the video properties on some systems; and then use cvGetCaptureProperty() to ascertain various important properties of the video stream. We then open a video file for writing, convert the frame to log-polar format, and write the frames to this new file one at a time until there are none left. Then we close up.

The call to cvCreateVideoWriter() contains several parameters that we should understand. The first is just the filename for the new file. The second is the *video codec* with which the video stream will be compressed. There are countless such codecs in circulation, but whichever codec you choose must be available on your machine (codecs are installed separately from OpenCV). In our case we choose the relatively popular *MJPG* codec; this is indicated to OpenCV by using the macro CV_FOURCC(), which takes four characters as arguments. These characters constitute the "four-character code" of the codec, and every codec has such a code. The four-character code for *motion jpeg* is MJPG, so we specify that as CV_FOURCC('M','J','P','G').

The next two arguments are the replay frame rate, and the size of the images we will be using. In our case, we set these to the values we got from the original (color) video.

Onward

Before moving on to the next chapter, we should take a moment to take stock of where we are and look ahead to what is coming. We have seen that the OpenCV API provides us with a variety of easy-to-use tools for loading still images from files, reading video from disk, or capturing video from cameras. We have also seen that the library contains primitive functions for manipulating these images. What we have not yet seen are the powerful elements of the library, which allow for more sophisticated manipulation of the entire set of abstract data types that are important to practical vision problem solving.

In the next few chapters we will delve more deeply into the basics and come to understand in greater detail both the interface-related functions and the image data types. We will investigate the primitive image manipulation operators and, later, some much more advanced ones. Thereafter, we will be ready to explore the many specialized services that the API provides for tasks as diverse as camera calibration, tracking, and recognition. Ready? Let's go!

Exercises

Download and install OpenCV if you have not already done so. Systematically go through the directory structure. Note in particular the *docs* directory; there you can load *index.htm*, which further links to the main documentation of the library. Further explore the main areas of the library. *Cvcore* contains the basic data structures and algorithms, *cv* contains the image processing and vision algorithms, *ml* includes algorithms for machine learning and clustering, and *otherlibs/highgui* contains the I/O functions. Check out the *_make* directory (containing the OpenCV build files) and also the samples directory, where example code is stored.

1. Go to the *.../opencv/_make* directory. On Windows, open the solution file *opencv .sln*; on Linux, open the appropriate makefile. Build the library in both the debug and the release versions. This may take some time, but you will need the resulting library and *dll* files.

2. Go to the *.../opencv/samples/c/* directory. Create a project or make file and then import and build *lkdemo.c* (this is an example motion tracking program). Attach a camera to your system and run the code. With the display window selected, type "r" to initialize tracking. You can add points by clicking on video positions with the mouse. You can also switch to watching only the points (and not the image) by typing "n". Typing "n" again will toggle between "night" and "day" views.

3. Use the capture and store code in Example 2-10, together with the doPyrDown() code of Example 2-5 to create a program that reads from a camera and stores downsampled color images to disk.

4. Modify the code in exercise 3 and combine it with the window display code in Example 2-1 to display the frames as they are processed.

5. Modify the program of exercise 4 with a slider control from Example 2-3 so that the user can dynamically vary the pyramid downsampling reduction level by factors of between 2 and 8. You may skip writing this to disk, but you should display the results.

CHAPTER 3

Getting to Know OpenCV

OpenCV Primitive Data Types

OpenCV has several primitive data types. These data types are not primitive from the point of view of C, but they are all simple structures, and we will regard them as atomic. You can examine details of the structures described in what follows (as well as other structures) in the *cxtypes.h* header file, which is in the *.../OpenCV/cxcore/include* directory of the OpenCV install.

The simplest of these types is CvPoint. CvPoint is a simple structure with two integer members, x and y. CvPoint has two siblings: CvPoint2D32f and CvPoint3D32f. The former has the same two members x and y, which are both floating-point numbers. The latter also contains a third element, z.

CvSize is more like a cousin to CvPoint. Its members are width and height, which are both integers. If you want floating-point numbers, use CvSize's cousin CvSize2D32f.

CvRect is another child of CvPoint and CvSize; it contains four members: x, y, width, and height. (In case you were worried, this child was adopted.)

Last but not least is CvScalar, which is a set of four double-precision numbers. When memory is not an issue, CvScalar is often used to represent one, two, or three real numbers (in these cases, the unneeded components are simply ignored). CvScalar has a single member val, which is a pointer to an array containing the four double-precision floating-point numbers.

All of these data types have constructor methods with names like cvSize() (generally* the constructor has the same name as the structure type but with the first character not capitalized). Remember that this is C and not C++, so these "constructors" are just inline functions that take a list of arguments and return the desired structure with the values set appropriately.

---

* We say "generally" here because there are a few oddballs. In particular, we have cvScalarAll(double) and cvRealScalar(double); the former returns a CvScalar with all four values set to the argument, while the latter returns a CvScalar with the first value set and the other values 0.

The inline constructors for the data types listed in Table 3-1—cvPointXXX(), cvSize(), cvRect(), and cvScalar()—are extremely useful because they make your code not only easier to write but also easier to read. Suppose you wanted to draw a white rectangle between (5, 10) and (20, 30); you could simply call:

```
cvRectangle(
    myImg,
    cvPoint(5,10),
    cvPoint(20,30),
    cvScalar(255,255,255)
);
```

Table 3-1. Structures for points, size, rectangles, and scalar tuples

Structure	Contains	Represents
CvPoint	int x, y	Point in image
CvPoint2D32f	float x, y	Points in $\Re^2$
CvPoint3D32f	float x, y, z	Points in $\Re^3$
CvSize	int width, height	Size of image
CvRect	int x, y, width, height	Portion of image
CvScalar	double val[4]	RGBA value

cvScalar() is a special case: it has three constructors. The first, called cvScalar(), takes one, two, three, or four arguments and assigns those arguments to the corresponding elements of val[]. The second constructor is cvRealScalar(); it takes one argument, which it assigns to val[0] while setting the other entries to 0. The final variant is cvScalarAll(), which takes a single argument but sets all four elements of val[] to that same argument.

Matrix and Image Types

Figure 3-1 shows the class or structure hierarchy of the three image types. When using OpenCV, you will repeatedly encounter the IplImage data type. You have already seen it many times in the previous chapter. IplImage is the basic structure used to encode what we generally call "images". These images may be grayscale, color, four-channel (RGB+alpha), and each channel may contain any of several types of integer or floating-point numbers. Hence, this type is more general than the ubiquitous three-channel 8-bit RGB image that immediately comes to mind.*

OpenCV provides a vast arsenal of useful operators that act on these images, including tools to resize images, extract individual channels, find the largest or smallest value of a particular channel, add two images, threshold an image, and so on. In this chapter we will examine these sorts of operators carefully.

---

* If you are especially picky, you can say that OpenCV is a design, implemented in C, that is not only object-oriented but also template-oriented.

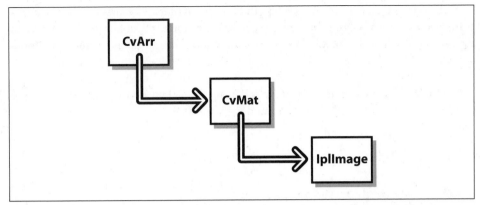

Figure 3-1. Even though OpenCV is implemented in C, the structures used in OpenCV have an object-oriented design; in effect, IplImage is derived from CvMat, which is derived from CvArr

Before we can discuss images in detail, we need to look at another data type: CvMat, the OpenCV matrix structure. Though OpenCV is implemented entirely in C, the relationship between CvMat and IplImage is akin to inheritance in C++. For all intents and purposes, an IplImage can be thought of as being derived from CvMat. Therefore, it is best to understand the (would-be) base class before attempting to understand the added complexities of the derived class. A third class, called CvArr, can be thought of as an abstract base class from which CvMat is itself derived. You will often see CvArr (or, more accurately, CvArr*) in function prototypes. When it appears, it is acceptable to pass CvMat* or IplImage* to the routine.

CvMat Matrix Structure

There are two things you need to know before we dive into the matrix business. First, there is no "vector" construct in OpenCV. Whenever we want a vector, we just use a matrix with one column (or one row, if we want a transpose or conjugate vector). Second, the concept of a matrix in OpenCV is somewhat more abstract than the concept you learned in your linear algebra class. In particular, the elements of a matrix need not themselves be simple numbers. For example, the routine that creates a new two-dimensional matrix has the following prototype:

```
CvMat* cvCreateMat ( int rows, int cols, int type );
```

Here type can be any of a long list of predefined types of the form: CV_<bit_depth>(S|U|F) C<number_of_channels>. Thus, the matrix could consist of 32-bit floats (CV_32FC1), of unsigned integer 8-bit triplets (CV_8UC3), or of countless other elements. An element of a CvMat is not necessarily a single number. Being able to represent multiple values for a single entry in the matrix allows us to do things like represent multiple color channels in an RGB image. For a simple image containing red, green and blue channels, most image operators will be applied to each channel separately (unless otherwise noted).

Internally, the structure of CvMat is relatively simple, as shown in Example 3-1 (you can see this for yourself by opening up *.../opencv/cxcore/include/cxtypes.h*). Matrices have a width, a height, a type, a step (the length of a row in bytes, not ints or floats), and a pointer to a data array (and some more stuff that we won't talk about just yet). You can access these members directly by de-referencing a pointer to CvMat or, for some more popular elements, by using supplied accessor functions. For example, to obtain the size of a matrix, you can get the information you want either by calling cvGetSize(CvMat*), which returns a CvSize structure, or by accessing the height and width independently with such constructs as matrix->height and matrix->width.

Example 3-1. CvMat structure: the matrix "header"

```
typedef struct CvMat {
    int type;
    int step;
    int* refcount;      // for internal use only
    union {
        uchar*  ptr;
        short*  s;
        int*    i;
        float*  fl;
        double* db;
    } data;
    union {
        int rows;
        int height;
    };
    union {
        int cols;
        int width;
    };
} CvMat;
```

This information is generally referred to as the *matrix header*. Many routines distinguish between the header and the data, the latter being the memory that the data element points to.

Matrices can be created in one of several ways. The most common way is to use cvCreateMat(), which is essentially shorthand for the combination of the more atomic functions cvCreateMatHeader() and cvCreateData(). cvCreateMatHeader() creates the CvMat structure without allocating memory for the data, while cvCreateData() handles the data allocation. Sometimes only cvCreateMatHeader() is required, either because you have already allocated the data for some other reason or because you are not yet ready to allocate it. The third method is to use the cvCloneMat(CvMat*), which creates a new matrix from an existing one.* When the matrix is no longer needed, it can be released by calling cvReleaseMat(CvMat**).

The list in Example 3-2 summarizes the functions we have just described as well as some others that are closely related.

---

* cvCloneMat() and other OpenCV functions containing the word "clone" not only create a new header that is identical to the input header, they also allocate a separate data area and copy the data from the source to the new object.

Example 3-2. Matrix creation and release

```
// Create a new rows by cols matrix of type 'type'.
//
CvMat* cvCreateMat( int rows, int cols, int type );

// Create only matrix header without allocating data
//
CvMat* cvCreateMatHeader( int rows, int cols, int type );

// Initialize header on existing CvMat structure
//
CvMat* cvInitMatHeader(
  CvMat* mat,
  int    rows,
  int    cols,
  int    type,
  void*  data = NULL,
  int    step = CV_AUTOSTEP
);

// Like cvInitMatHeader() but allocates CvMat as well.
//
CvMat cvMat(
  int    rows,
  int    cols,
  int    type,
  void*  data = NULL
);

// Allocate a new matrix just like the matrix 'mat'.
//
CvMat* cvCloneMat( const cvMat* mat );

// Free the matrix 'mat', both header and data.
//
void cvReleaseMat( CvMat** mat );
```

Analogously to many OpenCV structures, there is a constructor called cvMat() that creates a CvMat structure. This routine does not actually allocate memory; it only creates the header (this is similar to cvInitMatHeader()). These methods are a good way to take some data you already have lying around, package it by pointing the matrix header to it as in Example 3-3, and run it through routines that process OpenCV matrices.

Example 3-3. Creating an OpenCV matrix with fixed data

```
// Create an OpenCV Matrix containing some fixed data.
//
float vals[] = { 0.866025, -0.500000, 0.500000, 0.866025 };

CvMat rotmat;

cvInitMatHeader(
  &rotmat,
  2,
```

Example 3-3. Creating an OpenCV matrix with fixed data (continued)

```
  2,
  CV_32FC1,
  vals
);
```

Once we have a matrix, there are many things we can do with it. The simplest operations are querying aspects of the array definition and data access. To query the matrix, we have cvGetElemType(const CvArr* arr), cvGetDims(const CvArr* arr, int* sizes=NULL), and cvGetDimSize(const CvArr* arr, int index). The first returns an integer constant representing the type of elements stored in the array (this will be equal to something like CV_8UC1, CV_64FC4, etc). The second takes the array and an optional pointer to an integer; it returns the number of dimensions (two for the cases we are considering, but later on we will encounter *N*-dimensional matrixlike objects). If the integer pointer is not null then it will store the height and width (or *N* dimensions) of the supplied array. The last function takes an integer indicating the dimension of interest and simply returns the extent of the matrix in that dimension.*

Accessing Data in Your Matrix

There are three ways to access the data in your matrix: the easy way, the hard way, and the right way.

The easy way

The easiest way to get at a member element of an array is with the CV_MAT_ELEM() macro. This macro (see Example 3-4) takes the matrix, the type of element to be retrieved, and the row and column numbers and then returns the element.

Example 3-4. Accessing a matrix with the CV_MAT_ELEM() macro

```
CvMat* mat = cvCreateMat( 5, 5, CV_32FC1 );
float element_3_2 = CV_MAT_ELEM( *mat, float, 3, 2 );
```

"Under the hood" this macro is just calling the macro CV_MAT_ELEM_PTR(). CV_MAT_ELEM_PTR() (see Example 3-5) takes as arguments the matrix and the row and column of the desired element and returns (not surprisingly) a pointer to the indicated element. One important difference between CV_MAT_ELEM() and CV_MAT_ELEM_PTR() is that CV_MAT_ELEM() actually casts the pointer to the indicated type before de-referencing it. If you would like to set a value rather than just read it, you can call CV_MAT_ELEM_PTR() directly; in this case, however, you must cast the returned pointer to the appropriate type yourself.

Example 3-5. Setting a single value in a matrix using the CV_MAT_ELEM_PTR() macro

```
CvMat* mat = cvCreateMat( 5, 5, CV_32FC1 );
float element_3_2 = 7.7;
*( (float*)CV_MAT_ELEM_PTR( *mat, 3, 2 ) ) = element_3_2;
```

---

* For the regular two-dimensional matrices discussed here, dimension zero (0) is always the "width" and dimension one (1) is always the height.

Unfortunately, these macros recompute the pointer needed on every call. This means looking up the pointer to the base element of the data area of the matrix, computing an offset to get the address of the information you are interested in, and then adding that offset to the computed base. Thus, although these macros are easy to use, they may not be the best way to access a matrix. This is particularly true when you are planning to access all of the elements in a matrix sequentially. We will come momentarily to the best way to accomplish this important task.

The hard way

The two macros discussed in "The easy way" are suitable only for accessing one- and two-dimensional arrays (recall that one-dimensional arrays, or "vectors", are really just *n*-by-1 matrices). OpenCV provides mechanisms for dealing with multidimensional arrays. In fact OpenCV allows for a general *N*-dimensional matrix that can have as many dimensions as you like.

For accessing data in a general matrix, we use the family of functions cvPtr*D and cvGet*D… listed in Examples 3-6 and 3-7. The cvPtr*D family contains cvPtr1D(), cvPtr2D(), cvPtr3D(), and cvPtrND() .... Each of the first three takes a CvArr* matrix pointer argument followed by the appropriate number of integers for the indices, and an optional argument indicating the type of the output parameter. The routines return a pointer to the element of interest. With cvPtrND(), the second argument is a pointer to an array of integers containing the appropriate number of indices. We will return to this function later. (In the prototypes that follow, you will also notice some optional arguments; we will address those when we need them.)

Example 3-6. Pointer access to matrix structures

```
uchar* cvPtr1D(
  const CvArr* arr,
  int          idx0,
  int*         type = NULL
);

uchar* cvPtr2D(
  const CvArr* arr,
  int          idx0,
  int          idx1,
  int*         type = NULL
);

uchar* cvPtr3D(
  const CvArr* arr,
  int          idx0,
  int          idx1,
  int          idx2,
  int*         type = NULL
);

uchar* cvPtrND(
```

Example 3-6. Pointer access to matrix structures (continued)

```
    const CvArr* arr,
    int*         idx,
    int*         type          = NULL,
    int          create_node   = 1,
    unsigned*    precalc_hashval = NULL
);
```

For merely reading the data, there is another family of functions cvGet*D, listed in Example 3-7, that are analogous to those of Example 3-6 but return the actual value of the matrix element.

Example 3-7. CvMat and IplImage element functions

```
double cvGetReal1D( const CvArr* arr, int idx0 );
double cvGetReal2D( const CvArr* arr, int idx0, int idx1 );
double cvGetReal3D( const CvArr* arr, int idx0, int idx1, int idx2 );
double cvGetRealND( const CvArr* arr, int* idx );

CvScalar cvGet1D( const CvArr* arr, int idx0 );
CvScalar cvGet2D( const CvArr* arr, int idx0, int idx1 );
CvScalar cvGet3D( const CvArr* arr, int idx0, int idx1, int idx2 );
CvScalar cvGetND( const CvArr* arr, int* idx );
```

The return type of cvGet*D is double for four of the routines and CvScalar for the other four. This means that there can be some significant waste when using these functions. They should be used only where convenient and efficient; otherwise, it is better just to use cvPtr*D.

One reason it is better to use cvPtr*D() is that you can use these pointer functions to gain access to a particular point in the matrix and then use pointer arithmetic to move around in the matrix from there. It is important to remember that the channels are contiguous in a multichannel matrix. For example, in a three-channel two-dimensional matrix representing red, green, blue (RGB) bytes, the matrix data is stored: rgbrgbrgb .... Therefore, to move a pointer of the appropriate type to the next channel, we add 1. If we wanted to go to the next "pixel" or set of elements, we'd add and offset equal to the number of channels (in this case 3).

The other trick to know is that the step element in the matrix array (see Examples 3-1 and 3-3) is the length in bytes of a row in the matrix. In that structure, cols or width alone is not enough to move between matrix rows because, for machine efficiency, matrix or image allocation is done to the nearest four-byte boundary. Thus a matrix of width three bytes would be allocated four bytes with the last one ignored. For this reason, if we get a byte pointer to a data element then we add step to the pointer in order to step it to the next row directly below our point. If we have a matrix of integers or floating-point numbers and corresponding int or float pointers to a data element, we would step to the next row by adding step/4; for doubles, we'd add step/8 (this is just to take into account that C will automatically multiply the offsets we add by the data type's byte size).

Somewhat analogous to cvGet*D is cvSet*D in Example 3-8, which sets a matrix or image element with a single call, and the functions cvSetReal*D() and cvSet*D(), which can be used to set the values of elements of a matrix or image.

Example 3-8. Set element functions for CvMat or IplImage.

```
void cvSetReal1D( CvArr* arr, int idx0, double value );
void cvSetReal2D( CvArr* arr, int idx0, int idx1, double value );
void cvSetReal3D(
  CvArr* arr,
  int idx0,
  int idx1,
  int idx2,
  double value
);
void cvSetRealND( CvArr* arr, int* idx, double value );

void cvSet1D( CvArr* arr, int idx0, CvScalar value );
void cvSet2D( CvArr* arr, int idx0, int idx1, CvScalar value );
void cvSet3D(
  CvArr* arr,
  int idx0,
  int idx1,
  int idx2,
  CvScalar value
);
void cvSetND( CvArr* arr, int* idx, CvScalar value );
```

As an added convenience, we also have cvmSet() and cvmGet(), which are used when dealing with single-channel floating-point matrices. They are very simple:

```
double cvmGet( const CvMat* mat, int row, int col )
void cvmSet( CvMat* mat, int row, int col, double value )
```

So the call to the convenience function cvmSet(),

```
cvmSet( mat, 2, 2, 0.5000 );
```

is the same as the call to the equivalent cvSetReal2D function,

```
cvSetReal2D( mat, 2, 2, 0.5000 );
```

The right way

With all of those accessor functions, you might think that there's nothing more to say. In fact, you will rarely use any of the set and get functions. Most of the time, vision is a processor-intensive activity, and you will want to do things in the most efficient way possible. Needless to say, going through these interface functions is not efficient. Instead, you should do your own pointer arithmetic and simply de-reference your way into the matrix. Managing the pointers yourself is particularly important when you want to do something to every element in an array (assuming there is no OpenCV routine that can perform this task for you).

For direct access to the innards of a matrix, all you really need to know is that the data is stored sequentially in raster scan order, where columns ("x") are the fastest-running variable. Channels are interleaved, which means that, in the case of a multichannel matrix, they are a still faster-running ordinal. Example 3-9 shows an example of how this can be done.

Example 3-9. Summing all of the elements in a three-channel matrix

```
float sum( const CvMat* mat ) {

  float s = 0.0f;
  for(int row=0; row<mat->rows; row++ ) {
    const float* ptr = (const float*)(mat->data.ptr + row * mat->step);
    for( col=0; col<mat->cols; col++ ) {
      s += *ptr++;
    }
  }
  return( s );
}
```

When computing the pointer into the matrix, remember that the matrix element data is a union. Therefore, when de-referencing this pointer, you must indicate the correct element of the union in order to obtain the correct pointer type. Then, to offset that pointer, you must use the step element of the matrix. As noted previously, the step element is in bytes. To be safe, it is best to do your pointer arithmetic in bytes and then cast to the appropriate type, in this case float. Although the CVMat structure has the concept of height and width for compatibility with the older IplImage structure, we use the more up-to-date rows and cols instead. Finally, note that we recompute ptr for every row rather than simply starting at the beginning and then incrementing that pointer every read. This might seem excessive, but because the CvMat data pointer could just point to an ROI within a larger array, there is no guarantee that the data will be contiguous across rows.

Arrays of Points

One issue that will come up often—and that is important to understand—is the difference between a multidimensional array (or matrix) of multidimensional objects and an array of one higher dimension that contains only one-dimensional objects. Suppose, for example, that you have *n* points in three dimensions which you want to pass to some OpenCV function that takes an argument of type CvMat* (or, more likely, cvArr*). There are four obvious ways you could do this, and it is absolutely critical to remember that they are not necessarily equivalent. One method would be to use a two-dimensional array of type CV32FC1 with *n* rows and three columns (*n*-by-3). Similarly, you could use a two-dimensional array with three rows and *n* columns (3-by-*n*). You could also use an array with *n* rows and one column (*n*-by-1) of type CV32FC3 or an array with one row and *n* columns (3-by-1). Some of these cases can be freely converted from one to the other (meaning you can just pass one where the other is expected) but others cannot. To understand why, consider the memory layout shown in Figure 3-2.

As you can see in the figure, the points are mapped into memory in the same way for three of the four cases just described above but differently for the last. The situation is even

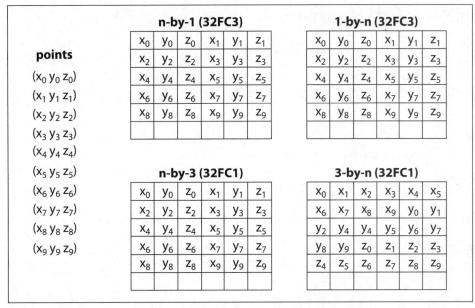

Figure 3-2. A set of ten points, each represented by three floating-point numbers, placed in four arrays that each use a slightly different structure; in three cases the resulting memory layout is identical, but one case is different

more complicated for the case of an N-dimensional array of c-dimensional points. The key thing to remember is that the location of any given point is given by the formula:

$$\delta = (\text{row}) \cdot N_{\text{cols}} \cdot N_{\text{channels}} + (\text{col}) \cdot N_{\text{channels}} + (\text{channel})$$

where N_{cols} and N_{channels} are the number of columns and channels, respectively.* From this formula one can see that, in general, an N-dimensional array of c-dimensional objects is not the same as an $(N + c)$-dimensional array of one-dimensional objects. In the special case of $N = 1$ (i.e., vectors represented either as n-by-1 or 1-by-n arrays), there is a special degeneracy (specifically, the equivalences shown in Figure 3-2) that can sometimes be taken advantage of for performance.

The last detail concerns the OpenCV data types such as CvPoint2D and CvPoint2D32f. These data types are defined as C structures and therefore have a strictly defined memory layout. In particular, the integers or floating-point numbers that these structures comprise are "channel" sequential. As a result, a one-dimensional C-style array of these objects has the same memory layout as an n-by-1 or a 1-by-n array of type CV32FC2. Similar reasoning applies for arrays of structures of the type CvPoint3D32f.

---

* In this context we use the term "channel" to refer to the fastest-running index. This index is the one associated with the C3 part of CV32FC3. Shortly, when we talk about images, the "channel" there will be exactly equivalent to our use of "channel" here.

IplImage Data Structure

With all of that in hand, it is now easy to discuss the IplImage data structure. In essence this object is a CvMat but with some extra goodies buried in it to make the matrix interpretable as an image. This structure was originally defined as part of Intel's Image Processing Library (IPL).* The exact definition of the IplImage structure is shown in Example 3-10.

Example 3-10. IplImage header structure

```
typedef struct _IplImage {
   int                   nSize;
   int                   ID;
   int                   nChannels;
   int                   alphaChannel;
   int                   depth;
   char                  colorModel[4];
   char                  channelSeq[4];
   int                   dataOrder;
   int                   origin;
   int                   align;
   int                   width;
   int                   height;
   struct _IplROI*       roi;
   struct _IplImage*     maskROI;
   void*                 imageId;
   struct _IplTileInfo*  tileInfo;
   int                   imageSize;
   char*                 imageData;
   int                   widthStep;
   int                   BorderMode[4];
   int                   BorderConst[4];
   char*                 imageDataOrigin;
} IplImage;
```

As crazy as it sounds, we want to discuss the function of several of these variables. Some are trivial, but many are very important to understanding how OpenCV interprets and works with images.

After the ubiquitous width and height, depth and nChannels are the next most crucial. The depth variable takes one of a set of values defined in *ipl.h*, which are (unfortunately) not exactly the values we encountered when looking at matrices. This is because for images we tend to deal with the depth and the number of channels separately (whereas in the matrix routines we tended to refer to them simultaneously). The possible depths are listed in Table 3-2.

---

* IPL was the predecessor to the more modern Intel Performance Primitives (IPP), discussed in Chapter 1. Many of the OpenCV functions are actually relatively thin wrappers around the corresponding IPL or IPP routines. This is why it is so easy for OpenCV to swap in the high-performance IPP library routines when available.

Table 3-2. OpenCV image types

Macro	Image pixel type
IPL_DEPTH_8U	Unsigned 8-bit integer (8u)
IPL_DEPTH_8S	Signed 8-bit integer (8s)
IPL_DEPTH_16S	Signed 16-bit integer (16s)
IPL_DEPTH_32S	Signed 32-bit integer (32s)
IPL_DEPTH_32F	32-bit floating-point single-precision (32f)
IPL_DEPTH_64F	64-bit floating-point double-precision (64f)

The possible values for nChannels are 1, 2, 3, or 4.

The next two important members are origin and dataOrder. The origin variable can take one of two values: IPL_ORIGIN_TL or IPL_ORIGIN_BL, corresponding to the origin of coordinates being located in either the upper-left or lower-left corners of the image, respectively. The lack of a standard origin (upper versus lower) is an important source of error in computer vision routines. In particular, depending on where an image came from, the operating system, codec, storage format, and so forth can all affect the location of the origin of the coordinates of a particular image. For example, you may think you are sampling pixels from a face in the top quadrant of an image when you are really sampling from a shirt in the bottom quadrant. It is best to check the system the first time through by drawing where you think you are operating on an image patch.

The dataOrder may be either IPL_DATA_ORDER_PIXEL or IPL_DATA_ORDER_PLANE.* This value indicates whether the data should be packed with multiple channels one after the other for each pixel (*interleaved,* the usual case), or rather all of the channels clustered into image planes with the planes placed one after another.

The parameter widthStep contains the number of bytes between points in the same column and successive rows (similar to the "step" parameter of CvMat discussed earlier). The variable width is not sufficient to calculate the distance because each row may be aligned with a certain number of bytes to achieve faster processing of the image; hence there may be some gaps between the end of *i*th row and the start of $(i + 1)$ row. The parameter imageData contains a pointer to the first row of image data. If there are several separate planes in the image (as when dataOrder = IPL_DATA_ORDER_PLANE) then they are placed consecutively as separate images with height*nChannels rows in total, but normally they are interleaved so that the number of rows is equal to height and with each row containing the interleaved channels in order.

Finally there is the practical and important *region of interest* (ROI), which is actually an instance of another IPL/IPP structure, IplROI. An IplROI contains an xOffset, a yOffset,

---

* We say that dataOrder may be either IPL_DATA_ORDER_PIXEL or IPL_DATA_ORDER_PLANE, but in fact only IPL_DATA_ORDER_PIXEL is supported by OpenCV. Both values are generally supported by IPL/IPP, but OpenCV always uses interleaved images.

a height, a width, and a coi, where COI stands for *channel of interest*.* The idea behind the ROI is that, once it is set, functions that would normally operate on the entire image will instead act only on the subset of the image indicated by the ROI. All OpenCV functions will use ROI if set. If the COI is set to a nonzero value then some operators will act only on the indicated channel.† Unfortunately, many OpenCV functions ignore this parameter.

Accessing Image Data

When working with image data we usually need to do so quickly and efficiently. This suggests that we should not subject ourselves to the overhead of calling accessor functions like cvSet*D or their equivalent. Indeed, we would like to access the data inside of the image in the most direct way possible. With our knowledge of the internals of the IplImage structure, we can now understand how best to do this.

Even though there are often well-optimized routines in OpenCV that accomplish many of the tasks we need to perform on images, there will always be tasks for which there is no prepackaged routine in the library. Consider the case of a three-channel HSV [Smith78] image‡ in which we want to set the saturation and value to 255 (their maximal values for an 8-bit image) while leaving the hue unmodified. We can do this best by handling the pointers into the image ourselves, much as we did with matrices in Example 3-9. However, there are a few minor differences that stem from the difference between the IplImage and CvMat structures. Example 3-11 shows the fastest way.

Example 3-11. Maxing out (saturating) only the "S" and "V" parts of an HSV image

```
void saturate_sv( IplImage* img ) {

  for( int y=0; y<img->height; y++ ) {
    uchar* ptr = (uchar*) (
      img->imageData + y * img->widthStep
    );
    for( int x=0; x<img->width; x++ ) {
      ptr[3*x+1] = 255;
      ptr[3*x+2] = 255;
    }
  }
}
```

We simply compute the pointer ptr directly as the head of the relevant row y. From there, we de-reference the saturation and value of the x column. Because this is a three-channel image, the location of channel c in column x is 3*x+c.

---

* Unlike other parts of the ROI, the COI is not respected by all OpenCV functions. More on this later, but for now you should keep in mind that COI is not as universally applied as the rest of the ROI.

† For the COI, the terminology is to indicate the channel as 1, 2, 3, or 4 and to reserve 0 for deactivating the COI all together (something like a "don't care").

‡ In OpenCV, an HSV image does not differ from an RGB image except in terms of how the channels are interpreted. As a result, constructing an HSV image from an RGB image actually occurs entirely within the "data" area; there is no representation in the header of what meaning is "intended" for the data channels.

One important difference between the `IplImage` case and the `CvMat` case is the behavior of `imageData`, compared to the element `data` of `CvMat`. The `data` element of `CvMat` is a union, so you must indicate which pointer type you want to use. The `imageData` pointer is a byte pointer (`uchar*`). We already know that the data pointed to is not necessarily of type `uchar`, which means that—when doing pointer arithmetic on images—you can simply add `widthStep` (also measured in bytes) without worrying about the actual data type until after the addition, when you cast the resultant pointer to the data type you need. To recap: when working with matrices, you must scale down the offset because the data pointer may be of nonbyte type; when working with images, you can use the offset "as is" because the data pointer is always of a byte type, so you can just cast the whole thing when you are ready to use it.

More on ROI and widthStep

ROI and `widthStep` have great practical importance, since in many situations they speed up computer vision operations by allowing the code to process only a small subregion of the image. Support for ROI and `widthStep` is universal in OpenCV:* every function allows operation to be limited to a subregion. To turn ROI on or off, use the `cvSetImageROI()` and `cvResetImageROI()` functions. Given a rectangular subregion of interest in the form of a `CvRect`, you may pass an image pointer and the rectangle to `cvSetImageROI()` to "turn on" ROI; "turn off" ROI by passing the image pointer to `cvResetImageROI()`.

```
void cvSetImageROI( IplImage* image, CvRect rect );
void cvResetImageROI( IplImage* image );
```

To see how ROI is used, let's suppose we want to load an image and modify some region of that image. The code in Example 3-12 reads an image and then sets the x, y, width, and height of the intended ROI and finally an integer value add to increment the ROI region with. The program then sets the ROI using the convenience of the inline `cvRect()` constructor. It's important to release the ROI with `cvResetImageROI()`, for otherwise the display will observe the ROI and dutifully display only the ROI region.

Example 3-12. Using ImageROI to increment all of the pixels in a region

```
// roi_add <image> <x> <y> <width> <height> <add>
#include <cv.h>
#include <highgui.h>

int main(int argc, char** argv)
{
    IplImage* src;
    if( argc == 7 && ((src=cvLoadImage(argv[1],1)) != 0 ))
    {
        int x = atoi(argv[2]);
        int y = atoi(argv[3]);
        int width = atoi(argv[4]);
        int height = atoi(argv[5]);
```

---

* Well, in theory at least. Any nonadherence to `widthStep` or ROI is considered a bug and may be posted as such to SourceForge, where it will go on a "to fix" list. This is in contrast with color channel of interest, "COI", which is supported only where explicitly stated.

```
        int add = atoi(argv[6]);
        cvSetImageROI(src, cvRect(x,y,width,height));
        cvAddS(src, cvScalar(add),src);
        cvResetImageROI(src);
        cvNamedWindow( "Roi_Add", 1 );
        cvShowImage( "Roi_Add", src );
        cvWaitKey();
    }
    return 0;
}
```

Figure 3-3 shows the result of adding 150 to the blue channel of the image of a cat with an ROI centered over its face, using the code from Example 3-12.

Figure 3-3. Result of adding 150 to the face ROI of a cat

We can achieve the same effect by clever use of widthStep. To do this, we create another image header and set its width and height equal to the interest_rect width and height. We also need to set the image origin (upper left or lower left) to be the same as the interest_img. Next we set the widthStep of this subimage to be the widthStep of the larger interest_ img; this way, stepping by rows in the subimage steps you to the appropriate place at the start of the next line of the subregion within the larger image. We finally set the subimage imageData pointer the start of the interest subregion, as shown in Example 3-13.

Example 3-13. Using alternate widthStep method to increment all of the pixels of interest_img by 1

```
// Assuming IplImage *interest_img; and
//   CvRect interest_rect;
//   Use widthStep to get a region of interest
//
// (Alternate method)
//
IplImage *sub_img = cvCreateImageHeader(
  cvSize(
      interest_rect.width,
      interest_rect.height
  ),
  interest_img->depth,
  interest_img->nChannels
);

sub_img->origin = interest_img->origin;

sub_img->widthStep = interest_img->widthStep;

sub_img->imageData = interest_img->imageData +
    interest_rect.y * interest_img->widthStep   +
    interest_rect.x * interest_img->nChannels;

cvAddS( sub_img, cvScalar(1), sub_img );

cvReleaseImageHeader(&sub_img);
```

So, why would you want to use the widthStep trick when setting and resetting ROI seem to be more convenient? The reason is that there are times when you want to set and perhaps keep multiple subregions of an image active during processing, but ROI can only be done serially and must be set and reset constantly.

Finally, a word should be said here about masks. The cvAddS() function used in the code examples allows the use of a fourth argument that defaults to NULL: const CvArr* mask=NULL. This is an 8-bit single-channel array that allows you to restrict processing to an arbitrarily shaped mask region indicated by nonzero pixels in the mask. If ROI is set along with a mask, processing will be restricted to the intersection of the ROI and the mask. Masks can be used only in functions that specify their use.

Matrix and Image Operators

Table 3-3 lists a variety of routines for matrix manipulation, most of which work equally well for images. They do all of the "usual" things, such as diagonalizing or transposing a matrix, as well as some more complicated operations, such as computing image statistics.

Table 3-3. Basic matrix and image operators

Function	Description
cvAbs	Absolute value of all elements in an array
cvAbsDiff	Absolute value of differences between two arrays
cvAbsDiffS	Absolute value of difference between an array and a scalar
cvAdd	Elementwise addition of two arrays
cvAddS	Elementwise addition of an array and a scalar
cvAddWeighted	Elementwise weighted addition of two arrays (alpha blending)
cvAvg	Average value of all elements in an array
cvAvgSdv	Absolute value and standard deviation of all elements in an array
cvCalcCovarMatrix	Compute covariance of a set of n-dimensional vectors
cvCmp	Apply selected comparison operator to all elements in two arrays
cvCmpS	Apply selected comparison operator to an array relative to a scalar
cvConvertScale	Convert array type with optional rescaling of the value
cvConvertScaleAbs	Convert array type after absolute value with optional rescaling
cvCopy	Copy elements of one array to another
cvCountNonZero	Count nonzero elements in an array
cvCrossProduct	Compute cross product of two three-dimensional vectors
cvCvtColor	Convert channels of an array from one color space to another
cvDet	Compute determinant of a square matrix
cvDiv	Elementwise division of one array by another
cvDotProduct	Compute dot product of two vectors
cvEigenVV	Compute eigenvalues and eigenvectors of a square matrix
cvFlip	Flip an array about a selected axis
cvGEMM	Generalized matrix multiplication
cvGetCol	Copy elements from column slice of an array
cvGetCols	Copy elements from multiple adjacent columns of an array
cvGetDiag	Copy elements from an array diagonal
cvGetDims	Return the number of dimensions of an array
cvGetDimSize	Return the sizes of all dimensions of an array
cvGetRow	Copy elements from row slice of an array
cvGetRows	Copy elements from multiple adjacent rows of an array
cvGetSize	Get size of a two-dimensional array and return as CvSize
cvGetSubRect	Copy elements from subregion of an array
cvInRange	Test if elements of an array are within values of two other arrays
cvInRangeS	Test if elements of an array are in range between two scalars
cvInvert	Invert a square matrix

Table 3-3. *Basic matrix and image operators* (continued)

Function	Description
cvMahalonobis	Compute Mahalonobis distance between two vectors
cvMax	Elementwise max operation on two arrays
cvMaxS	Elementwise max operation between an array and a scalar
cvMerge	Merge several single-channel images into one multichannel image
cvMin	Elementwise min operation on two arrays
cvMinS	Elementwise min operation between an array and a scalar
cvMinMaxLoc	Find minimum and maximum values in an array
cvMul	Elementwise multiplication of two arrays
cvNot	Bitwise inversion of every element of an array
cvNorm	Compute normalized correlations between two arrays
cvNormalize	Normalize elements in an array to some value
cvOr	Elementwise bit-level OR of two arrays
cvOrS	Elementwise bit-level OR of an array and a scalar
cvReduce	Reduce a two-dimensional array to a vector by a given operation
cvRepeat	Tile the contents of one array into another
cvSet	Set all elements of an array to a given value
cvSetZero	Set all elements of an array to 0
cvSetIdentity	Set all elements of an array to 1 for the diagonal and 0 otherwise
cvSolve	Solve a system of linear equations
cvSplit	Split a multichannel array into multiple single-channel arrays
cvSub	Elementwise subtraction of one array from another
cvSubS	Elementwise subtraction of a scalar from an array
cvSubRS	Elementwise subtraction of an array from a scalar
cvSum	Sum all elements of an array
cvSVD	Compute singular value decomposition of a two-dimensional array
cvSVBkSb	Compute singular value back-substitution
cvTrace	Compute the trace of an array
cvTranspose	Transpose all elements of an array across the diagonal
cvXor	Elementwise bit-level XOR between two arrays
cvXorS	Elementwise bit-level XOR between an array and a scalar
cvZero	Set all elements of an array to 0

cvAbs, cvAbsDiff, and cvAbsDiffS

```
void cvAbs(
    const CvArr* src,
    const        dst
);
```

```
void cvAbsDiff(
    const CvArr* src1,
    const CvArr* src2,
    const        dst
);
void cvAbsDiffS(
    const CvArr* src,
    CvScalar     value,
    const        dst
);
```

These functions compute the absolute value of an array or of the difference between the array and some reference. The cvAbs() function simply computes the absolute value of the elements in src and writes the result to dst; cvAbsDiff() first subtracts src2 from src1 and then writes the absolute value of the difference to dst. Note that cvAbsDiffS() is essentially the same as cvAbsDiff() except that the value subtracted from all of the elements of src is the constant scalar value.

cvAdd, cvAddS, cvAddWeighted, and alpha blending

```
void cvAdd(
    const CvArr* src1,
    const CvArr* src2,
    CvArr*       dst,
    const CvArr* mask = NULL
);
void cvAddS(
    const CvArr* src,
    CvScalar     value,
    CvArr*       dst,
    const CvArr* mask = NULL
);
void  cvAddWeighted(
    const CvArr* src1,
    double       alpha,
    const CvArr* src2,
    double       beta,
    double       gamma,
    CvArr*       dst
);
```

cvAdd() is a simple addition function: it adds all of the elements in src1 to the corresponding elements in src2 and puts the results in dst. If mask is not set to NULL, then any element of dst that corresponds to a zero element of mask remains unaltered by this operation. The closely related function cvAddS() does the same thing except that the constant scalar value is added to every element of src.

The function cvAddWeighted() is similar to cvAdd() except that the result written to dst is computed according to the following formula:

$$\mathrm{dst}_{x,y} = \alpha \cdot \mathrm{src1}_{x,y} + \beta \cdot \mathrm{src2}_{x,y} + \gamma$$

This function can be used to implement *alpha blending* [Smith79; Porter84]; that is, it can be used to blend one image with another. The form of this function is:

```
void  cvAddWeighted(
    const CvArr* src1,
    double       alpha,
    const CvArr* src2,
    double       beta,
    double       gamma,
    CvArr*       dst
);
```

In cvAddWeighted() we have two source images, src1 and src2. These images may be of any pixel type so long as both are of the same type. They may also be one or three channels (grayscale or color), again as long as they agree. The destination result image, dst, must also have the same pixel type as src1 and src2. These images may be of different sizes, but their ROIs must agree in size or else OpenCV will issue an error. The parameter alpha is the blending strength of src1, and beta is the blending strength of src2. The alpha blending equation is:

$$\text{dst}_{x,y} = \alpha \cdot \text{src1}_{x,y} + \beta \cdot \text{src2}_{x,y} + \gamma$$

You can convert to the standard alpha blend equation by choosing α between 0 and 1, setting $\beta = 1 - \alpha$, and setting γ to 0; this yields:

$$\text{dst}_{x,y} = \alpha \cdot \text{src1}_{x,y} + (1-\alpha) \cdot \text{src2}_{x,y}$$

However, cvAddWeighted() gives us a little more flexibility—both in how we weight the blended images and in the additional parameter γ, which allows for an additive offset to the resulting destination image. For the general form, you will probably want to keep alpha and beta at no less than 0 and their sum at no more than 1; gamma may be set depending on average or max image value to scale the pixels up. A program showing the use of alpha blending is shown in Example 3-14.

Example 3-14. Complete program to alpha blend the ROI starting at (0,0) in src2 with the ROI starting at (x,y) in src1

```
// alphablend <imageA> <image B> <x> <y> <width> <height>
//            <alpha> <beta>
#include <cv.h>
#include <highgui.h>

int main(int argc, char** argv)
{
    IplImage *src1, *src2;
    if( argc == 9 && ((src1=cvLoadImage(argv[1],1)) != 0
        )&&((src2=cvLoadImage(argv[2],1)) != 0 ))
    {
        int x = atoi(argv[3]);
        int y = atoi(argv[4]);
        int width = atoi(argv[5]);
```

```
        int height = atoi(argv[6]);
        double alpha = (double)atof(argv[7]);
        double beta  = (double)atof(argv[8]);
        cvSetImageROI(src1, cvRect(x,y,width,height));
        cvSetImageROI(src2, cvRect(0,0,width,height));
        cvAddWeighted(src1, alpha, src2, beta,0.0,src1);
        cvResetImageROI(src1);
        cvNamedWindow( "Alpha_blend", 1 );
        cvShowImage( "Alpha_blend", src1 );
        cvWaitKey();
    }
    return 0;
}
```

The code in Example 3-14 takes two source images: the primary one (src1) and the one to blend (src2). It reads in a rectangle ROI for src1 and applies an ROI of the same size to src2, this time located at the origin. It reads in alpha and beta levels but sets gamma to 0. Alpha blending is applied using cvAddWeighted(), and the results are put into src1 and displayed. Example output is shown in Figure 3-4, where the face of a child is blended onto the face and body of a cat. Note that the code took the same ROI as in the ROI addition example in Figure 3-3. This time we used the ROI as the target blending region.

cvAnd and cvAndS

```
void cvAnd(
    const CvArr* src1,
    const CvArr* src2,
    CvArr*       dst,
    const CvArr* mask = NULL
);
void cvAndS(
    const CvArr* src1,
    CvScalar     value,
    CvArr*       dst,
    const CvArr* mask = NULL
);
```

These two functions compute a bitwise AND operation on the array src1. In the case of cvAnd(), each element of dst is computed as the bitwise AND of the corresponding two elements of src1 and src2. In the case of cvAndS(), the bitwise AND is computed with the constant scalar value. As always, if mask is non-NULL then only the elements of dst corresponding to nonzero entries in mask are computed.

Though all data types are supported, src1 and src2 must have the same data type for cvAnd(). If the elements are of a floating-point type, then the bitwise representation of that floating-point number is used.

Figure 3-4. The face of a child is alpha blended onto the face of a cat

cvAvg

```
CvScalar cvAvg(
    const CvArr* arr,
    const CvArr* mask = NULL
);
```

cvAvg() computes the average value of the pixels in arr. If mask is non-NULL then the average will be computed only over those pixels for which the corresponding value of mask is nonzero.

This function has the now deprecated alias cvMean().

cvAvgSdv

```
cvAvgSdv(
    const CvArr* arr,
    CvScalar*    mean,
    CvScalar*    std_dev,
    const CvArr* mask     = NULL
);
```

This function is like cvAvg(), but in addition to the average it also computes the standard deviation of the pixels.

This function has the now deprecated alias cvMean_StdDev().

cvCalcCovarMatrix

```
void cvAdd(
    const CvArr**  vects,
    int            count,
    CvArr*         cov_mat,
    CvArr*         avg,
    int            flags
);
```

Given any number of vectors, cvCalcCovarMatrix() will compute the mean and *covariance matrix* for the Gaussian approximation to the distribution of those points. This can be used in many ways, of course, and OpenCV has some additional flags that will help in particular contexts (see Table 3-4). These flags may be combined by the standard use of the Boolean OR operator.

Table 3-4. Possible components of flags argument to cvCalcCovarMatrix()

Flag in flags argument	Meaning
CV_COVAR_NORMAL	Compute mean and covariance
CV_COVAR_SCRAMBLED	Fast PCA "scrambled" covariance
CV_COVAR_USE_AVERAGE	Use avg as input instead of computing it
CV_COVAR_SCALE	Rescale output covariance matrix

In all cases, the vectors are supplied in vects as an array of OpenCV arrays (i.e., a pointer to a list of pointers to arrays), with the argument count indicating how many arrays are being supplied. The results will be placed in cov_mat in all cases, but the exact meaning of avg depends on the flag values (see Table 3-4).

The flags CV_COVAR_NORMAL and CV_COVAR_SCRAMBLED are mutually exclusive; you should use one or the other but not both. In the case of CV_COVAR_NORMAL, the function will simply compute the mean and covariance of the points provided.

$$\Sigma^2_{\text{normal}} = z \begin{bmatrix} v_{0,0} - \bar{v}_0 & \cdots & v_{m,0} - \bar{v}_0 \\ \vdots & \ddots & \vdots \\ v_{0,n} - \bar{v}_n & \cdots & v_{m,n} - \bar{v}_n \end{bmatrix} \begin{bmatrix} v_{0,0} - \bar{v}_0 & \cdots & v_{m,0} - \bar{v}_0 \\ \vdots & \ddots & \vdots \\ v_{0,n} - \bar{v}_n & \cdots & v_{m,n} - \bar{v}_n \end{bmatrix}^T$$

Thus the normal covariance Σ^2_{normal} is computed from the m vectors of length n, where $\bar{v}_n$ is defined as the nth element of the average vector $\bar{v}$. The resulting covariance matrix is an n-by-n matrix. The factor z is an optional scale factor; it will be set to 1 unless the CV_COVAR_SCALE flag is used.

In the case of CV_COVAR_SCRAMBLED, cvCalcCovarMatrix() will compute the following:

$$\Sigma^2_{\text{scrambled}} = z \begin{bmatrix} v_{0,0} - \overline{v}_0 & \cdots & v_{m,0} - \overline{v}_0 \\ \vdots & \ddots & \vdots \\ v_{0,n} - \overline{v}_n & \cdots & v_{m,n} - \overline{v}_n \end{bmatrix}^{\text{T}} \begin{bmatrix} v_{0,0} - \overline{v}_0 & \cdots & v_{m,0} - \overline{v}_0 \\ \vdots & \ddots & \vdots \\ v_{0,n} - \overline{v}_n & \cdots & v_{m,n} - \overline{v}_n \end{bmatrix}$$

This matrix is not the usual covariance matrix (note the location of the transpose operator). This matrix is computed from the same m vectors of length n, but the resulting *scrambled covariance* matrix is an m-by-m matrix. This matrix is used in some specific algorithms such as fast PCA for very large vectors (as in the *eigenfaces* technique for face recognition).

The flag CV_COVAR_USE_AVG is used when the mean of the input vectors is already known. In this case, the argument avg is used as an input rather than an output, which reduces computation time.

Finally, the flag CV_COVAR_SCALE is used to apply a uniform scale to the covariance matrix calculated. This is the factor z in the preceding equations. When used in conjunction with the CV_COVAR_NORMAL flag, the applied scale factor will be $1.0/m$ (or, equivalently, $1.0/$ count). If instead CV_COVAR_SCRAMBLED is used, then the value of z will be $1.0/n$ (the inverse of the length of the vectors).

The input and output arrays to cvCalcCovarMatrix() should all be of the same floating-point type. The size of the resulting matrix cov_mat should be either n-by-n or m-by-m depending on whether the standard or scrambled covariance is being computed. It should be noted that the "vectors" input in vects do not actually have to be one-dimensional; they can be two-dimensional objects (e.g., images) as well.

cvCmp and cvCmpS

```
void cvCmp(
    const CvArr* src1,
    const CvArr* src2,
    CvArr*       dst,
    int          cmp_op
);
void cvCmpS(
    const CvArr* src,
    double       value,
    CvArr*       dst,
    int          cmp_op
);
```

Both of these functions make comparisons, either between corresponding pixels in two images or between pixels in one image and a constant scalar value. Both cvCmp() and cvCmpS() take as their last argument a comparison operator, which may be any of the types listed in Table 3-5.

*Table 3-5. Values of cmp_op used by cvCmp() and cvCmpS()
and the resulting comparison operation performed*

Value of cmp_op	Comparison
CV_CMP_EQ	(src1i == src2i)
CV_CMP_GT	(src1i > src2i)
CV_CMP_GE	(src1i >= src2i)
CV_CMP_LT	(src1i < src2i)
CV_CMP_LE	(src1i <= src2i)
CV_CMP_NE	(src1i != src2i)

All the listed comparisons are done with the same functions; you just pass in the appropriate argument to indicate what you would like done. These particular functions operate only on single-channel images.

These comparison functions are useful in applications where you employ some version of background subtraction and want to mask the results (e.g., looking at a video stream from a security camera) such that only novel information is pulled out of the image.

cvConvertScale

```
void cvConvertScale(
    const CvArr* src,
    CvArr*       dst,
    double       scale = 1.0,
    double       shift = 0.0
);
```

The cvConvertScale() function is actually several functions rolled into one; it will perform any of several functions or, if desired, all of them together. The first function is to convert the data type in the source image to the data type of the destination image. For example, if we have an 8-bit RGB grayscale image and would like to convert it to a 16-bit signed image, we can do that by calling cvConvertScale().

The second function of cvConvertScale() is to perform a linear transformation on the image data. After conversion to the new data type, each pixel value will be multiplied by the value scale and then have added to it the value shift.

It is critical to remember that, even though "Convert" precedes "Scale" in the function name, the actual order in which these operations is performed is the opposite. Specifically, multiplication by scale and the addition of shift occurs before the type conversion takes place.

When you simply pass the default values (scale = 1.0 and shift = 0.0), you need not have performance fears; OpenCV is smart enough to recognize this case and not waste processor time on useless operations. For clarity (if you think it adds any), OpenCV also provides the macro cvConvert(), which is the same as cvConvertScale() but is conventionally used when the scale and shift arguments will be left at their default values.

cvConvertScale() will work on all data types and any number of channels, but the number of channels in the source and destination images must be the same. (If you want to, say, convert from color to grayscale or vice versa, see cvCvtColor(), which is coming up shortly.)

cvConvertScaleAbs

```
void cvConvertScaleAbs(
    const CvArr* src,
    CvArr*       dst,
    double       scale = 1.0,
    double       shift = 0.0
);
```

cvConvertScaleAbs() is essentially identical to cvConvertScale() except that the dst image contains the absolute value of the resulting data. Specifically, cvConvertScaleAbs() first scales and shifts, then computes the absolute value, and finally performs the datatype conversion.

cvCopy

```
void cvCopy(
    const CvArr* src,
    CvArr*       dst,
    const CvArr* mask = NULL
);
```

This is how you copy one image to another. The cvCopy() function expects both arrays to have the same type, the same size, and the same number of dimensions. You can use it to copy sparse arrays as well, but for this the use of mask is not supported. For nonsparse arrays and images, the effect of mask (if non-NULL) is that only the pixels in dst that correspond to nonzero entries in mask will be altered.

cvCountNonZero

```
int cvCountNonZero( const CvArr* arr );
```

cvCountNonZero() returns the number of nonzero pixels in the array arr.

cvCrossProduct

```
void cvCrossProduct(
    const CvArr* src1,
    const CvArr* src2,
    CvArr*       dst
);
```

This function computes the vector cross product [Lagrange1773] of two three-dimensional vectors. It does not matter if the vectors are in row or column form (a little reflection reveals that, for single-channel objects, these two are really the same internally). Both src1 and src2 should be single-channel arrays, and dst should be single-channel and of length exactly 3.All three arrays should be of the same data type.

cvCvtColor

```
void cvCvtColor(
    const CvArr* src,
    CvArr*       dst,
    int          code
);
```

The previous functions were for converting from one data type to another, and they expected the number of channels to be the same in both source and destination images. The complementary function is cvCvtColor(), which converts from one color space (number of channels) to another [Wharton71] while expecting the data type to be the same. The exact conversion operation to be done is specified by the argument code, whose possible values are listed in Table 3-6.*

Table 3-6. Conversions available by means of cvCvtColor()

Conversion code	Meaning
CV_BGR2RGB CV_RGB2BGR CV_RGBA2BGRA CV_BGRA2RGBA	Convert between RGB and BGR color spaces (with or without alpha channel)
CV_RGB2RGBA CV_BGR2BGRA	Add alpha channel to RGB or BGR image
CV_RGBA2RGB CV_BGRA2BGR	Remove alpha channel from RGB or BGR image
CV_RGB2BGRA CV_RGBA2BGR CV_BGRA2RGB CV_BGR2RGBA	Convert RGB to BGR color spaces while adding or removing alpha channel
CV_RGB2GRAY CV_BGR2GRAY	Convert RGB or BGR color spaces to grayscale
CV_GRAY2RGB CV_GRAY2BGR CV_RGBA2GRAY CV_BGRA2GRAY	Convert grayscale to RGB or BGR color spaces (optionally removing alpha channel in the process)
CV_GRAY2RGBA CV_GRAY2BGRA	Convert grayscale to RGB or BGR color spaces and add alpha channel
CV_RGB2BGR565 CV_BGR2BGR565 CV_BGR5652RGB CV_BGR5652BGR CV_RGBA2BGR565 CV_BGRA2BGR565 CV_BGR5652RGBA CV_BGR5652BGRA	Convert from RGB or BGR color space to BGR565 color representation with optional addition or removal of alpha channel (16-bit images)
CV_GRAY2BGR565 CV_BGR5652GRAY	Convert grayscale to BGR565 color representation or vice versa (16-bit images)

---

* Long-time users of IPL should note that the function cvCvtColor() ignores the colorModel and chan-
  nelSeq fields of the IplImage header. The conversions are done exactly as implied by the code argument.

Table 3-6. Conversions available by means of cvCvtColor() (continued)

Conversion code	Meaning
CV_RGB2BGR555 CV_BGR2BGR555 CV_BGR5552RGB CV_BGR5552BGR CV_RGBA2BGR555 CV_BGRA2BGR555 CV_BGR5552RGBA CV_BGR5552BGRA	Convert from RGB or BGR color space to BGR555 color representation with optional addition or removal of alpha channel (16-bit images)
CV_GRAY2BGR555 CV_BGR5552GRAY	Convert grayscale to BGR555 color representation or vice versa (16-bit images)
CV_RGB2XYZ CV_BGR2XYZ CV_XYZ2RGB CV_XYZ2BGR	Convert RGB or BGR image to CIE XYZ representation or vice versa (Rec 709 with D65 white point)
CV_RGB2YCrCb CV_BGR2YCrCb CV_YCrCb2RGB CV_YCrCb2BGR	Convert RGB or BGR image to luma-chroma (aka YCC) color representation
CV_RGB2HSV CV_BGR2HSV CV_HSV2RGB CV_HSV2BGR	Convert RGB or BGR image to HSV (hue saturation value) color representation or vice versa
CV_RGB2HLS CV_BGR2HLS CV_HLS2RGB CV_HLS2BGR	Convert RGB or BGR image to HLS (hue lightness saturation) color representation or vice versa
CV_RGB2Lab CV_BGR2Lab CV_Lab2RGB CV_Lab2BGR	Convert RGB or BGR image to CIE Lab color representation or vice versa
CV_RGB2Luv CV_BGR2Luv CV_Luv2RGB CV_Luv2BGR	Convert RGB or BGR image to CIE Luv color representation
CV_BayerBG2RGB CV_BayerGB2RGB CV_BayerRG2RGB CV_BayerGR2RGB CV_BayerBG2BGR CV_BayerGB2BGR CV_BayerRG2BGR CV_BayerGR2BGR	Convert from Bayer pattern (single-channel) to RGB or BGR image

The details of many of these conversions are nontrivial, and we will not go into the subtleties of Bayer representations of the CIE color spaces here. For our purposes, it is sufficient to note that OpenCV contains tools to convert to and from these various color spaces, which are of importance to various classes of users.

The color-space conversions all use the conventions: 8-bit images are in the range 0–255, 16-bit images are in the range 0–65536, and floating-point numbers are in the range

0.0–1.0. When grayscale images are converted to color images, all components of the resulting image are taken to be equal; but for the reverse transformation (e.g., RGB or BGR to grayscale), the gray value is computed using the perceptually weighted formula:

$$Y = (0.299)R + (0.587)G + (0.114)B$$

In the case of HSV or HLS representations, hue is normally represented as a value from 0 to 360.* This can cause trouble in 8-bit representations and so, when converting to HSV, the hue is divided by 2 when the output image is an 8-bit image.

cvDet

```
double cvDet(
    const CvArr* mat
);
```

cvDet() computes the determinant (Det) of a square array. The array can be of any data type, but it must be single-channel. If the matrix is small then the determinant is directly computed by the standard formula. For large matrices, this is not particularly efficient and so the determinant is computed by *Gaussian elimination*.

It is worth noting that if you already know that a matrix is symmetric and has a positive determinant, you can also use the trick of solving via *singular value decomposition* (SVD). For more information see the section "cvSVD" to follow, but the trick is to set both **U** and **V** to NULL and then just take the products of the matrix **W** to obtain the determinant.

cvDiv

```
void cvDiv(
    const CvArr* src1,
    const CvArr* src2,
    CvArr*       dst,
    double       scale = 1
);
```

cvDiv() is a simple division function; it divides all of the elements in src1 by the corresponding elements in src2 and puts the results in dst. If mask is non-NULL, then any element of dst that corresponds to a zero element of mask is not altered by this operation. If you only want to invert all the elements in an array, you can pass NULL in the place of src1; the routine will treat this as an array full of 1s.

cvDotProduct

```
double cvDotProduct(
    const CvArr* src1,
    const CvArr* src2
);
```

* Excluding 360, of course.

This function computes the vector dot product [Lagrange1773] of two N-dimensional vectors.* As with the cross product (and for the same reason), it does not matter if the vectors are in row or column form. Both src1 and src2 should be single-channel arrays, and both arrays should be of the same data type.

cvEigenVV

```
double cvEigenVV(
    CvArr* mat,
    CvArr* evects,
    CvArr* evals,
    double eps    = 0
);
```

Given a symmetric matrix mat, cvEigenVV() will compute the *eigenvectors* and the corresponding *eigenvalues* of that matrix. This is done using *Jacobi's method* [Bronshtein97], so it is efficient for smaller matrices.† Jacobi's method requires a stopping parameter, which is the maximum size of the off-diagonal elements in the final matrix.‡ The optional argument eps sets this termination value. In the process of computation, the supplied matrix mat is used for the computation, so its values will be altered by the function. When the function returns, you will find your eigenvectors in evects in the form of subsequent rows. The corresponding eigenvalues are stored in evals. The order of the eigenvectors will always be in descending order of the magnitudes of the corresponding eigenvalues. The cvEigenVV() function requires all three arrays to be of floating-point type.

As with cvDet() (described previously), if the matrix in question is known to be symmetric and positive definite§ then it is better to use SVD to find the eigenvalues and eigenvectors of mat.

cvFlip

```
void cvFlip(
    const CvArr* src,
    CvArr*       dst       = NULL,
    int          flip_mode = 0
);
```

This function flips an image around the *x*-axis, the *y*-axis, or both. In particular, if the argument flip_mode is set to 0 then the image will be flipped around the *x*-axis.

---

* Actually, the behavior of cvDotProduct() is a little more general than described here. Given any pair of *n*-by-*m* matrices, cvDotProduct() will return the sum of the products of the corresponding elements.

† A good rule of thumb would be that matrices 10-by-10 or smaller are small enough for Jacobi's method to be efficient. If the matrix is larger than 20-by-20 then you are in a domain where this method is probably not the way to go.

‡ In principle, once the Jacobi method is complete then the original matrix is transformed into one that is diagonal and contains only the eigenvalues; however, the method can be terminated before the off-diagonal elements are all the way to zero in order to save on computation. In practice is it usually sufficient to set this value to DBL_EPSILON, or about 10^{-15}.

§ This is, for example, always the case for covariance matrices. See cvCalcCovarMatrix().

If `flip_mode` is set to a positive value (e.g., +1) the image will be flipped around the *y*-axis, and if set to a negative value (e.g., –1) the image will be flipped about both axes.

When video processing on Win32 systems, you will find yourself using this function often to switch between image formats with their origins at the upper-left and lower-left of the image.

cvGEMM

```
double cvGEMM(
    const CvArr* src1,
    const CvArr* src2,
    double       alpha,
    const CvArr* src3,
    double       beta,
    CvArr*       dst,
    int          tABC = 0
);
```

Generalized matrix multiplication (GEMM) in OpenCV is performed by `cvGEMM()`, which performs matrix multiplication, multiplication by a transpose, scaled multiplication, et cetera. In its most general form, `cvGEMM()` computes the following:

$$D = \alpha \cdot \mathrm{op}(\mathbf{A}) \cdot \mathrm{op}(\mathbf{B}) + \beta \cdot \mathrm{op}(\mathbf{C})$$

Where **A**, **B**, and **C** are (respectively) the matrices src1, src2, and src3, α and β are numerical coefficients, and op() is an optional transposition of the matrix enclosed. The argument src3 may be set to NULL, in which case it will not be added. The transpositions are controlled by the optional argument tABC, which may be 0 or any combination (by means of Boolean OR) of CV_GEMM_A_T, CV_GEMM_B_T, and CV_GEMM_C_T (with each flag indicating a transposition of the corresponding matrix).

In the distant past OpenCV contained the methods `cvMatMul()` and `cvMatMulAdd()`, but these were too often confused with `cvMul()`, which does something entirely different (i.e., element-by-element multiplication of two arrays). These functions continue to exist as macros for calls to `cvGEMM()`. In particular, we have the equivalences listed in Table 3-7.

Table 3-7. Macro aliases for common usages of cvGEMM()

cvMatMul(A, B, D)	cvGEMM(A, A, 1, NULL, 0, D, 0)
cvMatMulAdd(A, B, C, D)	cvGEMM(A, A, 1, C, 1, D, 0)

All matrices must be of the appropriate size for the multiplication, and all should be of floating-point type. The `cvGEMM()` function supports two-channel matrices, in which case it will treat the two channels as the two components of a single complex number.

cvGetCol and cvGetCols

```
CvMat* cvGetCol(
    const CvArr* arr,
```

```
    CvMat*      submat,
    int         col
);
CvMat* cvGetCols(
    const CvArr* arr,
    CvMat*      submat,
    int         start_col,
    int         end_col
);
```

The function cvGetCol() is used to pick a single column out of a matrix and return it as a vector (i.e., as a matrix with only one column). In this case the matrix header submat will be modified to point to a particular column in arr. It is important to note that such header modification does not include the allocation of memory or the copying of data. The contents of submat will simply be altered so that it correctly indicates the selected column in arr. All data types are supported.

cvGetCols() works precisely the same way, except that all columns from start_col to end_col are selected. With both functions, the return value is a pointer to a header corresponding to the particular specified column or column span (i.e., submat) selected by the caller.

cvGetDiag

```
CvMat* cvGetDiag(
    const CvArr* arr,
    CvMat*      submat,
    int         diag    = 0
);
```

cvGetDiag() is analogous to cvGetCol(); it is used to pick a single *diagonal* from a matrix and return it as a vector. The argument submat is a matrix header. The function cvGetDiag() will fill the components of this header so that it points to the correct information in arr. Note that the result of calling cvGetDiag() is that the header you supplied is correctly configured to point at the diagonal data in arr, but the data from arr is not copied. The optional argument diag specifies which diagonal is to be pointed to by submat. If diag is set to the default value of 0, the main diagonal will be selected. If diag is greater than 0, then the diagonal starting at (diag,0) will be selected; if diag is less than 0, then the diagonal starting at (0,-diag) will be selected instead. The cvGetDiag() function does not require the matrix arr to be square, but the array submat must have the correct length for the size of the input array. The final returned value is the same as the value of submat passed in when the function was called.

cvGetDims and cvGetDimSize

```
int cvGetDims(
    const CvArr* arr,
    int*        sizes=NULL
);
int cvGetDimSize(
    const CvArr* arr,
```

```
int          index
);
```

Recall that arrays in OpenCV can be of dimension much greater than two. The function cvGetDims() returns the number of array dimensions of a particular array and (optionally) the sizes of each of those dimensions. The sizes will be reported if the array sizes is non-NULL. If sizes is used, it should be a pointer to *n* integers, where *n* is the number of dimensions. If you do not know the number of dimensions in advance, you can allocate sizes to CV_MAX_DIM integers just to be safe.

The function cvGetDimSize() returns the size of a single dimension specified by index. If the array is either a matrix or an image, the number of dimensions returned will always be two.* For matrices and images, the order of sizes returned by cvGetDims() will always be the number of rows first followed by the number of columns.

cvGetRow and cvGetRows

```
CvMat* cvGetRow(
    const CvArr* arr,
    CvMat*       submat,
    int          row
);
CvMat* cvGetRows(
    const CvArr* arr,
    CvMat*       submat,
    int          start_row,
    int          end_row
);
```

cvGetRow() picks a single row out of a matrix and returns it as a vector (a matrix with only one row). As with cvGetRow(), the matrix header submat will be modified to point to a particular row in arr, and the modification of this header does not include the allocation of memory or the copying of data; the contents of submat will simply be altered such that it correctly indicates the selected column in arr. All data types are supported.

The function cvGetRows() works precisely the same way, except that all rows from start_row to end_row are selected. With both functions, the return value is a pointer to a header corresponding to the particular specified row or row span selected by the caller.

cvGetSize

```
CvSize cvGetSize( const CvArr* arr );
```

Closely related to cvGetDims(), cvGetSize() returns the size of an array. The primary difference is that cvGetSize() is designed to be used on matrices and images, which always have dimension two. The size can then be returned in the form of a CvSize structure, which is suitable to use when (for example) constructing a new matrix or image of the same size.

---

* Remember that OpenCV regards a "vector" as a matrix of size *n*-by-1 or 1-by-*n*.

cvGetSubRect

```
CvSize cvGetSubRect(
    const CvArr* arr,
    CvArr*       submat,
    CvRect       rect
);
```

cvGetSubRect() is similar to cvGetColumns() or cvGetRows() except that it selects some arbitrary subrectangle in the array specified by the argument rect. As with other routines that select subsections of arrays, submat is simply a header that will be filled by cvGetSubRect() in such a way that it correctly points to the desired submatrix (i.e., no memory is allocated and no data is copied).

cvInRange and cvInRangeS

```
void cvInRange(
    const CvArr* src,
    const CvArr* lower,
    const CvArr* upper,
    CvArr*       dst
);
void cvInRangeS(
    const CvArr* src,
    CvScalar     lower,
    CvScalar     upper,
    CvArr*       dst
);
```

These two functions can be used to check if the pixels in an image fall within a particular specified range. In the case of cvInRange(), each pixel of src is compared with the corresponding value in the images lower and upper. If the value in src is greater than or equal to the value in lower and also less than the value in upper, then the corresponding value in dst will be set to 0xff; otherwise, the value in dst will be set to 0.

The function cvInRangeS() works precisely the same way except that the image src is compared to the constant (CvScalar) values in lower and upper. For both functions, the image src may be of any type; if it has multiple channels then each channel will be handled separately. Note that dst must be of the same size and number of channels and also must be an 8-bit image.

cvInvert

```
double cvInvert(
    const CvArr* src,
    CvArr*       dst,
    Int          method = CV_LU
);
```

cvInvert() inverts the matrix in src and places the result in dst. This function supports several methods of computing the inverse matrix (see Table 3-8), but the default is Gaussian elimination. The return value depends on the method used.

Table 3-8. Possible values of method argument to cvInvert()

Value of method argument	Meaning
CV_LU	Gaussian elimination (LU Decomposition)
CV_SVD	Singular value decomposition (SVD)
CV_SVD_SYM	SVD for symmetric matrices

In the case of Gaussian elimination (method=CV_LU), the determinant of src is returned when the function is complete. If the determinant is 0, then the inversion is not actually performed and the array dst is simply set to all 0s.

In the case of CV_SVD or CV_SVD_SYM, the return value is the inverse condition number for the matrix (the ratio of the smallest to the largest eigenvalue). If the matrix src is singular, then cvInvert() in SVD mode will instead compute the pseudo-inverse.

cvMahalonobis

```
CvSize cvMahalonobis(
    const CvArr* vec1,
    const CvArr* vec2,
    CvArr*       mat
);
```

The *Mahalonobis distance* (Mahal) is defined as the vector distance measured between a point and the center of a Gaussian distribution; it is computed using the inverse covariance of that distribution as a metric. See Figure 3-5. Intuitively, this is analogous to the *z*-score in basic statistics, where the distance from the center of a distribution is measured in units of the variance of that distribution. The Mahalonobis distance is just a multivariable generalization of the same idea.

cvMahalonobis() computes the value:

$$r_{\text{Mahalonobis}} = \sqrt{(\mathbf{x} - \boldsymbol{\mu})^{\text{T}} \Sigma^{-1} (\mathbf{x} - \boldsymbol{\mu})}$$

The vector vec1 is presumed to be the point **x**, and the vector vec2 is taken to be the distribution's mean.* That matrix mat is the inverse covariance.

In practice, this covariance matrix will usually have been computed with cvCalcCovar Matrix() (described previously) and then inverted with cvInvert(). It is good programming practice to use the SV_SVD method for this inversion because someday you will encounter a distribution for which one of the eigenvalues is 0!

cvMax and cvMaxS

```
void cvMax(
    const CvArr* src1,
    const CvArr* src2,
```

---

* Actually, the Mahalonobis distance is more generally defined as the distance between any two vectors; in any case, the vector vec2 is subtracted from the vector vec1. Neither is there any fundamental connection between mat in cvMahalonobis() and the inverse covariance; any metric can be imposed here as appropriate.

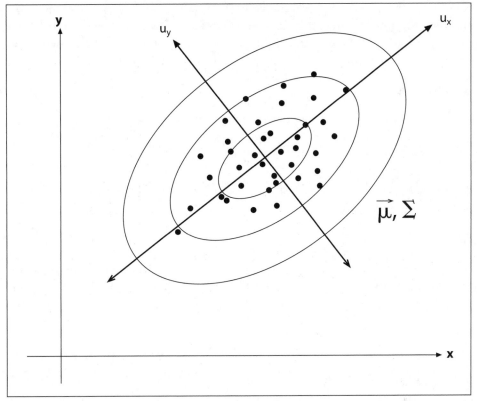

Figure 3-5. A distribution of points in two dimensions with superimposed ellipsoids representing Mahalonobis distances of 1.0, 2.0, and 3.0 from the distribution's mean

```
        CvArr* dst
    );
    void cvMaxS(
        const CvArr* src,
        double       value,
        CvArr*       dst
    );
```

cvMax() computes the maximum value of each corresponding pair of pixels in the arrays src1 and src2. With cvMaxS(), the src array is compared with the constant scalar value. As always, if mask is non-NULL then only the elements of dst corresponding to nonzero entries in mask are computed.

cvMerge

```
    void cvMerge(
        const CvArr* src0,
        const CvArr* src1,
        const CvArr* src2,
        const CvArr* src3,
        CvArr* dst
    );
```

cvMerge() is the inverse operation of cvSplit(). The arrays in src0, src1, src2, and src3 are combined into the array dst. Of course, dst should have the same data type and size as all of the source arrays, but it can have two, three, or four channels. The unused source images can be left set to NULL.

cvMin and cvMinS

```
void cvMin(
    const CvArr* src1,
    const CvArr* src2,
    CvArr* dst
);
void cvMinS(
    const CvArr* src,
    double value,
    CvArr* dst
);
```

cvMin() computes the minimum value of each corresponding pair of pixels in the arrays src1 and src2. With cvMinS(), the src arrays are compared with the constant scalar value. Again, if mask is non-NULL then only the elements of dst corresponding to nonzero entries in mask are computed.

cvMinMaxLoc

```
void cvMinMaxLoc(
    const CvArr* arr,
    double*      min_val,
    double*      max_val,
    CvPoint*     min_loc = NULL,
    CvPoint*     max_loc = NULL,
    const CvArr* mask    = NULL
);
```

This routine finds the minimal and maximal values in the array arr and (optionally) returns their locations. The computed minimum and maximum values are placed in min_val and max_val. Optionally, the locations of those extrema will also be written to the addresses given by min_loc and max_loc if those values are non-NULL.

As usual, if mask is non-NULL then only those portions of the image arr that correspond to nonzero pixels in mask are considered. The cvMinMaxLoc() routine handles only single-channel arrays, however, so if you have a multichannel array then you should use cvSetCOI() to set a particular channel for consideration.

cvMul

```
void cvMul(
    const CvArr* src1,
    const CvArr* src2,
    CvArr* dst,
    double scale=1
);
```

cvMul() is a simple multiplication function. It multiplies all of the elements in src1 by the corresponding elements in src2 and then puts the results in dst. If mask is non-NULL, then any element of dst that corresponds to a zero element of mask is not altered by this operation. There is no function cvMulS() because that functionality is already provided by cvScale() or cvCvtScale().

One further thing to keep in mind: cvMul() performs element-by-element multiplication. Someday, when you are multiplying some matrices, you may be tempted to reach for cvMul(). This will not work; remember that matrix multiplication is done with cvGEMM(), not cvMul().

cvNot

```
void(
    const CvArr* src,
    CvArr*       dst
);
```

The function cvNot() inverts every bit in every element of src and then places the result in dst. Thus, for an 8-bit image the value 0x00 would be mapped to 0xff and the value 0x83 would be mapped to 0x7c.

cvNorm

```
double cvNorm(
    const CvArr* arr1,
    const CvArr* arr2      = NULL,
    int          norm_type = CV_L2,
    const CvArr* mask      = NULL
);
```

This function can be used to compute the total norm of an array and also a variety of relative distance norms if two arrays are provided. In the former case, the norm computed is shown in Table 3-9.

Table 3-9. Norm computed by cvNorm() for different values of norm_type when arr2=NULL

norm_type	Result
CV_C	$\|arr1\|_C = \max_{x,y} abs(arr1_{x,y})$
CV_L1	$\|arr1\|_{L1} = \sum_{x,y} abs(arr1_{x,y})$
CV_L2	$\|arr1\|_{L2} = \sum_{x,y} arr1_{x,y}^2$

If the second array argument arr2 is non-NULL, then the norm computed is a difference norm—that is, something like the distance between the two arrays.* In the first three

---

* At least in the case of the L2 norm, there is an intuitive interpretation of the difference norm as a Euclidean distance in a space of dimension equal to the number of pixels in the images.

cases shown in Table 3-10, the norm is absolute; in the latter three cases it is rescaled by the magnitude of the second array arr2.

Table 3-10. Norm computed by cvNorm() for different values of norm_type when arr2 is non-NULL

norm_type	Result
CV_C	$\|arr1-arr2\|_c = \max_{x,y} abs(arr1_{x,y} - arr2_{x,y})$
CV_L1	$\|arr1-arr2\|_{L1} = \sum_{x,y} abs(arr1_{x,y} - arr2_{x,y})$
CV_L2	$\|arr1-arr2\|_{L2} = \sum_{x,y} (arr1_{x,y} - arr2_{x,y})^2$
CV_RELATIVE_C	$\dfrac{\|arr1-arr2\|_c}{\|arr2\|_c}$
CV_ RELATIVE_L1	$\dfrac{\|arr1-arr2\|_{L1}}{\|arr2\|_{L1}}$
CV_ RELATIVE_L2	$\dfrac{\|arr1-arr2\|_{L2}}{\|arr2\|_{L2}}$

In all cases, arr1 and arr2 must have the same size and number of channels. When there is more than one channel, the norm is computed over all of the channels together (i.e., the sums in Tables 3-9 and 3-10 are not only over *x* and *y* but also over the channels).

cvNormalize

```
cvNormalize(
    const CvArr*  src,
    CvArr*        dst,
    double        a         = 1.0,
    double        b         = 0.0,
    int           norm_type = CV_L2,
    const CvArr*  mask      = NULL
);
```

As with so many OpenCV functions, cvNormalize() does more than it might at first appear. Depending on the value of norm_type, image src is normalized or otherwise mapped into a particular range in dst. The possible values of norm_type are shown in Table 3-11.

Table 3-11. Possible values of norm_type argument to cvNormalize()

norm_type	Result
CV_C	$\|arr1\|_c = \max_{dst} abs(I_{x,y}) = a$
CV_L1	$\|arr1\|_{L1} = \sum_{dst} abs(I_{x,y}) = a$
CV_L2	$\|arr1\|_{L2} = \sum_{dst} I^2_{x\ y,} = a$
CV_MINMAX	Map into range [*a, b*]

In the case of the C norm, the array src is rescaled such that the magnitude of the absolute value of the largest entry is equal to a. In the case of the L1 or L2 norm, the array is rescaled so that the given norm is equal to the value of a. If norm_type is set to CV_MINMAX, then the values of the array are rescaled and translated so that they are linearly mapped into the interval between a and b (inclusive).

As before, if mask is non-NULL then only those pixels corresponding to nonzero values of the mask image will contribute to the computation of the norm—and only those pixels will be altered by cvNormalize().

cvOr and cvOrS

```
void cvOr(
    const CvArr* src1,
    const CvArr* src2,
    CvArr*       dst,
    const CvArr* mask=NULL
);
void cvOrS(
    const CvArr* src,
    CvScalar     value,
    CvArr*       dst,
    const CvArr* mask  = NULL
);
```

These two functions compute a bitwise OR operation on the array src1. In the case of cvOr(), each element of dst is computed as the bitwise OR of the corresponding two elements of src1 and src2. In the case of cvOrS(), the bitwise OR is computed with the constant scalar value. As usual, if mask is non-NULL then only the elements of dst corresponding to nonzero entries in mask are computed.

All data types are supported, but src1 and src2 must have the same data type for cvOr(). If the elements are of floating-point type, then the bitwise representation of that floating-point number is used.

cvReduce

```
CvSize cvReduce(
    const CvArr* src,
    CvArr*       dst,
    int          dim,
    int          op = CV_REDUCE_SUM
);
```

Reduction is the systematic transformation of the input matrix src into a vector dst by applying some combination rule op on each row (or column) and its neighbor until only one row (or column) remains (see Table 3-12).* The argument op controls how the reduction is done, as summarized in Table 3-13.

---

* Purists will note that averaging is not technically a proper *fold* in the sense implied here. OpenCV has a more practical view of reductions and so includes this useful operation in cvReduce.

Table 3-12. Argument op in cvReduce() selects the reduction operator

Value of op	Result
CV_REDUCE_SUM	Compute sum across vectors
CV_REDUCE_AVG	Compute average across vectors
CV_REDUCE_MAX	Compute maximum across vectors
CV_REDUCE_MIN	Compute minimum across vectors

Table 3-13. Argument dim in cvReduce() controls the direction of the reduction

Value of dim	Result
+1	Collapse to a single row
0	Collapse to a single column
−1	Collapse as appropriate for dst

cvReduce() supports multichannel arrays of floating-point type. It is also allowable to use a higher precision type in dst than appears in src. This is primarily relevant for CV_REDUCE_SUM and CV_REDUCE_AVG, where overflows and summation problems are possible.

cvRepeat

```
void cvRepeat(
    const CvArr* src,
    CvArr*       dst
);
```

This function copies the contents of src into dst, repeating as many times as necessary to fill dst. In particular, dst can be of any size relative to src. It may be larger or smaller, and it need not have an integer relationship between any of its dimensions and the corresponding dimensions of src.

cvScale

```
void cvScale(
    const CvArr* src,
    CvArr*       dst,
    double       scale
);
```

The function cvScale() is actually a macro for cvConvertScale() that sets the shift argument to 0.0. Thus, it can be used to rescale the contents of an array and to convert from one kind of data type to another.

cvSet and cvSetZero

```
void cvSet(
    CvArr*       arr,
    CvScalar     value,
    const CvArr* mask   = NULL
);
```

These functions set all values in all channels of the array to a specified value. The cvSet() function accepts an optional mask argument: if a mask is provided, then only those pixels in the image arr that correspond to nonzero values of the mask image will be set to the specified value. The function cvSetZero() is just a synonym for cvSet(0.0).

cvSetIdentity

```
void cvSetIdentity( CvArr* arr );
```

cvSetIdentity() sets all elements of the array to 0 except for elements whose row and column are equal; those elements are set to 1. cvSetIdentity() supports all data types and does not even require the array to be square.

cvSolve

```
int cvSolve(
    const CvArr* src1,
    const CvArr* src2,
    CvArr*       dst,
    int          method = CV_LU
);
```

The function cvSolve() provides a fast way to solve linear systems based on cvInvert(). It computes the solution to

$$C = \arg\min_X \left\| \mathbf{A} \cdot \mathbf{X} - \mathbf{B} \right\|$$

where **A** is a square matrix given by src1, **B** is the vector src2, and C is the solution computed by cvSolve() for the best vector **X** it could find. That best vector **X** is returned in dst. The same methods are supported as by cvInvert() (described previously); only floating-point data types are supported. The function returns an integer value where a nonzero return indicates that it could find a solution.

It should be noted that cvSolve() can be used to solve overdetermined linear systems. Overdetermined systems will be solved using something called the *pseudo-inverse*, which uses SVD methods to find the least-squares solution for the system of equations.

cvSplit

```
void cvSplit(
    const CvArr* src,
    CvArr*       dst0,
    CvArr*       dst1,
    CvArr*       dst2,
    CvArr*       dst3
);
```

There are times when it is not convenient to work with a multichannel image. In such cases, we can use cvSplit() to copy each channel separately into one of several supplied single-channel images. The cvSplit() function will copy the channels in src into the images dst0, dst1, dst2, and dst3 as needed. The destination images must match the source image in size and data type but, of course, should be single-channel images.

If the source image has fewer than four channels (as it often will), then the unneeded destination arguments to cvSplit() can be set to NULL.

cvSub

```
void cvSub(
    const CvArr* src1,
    const CvArr* src2,
    CvArr*       dst,
    const CvArr* mask = NULL
);
```

This function performs a basic element-by-element subtraction of one array src2 from another src1 and places the result in dst. If the array mask is non-NULL, then only those elements of dst corresponding to nonzero elements of mask are computed. Note that src1, src2, and dst must all have the same type, size, and number of channels; mask, if used, should be an 8-bit array of the same size and number of channels as dst.

cvSub, cvSubS, and cvSubRS

```
void cvSub(
    const CvArr* src1,
    const CvArr* src2,
    CvArr*       dst,
    const CvArr* mask  = NULL
);
void cvSubS(
    const CvArr* src,
    CvScalar     value,
    CvArr*       dst,
    const CvArr* mask  = NULL
);
void cvSubRS(
    const CvArr* src,
    CvScalar     value,
    CvArr*       dst,
    const CvArr* mask  = NULL
);
```

cvSub() is a simple subtraction function; it subtracts all of the elements in src2 from the corresponding elements in src1 and puts the results in dst. If mask is non-NULL, then any element of dst that corresponds to a zero element of mask is not altered by this operation. The closely related function cvSubS() does the same thing except that the constant scalar value is added to every element of src. The function cvSubRS() is the same as cvSubS() except that, rather than subtracting a constant from every element of src, it subtracts every element of src from the constant value.

cvSum

```
CvScalar cvSum(
    CvArr* arr
);
```

cvSum() sums all of the pixels in all of the channels of the array arr. Observe that the return value is of type CvScalar, which means that cvSum() can accommodate multi-channel arrays. In that case, the sum for each channel is placed in the corresponding component of the CvScalar return value.

cvSVD

```
void cvSVD(
    CvArr* A,
    CvArr* W,
    CvArr* U     = NULL,
    CvArr* V     = NULL,
    int    flags = 0
);
```

Singular value decomposition (SVD) is the decomposing of an *m*-by-*m* matrix **A** into the form:

$$A = U \cdot W \cdot V^T$$

where **W** is a diagonal matrix and **U** and **V** are *m*-by-*m* and *n*-by-*n* unitary matrices. Of course the matrix **W** is also an *m*-by-*n* matrix, so here "diagonal" means that any element whose row and column numbers are not equal is necessarily 0. Because **W** is necessarily diagonal, OpenCV allows it to be represented either by an *m*-by-*n* matrix or by an *n*-by-1 vector (in which case that vector will contain only the diagonal "singular" values).

The matrices U and V are optional to cvSVD(), and if they are left set to NULL then no value will be returned. The final argument flags can be any or all of the three options described in Table 3-14 (combined as appropriate with the Boolean OR operator).

Table 3-14. Possible flags for flags argument to cvSVD()

Flag	Result
CV_SVD_MODIFY_A	Allows modification of matrix A
CV_SVD_U_T	Return U^T instead of U
CV_SVD_V_T	Return V^T instead of V

cvSVBkSb

```
void cvSVBkSb(
    const CvArr* W,
    const CvArr* U,
    const CvArr* V,
    const CvArr* B,
    CvArr* X,
    int    flags = 0
);
```

This is a function that you are unlikely to call directly. In conjunction with cvSVD() (just described), it underlies the SVD-based methods of cvInvert() and cvSolve(). That being said, you may want to cut out the middleman and do your own matrix inversions (depending on the data source, this could save you from making a bunch of memory allocations for temporary matrices inside of cvInvert() or cvSolve()).

The function cvSVBkSb() computes the back-substitution for a matrix **A** that is represented in the form of a decomposition of matrices **U**, **W**, and **V** (e.g., an SVD). The result matrix **X** is given by the formula:

$$X = V \cdot W^* \cdot U^T \cdot B$$

The matrix **B** is optional, and if set to NULL it will be ignored. The matrix **W*** is a matrix whose diagonal elements are defined by $\lambda_i^* = \lambda_i^{-1}$ for $\lambda_i \geq \varepsilon$. This value ε is the *singularity threshold*, a very small number that is typically proportional to the sum of the diagonal elements of **W** (i.e., $\varepsilon \propto \sum_i \lambda_i$).

cvTrace

```
CvScalar cvTrace( const CvArr* mat );
```

The trace of a matrix (Trace) is the sum of all of the diagonal elements. The trace in OpenCV is implemented on top of the cvGetDiag() function, so it does not require the array passed in to be square. Multichannel arrays are supported, but the array mat should be of floating-point type.

cvTranspose and cvT

```
void cvTranspose(
  const CvArr* src,
  CvArr*       dst
);
```

cvTranspose() copies every element of src into the location in dst indicated by reversing the row and column index. This function does support multichannel arrays; however, if you are using multiple channels to represent complex numbers, remember that cvTranspose() does not perform complex conjugation (a fast way to accomplish this task is by means of the cvXorS() function, which can be used to directly flip the sign bits in the imaginary part of the array). The macro cvT() is simply shorthand for cvTranspose().

cvXor and cvXorS

```
void cvXor(
    const CvArr* src1,
    const CvArr* src2,
    CvArr* dst,
    const CvArr* mask=NULL
);
void cvXorS(
    const CvArr* src,
    CvScalar value,
    CvArr* dst,
    const CvArr* mask=NULL
);
```

These two functions compute a bitwise XOR operation on the array src1. In the case of cvXor(), each element of dst is computed as the bitwise XOR of the corresponding two elements of src1 and src2. In the case of cvXorS(), the bitwise XOR is computed with the constant scalar value. Once again, if mask is non-NULL then only the elements of dst corresponding to nonzero entries in mask are computed.

All data types are supported, but src1 and src2 must be of the same data type for cvXor(). For floating-point elements, the bitwise representation of that floating-point number is used.

cvZero

```
void cvZero( CvArr* arr );
```

This function sets all values in all channels of the array to 0.

Drawing Things

Something that frequently occurs is the need to draw some kind of picture or to draw something on top of an image obtained from somewhere else. Toward this end, OpenCV provides a menagerie of functions that will allow us to make lines, squares, circles, and the like.

Lines

The simplest of these routines just draws a line by the Bresenham algorithm [Bresenham65]:

```
void  cvLine(
   CvArr*   array,
   CvPoint  pt1,
   CvPoint  pt2,
   CvScalar color,
   int      thickness    = 1,
   int      connectivity = 8
);
```

The first argument to cvLine() is the usual CvArr*, which in this context typically means an IplImage* image pointer. The next two arguments are CvPoints. As a quick reminder, CvPoint is a simple structure containing only the integer members x and y. We can create a CvPoint "on the fly" with the routine cvPoint(int x, int y), which conveniently packs the two integers into a CvPoint structure for us.

The next argument, color, is of type CvScalar. CvScalars are also structures, which (you may recall) are defined as follows:

```
typdef struct {
   double val[4];
} CvScalar;
```

As you can see, this structure is just a collection of four doubles. In this case, the first three represent the red, green, and blue channels; the fourth is not used (it can be used for an alpha channel when appropriate). One typically makes use of the handy macro CV_RGB(r, g, b). This macro takes three numbers and packs them up into a CvScalar.

The next two arguments are optional. The thickness is the thickness of the line (in pixels), and connectivity sets the anti-aliasing mode. The default is "8 connected", which will give a nice, smooth, anti-aliased line. You can also set this to a "4 connected" line; diagonals will be blocky and chunky, but they will be drawn a lot faster.

At least as handy as cvLine() is cvRectangle(). It is probably unnecessary to tell you that cvRectangle() draws a rectangle. It has the same arguments as cvLine() except that there is no connectivity argument. This is because the resulting rectangles are always oriented with their sides parallel to the x- and y-axes. With cvRectangle(), we simply give two points for the opposite corners and OpenCV will draw a rectangle.

```
void  cvRectangle(
    CvArr*   array,
    CvPoint  pt1,
    CvPoint  pt2,
    CvScalar color,
    int      thickness = 1
);
```

Circles and Ellipses

Similarly straightforward is the method for drawing circles, which pretty much has the same arguments.

```
void  cvCircle (
    CvArr*   array,
    CvPoint  center,
    int      radius,
    CvScalar color,
    int      thickness    = 1,
    int      connectivity = 8
);
```

For circles, rectangles, and all of the other closed shapes to come, the thickness argument can also be set to CV_FILL, which is just an alias for –1; the result is that the drawn figure will be filled in the same color as the edges.

Only slightly more complicated than cvCircle() is the routine for drawing generalized ellipses:

```
void cvEllipse(
    CvArr*   img,
    CvPoint  center,
    CvSize   axes,
    double   angle,
    double   start_angle,
    double   end_angle,
    CvScalar color,
    int      thickness = 1,
    int      line_type = 8
);
```

In this case, the major new ingredient is the axes argument, which is of type CvSize. The function CvSize is very much like CvPoint and CvScalar; it is a simple structure, in this case containing only the members width and height. Like CvPoint and CvScalar, there is a convenient helper function cvSize(int height, int width) that will return a CvSize structure when we need one. In this case, the height and width arguments represent the length of the ellipse's major and minor axes.

The angle is the angle (in degrees) of the major axis, which is measured counterclockwise from horizontal (i.e., from the *x*-axis). Similarly the start_angle and end_angle indicate (also in degrees) the angle for the arc to start and for it to finish. Thus, for a complete ellipse you must set these values to 0 and 360, respectively.

An alternate way to specify the drawing of an ellipse is to use a bounding box:

```
void cvEllipseBox(
    CvArr*   img,
    CvBox2D  box,
    CvScalar color,
    int      thickness = 1,
    int      line_type = 8,
    int      shift     = 0
);
```

Here again we see another of OpenCV's helper structures, CvBox2D:

```
typdef struct {
    CvPoint2D32f center;
    CvSize2D32f  size;
    float        angle;
} CvBox2D;
```

CvPoint2D32f is the floating-point analogue of CvPoint, and CvSize2D32f is the floating-point analog of CvSize. These, along with the tilt angle, effectively specify the bounding box for the ellipse.

Polygons

Finally, we have a set of functions for drawing polygons:

```
void cvFillPoly(
    CvArr*    img,
    CvPoint** pts,
    int*      npts,
    int       contours,
    CvScalar  color,
    int       line_type = 8
);

void cvFillConvexPoly(
    CvArr*   img,
    CvPoint* pts,
    int      npts,
    CvScalar color,
    int      line_type = 8
```

```
    );

    void cvPolyLine(
        CvArr*      img,
        CvPoint**   pts,
        int*        npts,
        int         contours,
        int         is_closed,
        CvScalar    color,
        int         thickness = 1,
        int         line_type = 8
    );
```

All three of these are slight variants on the same idea, with the main difference being how the points are specified.

In cvFillPoly(), the points are provided as an array of CvPoint structures. This allows cvFillPoly() to draw many polygons in a single call. Similarly npts is an array of point counts, one for each polygon to be drawn. If the is_closed variable is set to true, then an additional segment will be drawn from the last to the first point for each polygon. cvFillPoly() is quite robust and will handle self-intersecting polygons, polygons with holes, and other such complexities. Unfortunately, this means the routine is comparatively slow.

cvFillConvexPoly() works like cvFillPoly() except that it draws only one polygon at a time and can draw only convex polygons.* The upside is that cvFillConvexPoly() runs much faster.

The third function, cvPolyLine(), takes the same arguments as cvFillPoly(); however, since only the polygon edges are drawn, self-intersection presents no particular complexity. Hence this function is much faster than cvFillPoly().

Fonts and Text

One last form of drawing that one may well need is to draw text. Of course, text creates its own set of complexities, but—as always with this sort of thing—OpenCV is more concerned with providing a simple "down and dirty" solution that will work for simple cases than a robust, complex solution (which would be redundant anyway given the capabilities of other libraries).

OpenCV has one main routine, called cvPutText() that just throws some text onto an image. The text indicated by text is printed with its lower-left corner of the text box at origin and in the color indicated by color.

```
    void cvPutText(
        CvArr*          img,
        const char*     text,
        CvPoint         origin,
        const CvFont*   font,
```

---

* Strictly speaking, this is not quite true; it can actually draw and fill any *monotone* polygon, which is a slightly larger class of polygons.

```
CvScalar      color
);
```

There is always some little thing that makes our job a bit more complicated than we'd like, and in this case it's the appearance of the pointer to CvFont.

In a nutshell, the way to get a valid CvFont* pointer is to call the function cvInitFont(). This function takes a group of arguments that configure some particular font for use on the screen. Those of you familiar with GUI programming in other environments will find cvInitFont() to be reminiscent of similar devices but with many fewer options.

In order to create a CvFont that we can pass to cvPutText(), we must first declare a CvFont variable; then we can pass it to cvInitFont().

```
void cvInitFont(
    CvFont* font,
    int     font_face,
    double  hscale,
    double  vscale,
    double  shear     = 0,
    int     thickness = 1,
    int     line_type = 8
);
```

Observe that this is a little different than how seemingly similar functions, such as cvCreateImage(), work in OpenCV. The call to cvInitFont() initializes an existing CvFont structure (which means that you create the variable and pass cvInitFont() a pointer to the variable you created). This is unlike cvCreateImage(), which creates the structure for you and returns a pointer.

The argument font_face is one of those listed in Table 3-15 (and pictured in Figure 3-6), and it may optionally be combined (by Boolean OR) with CV_FONT_ITALIC.

Table 3-15. Available fonts (all are variations of Hershey)

Identifier	Description
CV_FONT_HERSHEY_SIMPLEX	Normal size sanserif
CV_FONT_HERSHEY_PLAIN	Small size sanserif
CV_FONT_HERSHEY_DUPLEX	Normal size sanserif, more complex than CV_FONT_HERSHEY_SIMPLEX
CV_FONT_HERSHEY_COMPLEX	Normal size serif, more complex than CV_FONT_HERSHEY_DUPLEX
CV_FONT_HERSHEY_TRIPLEX	Normal size serif, more complex than CV_FONT_HERSHEY_COMPLEX
CV_FONT_HERSHEY_COMPLEX_SMALL	Smaller version of CV_FONT_HERSHEY_COMPLEX
CV_FONT_HERSHEY_SCRIPT_SIMPLEX	Handwriting style
CV_FONT_HERSHEY_SCRIPT_COMPLEX	More complex variant of CV_FONT_HERSHEY_SCRIPT_SIMPLEX

Figure 3-6. The eight fonts of Table 3-15 drawn with hscale = vscale = 1.0, with the origin of each line separated from the vertical by 30 pixels

Both hscale and vscale can be set to either 1.0 or 0.5 only. This causes the font to be rendered at full or half height (and width) relative to the basic definition of the particular font.

The shear function creates an italicized slant to the font; if set to 0.0, the font is not slanted. It can be set as large as 1.0, which sets the slope of the characters to approximately 45 degrees.

Both thickness and line_type are the same as defined for all the other drawing functions.

Data Persistence

OpenCV provides a mechanism for serializing and de-serializing its various data types to and from disk in either YAML or XML format. In the chapter on HighGUI, which addresses user interface functions, we will cover specific functions that store and recall our most common object: IplImages (these functions are cvSaveImage() and cvLoadImage()).

In addition, the HighGUI chapter will discuss read and write functions specific to movies: cvGrabFrame(), which reads from file or from camera; and cvCreateVideoWriter() and cvWriteFrame(). In this section, we will focus on general object persistence: reading and writing matrices, OpenCV structures, and configuration and log files.

First we start with specific and convenient functions that save and load OpenCV matrices. These functions are cvSave() and cvLoad(). Suppose you had a 5-by-5 identity matrix (0 everywhere except for 1s on the diagonal). Example 3-15 shows how to accomplish this.

Example 3-15. Saving and loading a CvMat
```
CvMat A = cvMat( 5, 5, CV_32F, the_matrix_data );

cvSave( "my_matrix.xml", &A );
. . .
// to load it then in some other program use …
CvMat* A1 = (CvMat*) cvLoad( "my_matrix.xml" );
```

The CxCore reference manual contains an entire section on data persistence. What you really need to know is that general data persistence in OpenCV consists of creating a CvFileStorage structure, as in Example 3-16, that stores memory objects in a tree structure. You can create and fill this structure by reading from disk via cvOpenFileStorage() with CV_STORAGE_READ, or you can create and open CvFileStorage via cvOpenFileStorage() with CV_STORAGE_WRITE for writing and then fill it using the appropriate data persistence functions. On disk, the data is stored in an XML or YAML format.

Example 3-16. CvFileStorage structure; data is accessed by CxCore data persistence functions
```
typedef struct CvFileStorage
{
    ...         // hidden fields
} CvFileStorage;
```

The internal data inside the CvFileStorage tree may consist of a hierarchical collection of scalars, CxCore objects (matrices, sequences, and graphs) and/or user-defined objects.

Let's say you have a configuration or logging file. For example, consider the case of a movie configuration file that tells us how many frames we want (10), what their size is (320 by 240) and a 3-by-3 color conversion matrix that should be applied. We want to call the file "cfg.xml" on disk. Example 3-17 shows how to do this.

Example 3-17. Writing a configuration file "cfg.xml" to disk
```
CvFileStorage* fs = cvOpenFileStorage(
  "cfg.xml",
  0,
  CV_STORAGE_WRITE
);
cvWriteInt( fs, "frame_count", 10 );
cvStartWriteStruct( fs, "frame_size", CV_NODE_SEQ );
cvWriteInt( fs, 0, 320 );
cvWriteInt( fs, 0, 200 );
```

Example 3-17. Writing a configuration file "cfg.xml" to disk (continued)

```
cvEndWriteStruct(fs);
cvWrite( fs, "color_cvt_matrix", cmatrix );
cvReleaseFileStorage( &fs );
```

Note some of the key functions in this example. We can give a name to integers that we write to the structure using cvWriteInt(). We can create an arbitrary structure, using cvStartWriteStruct(), which is also given an optional name (pass a 0 or NULL if there is no name). This structure has two ints that have no name and so we pass a 0 for them in the name field, after which we use cvEndWriteStruct() to end the writing of that structure. If there were more structures, we'd Start and End each of them similarly; the structures may be nested to arbitrary depth. We then use cvWrite() to write out the color conversion matrix. Contrast this fairly complex matrix write procedure with the simpler cvSave() in Example 3-15. The cvSave() function is just a convenient shortcut for cvWrite() when you have only one matrix to write. When we are finished writing the data, the CvFileStorage handle is released in cvReleaseFileStorage(). The output (here, in XML form) would look like Example 3-18.

Example 3-18. XML version of cfg.xml on disk

```
<?xml version="1.0"?>
<opencv_storage>
<frame_count>10</frame_count>
<frame_size>320 200</frame_size>
<color_cvt_matrix type_id="opencv-matrix">
  <rows>3</rows> <cols>3</cols>
  <dt>f</dt>
  <data>...</data></color_cvt_matrix>
</opencv_storage>
```

We may then read this configuration file as shown in Example 3-19.

Example 3-19. Reading cfg.xml from disk

```
CvFileStorage* fs = cvOpenFileStorage(
  "cfg.xml",
  0,
  CV_STORAGE_READ
);

int frame_count = cvReadIntByName(
  fs,
  0,
  "frame_count",
  5 /* default value */
);

CvSeq* s = cvGetFileNodeByName(fs,0,"frame_size")->data.seq;

int frame_width = cvReadInt(
  (CvFileNode*)cvGetSeqElem(s,0)
);
```

Example 3-19. Reading cfg.xml from disk (continued)

```
int frame_height = cvReadInt(
  (CvFileNode*)cvGetSeqElem(s,1)
);

CvMat* color_cvt_matrix = (CvMat*) cvReadByName(
  fs,
  0,
  "color_cvt_matrix"
);

cvReleaseFileStorage( &fs );
```

When reading, we open the XML configuration file with cvOpenFileStorage() as in Example 3-19. We then read the frame_count using cvReadIntByName(), which allows for a default value to be given if no number is read. In this case the default is 5. We then get the structure that we named "frame_size" using cvGetFileNodeByName(). From here, we read our two unnamed integers using cvReadInt(). Next we read our named color conversion matrix using cvReadByName().* Again, contrast this with the short form cvLoad() in Example 3-15. We can use cvLoad() if we only have one matrix to read, but we must use cvRead() if the matrix is embedded within a larger structure. Finally, we release the CvFileStorage structure.

The list of relevant data persistence functions associated with the CvFileStorage structure is shown in Table 3-16. See the CxCore manual for more details.

Table 3-16. Data persistence functions

Function	Description
Open and Release	
cvOpenFileStorage	Opens file storage for reading or writing
cvReleaseFileStorage	Releases data storage
Writing	
cvStartWriteStruct	Starts writing a new structure
cvEndWriteStruct	Ends writing a structure
cvWriteInt	Writes integer
cvWriteReal	Writes float
cvWriteString	Writes text string
cvWriteComment	Writes an XML or YAML comment string
cvWrite	Writes an object such as a CvMat
cvWriteRawData	Writes multiple numbers
cvWriteFileNode	Writes file node to another file storage

---

* One could also use cvRead() to read in the matrix, but it can only be called after the appropriate CvFileNode{} is located, e.g., using cvGetFileNodeByName().

Table 3-16. Data persistence functions (continued)

Function	Description
Reading	
cvGetRootFileNode	Gets the top-level nodes of the file storage
cvGetFileNodeByName	Finds node in the map or file storage
cvGetHashedKey	Returns a unique pointer for given name
cvGetFileNode	Finds node in the map or file storage
cvGetFileNodeName	Returns name of file node
cvReadInt	Reads unnamed int
cvReadIntByName	Reads named int
cvReadReal	Reads unnamed float
cvReadRealByName	Reads named float
cvReadString	Retrieves text string from file node
cvReadStringByName	Finds named file node and returns its value
cvRead	Decodes object and returns pointer to it
cvReadByName	Finds object and decodes it
cvReadRawData	Reads multiple numbers
cvStartReadRawData	Initializes file node sequence reader
cvReadRawDataSlice	Reads data from sequence reader above

Integrated Performance Primitives

Intel has a product called the Integrated Performance Primitives (IPP) library (IPP). This library is essentially a toolbox of high-performance kernels for handling multimedia and other processor-intensive operations in a manner that makes extensive use of the detailed architecture of their processors (and, to a lesser degree, other manufacturers' processors that have a similar architecture).

As discussed in Chapter 1, OpenCV enjoys a close relationship with IPP, both at a software level and at an organizational level inside of the company. As a result, OpenCV is designed to automatically* recognize the presence of the IPP library and to automatically "swap out" the lower-performance implementations of many core functionalities for their higher-performance counterparts in IPP. The IPP library allows OpenCV to take advantage of performance opportunities that arrive from SIMD instructions in a single processor as well as from modern multicore architectures.

With these basics in hand, we can perform a wide variety of basic tasks. Moving onward through the text, we will look at many more sophisticated capabilities of OpenCV,

---

* The one prerequisite to this automatic recognition is that the binary directory of IPP must be in the system path. So on a Windows system, for example, if you have IPP in *C:/Program Files/Intel/IPP* then you want to ensure that *C:/Program Files/Intel/IPP/bin* is in your system path.

almost all of which are built on these routines. It should be no surprise that image processing—which often requires doing the same thing to a whole lot of data, much of which is completely parallel—would realize a great benefit from any code that allows it to take advantage of parallel execution units of any form (MMX, SSE, SSE2, etc.).

Verifying Installation

The way to check and make sure that IPP is installed and working correctly is with the function cvGetModuleInfo(), shown in Example 3-20. This function will identify both the version of OpenCV you are currently running and the version and identity of any add-in modules.

Example 3-20. Using cvGetModuleInfo() to check for IPP

```
char* libraries;
char* modules;
cvGetModuleInfo( 0, &libraries, &modules );
printf("Libraries: %s/nModules: %s/n", libraries, modules );
```

The code in Example 3-20 will generate text strings which describe the installed libraries and modules. The output might look like this:

```
Libraries cxcore: 1.0.0
Modules: ippcv20.dll, ippi20.dll, ipps20.dll, ippvm20.dll
```

The modules listed in this output are the IPP modules used by OpenCV. Those modules are themselves actually proxies for even lower-level CPU-specific libraries. The details of how it all works are well beyond the scope of this book, but if you see the IPP libraries in the Modules string then you can be pretty confident that everything is working as expected. Of course, you could use this information to verify that IPP is running correctly on your own system. You might also use it to check for IPP on a machine on which your finished software is installed, perhaps then making some dynamic adjustments depending on whether IPP is available.

Summary

In this chapter we introduced some basic data structures that we will often encounter. In particular, we met the OpenCV matrix structure and the all-important OpenCV image structure, IplImage. We considered both in some detail and found that the matrix and image structures are very similar: the functions used for primitive manipulations in one work equally well in the other.

Exercises

In the following exercises, you may need to refer to the CxCore manual that ships with OpenCV or to the OpenCV Wiki on the Web for details of the functions outlined in this chapter.

1. Find and open *.../opencv/cxcore/include/cxtypes.h*. Read through and find the many conversion helper functions.

a. Choose a negative floating-point number. Take its absolute value, round it, and then take its ceiling and floor.

b. Generate some random numbers.

c. Create a floating point CvPoint2D32f and convert it to an integer CvPoint.

d. Convert a CvPoint to a CvPoint2D32f.

2. This exercise will accustom you to the idea of many functions taking matrix types. Create a two-dimensional matrix with three channels of type byte with data size 100-by-100. Set all the values to 0.

a. Draw a circle in the matrix using void cvCircle(CvArr* img, CvPoint center, intradius, CvScalar color, int thickness=1, int line_type=8, int shift=0).

b. Display this image using methods described in Chapter 2.

3. Create a two-dimensional matrix with three channels of type byte with data size 100-by-100, and set all the values to 0. Use the pointer element access function cvPtr2D to point to the middle ("green") channel. Draw a green rectangle between (20, 5) and (40, 20).

4. Create a three-channel RGB image of size 100-by-100. Clear it. Use pointer arithmetic to draw a green square between (20, 5) and (40, 20).

5. Practice using region of interest (ROI). Create a 210-by-210 single-channel byte image and zero it. Within the image, build a pyramid of increasing values using ROI and cvSet(). That is: the outer border should be 0, the next inner border should be 20, the next inner border should be 40, and so on until the final innermost square is set to value 200; all borders should be 10 pixels wide. Display the image.

6. Use multiple image headers for one image. Load an image that is at least 100-by-100. Create two additional image headers and set their origin, depth, number of channels, and widthstep to be the same as the loaded image. In the new image headers, set the width at 20 and the height at 30. Finally, set their imageData pointers to point to the pixel at (5, 10) and (50, 60), respectively. Pass these new image subheaders to cvNot(). Display the loaded image, which should have two inverted rectangles within the larger image.

7. Create a mask using cvCmp(). Load a real image. Use cvSplit() to split the image into red, green, and blue images.

a. Find and display the green image.

b. Clone this green plane image twice (call these clone1 and clone2).

c. Find the green plane's minimum and maximum value.

d. Set clone1's values to *thresh = (unsigned char)((maximum - minimum)/2.0)*.

e. Set clone2 to 0 and use cvCmp(green_image, clone1, clone2, CV_CMP_GE). Now clone2 will have a mask of where the value exceeds thresh in the green image.

f. Finally, use cvSubS(green_image,thresh/2, green_image, clone2) and display the results.

8. Create a structure of an integer, a CvPoint and a CvRect; call it "my_struct".

  a. Write two functions: void write_my_struct(CvFileStorage * fs, const char * name, my_struct *ms) and void read_my_struct(CvFileStorage* fs, CvFileNode* ms_node, my_struct* ms). Use them to write and read my_struct.

  b. Write and read an array of 10 my_struct structures.

CHAPTER 4

HighGUI

A Portable Graphics Toolkit

The OpenCV functions that allow us to interact with the operating system, the file system, and hardware such as cameras are collected into a library called HighGUI (which stands for "high-level graphical user interface"). HighGUI allows us to open windows, to display images, to read and write graphics-related files (both images and video), and to handle simple mouse, pointer, and keyboard events. We can also use it to create other useful doodads like sliders and then add them to our windows. If you are a GUI guru in your window environment of choice, then you might find that much of what HighGUI offers is redundant. Yet even so you might find that the benefit of cross-platform portability is itself a tempting morsel.

From our initial perspective, the HighGUI library in OpenCV can be divided into three parts: the hardware part, the file system part, and the GUI part.* We will take a moment to overview what is in each part before we really dive in.

The hardware part is primarily concerned with the operation of cameras. In most operating systems, interaction with a camera is a tedious and painful task. HighGUI allows an easy way to query a camera and retrieve the latest image from the camera. It hides all of the nasty stuff, and that keeps us happy.

The file system part is concerned primarily with loading and saving images. One nice feature of the library is that it allows us to read video using the same methods we would use to read a camera. We can therefore abstract ourselves away from the particular device we're using and get on with writing interesting code. In a similar spirit, HighGUI provides us with a (relatively) universal pair of functions to load and save still images. These functions simply rely on the filename extension and automatically handle all of the decoding or encoding that is necessary.

---

* Under the hood, the architectural organization is a bit different from what we described, but the breakdown into hardware, file system, and GUI is an easier way to organize things conceptually. The actual HighGUI functions are divided into "video IO", "image IO", and "GUI tools". These categories are represented by the *cvcap**, *grfmt**, and *window** source files, respectively.

The third part of HighGUI is the window system (or GUI). The library provides some simple functions that will allow us to open a window and throw an image into that window. It also allows us to register and respond to mouse and keyboard events on that window. These features are most useful when trying to get off of the ground with a simple application. Tossing in some slider bars, which we can also use as switches,* we find ourselves able to prototype a surprising variety of applications using only the HighGUI library.

As we proceed in this chapter, we will not treat these three segments separately; rather, we will start with some functions of highest immediate utility and work our way to the subtler points thereafter. In this way you will learn what you need to get going as soon as possible.

Creating a Window

First, we want to show an image on the screen using HighGUI. The function that does this for us is cvNamedWindow(). The function expects a name for the new window and one flag. The name appears at the top of the window, and the name is also used as a handle for the window that can be passed to other HighGUI functions. The flag indicates if the window should autosize itself to fit an image we put into it. Here is the full prototype:

```
int cvNamedWindow(
    const char* name,
    int         flags = CV_WINDOW_AUTOSIZE
);
```

Notice the parameter flags. For now, the only valid options available are to set flags to 0 or to use the default setting, CV_WINDOW_AUTOSIZE. If CV_WINDOW_AUTOSIZE is set, then HighGUI resizes the window to fit the image. Thereafter, the window will automatically resize itself if a new image is loaded into the window but cannot be resized by the user. If you don't want autosizing, you can set this argument to 0; then users can resize the window as they wish.

Once we create a window, we usually want to put something into it. But before we do that, let's see how to get rid of the window when it is no longer needed. For this we use cvDestroyWindow(), a function whose argument is a string: the name given to the window when it was created. In OpenCV, windows are referenced by name instead of by some unfriendly (and invariably OS-dependent) "handle". Conversion between handles and names happens under the hood of HighGUI, so you needn't worry about it.

Having said that, some people do worry about it, and that's OK, too. For those people, HighGUI provides the following functions:

```
void*      cvGetWindowHandle( const char* name );
const char* cvGetWindowName( void* window_handle );
```

---

* OpenCV HighGUI does not provide anything like a button. The common trick is to use a two-position slider to achieve this functionality (more on this later).

These functions allow us to convert back and forth between the human-readable names preferred by OpenCV and the "handle" style of reference used by different window systems.*

To resize a window, call (not surprisingly) cvResizeWindow():

```
void cvResizeWindow(
    const char* name,
    int         width,
    int         height
);
```

Here the width and height are in pixels and give the size of the drawable part of the window (which are probably the dimensions you actually care about).

Loading an Image

Before we can display an image in our window, we'll need to know how to load an image from disk. The function for this is cvLoadImage():

```
IplImage* cvLoadImage(
    const char* filename,
    int         iscolor = CV_LOAD_IMAGE_COLOR
);
```

When opening an image, cvLoadImage() does not look at the file extension. Instead, cvLoadImage() analyzes the first few bytes of the file (aka its *signature* or "magic sequence") and determines the appropriate codec using that. The second argument iscolor can be set to one of several values. By default, images are loaded as three-channel images with 8 bits per channel; the optional flag CV_LOAD_IMAGE_ANYDEPTH can be added to allow loading of non-8-bit images. By default, the number of channels will be three because the iscolor flag has the default value of CV_LOAD_IMAGE_COLOR. This means that, regardless of the number of channels in the image file, the image will be converted to three channels if needed. The alternatives to CV_LOAD_IMAGE_COLOR are CV_LOAD_IMAGE_GRAYSCALE and CV_LOAD_IMAGE_ANYCOLOR. Just as CV_LOAD_IMAGE_COLOR forces any image into a three-channel image, CV_LOAD_IMAGE_GRAYSCALE automatically converts any image into a single-channel image. CV_LOAD_IMAGE_ANYCOLOR will simply load the image as it is stored in the file. Thus, to load a 16-bit color image you would use CV_LOAD_IMAGE_COLOR | CV_LOAD_IMAGE_ANYDEPTH. If you want both the color and depth to be loaded exactly "as is", you could instead use the all-purpose flag CV_LOAD_IMAGE_UNCHANGED. Note that cvLoadImage() does not signal a runtime error when it fails to load an image; it simply returns a null pointer.

The obvious complementary function to cvLoadImage() is cvSaveImage(), which takes two arguments:

```
int cvSaveImage(
    const char*  filename,
    const CvArr* image
);
```

---

* For those who know what this means: the window handle returned is a HWND on Win32 systems, a Carbon WindowRef on Mac OS X, and a Widget* pointer on systems (e.g., GtkWidget) of X Window type.

The first argument gives the filename, whose extension is used to determine the format in which the file will be stored. The second argument is the name of the image to be stored. Recall that CvArr is kind of a C-style way of creating something equivalent to a base-class in an object-oriented language; wherever you see CvArr*, you can use an IplImage*. The cvSaveImage() function will store only 8-bit single- or three-channel images for most file formats. Newer back ends for flexible image formats like PNG, TIFF or JPEG2000 allow storing 16-bit or even float formats and some allow four-channel images (BGR plus alpha) as well. The return value will be 1 if the save was successful and should be 0 if the save was not.*

Displaying Images

Now we are ready for what we really want to do, and that is to load an image and to put it into the window where we can view it and appreciate its profundity. We do this via one simple function, cvShowImage():

```
void cvShowImage(
    const char*  name,
    const CvArr* image
);
```

The first argument here is the name of the window within which we intend to draw. The second argument is the image to be drawn.

Let's now put together a simple program that will display an image on the screen. We can read a filename from the command line, create a window, and put our image in the window in 25 lines, including comments and tidily cleaning up our memory allocations!

```
int main(int argc, char** argv)
{

    // Create a named window with the name of the file.
    cvNamedWindow( argv[1], 1 );

    // Load the image from the given file name.
    IplImage* img = cvLoadImage( argv[1] );

    // Show the image in the named window
    cvShowImage( argv[1], img );

    // Idle until the user hits the "Esc" key.
    while( 1 ) {
        if( cvWaitKey( 100 ) == 27 ) break;
    }

    // Clean up and don't be piggies
    cvDestroyWindow( argv[1] );
    cvReleaseImage( &img );
```

---

* The reason we say "should" is that, in some OS environments, it is possible to issue save commands that will actually cause the operating system to throw an exception. Normally, however, a zero value will be returned to indicate failure.

```
    exit(0);
  }
```

For convenience we have used the filename as the window name. This is nice because OpenCV automatically puts the window name at the top of the window, so we can tell which file we are viewing (see Figure 4-1). Easy as cake.

Figure 4-1. A simple image displayed with cvShowImage()

Before we move on, there are a few other window-related functions you ought to know about. They are:

```
void cvMoveWindow( const char* name, int x, int y );
void cvDestroyAllWindows( void );
int  cvStartWindowThread( void );
```

cvMoveWindow() simply moves a window on the screen so that its upper left corner is positioned at x,y.

cvDestroyAllWindows() is a useful cleanup function that closes all of the windows and de-allocates the associated memory.

On Linux and MacOS, cvStartWindowThread() tries to start a thread that updates the window automatically and handles resizing and so forth. A return value of 0 indicates that no thread could be started—for example, because there is no support for this feature in the version of OpenCV that you are using. Note that, if you do not start a separate window thread, OpenCV can react to user interface actions only when it is explicitly given time to do so (this happens when your program invokes cvWaitKey(), as described next).

WaitKey

Observe that inside the `while` loop in our window creation example there is a new function we have not seen before: `cvWaitKey()`. This function causes OpenCV to wait for a specified number of milliseconds for a user keystroke. If the key is pressed within the allotted time, the function returns the key pressed;* otherwise, it returns 0. With the construction:

```
while( 1 ) {
  if( cvWaitKey(100)==27 ) break;
}
```

we tell OpenCV to wait 100 ms for a key stroke. If there is no keystroke, then repeat ad infinitum. If there is a keystroke and it happens to have ASCII value 27 (the Escape key), then break out of that loop. This allows our user to leisurely peruse the image before ultimately exiting the program by hitting Escape.

As long as we're introducing `cvWaitKey()`, it is worth mentioning that `cvWaitKey()` can also be called with 0 as an argument. In this case, `cvWaitKey()` will wait indefinitely until a keystroke is received and then return that key. Thus, in our example we could just as easily have used `cvWaitKey(0)`. The difference between these two options would be more apparent if we were displaying a video, in which case we would want to take an action (i.e., display the next frame) if the user supplied no keystroke.

Mouse Events

Now that we can display an image to a user, we might also want to allow the user to interact with the image we have created. Since we are working in a window environment and since we already learned how to capture single keystrokes with `cvWaitKey()`, the next logical thing to consider is how to "listen to" and respond to mouse events.

Unlike keyboard events, mouse events are handled by a more typical callback mechanism. This means that, to enable response to mouse clicks, we must first write a callback routine that OpenCV can call whenever a mouse event occurs. Once we have done that, we must register the callback with OpenCV, thereby informing OpenCV that this is the correct function to use whenever the user does something with the mouse over a particular window.

Let's start with the callback. For those of you who are a little rusty on your event-driven program lingo, the *callback* can be any function that takes the correct set of arguments and returns the correct type. Here, we must be able to tell the function to be used as a

---

* The careful reader might legitimately ask exactly what this means. The short answer is "an ASCII value", but the long answer depends on the operating system. In Win32 environments, `cvWaitKey()` is actually waiting for a message of type WM_CHAR and, after receiving that message, returns the wParam field from the message (wParam is not actually type char at all!). On Unix-like systems, `cvWaitKey()` is using GTK; the return value is (event->keyval | (event->state<<16)), where event is a GdkEventKey structure. Again, this is not really a char. That state information is essentially the state of the Shift, Control, etc. keys at the time of the key press. This means that, if you are expecting (say) a capital Q, then you should either cast the return of `cvWaitKey()` to type char or AND with 0xff, because the shift key will appear in the upper bits (e.g., Shift-Q will return 0x10051).

callback exactly what kind of event occurred and where it occurred. The function must also be told if the user was pressing such keys as Shift or Alt when the mouse event occurred. Here is the exact prototype that your callback function must match:

```
void CvMouseCallback(
    int    event,
    int    x,
    int    y,
    int    flags,
    void* param
);
```

Now, whenever your function is called, OpenCV will fill in the arguments with their appropriate values. The first argument, called the event, will have one of the values shown in Table 4-1.

Table 4-1. Mouse event types

Event	Numerical value
CV_EVENT_MOUSEMOVE	0
CV_EVENT_LBUTTONDOWN	1
CV_EVENT_RBUTTONDOWN	2
CV_EVENT_MBUTTONDOWN	3
CV_EVENT_LBUTTONUP	4
CV_EVENT_RBUTTONUP	5
CV_EVENT_MBUTTONUP	6
CV_EVENT_LBUTTONDBLCLK	7
CV_EVENT_RBUTTONDBLCLK	8
CV_EVENT_MBUTTONDBLCLK	9

The second and third arguments will be set to the x and y coordinates of the mouse event. It is worth noting that these coordinates represent the pixel in the image independent of the size of the window (in general, this is not the same as the pixel coordinates of the event).

The fourth argument, called flags, is a bit field in which individual bits indicate special conditions present at the time of the event. For example, CV_EVENT_FLAG_SHIFTKEY has a numerical value of 16 (i.e., the fifth bit) and so, if we wanted to test whether the shift key were down, we could AND the flags variable with the bit mask (1<<4). Table 4-2 shows a complete list of the flags.

Table 4-2. Mouse event flags

Flag	Numerical value
CV_EVENT_FLAG_LBUTTON	1
CV_EVENT_FLAG_RBUTTON	2
CV_EVENT_FLAG_MBUTTON	4

Table 4-2. Mouse event flags (continued)

Flag	Numerical value
CV_EVENT_FLAG_CTRLKEY	8
CV_EVENT_FLAG_SHIFTKEY	16
CV_EVENT_FLAG_ALTKEY	32

The final argument is a void pointer that can be used to have OpenCV pass in any additional information in the form of a pointer to whatever kind of structure you need. A common situation in which you will want to use the param argument is when the callback itself is a static member function of a class. In this case, you will probably find yourself wanting to pass the this pointer and so indicate which class object instance the callback is intended to affect.

Next we need the function that registers the callback. That function is called cvSetMouseCallback(), and it requires three arguments.

```
void cvSetMouseCallback(
  const char*     window_name,
  CvMouseCallback on_mouse,
  void*           param       = NULL
);
```

The first argument is the name of the window to which the callback will be attached. Only events in that particular window will trigger this specific callback. The second argument is your callback function. Finally, the third param argument allows us to specify the param information that should be given to the callback whenever it is executed. This is, of course, the same param we were just discussing in regard to the callback prototype.

In Example 4-1 we write a small program to draw boxes on the screen with the mouse. The function my_mouse_callback() is installed to respond to mouse events, and it uses the event to determine what to do when it is called.

Example 4-1. Toy program for using a mouse to draw boxes on the screen

```
// An example program in which the
// user can draw boxes on the screen.
//
#include <cv.h>
#include <highgui.h>

// Define our callback which we will install for
// mouse events.
//
void my_mouse_callback(
   int event, int x, int y, int flags, void* param
);

CvRect box;
bool drawing_box = false;

// A litte subroutine to draw a box onto an image
```

Example 4-1. Toy program for using a mouse to draw boxes on the screen (continued)

```
//
void draw_box( IplImage* img, CvRect rect ) {
  cvRectangle (
    img,
    cvPoint(box.x,box.y),
    cvPoint(box.x+box.width,box.y+box.height),
    cvScalar(0xff,0x00,0x00)    /* red */
  );
}

int main( int argc, char* argv[] ) {

  box = cvRect(-1,-1,0,0);

  IplImage* image = cvCreateImage(
    cvSize(200,200),
    IPL_DEPTH_8U,
    3
  );
  cvZero( image );
  IplImage* temp = cvCloneImage( image );

  cvNamedWindow( "Box Example" );

  // Here is the crucial moment that we actually install
  // the callback. Note that we set the value 'param' to
  // be the image we are working with so that the callback
  // will have the image to edit.
  //
  cvSetMouseCallback(
    "Box Example",
    my_mouse_callback,
    (void*) image
  );

  // The main program loop. Here we copy the working image
  // to the 'temp' image, and if the user is drawing, then
  // put the currently contemplated box onto that temp image.
  // display the temp image, and wait 15ms for a keystroke,
  // then repeat...
  //
  while( 1 ) {

    cvCopyImage( image, temp );
    if( drawing_box ) draw_box( temp, box );
    cvShowImage( "Box Example", temp );

    if( cvWaitKey( 15 )==27 ) break;
  }

  // Be tidy
  //
  cvReleaseImage( &image );
```

```
    cvReleaseImage( &temp );
    cvDestroyWindow( "Box Example" );
}

// This is our mouse callback. If the user
// presses the left button, we start a box.
// when the user releases that button, then we
// add the box to the current image. When the
// mouse is dragged (with the button down) we
// resize the box.
//
void my_mouse_callback(
    int event, int x, int y, int flags, void* param
) {

  IplImage* image = (IplImage*) param;

  switch( event ) {
    case CV_EVENT_MOUSEMOVE: {
      if( drawing_box ) {
        box.width  = x-box.x;
        box.height = y-box.y;
      }
    }
    break;
    case CV_EVENT_LBUTTONDOWN: {
      drawing_box = true;
      box = cvRect(x, y, 0, 0);
    }
    break;
    case CV_EVENT_LBUTTONUP: {
      drawing_box = false;
      if(box.width<0) {
        box.x+=box.width;
        box.width *=-1;
      }
      if(box.height<0) {
        box.y+=box.height;
        box.height*=-1;
      }
      draw_box(image, box);
    }
    break;
  }
}
```

Sliders, Trackbars, and Switches

HighGUI provides a convenient slider element. In HighGUI, sliders are called *trackbars*. This is because their original (historical) intent was for selecting a particular frame in the playback of a video. Of course, once added to HighGUI, people began to use trackbars for all of the usual things one might do with a slider as well as many unusual ones (see the next section, "No Buttons")!

As with the parent window, the slider is given a unique name (in the form of a character string) and is thereafter always referred to by that name. The HighGUI routine for creating a trackbar is:

```
int cvCreateTrackbar(
    const char*          trackbar_name,
    const char*          window_name,
    int*                 value,
    int                  count,
    CvTrackbarCallback on_change
);
```

The first two arguments are the name for the trackbar itself and the name of the parent window to which the trackbar will be attached. When the trackbar is created it is added to either the top or the bottom of the parent window;* it will not occlude any image that is already in the window.

The next two arguments are value, a pointer to an integer that will be set automatically to the value to which the slider has been moved, and count, a numerical value for the maximum value of the slider.

The last argument is a pointer to a callback function that will be automatically called whenever the slider is moved. This is exactly analogous to the callback for mouse events. If used, the callback function must have the form CvTrackbarCallback, which is defined as:

```
void (*callback)( int position )
```

This callback is not actually required, so if you don't want a callback then you can simply set this value to NULL. Without a callback, the only effect of the user moving the slider will be the value of *value being changed.

Finally, here are two more routines that will allow you to programmatically set or read the value of a trackbar if you know its name:

```
int cvGetTrackbarPos(
    const char* trackbar_name,
    const char* window_name
);

void cvSetTrackbarPos(
    const char* trackbar_name,
    const char* window_name,
    int         pos
);
```

These functions allow you to set or read the value of a trackbar from anywhere in your program.

---

* Whether it is added to the top or bottom depends on the operating system, but it will always appear in the same place on any given platform.

No Buttons

Unfortunately, HighGUI does not provide any explicit support for buttons. It is thus common practice, among the particularly lazy,* to instead use sliders with only two positions. Another option that occurs often in the OpenCV samples in *…/opencv/ samples/c/* is to use keyboard shortcuts instead of buttons (see, e.g., the *floodfill* demo in the OpenCV source-code bundle).

Switches are just sliders (trackbars) that have only two positions, "on" (1) and "off" (0) (i.e., count has been set to 1). You can see how this is an easy way to obtain the functionality of a button using only the available trackbar tools. Depending on exactly how you want the switch to behave, you can use the trackbar callback to automatically reset the button back to 0 (as in Example 4-2; this is something like the standard behavior of most GUI "buttons") or to automatically set other switches to 0 (which gives the effect of a "radio button").

Example 4-2. Using a trackbar to create a "switch" that the user can turn on and off

```
// We make this value global so everyone can see it.
//
int g_switch_value = 0;

// This will be the callback that we give to the
// trackbar.
//
void switch_callback( int position ) {
  if( position == 0 ) {
    switch_off_function();
  } else {
    switch_on_function();
  }
}

int main( int argc, char* argv[] ) {

  // Name the main window
  //
  cvNamedWindow( "Demo Window", 1 );

  // Create the trackbar. We give it a name,
  // and tell it the name of the parent window.
  //
  cvCreateTrackbar(
    "Switch",
    "Demo Window",
    &g_switch_value,
    1,
```

---

* For the less lazy, another common practice is to compose the image you are displaying with a "control panel" you have drawn and then use the mouse event callback to test for the mouse's location when the event occurs. When the (x, y) location is within the area of a button you have drawn on your control panel, the callback is set to perform the button action. In this way, all "buttons" are internal to the mouse event callback routine associated with the parent window.

```
    Switch_callback
    );

    // This will just cause OpenCV to idle until
    // someone hits the "Escape" key.
    //
    while( 1 ) {
        if( cvWaitKey(15)==27 ) break;
    }

}
```

You can see that this will turn on and off just like a light switch. In our example, whenever the trackbar "switch" is set to 0, the callback executes the function switch_off_function(), and whenever it is switched on, the switch_on_function() is called.

Working with Video

When working with video we must consider several functions, including (of course) how to read and write video files. We must also think about how to actually play back such files on the screen.

The first thing we need is the CvCapture device. This structure contains the information needed for reading frames from a camera or video file. Depending on the source, we use one of two different calls to create and initialize a CvCapture structure.

```
    CvCapture* cvCreateFileCapture( const char* filename );
    CvCapture* cvCreateCameraCapture( int index );
```

In the case of cvCreateFileCapture(), we can simply give a filename for an MPG or AVI file and OpenCV will open the file and prepare to read it. If the open is successful and we are able to start reading frames, a pointer to an initialized CvCapture structure will be returned.

A lot of people don't always check these sorts of things, thinking that nothing will go wrong. Don't do that here. The returned pointer will be NULL if for some reason the file could not be opened (e.g., if the file does not exist), but cvCreateFileCapture() will also return a NULL pointer if the codec with which the video is compressed is not known. The subtleties of compression codecs are beyond the scope of this book, but in general you will need to have the appropriate library already resident on your computer in order to successfully read the video file. For example, if you want to read a file encoded with DIVX or MPG4 compression on a Windows machine, there are specific DLLs that provide the necessary resources to decode the video. This is why it is always important to check the return value of cvCreateFileCapture(), because even if it works on one machine (where the needed DLL is available) it might not work on another machine (where that codec DLL is missing). Once we have the CvCapture structure, we can begin reading frames and do a number of other things. But before we get into that, let's take a look at how to capture images from a camera.

The routine cvCreateCameraCapture() works very much like cvCreateFileCapture() except without the headache from the codecs.* In this case we give an *identifier* that indicates which camera we would like to access and how we expect the operating system to talk to that camera. For the former, this is just an identification number that is zero (0) when we only have one camera, and increments upward when there are multiple cameras on the same system. The other part of the identifier is called the *domain* of the camera and indicates (in essence) what type of camera we have. The domain can be any of the predefined constants shown in Table 4-3.

Table 4-3. Camera "domain" indicates where HighGUI should look for your camera

Camera capture constant	Numerical value
CV_CAP_ANY	0
CV_CAP_MIL	100
CV_CAP_VFW	200
CV_CAP_V4L	200
CV_CAP_V4L2	200
CV_CAP_FIREWIRE	300
CV_CAP_IEEE1394	300
CV_CAP_DC1394	300
CV_CAP_CMU1394	300

When we call cvCreateCameraCapture(), we pass in an identifier that is just the sum of the domain index and the camera index. For example:

```
CvCapture* capture = cvCreateCameraCapture( CV_CAP_FIREWIRE );
```

In this example, cvCreateCameraCapture() will attempt to open the first (i.e., number-zero) Firewire camera. In most cases, the domain is unnecessary when we have only one camera; it is sufficient to use CV_CAP_ANY (which is conveniently equal to 0, so we don't even have to type that in). One last useful hint before we move on: you can pass -1 to cvCreateCameraCapture(), which will cause OpenCV to open a window that allows you to select the desired camera.

Reading Video

```
int       cvGrabFrame( CvCapture* capture );
IplImage* cvRetrieveFrame( CvCapture* capture );
IplImage* cvQueryFrame( CvCapture* capture );
```

Once you have a valid CvCapture object, you can start grabbing frames. There are two ways to do this. One way is to call cvGrabFrame(), which takes the CvCapture* pointer and returns an integer. This integer will be 1 if the grab was successful and 0 if the grab

---

* Of course, to be completely fair, we should probably confess that the headache caused by different codecs has been replaced by the analogous headache of determining which cameras are (or are not) supported on our system.

failed. The cvGrabFrame() function copies the captured image to an internal buffer that is invisible to the user. Why would you want OpenCV to put the frame somewhere you can't access it? The answer is that this grabbed frame is unprocessed, and cvGrabFrame() is designed simply to get it onto the computer as quickly as possible.

Once you have called cvGrabFrame(), you can then call cvRetrieveFrame(). This function will do any necessary processing on the frame (such as the decompression stage in the codec) and then return an IplImage* pointer that points to another internal buffer (so do not rely on this image, because it will be overwritten the next time you call cvGrabFrame()). If you want to do anything in particular with this image, copy it elsewhere first. Because this pointer points to a structure maintained by OpenCV itself, you are not required to release the image and can expect trouble if you do so.

Having said all that, there is a somewhat simpler method called cvQueryFrame(). This is, in effect, a combination of cvGrabFrame() and cvRetrieveFrame(); it also returns the same IplImage* pointer as cvRetrieveFrame() did.

It should be noted that, with a video file, the frame is automatically advanced whenever a cvGrabFrame() call is made. Hence a subsequent call will retrieve the next frame automatically.

Once you are done with the CvCapture device, you can release it with a call to cvReleaseCapture(). As with most other de-allocators in OpenCV, this routine takes a pointer to the CvCapture* pointer:

```
void cvReleaseCapture( CvCapture** capture );
```

There are many other things we can do with the CvCapture structure. In particular, we can check and set various properties of the video source:

```
double cvGetCaptureProperty(
  CvCapture* capture,
  int        property_id
);

int cvSetCaptureProperty(
  CvCapture* capture,
  int        property_id,
  double     value
);
```

The routine cvGetCaptureProperty() accepts any of the property IDs shown in Table 4-4.

Table 4-4. Video capture properties used by cvGetCaptureProperty() and cvSetCaptureProperty()

Video capture property	Numerical value
CV_CAP_PROP_POS_MSEC	0
CV_CAP_PROP_POS_FRAME	1
CV_CAP_PROP_POS_AVI_RATIO	2
CV_CAP_PROP_FRAME_WIDTH	3
CV_CAP_PROP_FRAME_HEIGHT	4

*Table 4-4. Video capture properties used by cvGetCaptureProperty()
and cvSetCaptureProperty() (continued)*

Video capture property	Numerical value
CV_CAP_PROP_FPS	5
CV_CAP_PROP_FOURCC	6
CV_CAP_PROP_FRAME_COUNT	7

Most of these properties are self explanatory. POS_MSEC is the current position in a video file, measured in milliseconds. POS_FRAME is the current position in frame number. POS_AVI_RATIO is the position given as a number between 0 and 1 (this is actually quite useful when you want to position a trackbar to allow folks to navigate around your video). FRAME_WIDTH and FRAME_HEIGHT are the dimensions of the individual frames of the video to be read (or to be captured at the camera's current settings). FPS is specific to video files and indicates the number of frames per second at which the video was captured; you will need to know this if you want to play back your video and have it come out at the right speed. FOURCC is the four-character code for the compression codec to be used for the video you are currently reading. FRAME_COUNT should be the total number of frames in the video, but this figure is not entirely reliable.

All of these values are returned as type double, which is perfectly reasonable except for the case of FOURCC (FourCC) [FourCC85]. Here you will have to recast the result in order to interpret it, as described in Example 4-3.

Example 4-3. Unpacking a four-character code to identify a video codec

```
double f = cvGetCaptureProperty(
  capture,
  CV_CAP_PROP_FOURCC
);

char* fourcc = (char*) (&f);
```

For each of these video capture properties, there is a corresponding cvSetCaptureProperty() function that will attempt to set the property. These are not all entirely meaningful; for example, you should not be setting the FOURCC of a video you are currently reading. Attempting to move around the video by setting one of the position properties will work, but only for some video codecs (we'll have more to say about video codecs in the next section).

Writing Video

The other thing we might want to do with video is writing it out to disk. OpenCV makes this easy; it is essentially the same as reading video but with a few extra details.

First we must create a CvVideoWriter device, which is the video writing analogue of CvCapture. This device will incorporate the following functions.

```
CvVideoWriter* cvCreateVideoWriter(
  const char* filename,
```

```
    int          fourcc,
    double       fps,
    CvSize       frame_size,
    int          is_color  = 1
);
int cvWriteFrame(
    CvVideoWriter*  writer,
    const IplImage* image
);
void cvReleaseVideoWriter(
    CvVideoWriter** writer
);
```

You will notice that the video writer requires a few extra arguments. In addition to the filename, we have to tell the writer what codec to use, what the frame rate is, and how big the frames will be. Optionally we can tell OpenCV if the frames are black and white or color (the default is color).

Here, the codec is indicated by its four-character code. (For those of you who are not experts in compression codecs, they all have a unique four-character identifier associated with them). In this case the int that is named fourcc in the argument list for cvCreateVideoWriter() is actually the four characters of the fourcc packed together. Since this comes up relatively often, OpenCV provides a convenient macro CV_FOURCC(c0,c1,c2,c3) that will do the bit packing for you.

Once you have a video writer, all you have to do is call cvWriteFrame() and pass in the CvVideoWriter* pointer and the IplImage* pointer for the image you want to write out.

Once you are finished, you must call CvReleaseVideoWriter() in order to close the writer and the file you were writing to. Even if you are normally a bit sloppy about de-allocating things at the end of a program, do not be sloppy about this. Unless you explicitly release the video writer, the video file to which you are writing may be corrupted.

ConvertImage

For purely historical reasons, there is one orphan routine in the HighGUI that fits into none of the categories described above. It is so tremendously useful, however, that you should know about it and what it does. The function is called cvConvertImage().

```
void cvConvertImage(
    const CvArr* src,
    CvArr*       dst,
    int          flags = 0
);
```

cvConvertImage() is used to perform common conversions between image formats. The formats are specified in the headers of the src and dst images or arrays (the function prototype allows the more general CvArr type that works with IplImage).

The source image may be one, three, or four channels with either 8-bit or floating-point pixels. The destination must be 8 bits with one or three channels. This function can also convert color to grayscale or one-channel grayscale to three-channel grayscale (color).

Finally, the flag (if set) will flip the image vertically. This is useful because sometimes camera formats and display formats are reversed. Setting this flag actually flips the pixels in memory.

Exercises

1. This chapter completes our introduction to basic I/O programming and data structures in OpenCV. The following exercises build on this knowledge and create useful utilities for later use.

   a. Create a program that (1) reads frames from a video, (2) turns the result to grayscale, and (3) performs Canny edge detection on the image. Display all three stages of processing in three different windows, with each window appropriately named for its function.

   b. Display all three stages of processing in one image.

      > Hint: Create another image of the same height but three times the width as the video frame. Copy the images into this, either by using pointers or (more cleverly) by creating three new image headers that point to the beginning of and to one-third and two-thirds of the way into the imageData. Then use cvCopy().

   c. Write appropriate text labels describing the processing in each of the three slots.

2. Create a program that reads in and displays an image. When the user's mouse clicks on the image, read in the corresponding pixel (blue, green, red) values and write those values as text to the screen at the mouse location.

   a. For the program of exercise 1b, display the mouse coordinates of the individual image when clicking anywhere within the three-image display.

3. Create a program that reads in and displays an image.

   a. Allow the user to select a rectangular region in the image by drawing a rectangle with the mouse button held down, and highlight the region when the mouse button is released. Be careful to save an image copy in memory so that your drawing into the image does not destroy the original values there. The next mouse click should start the process all over again from the original image.

   b. In a separate window, use the drawing functions to draw a graph in blue, green, and red for how many pixels of each value were found in the selected box. This is the *color histogram* of that color region. The x-axis should be eight bins that represent pixel values falling within the ranges 0–31, 32–63, . . ., 223–255. The y-axis should be counts of the number of pixels that were found in that bin range. Do this for each color channel, BGR.

4. Make an application that reads and displays a video and is controlled by sliders. One slider will control the position within the video from start to end in 10

increments; another binary slider should control pause/unpause. Label both sliders appropriately.

5. Create your own simple paint program.

   a. Write a program that creates an image, sets it to 0, and then displays it. Allow the user to draw lines, circles, ellipses, and polygons on the image using the left mouse button. Create an eraser function when the right mouse button is held down.

   b. Allow "logical drawing" by allowing the user to set a slider setting to AND, OR, and XOR. That is, if the setting is AND then the drawing will appear only when it crosses pixels greater than 0 (and so on for the other logical functions).

6. Write a program that creates an image, sets it to 0, and then displays it. When the user clicks on a location, he or she can type in a label there. Allow Backspace to edit and provide for an abort key. Hitting Enter should fix the label at the spot it was typed.

7. Perspective transform.

   a. Write a program that reads in an image and uses the numbers 1–9 on the keypad to control a perspective transformation matrix (refer to our discussion of the cvWarpPerspective() in the Dense Perspective Transform section of Chapter 6). Tapping any number should increment the corresponding cell in the perspective transform matrix; tapping with the Shift key depressed should decrement the number associated with that cell (stopping at 0). Each time a number is changed, display the results in two images: the raw image and the transformed image.

   b. Add functionality to zoom in or out?

   c. Add functionality to rotate the image?

8. Face fun. Go to the */samples/c/* directory and build the *facedetect.c* code. Draw a skull image (or find one on the Web) and store it to disk. Modify the *facedetect* program to load in the image of the skull.

   a. When a face rectangle is detected, draw the skull in that rectangle.

      Hint: cvConvertImage() can convert the size of the image, or you could look up the cvResize function. One may then set the ROI to the rectangle and use cvCopy() to copy the properly resized image there.

   b. Add a slider with 10 settings corresponding to 0.0 to 1.0. Use this slider to alpha blend the skull over the face rectangle using the cvAddWeighted function.

9. Image stabilization. Go to the */samples/c/* directory and build the *lkdemo* code (the motion tracking or *optical flow* code). Create and display a video image in a much larger window image. Move the camera slightly but use the optical flow vectors to display the image in the same place within the larger window. This is a rudimentary image stabilization technique.

CHAPTER 5

Image Processing

Overview

At this point we have all of the basics at our disposal. We understand the structure of the library as well as the basic data structures it uses to represent images. We understand the HighGUI interface and can actually run a program and display our results on the screen. Now that we understand these primitive methods required to manipulate image structures, we are ready to learn some more sophisticated operations.

We will now move on to higher-level methods that treat the images as images, and not just as arrays of colored (or grayscale) values. When we say "image processing", we mean just that: using higher-level operators that are defined on image structures in order to accomplish tasks whose meaning is naturally defined in the context of graphical, visual images.

Smoothing

Smoothing, also called *blurring*, is a simple and frequently used image processing operation. There are many reasons for smoothing, but it is usually done to reduce noise or camera artifacts. Smoothing is also important when we wish to reduce the resolution of an image in a principled way (we will discuss this in more detail in the "Image Pyramids" section of this chapter).

OpenCV offers five different smoothing operations at this time. All of them are supported through one function, cvSmooth(),* which takes our desired form of smoothing as an argument.

```
void cvSmooth(
   const CvArr*    src,
   CvArr*          dst,
   int             smoothtype  = CV_GAUSSIAN,
   int             param1      = 3,
```

---

* Note that—unlike in, say, Matlab—the filtering operations in OpenCV (e.g., cvSmooth(), cvErode(), cvDilate()) produce output images of the same size as the input. To achieve that result, OpenCV creates "virtual" pixels outside of the image at the borders. By default, this is done by replication at the border, i.e., input(-dx,y)=input(0,y), input(w+dx,y)=input(w-1,y), and so forth.

```
    int         param2      = 0,
    double      param3      = 0,
    double      param4      = 0
);
```

The `src` and `dst` arguments are the usual source and destination for the smooth operation. The `cv_Smooth()` function has four parameters with the particularly uninformative names of param1, param2, param3, and param4. The meaning of these parameters depends on the value of smoothtype, which may take any of the five values listed in Table 5-1.*
(Please notice that for some values of ST, "in place operation", in which `src` and `dst` indicate the same image, is not allowed.)

Table 5-1. Types of smoothing operations

Smooth type	Name	In place?	Nc	Depth of src	Depth of dst	Brief description
CV_BLUR	Simple blur	Yes	1,3	8u, 32f	8u, 32f	Sum over a param1×param2 neighborhood with subsequent scaling by 1/(param1×param2).
CV_BLUR_NO_SCALE	Simple blur with no scaling	No	1	8u	16s (for 8u source) or 32f (for 32f source)	Sum over a param1×param2 neighborhood.
CV_MEDIAN	Median blur	No	1,3	8u	8u	Find median over a param1×param1 square neighborhood.
CV_GAUSSIAN	Gaussian blur	Yes	1,3	8u, 32f	8u (for 8u source) or 32f (for 32f source)	Sum over a param1×param2 neighborhood.
CV_BILATERAL	Bilateral filter	No	1,3	8u	8u	Apply bilateral 3-by-3 filtering with color sigma=param1 and a space sigma=param2.

The *simple blur* operation, as exemplified by CV_BLUR in Figure 5-1, is the simplest case. Each pixel in the output is the simple mean of all of the pixels in a window around the corresponding pixel in the input. Simple blur supports 1–4 image channels and works on 8-bit images or 32-bit floating-point images.

Not all of the smoothing operators act on the same sorts of images. CV_BLUR_NO_SCALE (*simple blur without scaling*) is essentially the same as simple blur except that there is no division performed to create an average. Hence the source and destination images must have different numerical precision so that the blurring operation will not result in an overflow. Simple blur without scaling may be performed on 8-bit images, in which case the destination image should have IPL_DEPTH_16S (CV_16S) or IPL_DEPTH_32S (CV_32S)

---

* Here and elsewhere we sometimes use 8u as shorthand for 8-bit unsigned image depth (IPL_DEPTH_8U). See Table 3-2 for other shorthand notation.

Figure 5-1. Image smoothing by block averaging: on the left are the input images; on the right, the output images

data types. The same operation may also be performed on 32-bit floating-point images, in which case the destination image may also be a 32-bit floating-point image. Simple blur without scaling cannot be done in place: the source and destination images must be different. (This requirement is obvious in the case of 8 bits to 16 bits, but it applies even when you are using a 32-bit image). Simple blur without scaling is sometimes chosen because it is a little faster than blurring with scaling.

The *median filter* (CV_MEDIAN) [Bardyn84] replaces each pixel by the median or "middle" pixel (as opposed to the mean pixel) value in a square neighborhood around the center pixel. Median filter will work on single-channel or three-channel or four-channel 8-bit images, but it cannot be done in place. Results of median filtering are shown in Figure 5-2. Simple blurring by averaging can be sensitive to noisy images, especially images with large isolated outlier points (sometimes called "shot noise"). Large differences in even a small number of points can cause a noticeable movement in the average value. Median filtering is able to ignore the outliers by selecting the middle points.

The next smoothing filter, the *Gaussian filter* (CV_GAUSSIAN), is probably the most useful though not the fastest. Gaussian filtering is done by convolving each point in the input array with a Gaussian kernel and then summing to produce the output array.

Figure 5-2. Image blurring by taking the median of surrounding pixels

For the Gaussian blur (Figure 5-3), the first two parameters give the width and height of the filter window; the (optional) third parameter indicates the sigma value (half width at half max) of the Gaussian kernel. If the third parameter is not specified, then the Gaussian will be automatically determined from the window size using the following formulae:

$$\sigma_x = \left(\frac{n_x}{2} - 1\right) \cdot 0.30 + 0.80, \quad n_x = \text{param1}$$

$$\sigma_y = \left(\frac{n_y}{2} - 1\right) \cdot 0.30 + 0.80, \quad n_y = \text{param2}$$

If you wish the kernel to be asymmetric, then you may also (optionally) supply a fourth parameter; in this case, the third and fourth parameters will be the values of sigma in the horizontal and vertical directions, respectively.

If the third and fourth parameters are given but the first two are set to 0, then the size of the window will be automatically determined from the value of sigma.

The OpenCV implementation of Gaussian smoothing also provides a higher performance optimization for several common kernels. 3-by-3, 5-by-5 and 7-by-7 with

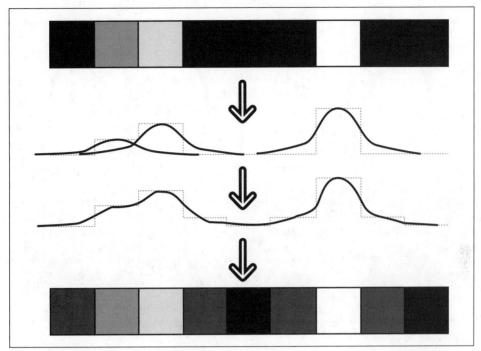

Figure 5-3. Gaussian blur on 1D pixel array

the "standard" sigma (i.e., param3 = 0.0) give better performance than other kernels. Gaussian blur supports single- or three-channel images in either 8-bit or 32-bit floating-point formats, and it can be done in place. Results of Gaussian blurring are shown in Figure 5-4.

The fifth and final form of smoothing supported by OpenCV is called *bilateral filtering* [Tomasi98], an example of which is shown in Figure 5-5. Bilateral filtering is one operation from a somewhat larger class of image analysis operators known as *edge-preserving smoothing*. Bilateral filtering is most easily understood when contrasted to Gaussian smoothing. A typical motivation for Gaussian smoothing is that pixels in a real image should vary slowly over space and thus be correlated to their neighbors, whereas random noise can be expected to vary greatly from one pixel to the next (i.e., noise is not spatially correlated). It is in this sense that Gaussian smoothing reduces noise while preserving signal. Unfortunately, this method breaks down near edges, where you do expect pixels to be uncorrelated with their neighbors. Thus Gaussian smoothing smoothes away the edges. At the cost of a little more processing time, bilateral filtering provides us a means of smoothing an image without smoothing away the edges.

Like Gaussian smoothing, bilateral filtering constructs a weighted average of each pixel and its neighboring components. The weighting has two components, the first of which is the same weighting used by Gaussian smoothing. The second component is also a Gaussian weighting but is based not on the spatial distance from the center pixel

Figure 5-4. Gaussian blurring

but rather on the difference in intensity[*] from the center pixel.[†] You can think of bilateral filtering as Gaussian smoothing that weights more similar pixels more highly than less similar ones. The effect of this filter is typically to turn an image into what appears to be a watercolor painting of the same scene.[‡] This can be useful as an aid to segmenting the image.

Bilateral filtering takes two parameters. The first is the width of the Gaussian kernel used in the spatial domain, which is analogous to the sigma parameters in the Gaussian filter. The second is the width of the Gaussian kernel in the color domain. The larger this second parameter is, the broader is the range of intensities (or colors) that will be included in the smoothing (and thus the more extreme a discontinuity must be in order to be preserved).

---

[*] In the case of multichannel (i.e., color) images, the difference in intensity is replaced with a weighted sum over colors. This weighting is chosen to enforce a Euclidean distance in the CIE color space.

[†] Technically, the use of Gaussian distribution functions is not a necessary feature of bilateral filtering. The implementation in OpenCV uses Gaussian weighting even though the method is general to many possible weighting functions.

[‡] This effect is particularly pronounced after multiple iterations of bilateral filtering.

Figure 5-5. Results of bilateral smoothing

Image Morphology

OpenCV provides a fast, convenient interface for doing *morphological transformations* [Serra83] on an image. The basic morphological transformations are called *dilation* and *erosion*, and they arise in a wide variety of contexts such as removing noise, isolating individual elements, and joining disparate elements in an image. Morphology can also be used to find intensity bumps or holes in an image and to find image gradients.

Dilation and Erosion

Dilation is a convolution of some image (or region of an image), which we will call A, with some *kernel*, which we will call B. The kernel, which can be any shape or size, has a single defined *anchor point*. Most often, the kernel is a small solid square or disk with the anchor point at the center. The kernel can be thought of as a template or mask, and its effect for dilation is that of a *local maximum* operator. As the kernel B is scanned over the image, we compute the maximal pixel value overlapped by B and replace the image pixel under the anchor point with that maximal value. This causes bright regions within an image to grow as diagrammed in Figure 5-6. This growth is the origin of the term "dilation operator".

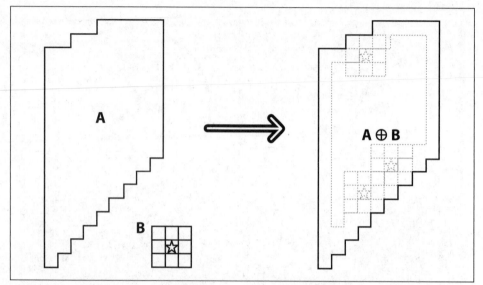

Figure 5-6. Morphological dilation: take the maximum under the kernel B

Erosion is the converse operation. The action of the erosion operator is equivalent to computing a *local minimum* over the area of the kernel. Erosion generates a new image from the original using the following algorithm: as the kernel B is scanned over the image, we compute the minimal pixel value overlapped by B and replace the image pixel under the anchor point with that minimal value.* Erosion is diagrammed in Figure 5-7.

> Image morphology is often done on binary images that result from thresholding. However, because dilation is just a max operator and erosion is just a min operator, morphology may be used on intensity images as well.

In general, whereas dilation expands region A, erosion reduces region A. Moreover, dilation will tend to smooth concavities and erosion will tend to smooth away protrusions. Of course, the exact result will depend on the kernel, but these statements are generally true for the filled convex kernels typically used.

In OpenCV, we effect these transformations using the cvErode() and cvDilate() functions:

```
void cvErode(
    IplImage*       src,
    IplImage*       dst,
    IplConvKernel*  B           = NULL,
    int             iterations = 1
);
```

---

* To be precise, the pixel in the destination image is set to the value equal to the minimal value of the pixels under the kernel in the source image.

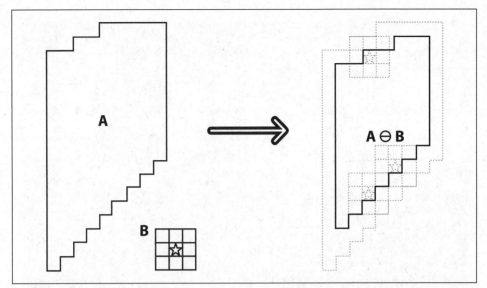

Figure 5-7. Morphological erosion: take the minimum under the kernel B

```
void cvDilate(
    IplImage*       src,
    IplImage*       dst,
    IplConvKernel*  B         = NULL,
    int             iterations = 1
);
```

Both cvErode() and cvDilate() take a source and destination image, and both support "in place" calls (in which the source and destination are the same image). The third argument is the kernel, which defaults to NULL. In the NULL case, the kernel used is a 3-by-3 kernel with the anchor at its center (we will discuss shortly how to create your own kernels). Finally, the fourth argument is the number of iterations. If not set to the default value of 1, the operation will be applied multiple times during the single call to the function. The results of an erode operation are shown in Figure 5-8 and those of a dilation operation in Figure 5-9. The erode operation is often used to eliminate "speckle" noise in an image. The idea here is that the speckles are eroded to nothing while larger regions that contain visually significant content are not affected. The dilate operation is often used when attempting to find *connected components* (i.e., large discrete regions of similar pixel color or intensity). The utility of dilation arises because in many cases a large region might otherwise be broken apart into multiple components as a result of noise, shadows, or some other similar effect. A small dilation will cause such components to "melt" together into one.

To recap: when OpenCV processes the cvErode() function, what happens beneath the hood is that the value of some point p is set to the minimum value of all of the points covered by the kernel when aligned at p; for the dilation operator, the equation is the same except that max is considered rather than min:

Figure 5-8. Results of the erosion, or "min", operator: bright regions are isolated and shrunk

$$\text{erode}(x, y) = \min_{(x', y') \in \text{kernel}} \text{src}(x + x', y + y')$$

$$\text{dilate}(x, y) = \max_{(x', y') \in \text{kernel}} \text{src}(x + x', y + y')$$

You might be wondering why we need a complicated formula when the earlier heuristic description was perfectly sufficient. Some readers actually prefer such formulas but, more importantly, the formulas capture some generality that isn't apparent in the qualitative description. Observe that if the image is not binary then the min and max operators play a less trivial role. Take another look at Figures 5-8 and 5-9, which show the erosion and dilation operators applied to two real images.

Making Your Own Kernel

You are not limited to the simple 3-by-3 square kernel. You can make your own custom morphological kernels (our previous "kernel B") using `IplConvKernel`. Such kernels are allocated using `cvCreateStructuringElementEx()` and are released using `cvReleaseStructuringElement()`.

```
IplConvKernel* cvCreateStructuringElementEx(
    int         cols,
    int         rows,
```

Figure 5-9. Results of the dilation, or "max", operator: bright regions are expanded and often joined

```
    int         anchor_x,
    int         anchor_y,
    int         shape,
    int*        values=NULL
);
```

```
void cvReleaseStructuringElement( IplConvKernel** element );
```

A morphological kernel, unlike a convolution kernel, doesn't require any numerical values. The elements of the kernel simply indicate where the max or min computations take place as the kernel moves around the image. The anchor point indicates how the kernel is to be aligned with the source image and also where the result of the computation is to be placed in the destination image. When creating the kernel, cols and rows indicate the size of the rectangle that holds the structuring element. The next parameters, anchor_x and anchor_y, are the (*x*, *y*) coordinates of the anchor point within the enclosing rectangle of the kernel. The fifth parameter, shape, can take on values listed in Table 5-2. If CV_SHAPE_CUSTOM is used, then the integer vector values is used to define a custom shape of the kernel within the rows-by-cols enclosing rectangle. This vector is read in raster scan order with each entry representing a different pixel in the enclosing rectangle. Any nonzero value is taken to indicate that the corresponding pixel should be included in the kernel. If values is NULL then the custom shape is interpreted to be all nonzero, resulting in a rectangular kernel.*

Table 5-2. Possible IplConvKernel shape values

Shape value	Meaning
CV_SHAPE_RECT	The kernel is rectangular
CV_SHAPE_CROSS	The kernel is cross shaped
CV_SHAPE_ELLIPSE	The kernel is elliptical
CV_SHAPE_CUSTOM	The kernel is user-defined via values

More General Morphology

When working with Boolean images and image masks, the basic erode and dilate operations are usually sufficient. When working with grayscale or color images, however, a number of additional operations are often helpful. Several of the more useful operations can be handled by the multi-purpose cvMorphologyEx() function.

```
void cvMorphologyEx(
    const CvArr*    src,
    CvArr*          dst,
    CvArr*          temp,
    IplConvKernel*  element,
    int             operation,
    int             iterations   = 1
);
```

In addition to the arguments src, dst, element, and iterations, which we used with previous operators, cvMorphologyEx() has two new parameters. The first is the temp array, which is required for some of the operations (see Table 5-3). When required, this array should be the same size as the source image. The second new argument—the really interesting one—is operation, which selects the morphological operation that we will do.

Table 5-3. cvMorphologyEx() operation options

Value of operation	Morphological operator	Requires temp image?
CV_MOP_OPEN	Opening	No
CV_MOP_CLOSE	Closing	No
CV_MOP_GRADIENT	Morphological gradient	Always
CV_MOP_TOPHAT	Top Hat	For in-place only (src = dst)
CV_MOP_BLACKHAT	Black Hat	For in-place only (src = dst)

Opening and closing

The first two operations in Table 5-3, *opening* and *closing*, are combinations of the erosion and dilation operators. In the case of opening, we erode first and then dilate (Figure 5-10).

---

* If the use of this strange integer vector strikes you as being incongruous with other OpenCV functions, you are not alone. The origin of this syntax is the same as the origin of the IPL prefix to this function—another instance of archeological code relics.

Opening is often used to count regions in a binary image. For example, if we have thresholded an image of cells on a microscope slide, we might use opening to separate out cells that are near each other before counting the regions. In the case of closing, we dilate first and then erode (Figure 5-12). Closing is used in most of the more sophisticated connected-component algorithms to reduce unwanted or noise-driven segments. For connected components, usually an erosion or closing operation is performed first to eliminate elements that arise purely from noise and then an opening operation is used to connect nearby large regions. (Notice that, although the end result of using open or close is similar to using erode or dilate, these new operations tend to preserve the area of connected regions more accurately.)

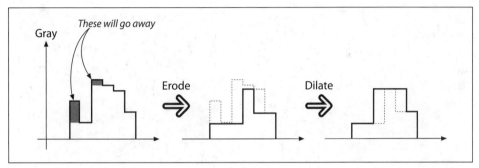

Figure 5-10. Morphological opening operation: the upward outliers are eliminated as a result

Both the opening and closing operations are approximately area-preserving: the most prominent effect of closing is to eliminate lone outliers that are lower than their neighbors whereas the effect of opening is to eliminate lone outliers that are higher than their neighbors. Results of using the opening operator are shown in Figure 5-11, and of the closing operator in Figure 5-13.

One last note on the opening and closing operators concerns how the iterations argument is interpreted. You might expect that asking for two iterations of closing would yield something like dilate-erode-dilate-erode. It turns out that this would not be particularly useful. What you really want (and what you get) is dilate-dilate-erode-erode. In this way, not only the single outliers but also neighboring pairs of outliers will disappear.

Morphological gradient

Our next available operator is the *morphological gradient*. For this one it is probably easier to start with a formula and then figure out what it means:

$$\text{gradient}(src) = \text{dilate}(src) - \text{erode}(src)$$

The effect of this operation on a Boolean image would be simply to isolate perimeters of existing blobs. The process is diagrammed in Figure 5-14, and the effect of this operator on our test images is shown in Figure 5-15.

Figure 5-11. Results of morphological opening on an image: small bright regions are removed, and the remaining bright regions are isolated but retain their size

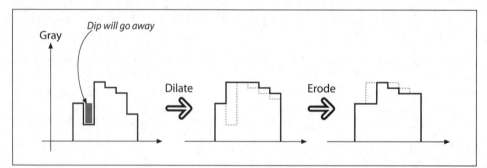

Figure 5-12. Morphological closing operation: the downward outliers are eliminated as a result

With a grayscale image we see that the value of the operator is telling us something about how fast the image brightness is changing; this is why the name "morphological gradient" is justified. Morphological gradient is often used when we want to isolate the perimeters of bright regions so we can treat them as whole objects (or as whole parts of objects). The complete perimeter of a region tends to be found because an expanded version is subtracted from a contracted version of the region, leaving a complete perimeter

Figure 5-13. Results of morphological closing on an image: bright regions are joined but retain their basic size

edge. This differs from calculating a gradient, which is much less likely to work around the full perimeter of an object.*

Top Hat and Black Hat

The last two operators are called *Top Hat* and *Black Hat* [Meyer78]. These operators are used to isolate patches that are, respectively, brighter or dimmer than their immediate neighbors. You would use these when trying to isolate parts of an object that exhibit brightness changes relative only to the object to which they are attached. This often occurs with microscope images of organisms or cells, for example. Both operations are defined in terms of the more primitive operators, as follows:

$$TopHat(src) = src - open(src)$$

$$BlackHat(src) = close(src) - src$$

As you can see, the Top Hat operator subtracts the opened form of A from A. Recall that the effect of the open operation was to exaggerate small cracks or local drops. Thus,

---

* We will return to the topic of gradients when we introduce the Sobel and Scharr operators in the next chapter.

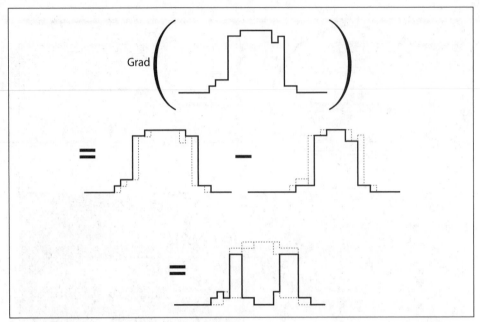

Figure 5-14. Morphological gradient applied to a grayscale image: as expected, the operator has its highest values where the grayscale image is changing most rapidly

subtracting open(A) from A should reveal areas that are lighter then the surrounding region of A, relative to the size of the kernel (see Figure 5-16); conversely, the Black Hat operator reveals areas that are darker than the surrounding region of A (Figure 5-17). Summary results for all the morphological operators discussed in this chapter are assembled in Figure 5-18.*

Flood Fill

Flood fill [Heckbert00; Shaw04; Vandevenne04] is an extremely useful function that is often used to mark or isolate portions of an image for further processing or analysis. Flood fill can also be used to derive, from an input image, masks that can be used for subsequent routines to speed or restrict processing to only those pixels indicated by the mask. The function cvFloodFill() itself takes an optional mask that can be further used to control where filling is done (e.g., when doing multiple fills of the same image).

In OpenCV, flood fill is a more general version of the sort of fill functionality which you probably already associate with typical computer painting programs. For both, a *seed point* is selected from an image and then all similar neighboring points are colored with a uniform color. The difference here is that the neighboring pixels need not all be

---

* Both of these operations (Top Hat and Black Hat) make more sense in grayscale morphology, where the structuring element is a matrix of real numbers (not just a binary mask) and the matrix is added to the current pixel neighborhood before taking a minimum or maximum. Unfortunately, this is not yet implemented in OpenCV.

Figure 5-15. Results of the morphological gradient operator: bright perimeter edges are identified

identical in color.* The result of a flood fill operation will always be a single contiguous region. The cvFloodFill() function will color a neighboring pixel if it is within a specified range (loDiff to upDiff) of either the current pixel or if (depending on the settings of flags) the neighboring pixel is within a specified range of the original seedPoint value. Flood filling can also be constrained by an optional mask argument. The prototype for the flood fill routine is:

```
void cvFloodFill(
    IplImage*        img,
    CvPoint          seedPoint,
    CvScalar         newVal,
    CvScalar         loDiff    = cvScalarAll(0),
    CvScalar         upDiff    = cvScalarAll(0),
    CvConnectedComp* comp      = NULL,
    int              flags     = 4,
    CvArr*           mask      = NULL
);
```

The parameter img is the input image, which can be 8-bit or floating-point and one-channel or three-channel. We start the flood filling from seedPoint, and newVal is the

---

* Users of contemporary painting and drawing programs should note that most now employ a filling algorithm very much like cvFloodFill().

Figure 5-16. Results of morphological Top Hat operation: bright local peaks are isolated

value to which colorized pixels are set. A pixel will be colorized if its intensity is not less than a colorized neighbor's intensity minus loDiff and not greater than the colorized neighbor's intensity plus upDiff. If the flags argument includes CV_FLOODFILL_FIXED_RANGE, then a pixel will be compared to the original seed point rather than to its neighbors. If non-NULL, comp is a CvConnectedComp structure that will hold statistics about the areas filled.* The flags argument (to be discussed shortly) is a little tricky; it controls the connectivity of the fill, what the fill is relative to, whether we are filling only a mask, and what values are used to fill the mask. Our first example of flood fill is shown in Figure 5-19.

The argument mask indicates a mask that can function both as input to cvFloodFill() (in which case it constrains the regions that can be filled) and as output from cvFloodFill() (in which case it will indicate the regions that actually were filled). If set to a non-NULL value, then mask must be a one-channel, 8-bit image whose size is exactly two pixels larger in width and height than the source image (this is to make processing easier and faster for the internal algorithm). Pixel $(x + 1, y + 1)$ in the mask image corresponds to image pixel (x, y) in the source image. Note that cvFloodFill() will not flood across

---

* We will address the specifics of a "connected component" in the section "Image Pyramids". For now, just think of it as being similar to a mask that identifies some subsection of an image.

Figure 5-17. Results of morphological Black Hat operation: dark holes are isolated

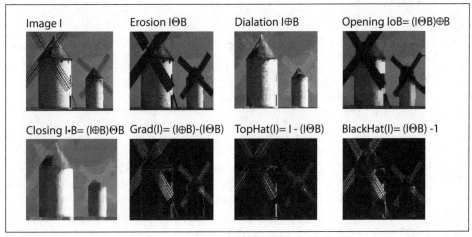

Figure 5-18. Summary results for all morphology operators

nonzero pixels in the mask, so you should be careful to zero it before use if you don't want masking to block the flooding operation. Flood fill can be set to colorize either the source image img or the mask image mask.

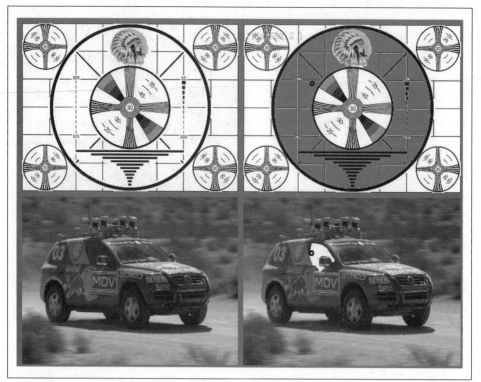

Figure 5-19. Results of flood fill (top image is filled with gray, bottom image with white) from the dark circle located just off center in both images; in this case, the hiDiff and loDiff parameters were each set to 7.0

If the flood-fill mask is set to be marked, then it is marked with the values set in the middle bits (8–15) of the flags value (see text). If these bits are not set then the mask is set to 1 as the default value. Don't be confused if you fill the mask and see nothing but black upon display; the filled values (if the middle bits of the flag weren't set) are 1s, so the mask image needs to be rescaled if you want to display it visually.

It's time to clarify the flags argument, which is tricky because it has three parts. The *low* 8 bits (0–7) can be set to 4 or 8. This controls the connectivity considered by the filling algorithm. If set to 4, only horizontal and vertical neighbors to the current pixel are considered in the filling process; if set to 8, flood fill will additionally include diagonal neighbors. The *high* 8 bits (16–23) can be set with the flags CV_FLOODFILL_FIXED_RANGE (fill relative to the seed point pixel value; otherwise, fill relative to the neighbor's value), and/or CV_FLOODFILL_MASK_ONLY (fill the mask location instead of the source image location). Obviously, you must supply an appropriate mask if CV_FLOODFILL_MASK_ONLY is set. The *middle* bits (8–15) of flags can be set to the value with which you want the mask to be filled. If the middle bits of flags are 0s, the mask will be filled with 1s. All these flags may be linked together via OR. For example, if you want an 8-way connectivity fill, filling only a fixed range, filling the mask not the image, and filling using a value of 47, then the parameter to pass in would be:

```
flags = 8
    | CV_FLOODFILL_MASK_ONLY
    | CV_FLOODFILL_FIXED_RANGE
    | (47<<8);
```

Figure 5-20 shows flood fill in action on a sample image. Using CV_FLOODFILL_FIXED_RANGE with a wide range resulted in most of the image being filled (starting at the center). We should note that newVal, loDiff, and upDiff are prototyped as type CvScalar so they can be set for three channels at once (i.e., to encompass the RGB colors specified via CV_RGB()). For example, lowDiff = CV_RGB(20,30,40) will set lowDiff thresholds of 20 for red, 30 for green, and 40 for blue.

Figure 5-20. Results of flood fill (top image is filled with gray, bottom image with white) from the dark circle located just off center in both images; in this case, flood fill was done with a fixed range and with a high and low difference of 25.0

Resize

We often encounter an image of some size that we would like to convert to an image of some other size. We may want to upsize (zoom in) or downsize (zoom out) the image; we can accomplish either task by using cvResize(). This function will fit the source image exactly to the destination image size. If the ROI is set in the source image then that ROI will be resized to fit in the destination image. Likewise, if an ROI is set in the destination image then the source will be resized to fit into the ROI.

```
void cvResize(
    const CvArr*    src,
    CvArr*          dst,
    int             interpolation = CV_INTER_LINEAR
);
```

The last argument is the interpolation method, which defaults to linear interpolation. The other available options are shown in Table 5-4.

Table 5-4. cvResize() interpolation options

Interpolation	Meaning
CV_INTER_NN	Nearest neighbor
CV_INTER_LINEAR	Bilinear
CV_INTER_AREA	Pixel area re-sampling
CV_INTER_CUBIC	Bicubic interpolation

In general, we would like the mapping from the source image to the resized destination image to be as smooth as possible. The argument interpolation controls exactly how this will be handled. Interpolation arises when we are shrinking an image and a pixel in the destination image falls in between pixels in the source image. It can also occur when we are expanding an image and need to compute values of pixels that do not directly correspond to any pixel in the source image. In either case, there are several options for computing the values of such pixels. The easiest approach is to take the resized pixel's value from its closest pixel in the source image; this is the effect of choosing the interpolation value CV_INTER_NN. Alternatively, we can linearly weight the 2-by-2 surrounding source pixel values according to how close they are to the destination pixel, which is what CV_INTER_LINEAR does. We can also virtually place the new resized pixel over the old pixels and then average the covered pixel values, as done with CV_INTER_AREA.* Finally, we have the option of fitting a cubic spline between the 4-by-4 surrounding pixels in the source image and then reading off the corresponding destination value from the fitted spline; this is the result of choosing the CV_INTER_CUBIC interpolation method.

Image Pyramids

Image pyramids [Adelson84] are heavily used in a wide variety of vision applications. An image pyramid is a collection of images—all arising from a single original image— that are successively downsampled until some desired stopping point is reached. (Of course, this stopping point could be a single-pixel image!)

---

* At least that's what happens when cvResize() shrinks an image. When it expands an image, CV_INTER_AREA amounts to the same thing as CV_INTER_NN.

There are two kinds of image pyramids that arise often in the literature and in application: the Gaussian [Rosenfeld80] and Laplacian [Burt83] pyramids [Adelson84]. The *Gaussian pyramid* is used to downsample images, and the Laplacian pyramid (to be discussed shortly) is required when we want to reconstruct an upsampled image from an image lower in the pyramid.

To produce layer (i+1) in the Gaussian pyramid (we denote this layer G_{i+1}) from layer G_i of the pyramid, we first convolve G_i with a Gaussian kernel and then remove every even-numbered row and column. Of course, from this it follows immediately that each image is exactly one-quarter the area of its predecessor. Iterating this process on the input image G_0 produces the entire pyramid. OpenCV provides us with a method for generating each pyramid stage from its predecessor:

```
void cvPyrDown(
    IplImage*    src,
    IplImage*    dst,
    IplFilter    filter = IPL_GAUSSIAN_5x5
);
```

Currently, the last argument `filter` supports only the single (default) option of a 5-by-5 Gaussian kernel.

Similarly, we can convert an existing image to an image that is twice as large in each direction by the following analogous (but not inverse!) operation:

```
void cvPyrUp(
    IplImage*    src,
    IplImage*    dst,
    IplFilter    filter = IPL_GAUSSIAN_5x5
);
```

In this case the image is first upsized to twice the original in each dimension, with the new (even) rows filled with 0s. Thereafter, a convolution is performed with the given filter (actually, a filter twice as large in each dimension than that specified*) to approximate the values of the "missing" pixels.

We noted previously that the operator `PyrUp()` is not the inverse of `PyrDown()`. This should be evident because `PyrDown()` is an operator that loses information. In order to restore the original (higher-resolution) image, we would require access to the information that was discarded by the downsampling. This data forms the *Laplacian pyramid*. The ith layer of the Laplacian pyramid is defined by the relation:

$$L_i = G_i - \text{UP}(G_{i+1}) \otimes \mathcal{G}_{5\times5}$$

Here the operator $\text{UP}()$ upsizes by mapping each pixel in location (x, y) in the original image to pixel $(2x + 1, 2y + 1)$ in the destination image; the $\otimes$ symbol denotes convolution; and $\mathcal{G}_{5\times5}$ is a 5-by-5 Gaussian kernel. Of course, $G_i - \text{UP}(G_{i+1}) \otimes \mathcal{G}_{5\times5}$ is the definition

---

* This filter is also normalized to four, rather than to one. This is appropriate because the inserted rows have 0s in all of their pixels before the convolution.

of the PyrUp() operator provided by OpenCv. Hence, we can use OpenCv to compute the Laplacian operator directly as:

$$L_i = G_i - \text{PyrUp}(G_{i+1})$$

The Gaussian and Laplacian pyramids are shown diagrammatically in Figure 5-21, which also shows the inverse process for recovering the original image from the sub-images. Note how the Laplacian is really an approximation that uses the difference of Gaussians, as revealed in the preceding equation and diagrammed in the figure.

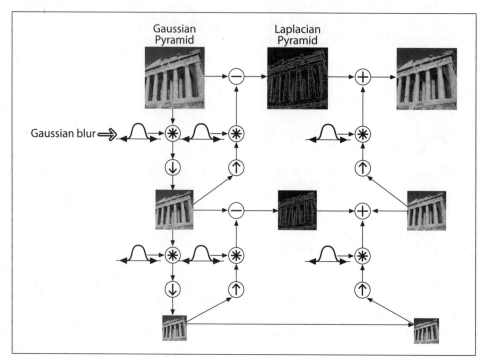

Figure 5-21. The Gaussian pyramid and its inverse, the Laplacian pyramid

There are many operations that can make extensive use of the Gaussian and Laplacian pyramids, but a particularly important one is image segmentation (see Figure 5-22). In this case, one builds an image pyramid and then associates to it a system of parent–child relations between pixels at level G_{i+1} and the corresponding reduced pixel at level G_i. In this way, a fast initial segmentation can be done on the low-resolution images high in the pyramid and then can be refined and further differentiated level by level.

This algorithm (due to B. Jaehne [Jaehne95; Antonisse82]) is implemented in OpenCV as cvPyrSegmentation():

```
void cvPyrSegmentation(
    IplImage*     src,
    IplImage*     dst,
```

Figure 5-22. Pyramid segmentation with threshold1 set to 150 and threshold2 set to 30; the images on the right contain only a subsection of the images on the left because pyramid segmentation requires images that are N-times divisible by 2, where N is the number of pyramid layers to be computed (these are 512-by-512 areas from the original images)

```
    CvMemStorage*    storage,
    CvSeq**          comp,
    int              level,
    double           threshold1,
    double           threshold2
);
```

As usual, src and dst are the source and destination images, which must both be 8-bit, of the same size, and of the same number of channels (one or three). You might be wondering, "What destination image?" Not an unreasonable question, actually. The destination image dst is used as scratch space for the algorithm and also as a return visualization of the segmentation. If you view this image, you will see that each segment is colored in a single color (the color of some pixel in that segment). Because this image is the algorithm's scratch space, you cannot simply set it to NULL. Even if you do not want the result, you must provide an image. One important word of warning about src and dst: because all levels of the image pyramid must have integer sizes in both dimensions, the starting images must be divisible by two as many times as there are levels in the pyramid. For example, for a four-level pyramid, a height or width of 80 ($2 \times 2 \times 2 \times 5$) would be acceptable, but a value of 90 ($2 \times 3 \times 3 \times 5$) would not.[*]

The pointer storage is for an OpenCV memory storage area. In Chapter 8 we will discuss such areas in more detail, but for now you should know that such a storage area is allocated with a command like[†]

```
CvMemStorage* storage = cvCreateMemStorage();
```

The argument comp is a location for storing further information about the resulting segmentation: a sequence of connected components is allocated from this memory storage. Exactly how this works will be detailed in Chapter 8, but for convenience here we briefly summarize what you'll need in the context of cvPyrSegmentation().

First of all, a *sequence* is essentially a list of structures of a particular kind. Given a sequence, you can obtain the number of elements as well as a particular element if you know both its type and its number in the sequence. Take a look at the Example 5-1 approach to accessing a sequence.

Example 5-1. Doing something with each element in the sequence of connected components returned by cvPyrSegmentation()

```
void f(
  IplImage* src,
  IplImage* dst
) {
  CvMemStorage* storage = cvCreateMemStorage(0);
  CvSeq* comp = NULL;
  cvPyrSegmentation( src, dst, storage, &comp, 4, 200, 50 );
  int n_comp = comp->total;
  for( int i=0; i<n_comp; i++ ) {
    CvConnectedComp* cc = (CvConnectedComp*) cvGetSeqElem( comp, i );
    do_something_with( cc );
  }
  cvReleaseMemStorage( &storage );
}
```

There are several things you should notice in this example. First, observe the allocation of a *memory storage*; this is where cvPyrSegmentation() will get the memory it needs for the connected components it will have to create. Then the pointer comp is allocated as type CvSeq*. It is initialized to NULL because its current value means nothing. We will pass to cvPyrSegmentation() a pointer to comp so that comp can be set to the location of the sequence created by cvPyrSegmentation(). Once we have called the segmentation, we can figure out how many elements there are in the sequence with the member element total. Thereafter we can use the generic cvGetSeqElem() to obtain the *i*th element of comp; however, because cvGetSeqElem() is generic and returns only a void pointer, we must cast the return pointer to the appropriate type (in this case, CvConnectedComp*).

---

[*] Heed this warning! Otherwise, you will get a totally useless error message and probably waste hours trying to figure out what's going on.

[†] Actually, the current implementation of cvPyrSegmentation() is a bit incomplete in that it returns not the computed segments but only the bounding rectangles (as CvSeq<CvConnectedComp>).

Finally, we need to know that a connected component is one of the basic structure types in OpenCV. You can think of it as a way of describing a "blob" in an image. It has the following definition:

```
typedef struct CvConnectedComponent {
    double   area;
    CvScalar value;
    CvRect   rect;
    CvSeq*   contour;
};
```

The area is the area of the component. The value is the average color* over the area of the component and rect is a bounding box for the component (defined in the coordinates of the parent image). The final element, contour, is a pointer to another sequence. This sequence can be used to store a representation of the boundary of the component, typically as a sequence of points (type CvPoint).

In the specific case of cvPyrSegmentation(), the contour member is not set. Thus, if you want some specific representation of the component's pixels then you will have to compute it yourself. The method to use depends, of course, on the representation you have in mind. Often you will want a Boolean mask with nonzero elements wherever the component was located. You can easily generate this by using the rect portion of the connected component as a mask and then using cvFloodFill() to select the desired pixels inside of that rectangle.

Threshold

Frequently we have done many layers of processing steps and want either to make a final decision about the pixels in an image or to categorically reject those pixels below or above some value while keeping the others. The OpenCV function cvThreshold() accomplishes these tasks (see survey [Sezgin04]). The basic idea is that an array is given, along with a threshold, and then something happens to every element of the array depending on whether it is below or above the threshold.

```
double cvThreshold(
    CvArr*      src,
    CvArr*      dst,
    double      threshold,
    double      max_value,
    int         threshold_type
);
```

As shown in Table 5-5, each threshold type corresponds to a particular comparison operation between the ith source pixel (src_i) and the threshold (denoted in the table by T). Depending on the relationship between the source pixel and the threshold, the destination pixel dst_i may be set to 0, the src_i, or the max_value (denoted in the table by M).

---

* Actually the meaning of value is context dependant and could be just about anything, but it is typically a color associated with the component. In the case of cvPyrSegmentation(), value is the average color over the segment.

Table 5-5. cvThreshold() threshold_type options

Threshold type	Operation
CV_THRESH_BINARY	$dst_i = (src_i > T) ? M : 0$
CV_THRESH_BINARY_INV	$dst_i = (src_i > T) ? 0 : M$
CV_THRESH_TRUNC	$dst_i = (src_i > T) ? M : src_i$
CV_THRESH_TOZERO_INV	$dst_i = (src_i > T) ? 0 : src_i$
CV_THRESH_TOZERO	$dst_i = (src_i > T) ? src_i : 0$

Figure 5-23 should help to clarify the exact implications of each threshold type.

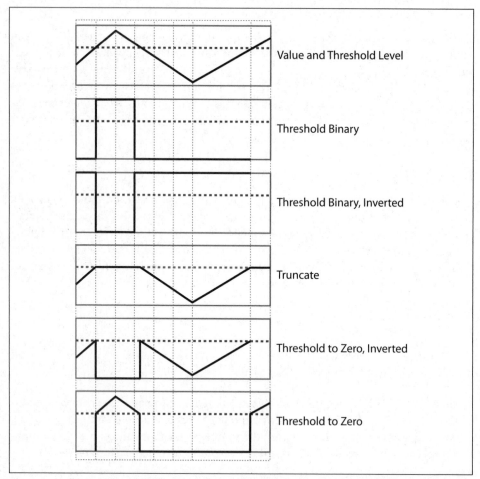

Figure 5-23. Results of varying the threshold type in cvThreshold(). The horizontal line through each chart represents a particular threshold level applied to the top chart and its effect for each of the five types of threshold operations below

Let's look at a simple example. In Example 5-2 we sum all three channels of an image and then clip the result at 100.

Example 5-2. Example code making use of cvThreshold()

```
#include <stdio.h>
#include <cv.h>
#include <highgui.h>
void sum_rgb( IplImage* src, IplImage* dst ) {

  // Allocate individual image planes.
  IplImage* r = cvCreateImage( cvGetSize(src), IPL_DEPTH_8U, 1 );
  IplImage* g = cvCreateImage( cvGetSize(src), IPL_DEPTH_8U, 1 );
  IplImage* b = cvCreateImage( cvGetSize(src), IPL_DEPTH_8U, 1 );

  // Split image onto the color planes.
  cvSplit( src, r, g, b, NULL );

  // Temporary storage.
  IplImage* s = cvCreateImage( cvGetSize(src), IPL_DEPTH_8U, 1 );

  // Add equally weighted rgb values.
  cvAddWeighted( r, 1./3., g, 1./3., 0.0, s );
  cvAddWeighted( s, 2./3., b, 1./3., 0.0, s );

  // Truncate values above 100.
  cvThreshold( s, dst, 100, 100, CV_THRESH_TRUNC );

  cvReleaseImage( &r );
  cvReleaseImage( &g );
  cvReleaseImage( &b );
  cvReleaseImage( &s );
}

int main(int argc, char** argv)
{

  // Create a named window with the name of the file.
  cvNamedWindow( argv[1], 1 );

  // Load the image from the given file name.
  IplImage* src = cvLoadImage( argv[1] );
  IplImage* dst = cvCreateImage( cvGetSize(src), src->depth, 1);
  sum_rgb( src, dst);

  // Show the image in the named window
  cvShowImage( argv[1], dst );

  // Idle until the user hits the "Esc" key.
  while( 1 ) { if( (cvWaitKey( 10 )&0x7f) == 27 ) break; }

  // Clean up and don't be piggies
  cvDestroyWindow( argv[1] );
```

Example 5-2. Example code making use of cvThreshold() (continued)

```
    cvReleaseImage( &src );
    cvReleaseImage( &dst );

}
```

Some important ideas are shown here. One thing is that we don't want to add into an 8-bit array because the higher bits will overflow. Instead, we use equally weighted addition of the three color channels (cvAddWeighted()); then the results are truncated to saturate at the value of 100 for the return. The cvThreshold() function handles only 8-bit or floating-point grayscale source images. The destination image must either match the source image or be an 8-bit image. In fact, cvThreshold() also allows the source and destination images to be the same image. Had we used a floating-point temporary image s in Example 5-2, we could have substituted the code shown in Example 5-3. Note that cvAcc() can accumulate 8-bit integer image types into a floating-point image; however, cvADD() cannot add integer bytes into floats.

Example 5-3. Alternative method to combine and threshold image planes

```
IplImage* s = cvCreateImage(cvGetSize(src), IPL_DEPTH_32F, 1);
cvZero(s);
cvAcc(b,s);
cvAcc(g,s);
cvAcc(r,s);
cvThreshold( s, s, 100, 100, CV_THRESH_TRUNC );
cvConvertScale( s, dst, 1, 0 );
```

Adaptive Threshold

There is a modified threshold technique in which the threshold level is itself variable. In OpenCV, this method is implemented in the cvAdaptiveThreshold() [Jain86] function:

```
    void cvAdaptiveThreshold(
        CvArr*          src,
        CvArr*          dst,
        double          max_val,
        int             adaptive_method = CV_ADAPTIVE_THRESH_MEAN_C
        int             threshold_type  = CV_THRESH_BINARY,
        int             block_size      = 3,
        double          param1          = 5
    );
```

cvAdaptiveThreshold() allows for two different adaptive threshold types depending on the settings of adaptive_method. In both cases the *adaptive threshold* $T(x, y)$ is set on a pixel-by-pixel basis by computing a weighted average of the *b*-by-*b* region around each pixel location minus a constant, where *b* is given by block_size and the constant is given by param1. If the method is set to CV_ADAPTIVE_THRESH_MEAN_C, then all pixels in the area are weighted equally. If it is set to CV_ADAPTIVE_THRESH_GAUSSIAN_C, then the pixels in the region around (x, y) are weighted according to a Gaussian function of their distance from that center point.

Finally, the parameter threshold_type is the same as for cvThreshold() shown in Table 5-5.

The adaptive threshold technique is useful when there are strong illumination or reflectance gradients that you need to threshold relative to the general intensity gradient. This function handles only single-channel 8-bit or floating-point images, and it requires that the source and destination images be distinct.

Source code for comparing cvAdaptiveThreshold() and cvThreshold() is shown in Example 5-4. Figure 5-24 displays the result of processing an image that has a strong lighting gradient across it. The lower-left portion of the figure shows the result of using a single global threshold as in cvThreshold(); the lower-right portion shows the result of adaptive local threshold using cvAdaptiveThreshold(). We get the whole checkerboard via adaptive threshold, a result that is impossible to achieve when using a single threshold. Note the calling-convention comments at the top of the code in Example 5-4; the parameters used for Figure 5-24 were:

```
./adaptThresh 15 1 1 71 15 ../Data/cal3-L.bmp
```

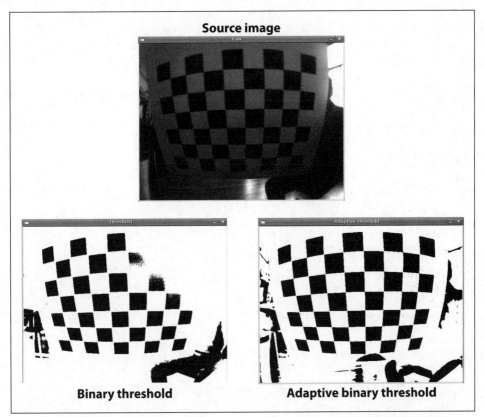

Figure 5-24. Binary threshold versus adaptive binary threshold: the input image (top) was turned into a binary image using a global threshold (lower left) and an adaptive threshold (lower right); raw image courtesy of Kurt Konolidge

Example 5-4. Threshold versus adaptive threshold

```
// Compare thresholding with adaptive thresholding
// CALL:
// ./adaptThreshold Threshold 1binary 1adaptivemean \
//                  blocksize offset filename
#include "cv.h"
#include "highgui.h"
#include "math.h"
IplImage *Igray=0, *It = 0, *Iat;
int main( int argc, char** argv )
{
    if(argc != 7){return -1;          }

    //Command line
    double threshold = (double)atof(argv[1]);
    int threshold_type =  atoi(argv[2]) ?
            CV_THRESH_BINARY : CV_THRESH_BINARY_INV;
    int adaptive_method = atoi(argv[3]) ?
            CV_ADAPTIVE_THRESH_MEAN_C : CV_ADAPTIVE_THRESH_GAUSSIAN_C;
    int block_size = atoi(argv[4]);
    double offset = (double)atof(argv[5]);

    //Read in gray image
    if((Igray = cvLoadImage( argv[6], CV_LOAD_IMAGE_GRAYSCALE)) == 0){
        return       -1;}

    // Create the grayscale output images
    It = cvCreateImage(cvSize(Igray->width,Igray->height),
                        IPL_DEPTH_8U, 1);
    Iat = cvCreateImage(cvSize(Igray->width,Igray->height),
                        IPL_DEPTH_8U, 1);
    //Threshold
    cvThreshold(Igray,It,threshold,255,threshold_type);
    cvAdaptiveThreshold(Igray, Iat, 255, adaptive_method,
                        threshold_type, block_size, offset);
    //PUT UP 2 WINDOWS
    cvNamedWindow("Raw",1);
    cvNamedWindow("Threshold",1);
    cvNamedWindow("Adaptive Threshold",1);

    //Show the results
    cvShowImage("Raw",Igray);
    cvShowImage("Threshold",It);
    cvShowImage("Adaptive Threshold",Iat);

    cvWaitKey(0);

    //Clean up
    cvReleaseImage(&Igray);
    cvReleaseImage(&It);
    cvReleaseImage(&Iat);
    cvDestroyWindow("Raw");
    cvDestroyWindow("Threshold");
```

Example 5-4. Threshold versus adaptive threshold (continued)

```
    cvDestroyWindow("Adaptive Threshold");
    return(0);
}
```

Exercises

1. Load an image with interesting textures. Smooth the image in several ways using cvSmooth() with smoothtype=CV_GAUSSIAN.

   a. Use a symmetric 3-by-3, 5-by-5, 9-by-9 and 11-by-11 smoothing window size and display the results.

   b. Are the output results nearly the same by smoothing the image twice with a 5-by-5 Gaussian filter as when you smooth once with two 11-by-11 filters? Why or why not?

2. Display the filter, creating a 100-by-100 single-channel image. Clear it and set the center pixel equal to 255.

   a. Smooth this image with a 5-by-5 Gaussian filter and display the results. What did you find?

   b. Do this again but now with a 9-by-9 Gaussian filter.

   c. What does it look like if you start over and smooth the image twice with the 5-by-5 filter? Compare this with the 9-by-9 results. Are they nearly the same? Why or why not?

3. Load an interesting image. Again, blur it with cvSmooth() using a Gaussian filter.

   a. Set param1=param2=9. Try several settings of param3 (e.g., 1, 4, and 6). Display the results.

   b. This time, set param1=param2=0 before setting param3 to 1, 4, and 6. Display the results. Are they different? Why?

   c. Again use param1=param2=0 but now set param3=1 and param4=9. Smooth the picture and display the results.

   d. Repeat part c but with param3=9 and param4=1. Display the results.

   e. Now smooth the image once with the settings of part c and once with the settings of part d. Display the results.

   f. Compare the results in part e with smoothings that use param3=param4=9 and param3=param4=0 (i.e., a 9-by-9 filter). Are the results the same? Why or why not?

4. Use a camera to take two pictures of the same scene while moving the camera as little as possible. Load these images into the computer as src1 and src1.

   a. Take the absolute value of src1 minus src1 (subtract the images); call it diff12 and display. If this were done perfectly, diff12 would be black. Why isn't it?

b. Create `cleandiff` by using `cvErode()` and then `cvDilate()` on `diff12`. Display the results.

c. Create `dirtydiff` by using `cvDilate()` and then `cvErode()` on `diff12` and then display.

d. Explain the difference between `cleandiff` and `dirtydiff`.

5. Take a picture of a scene. Then, without moving the camera, put a coffee cup in the scene and take a second picture. Load these images and convert both to 8-bit grayscale images.

a. Take the absolute value of their difference. Display the result, which should look like a noisy mask of a coffee mug.

b. Do a binary threshold of the resulting image using a level that preserves most of the coffee mug but removes some of the noise. Display the result. The "on" values should be set to 255.

c. Do a `CV_MOP_OPEN` on the image to further clean up noise.

6. Create a clean mask from noise. After completing exercise 5, continue by keeping only the largest remaining shape in the image. Set a pointer to the upper left of the image and then traverse the image. When you find a pixel of value 255 ("on"), store the location and then flood fill it using a value of 100. Read the connected component returned from flood fill and record the area of filled region. If there is another larger region in the image, then flood fill the smaller region using a value of 0 and delete its recorded area. If the new region is larger than the previous region, then flood fill the previous region using the value 0 and delete its location. Finally, fill the remaining largest region with 255. Display the results. We now have a single, solid mask for the coffee mug.

7. For this exercise, use the mask created in exercise 6 or create another mask of your own (perhaps by drawing a digital picture, or simply use a square). Load an outdoor scene. Now use this mask with `cvCopy()`, to copy an image of a mug into the scene.

8. Create a low-variance random image (use a random number call such that the numbers don't differ by much more than 3 and most numbers are near 0). Load the image into a drawing program such as PowerPoint and then draw a wheel of lines meeting at a single point. Use bilateral filtering on the resulting image and explain the results.

9. Load an image of a scene and convert it to grayscale.

a. Run the morphological Top Hat operation on your image and display the results.

b. Convert the resulting image into an 8-bit mask.

c. Copy a grayscale value into the Top Hat pieces and display the results.

10. Load an image with many details.

a. Use cvResize() to reduce the image by a factor of 2 in each dimension (hence the image will be reduced by a factor of 4). Do this three times and display the results.

b. Now take the original image and use cvPyrDown() to reduce it three times and then display the results.

c. How are the two results different? Why are the approaches different?

11. Load an image of a scene. Use cvPyrSegmentation() and display the results.

12. Load an image of an interesting or sufficiently "rich" scene. Using cvThreshold(), set the threshold to 128. Use each setting type in Table 5-5 on the image and display the results. You should familiarize yourself with thresholding functions because they will prove quite useful.

a. Repeat the exercise but use cvAdaptiveThreshold() instead. Set param1=5.

b. Repeat part a using param1=0 and then param1=-5.

CHAPTER 6

Image Transforms

Overview

In the previous chapter we covered a lot of different things you could do with an image. The majority of the operators presented thus far are used to enhance, modify, or otherwise "process" one image into a similar but new image.

In this chapter we will look at *image transforms*, which are methods for changing an image into an alternate representation of the data entirely. Perhaps the most common example of a transform would be a something like a *Fourier transform*, in which the image is converted to an alternate representation of the data in the original image. The result of this operation is still stored in an OpenCV "image" structure, but the individual "pixels" in this new image represent spectral components of the original input rather than the spatial components we are used to thinking about.

There are a number of useful transforms that arise repeatedly in computer vision. OpenCV provides complete implementations of some of the more common ones as well as building blocks to help you implement your own image transforms.

Convolution

Convolution is the basis of many of the transformations that we discuss in this chapter. In the abstract, this term means something we do to every part of an image. In this sense, many of the operations we looked at in Chapter 5 can also be understood as special cases of the more general process of convolution. What a particular convolution "does" is determined by the form of the *Convolution kernel* being used. This kernel is essentially just a fixed size array of numerical coefficients along with an *anchor point* in that array, which is typically located at the center. The size of the array* is called the *support* of the kernel.

Figure 6-1 depicts a 3-by-3 convolution kernel with the anchor located at the center of the array. The value of the convolution at a particular point is computed by first placing

---

* For technical purists, the support of the kernel actually consists of only the nonzero portion of the kernel array.

the kernel anchor on top of a pixel on the image with the rest of the kernel overlaying the corresponding local pixels in the image. For each kernel point, we now have a value for the kernel at that point and a value for the image at the corresponding image point. We multiply these together and sum the result; this result is then placed in the resulting image at the location corresponding to the location of the anchor in the input image. This process is repeated for every point in the image by scanning the kernel over the entire image.

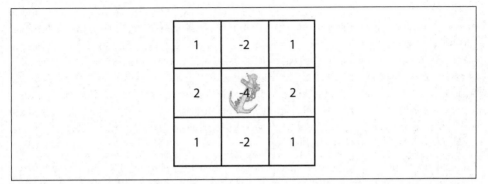

Figure 6-1. A 3-by-3 kernel for a Sobel derivative; note that the anchor point is in the center of the kernel

We can, of course, express this procedure in the form of an equation. If we define the image to be $I(x, y)$, the kernel to be $G(i, j)$ (where $0 < i < M_i - 1$ and $0 < j < M_j - 1$), and the anchor point to be located at (a_i, a_j) in the coordinates of the kernel, then the convolution $H(x, y)$ is defined by the following expression:

$$H(x, y) = \sum_{i=0}^{M_i-1} \sum_{j=0}^{M_j-1} I(x+i-a_i, y+j-a_j) G(i, j)$$

Observe that the number of operations, at least at first glance, seems to be the number of pixels in the image multiplied by the number of pixels in the kernel.* This can be a lot of computation and so is not something you want to do with some "for" loop and a lot of pointer de-referencing. In situations like this, it is better to let OpenCV do the work for you and take advantage of the optimizations already programmed into OpenCV. The OpenCV way to do this is with cvFilter2D():

```
void cvFilter2D(
    const CvArr*      src,
    CvArr*            dst,
    const CvMat*      kernel,
```

---

* We say "at first glance" because it is also possible to perform convolutions in the frequency domain. In this case, for an N-by-N image and an M-by-M kernel with N > M, the computational time will be proportional to $N^2 \log(N)$ and not to the $N^2 M^2$ that is expected for computations in the spatial domain. Because the frequency domain computation is independent of the size of the kernel, it is more efficient for large kernels. OpenCV automatically decides whether to do the convolution in the frequency domain based on the size of the kernel.

```
    CvPoint        anchor = cvPoint(-1,-1)
);
```

Here we create a matrix of the appropriate size, fill it with the coefficients, and then pass it together with the source and destination images into cvFilter2D(). We can also optionally pass in a CvPoint to indicate the location of the center of the kernel, but the default value (equal to cvPoint(-1,-1)) is interpreted as indicating the center of the kernel. The kernel can be of even size if its anchor point is defined; otherwise, it should be of odd size.

The src and dst images should be the same size. One might think that the src image should be larger than the dst image in order to allow for the extra width and length of the convolution kernel. But the sizes of the src and dst can be the same in OpenCV because, by default, prior to convolution OpenCV creates virtual pixels via replication past the border of the src image so that the border pixels in dst can be filled in. The replication is done as input($-dx$, y) = input(0, y), input($w + dx$, y) = input($w - 1$, y), and so forth. There are some alternatives to this default behavior; we will discuss them in the next section.

We remark that the coefficients of the convolution kernel should always be floating-point numbers. This means that you should use CV_32FC1 when allocating that matrix.

Convolution Boundaries

One problem that naturally arises with convolutions is how to handle the boundaries. For example, when using the convolution kernel just described, what happens when the point being convolved is at the edge of the image? Most of OpenCV's built-in functions that make use of cvFilter2D() must handle this in one way or another. Similarly, when doing your own convolutions, you will need to know how to deal with this efficiently.

The solution comes in the form of the cvCopyMakeBorder() function, which copies a given image onto another slightly larger image and then automatically pads the boundary in one way or another:

```
void cvCopyMakeBorder(
    const CvArr*    src,
    CvArr*          dst,
    CvPoint         offset,
    int             bordertype,
    CvScalar        value    = cvScalarAll(0)
);
```

The offset argument tells cvCopyMakeBorder() where to place the copy of the original image within the destination image. Typically, if the kernel is N-by-N (for odd N) then you will want a boundary that is $(N - 1)/2$ wide on all sides or, equivalently, an image that is $N - 1$ wider and taller than the original. In this case you would set the offset to cvPoint((N-1)/2,(N-1)/2) so that the boundary would be even on all sides.*

---

* Of course, the case of N-by-N with N odd and the anchor located at the center is the simplest case. In general, if the kernel is N-by-M and the anchor is located at (a_x, a_y), then the destination image will have to be $N - 1$ pixels wider and $M - 1$ pixels taller than the source image. The offset will simply be (a_x, a_y).

The border type can be either IPL_BORDER_CONSTANT or IPL_BORDER_REPLICATE (see Figure 6-2). In the first case, the value argument will be interpreted as the value to which all pixels in the boundary should be set. In the second case, the row or column at the very edge of the original is replicated out to the edge of the larger image. Note that the border of the test pattern image is somewhat subtle (examine the upper right image in Figure 6-2); in the test pattern image, there's a one-pixel-wide dark border except where the circle patterns come near the border where it turns white. There are two other border types defined, IPL_BORDER_REFLECT and IPL_BORDER_WRAP, which are not implemented at this time in OpenCV but may be supported in the future.

Figure 6-2. Expanding the image border. The left column shows IPL_BORDER_CONSTANT where a zero value is used to fill out the borders. The right column shows IPL_BORDER_REPLICATE where the border pixels are replicated in the horizontal and vertical directions

We mentioned previously that, when you make calls to OpenCV library functions that employ convolution, those library functions call cvCopyMakeBorder() to get their work done. In most cases the border type called is IPL_BORDER_REPLICATE, but sometimes you will not want it to be done that way. This is another occasion where you might want to use cvCopyMakeBorder(). You can create a slightly larger image with the border you want, call whatever routine on that image, and then clip back out the part you were originally interested in. This way, OpenCV's automatic bordering will not affect the pixels you care about.

Gradients and Sobel Derivatives

One of the most basic and important convolutions is the computation of derivatives (or approximations to them). There are many ways to do this, but only a few are well suited to a given situation.

In general, the most common operator used to represent differentiation is the *Sobel derivative* [Sobel68] operator (see Figures 6-3 and 6-4). Sobel operators exist for any order of derivative as well as for mixed partial derivatives (e.g., $\partial^2/\partial x \partial y$).

Figure 6-3. *The effect of the Sobel operator when used to approximate a first derivative in the x-dimension*

```
cvSobel(
    const CvArr*    src,
    CvArr*          dst,
    int             xorder,
    int             yorder,
    int             aperture_size = 3
);
```

Here, src and dst are your image input and output, and xorder and yorder are the orders of the derivative. Typically you'll use 0, 1, or at most 2; a 0 value indicates no derivative

Figure 6-4. The effect of the Sobel operator when used to approximate a first derivative in the y-dimension

in that direction.* The aperture_size parameter should be odd and is the width (and the height) of the square filter. Currently, aperture_sizes of 1, 3, 5, and 7 are supported. If src is 8-bit then the dst must be of depth IPL_DEPTH_16S to avoid overflow.

Sobel derivatives have the nice property that they can be defined for kernels of any size, and those kernels can be constructed quickly and iteratively. The larger kernels give a better approximation to the derivative because the smaller kernels are very sensitive to noise.

To understand this more exactly, we must realize that a Sobel derivative is not really a derivative at all. This is because the Sobel operator is defined on a discrete space. What the Sobel operator actually represents is a fit to a polynomial. That is, the Sobel derivative of second order in the *x*-direction is not really a second derivative; it is a local fit to a parabolic function. This explains why one might want to use a larger kernel: that larger kernel is computing the fit over a larger number of pixels.

---

* Either xorder or yorder must be nonzero.

Scharr Filter

In fact, there are many ways to approximate a derivative in the case of a discrete grid. The downside of the approximation used for the Sobel operator is that it is less accurate for small kernels. For large kernels, where more points are used in the approximation, this problem is less significant. This inaccuracy does not show up directly for the X and Y filters used in cvSobel(), because they are exactly aligned with the x- and y-axes. The difficulty arises when you want to make image measurements that are approximations of *directional derivatives* (i.e., direction of the image gradient by using the arctangent of the y/x filter responses).

To put this in context, a concrete example of where you may want image measurements of this kind would be in the process of collecting shape information from an object by assembling a histogram of gradient angles around the object. Such a histogram is the basis on which many common shape classifiers are trained and operated. In this case, inaccurate measures of gradient angle will decrease the recognition performance of the classifier.

For a 3-by-3 Sobel filter, the inaccuracies are more apparent the further the gradient angle is from horizontal or vertical. OpenCV addresses this inaccuracy for small (but fast) 3-by-3 Sobel derivative filters by a somewhat obscure use of the special aperture_size value CV_SCHARR in the cvSobel() function. The Scharr filter is just as fast but more accurate than the Sobel filter, so it should always be used if you want to make image measurements using a 3-by-3 filter. The filter coefficients for the Scharr filter are shown in Figure 6-5 [Scharr00].

-3	0	3
-10	0	10
-3	0	3

-3	-10	-3
0	0	0
3	10	3

Figure 6-5. The 3-by-3 Scharr filter using flag CV_SHARR

Laplace

The OpenCV *Laplacian* function (first used in vision by Marr [Marr82]) implements a discrete analog of the Laplacian operator:*

---

* Note that the Laplacian operator is completely distinct from the Laplacian pyramid of Chapter 5.

$$\text{Laplace}(f) \equiv \frac{\partial^2 f}{\partial x^2} + \frac{\partial^2 f}{\partial y^2}$$

Because the Laplacian operator can be defined in terms of second derivatives, you might well suppose that the discrete implementation works something like the second-order Sobel derivative. Indeed it does, and in fact the OpenCV implementation of the Laplacian operator uses the Sobel operators directly in its computation.

```
void cvLaplace(
  const CvArr* src,
  CvArr*       dst,
  int          apertureSize = 3
);
```

The cvLaplace() function takes the usual source and destination images as arguments as well as an aperture size. The source can be either an 8-bit (unsigned) image or a 32-bit (floating-point) image. The destination must be a 16-bit (signed) image or a 32-bit (floating-point) image. This aperture is precisely the same as the aperture appearing in the Sobel derivatives and, in effect, gives the size of the region over which the pixels are sampled in the computation of the second derivatives.

The Laplace operator can be used in a variety of contexts. A common application is to detect "blobs." Recall that the form of the Laplacian operator is a sum of second derivatives along the x-axis and y-axis. This means that a single point or any small blob (smaller than the aperture) that is surrounded by higher values will tend to maximize this function. Conversely, a point or small blob that is surrounded by lower values will tend to maximize the negative of this function.

With this in mind, the Laplace operator can also be used as a kind of edge detector. To see how this is done, consider the first derivative of a function, which will (of course) be large wherever the function is changing rapidly. Equally important, it will grow rapidly as we approach an edge-like discontinuity and shrink rapidly as we move past the discontinuity. Hence the derivative will be at a local maximum somewhere within this range. Therefore we can look to the 0s of the second derivative for locations of such local maxima. Got that? Edges in the original image will be 0s of the Laplacian. Unfortunately, both substantial and less meaningful edges will be 0s of the Laplacian, but this is not a problem because we can simply filter out those pixels that also have larger values of the first (Sobel) derivative. Figure 6-6 shows an example of using a Laplacian on an image together with details of the first and second derivatives and their zero crossings.

Canny

The method just described for finding edges was further refined by J. Canny in 1986 into what is now commonly called the *Canny edge detector* [Canny86]. One of the differences between the Canny algorithm and the simpler, Laplace-based algorithm from the previous section is that, in the Canny algorithm, the first derivatives are computed in x and y and then combined into four directional derivatives. The points where these directional derivatives are local maxima are then candidates for assembling into edges.

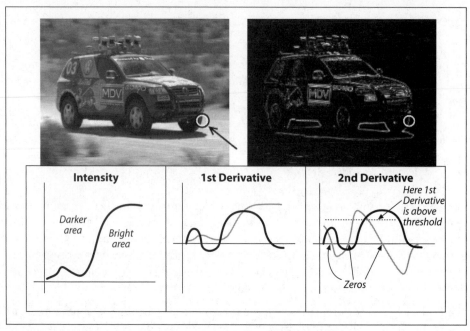

Figure 6-6. Laplace transform (upper right) of the racecar image: zooming in on the tire (circled in white) and considering only the x-dimension, we show a (qualitative) representation of the brightness as well as the first and second derivative (lower three cells); the 0s in the second derivative correspond to edges, and the 0 corresponding to a large first derivative is a strong edge

However, the most significant new dimension to the Canny algorithm is that it tries to assemble the individual edge candidate pixels into *contours*.* These contours are formed by applying an *hysteresis threshold* to the pixels. This means that there are two thresholds, an upper and a lower. If a pixel has a gradient larger than the upper threshold, then it is accepted as an edge pixel; if a pixel is below the lower threshold, it is rejected. If the pixel's gradient is between the thresholds, then it will be accepted only if it is connected to a pixel that is above the high threshold. Canny recommended a ratio of high:low threshold between 2:1 and 3:1. Figures 6-7 and 6-8 show the results of applying cvCanny() to a test pattern and a photograph using high:low hysteresis threshold ratios of 5:1 and 3:2, respectively.

```
void cvCanny(
    const CvArr* img,
    CvArr*       edges,
    double       lowThresh,
    double       highThresh,
    int          apertureSize = 3
);
```

---

* We'll have much more to say about contours later. As you await those revelations, though, keep in mind that the cvCanny() routine does not actually return objects of type CvContour; we will have to build those from the output of cvCanny() if we want them by using cvFindContours(). Everything you ever wanted to know about contours will be covered in Chapter 8.

Figure 6-7. Results of Canny edge detection for two different images when the high and low thresholds are set to 50 and 10, respectively

The cvCanny() function expects an input image, which must be grayscale, and an output image, which must also be grayscale (but which will actually be a Boolean image). The next two arguments are the low and high thresholds, and the last argument is another aperture. As usual, this is the aperture used by the Sobel derivative operators that are called inside of the implementation of cvCanny().

Hough Transforms

The *Hough transform** is a method for finding lines, circles, or other simple forms in an image. The original Hough transform was a line transform, which is a relatively fast way of searching a binary image for straight lines. The transform can be further generalized to cases other than just simple lines.

Hough Line Transform

The basic theory of the Hough line transform is that any point in a binary image could be part of some set of possible lines. If we parameterize each line by, for example, a

---

* Hough developed the transform for use in physics experiments [Hough59]; its use in vision was introduced by Duda and Hart [Duda72].

Figure 6-8. Results of Canny edge detection for two different images when the high and low thresholds are set to 150 and 100, respectively

slope *a* and an intercept *b*, then a point in the original image is transformed to a locus of points in the (*a, b*) plane corresponding to all of the lines passing through that point (see Figure 6-9). If we convert every nonzero pixel in the input image into such a set of points in the output image and sum over all such contributions, then lines that appear in the input (i.e., (*x, y*) plane) image will appear as local maxima in the output (i.e., (*a, b*) plane) image. Because we are summing the contributions from each point, the (*a, b*) plane is commonly called the *accumulator plane*.

It might occur to you that the slope-intercept form is not really the best way to represent all of the lines passing through a point (because of the considerably different density of lines as a function of the slope, and the related fact that the interval of possible slopes goes from $-\infty$ to $+\infty$). It is for this reason that the actual parameterization of the transform image used in numerical computation is somewhat different. The preferred parameterization represents each line as a point in polar coordinates (ρ, θ), with the implied line being the line passing through the indicated point but perpendicular to the radial from the origin to that point (see Figure 6-10). The equation for such a line is:

$$\rho = x\cos\theta + y\sin\theta$$

Figure 6-9. *The Hough line transform finds many lines in each image; some of the lines found are expected, but others may not be*

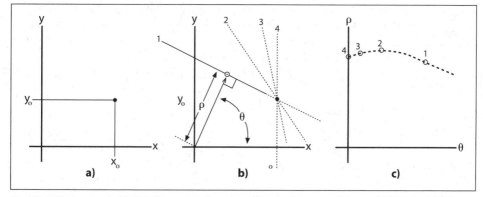

Figure 6-10. *A point (x_0, y_0) in the image plane (panel a) implies many lines each parameterized by a different ρ and θ (panel b); these lines each imply points in the (ρ, θ) plane, which taken together form a curve of characteristic shape (panel c)*

The OpenCV Hough transform algorithm does not make this computation explicit to the user. Instead, it simply returns the local maxima in the (ρ, θ) plane. However, you will need to understand this process in order to understand the arguments to the OpenCV Hough line transform function.

OpenCV supports two different kinds of Hough line transform: the *standard Hough transform* (SHT) [Duda72] and the *progressive probabilistic Hough transform* (PPHT).* The SHT is the algorithm we just looked at. The PPHT is a variation of this algorithm that, among other things, computes an extent for individual lines in addition to the orientation (as shown in Figure 6-11). It is "probabilistic" because, rather than accumulating every possible point in the accumulator plane, it accumulates only a fraction of them. The idea is that if the peak is going to be high enough anyhow, then hitting it only a fraction of the time will be enough to find it; the result of this conjecture can be a substantial reduction in computation time. Both of these algorithms are accessed with the same OpenCV function, though the meanings of some of the arguments depend on which method is being used.

```
CvSeq* cvHoughLines2(
  CvArr* image,
  void*  line_storage,
  int    method,
  double rho,
  double theta,
  int    threshold,
  double param1      = 0,
  double param2      = 0
);
```

The first argument is the input image. It must be an 8-bit image, but the input is treated as binary information (i.e., all nonzero pixels are considered to be equivalent). The second argument is a pointer to a place where the results can be stored, which can be either a memory storage (see CvMemoryStorage in Chapter 8) or a plain N-by-1 matrix array (the number of rows, N, will serve to limit the maximum number of lines returned). The next argument, method, can be CV_HOUGH_STANDARD, CV_HOUGH_PROBABILISTIC, or CV_HOUGH_MULTI_SCALE for (respectively) SHT, PPHT, or a multiscale variant of SHT.

The next two arguments, rho and theta, set the resolution desired for the lines (i.e., the resolution of the accumulator plane). The units of rho are pixels and the units of theta are radians; thus, the accumulator plane can be thought of as a two-dimensional histogram with cells of dimension rho pixels by theta radians. The threshold value is the value in the accumulator plane that must be reached for the routine to report a line. This last argument is a bit tricky in practice; it is not normalized, so you should expect to scale it up with the image size for SHT. Remember that this argument is, in effect, indicating the number of points (in the edge image) that must support the line for the line to be returned.

---

* The "probablistic Hough transform" (PHT) was introduced by Kiryati, Eldar, and Bruckshtein in 1991 [Kiryati91]; the PPHT was introduced by Matas, Galambosy, and Kittler in 1999 [Matas00].

Figure 6-11. The Canny edge detector (param1=50, param2=150) is run first, with the results shown in gray, and the progressive probabilistic Hough transform (param1=50, param2=10) is run next, with the results overlayed in white; you can see that the strong lines are generally picked up by the Hough transform

The param1 and param2 arguments are not used by the SHT. For the PPHT, param1 sets the minimum length of a line segment that will be returned, and param2 sets the separation between collinear segments required for the algorithm not to join them into a single longer segment. For the multiscale HT, the two parameters are used to indicate higher resolutions to which the parameters for the lines should be computed. The multiscale HT first computes the locations of the lines to the accuracy given by the rho and theta parameters and then goes on to refine those results by a factor of param1 and param2, respectively (i.e., the final resolution in rho is rho divided by param1 and the final resolution in theta is theta divided by param2).

What the function returns depends on how it was called. If the line_storage value was a matrix array, then the actual return value will be NULL. In this case, the matrix should be of type CV_32FC2 if the SHT or multi-scale HT is being used and should be CV_32SC4 if the PPHT is being used. In the first two cases, the ρ- and θ-values for each line will be placed in the two channels of the array. In the case of the PPHT, the four channels will hold the *x*- and *y*-values of the start and endpoints of the returned segments. In all of these cases, the number of rows in the array will be updated by cvHoughLines2() to correctly reflect the number of lines returned.

If the line_storage value was a pointer to a memory store,* then the return value will be a pointer to a CvSeq sequence structure. In that case, you can get each line or line segment from the sequence with a command like

```
float* line = (float*) cvGetSeqElem( lines , i );
```

where lines is the return value from cvHoughLines2() and i is index of the line of interest. In this case, line will be a pointer to the data for that line, with line[0] and line[1] being the floating-point values ρ and θ (for SHT and MSHT) or CvPoint structures for the endpoints of the segments (for PPHT).

Hough Circle Transform

The *Hough circle transform* [Kimme75] (see Figure 6-12) works in a manner roughly analogous to the Hough line transforms just described. The reason it is only "roughly" is that—if one were to try doing the exactly analogous thing—the accumulator *plane* would have to be replaced with an accumulator *volume* with three dimensions: one for *x*, one for *y*, and another for the circle radius *r*. This would mean far greater memory requirements and much slower speed. The implementation of the circle transform in OpenCV avoids this problem by using a somewhat more tricky method called the *Hough gradient method*.

The Hough gradient method works as follows. First the image is passed through an edge detection phase (in this case, cvCanny()). Next, for every nonzero point in the edge image, the local gradient is considered (the gradient is computed by first computing the first-order Sobel *x*- and *y*-derivatives via cvSobel()). Using this gradient, every point along the line indicated by this slope—from a specified minimum to a specified maximum distance—is incremented in the accumulator. At the same time, the location of every one of these nonzero pixels in the edge image is noted. The candidate centers are then selected from those points in this (two-dimensional) accumulator that are both above some given threshold and larger than all of their immediate neighbors. These candidate centers are sorted in descending order of their accumulator values, so that the centers with the most supporting pixels appear first. Next, for each center, all of the nonzero pixels (recall that this list was built earlier) are considered. These pixels are sorted according to their distance from the center. Working out from the smallest distances to the maximum radius, a single radius is selected that is best supported by the nonzero pixels. A center is kept if it has sufficient support from the nonzero pixels in the edge image *and* if it is a sufficient distance from any previously selected center.

This implementation enables the algorithm to run much faster and, perhaps more importantly, helps overcome the problem of the otherwise sparse population of a three-dimensional accumulator, which would lead to a lot of noise and render the results unstable. On the other hand, this algorithm has several shortcomings that you should be aware of.

---

* We have not yet introduced the concept of a memory store or a sequence, but Chapter 8 is devoted to this topic.

Figure 6-12. The Hough circle transform finds some of the circles in the test pattern and (correctly) finds none in the photograph

First, the use of the Sobel derivatives to compute the local gradient—and the attendant assumption that this can be considered equivalent to a local tangent—is not a numerically stable proposition. It might be true "most of the time," but you should expect this to generate some noise in the output.

Second, the entire set of nonzero pixels in the edge image is considered for every candidate center; hence, if you make the accumulator threshold too low, the algorithm will take a long time to run. Third, because only one circle is selected for every center, if there are concentric circles then you will get only one of them.

Finally, because centers are considered in ascending order of their associated accumulator value and because new centers are not kept if they are too close to previously accepted centers, there is a bias toward keeping the larger circles when multiple circles are concentric or approximately concentric. (It is only a "bias" because of the noise arising from the Sobel derivatives; in a smooth image at infinite resolution, it would be a certainty.)

With all of that in mind, let's move on to the OpenCV routine that does all this for us:

```
CvSeq* cvHoughCircles(
  CvArr* image,
```

```
    void*  circle_storage,
    int    method,
    double dp,
    double min_dist,
    double param1    = 100,
    double param2    = 300,
    int    min_radius = 0,
    int    max_radius = 0
);
```

The Hough circle transform function cvHoughCircles() has similar arguments to the line transform. The input image is again an 8-bit image. One significant difference between cvHoughCircles() and cvHoughLines2() is that the latter requires a binary image. The cvHoughCircles() function will internally (automatically) call cvSobel()* for you, so you can provide a more general grayscale image.

The circle_storage can be either an array or memory storage, depending on how you would like the results returned. If an array is used, it should be a single column of type CV_32FC3; the three channels will be used to encode the location of the circle and its radius. If memory storage is used, then the circles will be made into an OpenCV sequence and a pointer to that sequence will be returned by cvHoughCircles(). (Given an array pointer value for circle_storage, the return value of cvHoughCircles() is NULL.) The method argument must always be set to CV_HOUGH_GRADIENT.

The parameter dp is the resolution of the accumulator image used. This parameter allows us to create an accumulator of a lower resolution than the input image. (It makes sense to do this because there is no reason to expect the circles that exist in the image to fall naturally into the same number of categories as the width or height of the image itself.) If dp is set to 1 then the resolutions will be the same; if set to a larger number (e.g., 2), then the accumulator resolution will be smaller by that factor (in this case, half). The value of dp cannot be less than 1.

The parameter min_dist is the minimum distance that must exist between two circles in order for the algorithm to consider them distinct circles.

For the (currently required) case of the method being set to CV_HOUGH_GRADIENT, the next two arguments, param1 and param2, are the edge (Canny) threshold and the accumulator threshold, respectively. You may recall that the Canny edge detector actually takes two different thresholds itself. When cvCanny() is called internally, the first (higher) threshold is set to the value of param1 passed into cvHoughCircles(), and the second (lower) threshold is set to exactly half that value. The parameter param2 is the one used to threshold the accumulator and is exactly analogous to the threshold argument of cvHoughLines().

The final two parameters are the minimum and maximum radius of circles that can be found. This means that these are the radii of circles for which the accumulator has a representation. Example 6-1 shows an example program using cvHoughCircles().

---

* The function cvSobel(), not cvCanny(), is called internally. The reason is that cvHoughCircles() needs to estimate the orientation of a gradient at each pixel, and this is difficult to do with binary edge map.

Example 6-1. Using cvHoughCircles to return a sequence of circles found in a grayscale image

```
#include <cv.h>
#include <highgui.h>
#include <math.h>

int main(int argc, char** argv) {
  IplImage* image = cvLoadImage(
    argv[1],
    CV_LOAD_IMAGE_GRAYSCALE
  );

  CvMemStorage* storage = cvCreateMemStorage(0);
  cvSmooth(image, image, CV_GAUSSIAN, 5, 5 );
  CvSeq* results = cvHoughCircles(
    image,
    storage,
    CV_HOUGH_GRADIENT,
    2,
    image->width/10
  );

  for( int i = 0; i < results->total; i++ ) {
    float* p = (float*) cvGetSeqElem( results, i );
    CvPoint pt = cvPoint( cvRound( p[0] ), cvRound( p[1] ) );
    cvCircle(
      image,
      pt,
      cvRound( p[2] ),
      CV_RGB(0xff,0xff,0xff)
    );
  }
  cvNamedWindow( "cvHoughCircles", 1 );
  cvShowImage( "cvHoughCircles", image);
  cvWaitKey(0);
}
```

It is worth reflecting momentarily on the fact that, no matter what tricks we employ, there is no getting around the requirement that circles be described by three degrees of freedom (x, y, and r), in contrast to only two degrees of freedom (ρ and θ) for lines. The result will invariably be that any circle-finding algorithm requires more memory and computation time than the line-finding algorithms we looked at previously. With this in mind, it's a good idea to bound the radius parameter as tightly as circumstances allow in order to keep these costs under control.* The Hough transform was extended to arbitrary shapes by Ballard in 1981 [Ballard81] basically by considering objects as collections of gradient edges.

---

* Although cvHoughCircles() catches centers of the circles quite well, it sometimes fails to find the correct radius. Therefore, in an application where only a center must be found (or where some different technique can be used to find the actual radius), the radius returned by cvHoughCircles() can be ignored.

Remap

Under the hood, many of the transformations to follow have a certain common element. In particular, they will be taking pixels from one place in the image and mapping them to another place. In this case, there will always be some smooth mapping, which will do what we need, but it will not always be a one-to-one pixel correspondence.

We sometimes want to accomplish this interpolation programmatically; that is, we'd like to apply some known algorithm that will determine the mapping. In other cases, however, we'd like to do this mapping ourselves. Before diving into some methods that will compute (and apply) these mappings for us, let's take a moment to look at the function responsible for applying the mappings that these other methods rely upon. The OpenCV function we want is called cvRemap():

```
void cvRemap(
    const CvArr* src,
    CvArr*       dst,
    const CvArr* mapx,
    const CvArr* mapy,
    int          flags  = CV_INTER_LINEAR | CV_WARP_FILL_OUTLIERS,
    CvScalar     fillval = cvScalarAll(0)
);
```

The first two arguments of cvRemap() are the source and destination images, respectively. Obviously, these should be of the same size and number of channels, but they can have any data type. It is important to note that the two may not be the same image.* The next two arguments, mapx and mapy, indicate where any particular pixel is to be relocated. These should be the same size as the source and destination images, but they are single-channel and usually of data type float (IPL_DEPTH_32F). Noninteger mappings are OK, and cvRemap() will do the interpolation calculations for you automatically. One common use of cvRemap() is to rectify (correct distortions in) calibrated and stereo images. We will see functions in Chapters 11 and 12 that convert calculated camera distortions and alignments into mapx and mapy parameters. The next argument contains flags that tell cvRemap() exactly how that interpolation is to be done. Any one of the values listed in Table 6-1 will work.

Table 6-1. cvWarpAffine() additional flags values

flags values	Meaning
CV_INTER_NN	Nearest neighbor
CV_INTER_LINEAR	Bilinear (default)
CV_INTER_AREA	Pixel area resampling
CV_INTER_CUBIC	Bicubic interpolation

---

* A moment's thought will make it clear why the most efficient remapping strategy is incompatible with writing onto the source image. After all, if you move pixel A to location B then, when you get to location B and want to move it to location C, you will find that you've already written over the original value of B with A!

Interpolation is an important issue here. Pixels in the source image sit on an integer grid; for example, we can refer to a pixel at location (20, 17). When these integer locations are mapped to a new image, there can be gaps—either because the integer source pixel locations are mapped to float locations in the destination image and must be rounded to the nearest integer pixel location or because there are some locations to which no pixels at all are mapped (think about doubling the image size by stretching it; then every other destination pixel would be left blank). These problems are generally referred to as *forward projection* problems. To deal with such rounding problems and destination gaps, we actually solve the problem backwards: we step through each pixel of the destination image and ask, "Which pixels in the source are needed to fill in this destination pixel?" These source pixels will almost always be on fractional pixel locations so we must interpolate the source pixels to derive the correct value for our destination value. The default method is bilinear interpolation, but you may choose other methods (as shown in Table 6-1).

You may also add (using the OR operator) the flag CV_WARP_FILL_OUTLIERS, whose effect is to fill pixels in the destination image that are not the destination of any pixel in the input image with the value indicated by the final argument fillval. In this way, if you map all of your image to a circle in the center then the outside of that circle would automatically be filled with black (or any other color that you fancy).

Stretch, Shrink, Warp, and Rotate

In this section we turn to geometric manipulations of images.* Such manipulations include stretching in various ways, which includes both uniform and nonuniform resizing (the latter is known as *warping*). There are many reasons to perform these operations: for example, warping and rotating an image so that it can be superimposed on a wall in an existing scene, or artificially enlarging a set of training images used for object recognition.[†] The functions that can stretch, shrink, warp, and/or rotate an image are called *geometric transforms* (for an early exposition, see [Semple79]). For planar areas, there are two flavors of geometric transforms: transforms that use a 2-by-3 matrix, which are called *affine transforms*; and transforms based on a 3-by-3 matrix, which are called *perspective transforms* or *homographies*. You can think of the latter transformation as a method for computing the way in which a plane in three dimensions is perceived by a particular observer, who might not be looking straight on at that plane.

An affine transformation is any transformation that can be expressed in the form of a matrix multiplication followed by a vector addition. In OpenCV the standard style of representing such a transformation is as a 2-by-3 matrix. We define:

---

* We will cover these transformations in detail here; we will return to them when we discuss (in Chapter 11) how they can be used in the context of three-dimensional vision techniques.

† This activity might seem a bit dodgy; after all, wouldn't it be better just to use a recognition method that's invariant to local affine distortions? Nonetheless, this method has a long history and still can be quite useful in practice.

$$\mathbf{A} \equiv \begin{bmatrix} a_{00} & a_{01} \\ a_{10} & a_{11} \end{bmatrix} \quad \mathbf{B} \equiv \begin{bmatrix} b_0 \\ b_1 \end{bmatrix} \quad \mathbf{T} \equiv \begin{bmatrix} A & B \end{bmatrix} \quad \mathbf{X} \equiv \begin{bmatrix} x \\ y \end{bmatrix} \quad \mathbf{X}' \equiv \begin{bmatrix} x \\ y \\ 1 \end{bmatrix}$$

It is easily seen that the effect of the affine transformation $\mathbf{A} \cdot \mathbf{X} + \mathbf{B}$ is exactly equivalent to extending the vector $\mathbf{X}$ into the vector $\mathbf{X}'$ and simply left-multiplying $\mathbf{X}'$ by $\mathbf{T}$.

Affine transformations can be visualized as follows. Any parallelogram *ABCD* in a plane can be mapped to any other parallelogram *A'B'C'D'* by some affine transformation. If the areas of these parallelograms are nonzero, then the implied affine transformation is defined uniquely by (three vertices of) the two parallelograms. If you like, you can think of an affine transformation as drawing your image into a big rubber sheet and then deforming the sheet by pushing or pulling* on the corners to make different kinds of parallelograms.

When we have multiple images that we know to be slightly different views of the same object, we might want to compute the actual transforms that relate the different views. In this case, affine transformations are often used to model the views because, having fewer parameters, they are easier to solve for. The downside is that true perspective distortions can only be modeled by a homography,† so affine transforms yield a representation that cannot accommodate all possible relationships between the views. On the other hand, for small changes in viewpoint the resulting distortion is affine, so in some circumstances an affine transformation may be sufficient.

Affine transforms can convert rectangles to parallelograms. They can squash the shape but must keep the sides parallel; they can rotate it and/or scale it. Perspective transformations offer more flexibility; a perspective transform can turn a rectangle into a trapezoid. Of course, since parallelograms are also trapezoids, affine transformations are a subset of perspective transformations. Figure 6-13 shows examples of various affine and perspective transformations.

Affine Transform

There are two situations that arise when working with affine transformations. In the first case, we have an image (or a region of interest) we'd like to transform; in the second case, we have a list of points for which we'd like to compute the result of a transformation.

Dense affine transformations

In the first case, the obvious input and output formats are images, and the implicit requirement is that the warping assumes the pixels are a *dense representation* of the

---

* One can even pull in such a manner as to invert the parallelogram.

† "Homography" is the mathematical term for mapping points on one surface to points on another. In this sense it is a more general term than as used here. In the context of computer vision, homography almost always refers to mapping between points on two image planes that correspond to the same location on a planar object in the real world. It can be shown that such a mapping is representable by a single 3-by-3 orthogonal matrix (more on this in Chapter 11).

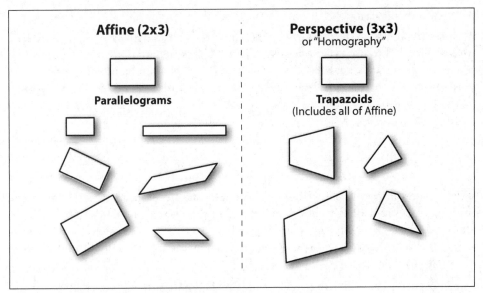

Figure 6-13. Affine and perspective transformations

underlying image. This means that image warping must necessarily handle interpolations so that the output images are smooth and look natural. The affine transformation function provided by OpenCV for dense transformations is cvWarpAffine().

```
void cvWarpAffine(
    const CvArr*  src,
    CvArr*        dst,
    const CvMat*  map_matrix,
    int           flags    = CV_INTER_LINEAR | CV_WARP_FILL_OUTLIERS,
    CvScalar      fillval  = cvScalarAll(0)
);
```

Here src and dst refer to an array or image, which can be either one or three channels and of any type (provided they are the *same* type and size).* The map_matrix is the 2-by-3 matrix we introduced earlier that quantifies the desired transformation. The next-to-last argument, flags, controls the interpolation method as well as either or both of the following additional options (as usual, combine with Boolean OR).

CV_WARP_FILL_OUTLIERS

Often, the transformed src image does not fit neatly into the dst image—there are pixels "mapped" there from the source file that don't actually exist. If this flag is set, then those missing values are filled with fillval (described previously).

CV_WARP_INVERSE_MAP

This flag is for convenience to allow inverse warping from dst to src instead of from src to dst.

---

* Since rotating an image will usually make its bounding box larger, the result will be a clipped image. You can circumvent this either by shrinking the image (as in the example code) or by copying the first image to a central ROI within a larger source image prior to transformation.

cVWarpAffine performance

It is worth knowing that cvWarpAffine() involves substantial associated overhead. An alternative is to use cvGetQuadrangleSubPix(). This function has fewer options but several advantages. In particular, it has less overhead and can handle the special case of when the source image is 8-bit and the destination image is a 32-bit floating-point image. It will also handle multichannel images.

```
void cvGetQuadrangleSubPix(
    const CvArr* src,
    CvArr*       dst,
    const CvMat* map_matrix
);
```

What cvGetQuadrangleSubPix() does is compute all the points in dst by mapping them (with interpolation) from the points in src that were computed by applying the affine transformation implied by multiplication by the 2-by-3 map_matrix. (Conversion of the locations in dst to homogeneous coordinates for the multiplication is done automatically.)

One idiosyncrasy of cvGetQuadrangleSubPix() is that there is an additional mapping applied by the function. In particular, the result points in dst are computed according to the formula:

$$dst(x, y) = src(a_{00}x'' + a_{01}y'' + b_0, \ a_{10}x'' + a_{11}y'' + b_1)$$

where:

$$M_{map} \equiv \begin{bmatrix} a_{00} & a_{01} & b_0 \\ a_{10} & a_{11} & b_1 \end{bmatrix} \quad \text{and} \quad \begin{bmatrix} x'' \\ y'' \end{bmatrix} = \begin{bmatrix} x - \dfrac{(width(dst)-1)}{2} \\ y - \dfrac{(height(dst)-1)}{2} \end{bmatrix}$$

Observe that the mapping from (x, y) to (x'', y'') has the effect that—even if the mapping M is an identity mapping—the points in the destination image at the center will be taken from the source image at the origin. If cvGetQuadrangleSubPix() needs points from outside the image, it uses replication to reconstruct those values.

Computing the affine map matrix

OpenCV provides two functions to help you generate the map_matrix. The first is used when you already have two images that you know to be related by an affine transformation or that you'd like to approximate in that way:

```
CvMat* cvGetAffineTransform(
    const CvPoint2D32f* pts_src,
    const CvPoint2D32f* pts_dst,
    CvMat*              map_matrix
);
```

Here src and dst are arrays containing three two-dimensional (*x*, *y*) points, and the map_matrix is the affine transform computed from those points.

The pts_src and pts_dst in cvGetAffineTransform() are just arrays of three points defining two parallelograms. The simplest way to define an affine transform is thus to set pts_src to three* corners in the source image—for example, the upper and lower left together with the upper right of the source image. The mapping from the source to destination image is then entirely defined by specifying pts_dst, the locations to which these three points will be mapped in that destination image. Once the mapping of these three independent corners (which, in effect, specify a "representative" parallelogram) is established, all the other points can be warped accordingly.

Example 6-2 shows some code that uses these functions. In the example we obtain the cvWarpAffine() matrix parameters by first constructing two three-component arrays of points (the corners of our representative parallelogram) and then convert that to the actual transformation matrix using cvGetAffineTransform(). We then do an affine warp followed by a rotation of the image. For our array of representative points in the source image, called srcTri[], we take the three points: (0,0), (0,height-1), and (width-1,0). We then specify the locations to which these points will be mapped in the corresponding array srcTri[].

Example 6-2. An affine transformation

```
// Usage: warp_affine <image>
//
#include <cv.h>
#include <highgui.h>

int main(int argc, char** argv)
{
  CvPoint2D32f srcTri[3], dstTri[3];
  CvMat*       rot_mat = cvCreateMat(2,3,CV_32FC1);
  CvMat*       warp_mat = cvCreateMat(2,3,CV_32FC1);
  IplImage     *src, *dst;

  if( argc == 2 && ((src=cvLoadImage(argv[1],1)) != 0 )) {

    dst = cvCloneImage( src );
    dst->origin = src->origin;
    cvZero( dst );

    // Compute warp matrix
    //
    srcTri[0].x = 0;                    //src Top left
    srcTri[0].y = 0;
    srcTri[1].x = src->width - 1;       //src Top right
    srcTri[1].y = 0;
    srcTri[2].x = 0;                    //src Bottom left offset
    srcTri[2].y = src->height - 1;
```

---

* We need just three points because, for an affine transformation, we are only representing a parallelogram. We will need four points to represent a general trapezoid when we address perspective transformations.

Example 6-2. An affine transformation (continued)

```
      dstTri[0].x = src->width*0.0;      //dst Top left
      dstTri[0].y = src->height*0.33;
      dstTri[1].x = src->width*0.85;     //dst Top right
      dstTri[1].y = src->height*0.25;
      dstTri[2].x = src->width*0.15;     //dst Bottom left offset
      dstTri[2].y = src->height*0.7;

      cvGetAffineTransform( srcTri, dstTri, warp_mat );
      cvWarpAffine( src, dst, warp_mat );
      cvCopy( dst, src );

      // Compute rotation matrix
      //
      CvPoint2D32f center = cvPoint2D32f(
        src->width/2,
        src->height/2
      );
      double angle = -50.0;
      double scale = 0.6;
      cv2DRotationMatrix( center, angle, scale, rot_mat );

      // Do the transformation
      //
      cvWarpAffine( src, dst, rot_mat );

      cvNamedWindow( "Affine_Transform", 1 );
        cvShowImage( "Affine_Transform", dst );
        cvWaitKey();
      }
      cvReleaseImage( &dst );
      cvReleaseMat( &rot_mat );
      cvReleaseMat( &warp_mat );
      return 0;
  }
}
```

The second way to compute the map_matrix is to use cv2DRotationMatrix(), which computes the map matrix for a rotation around some arbitrary point, combined with an optional rescaling. This is just one possible kind of affine transformation, but it represents an important subset that has an alternative (and more intuitive) representation that's easier to work with in your head:

```
CvMat* cv2DRotationMatrix(
    CvPoint2D32f center,
    double        angle,
    double        scale,
    CvMat*        map_matrix
);
```

The first argument, center, is the center point of the rotation. The next two arguments give the magnitude of the rotation and the overall rescaling. The final argument is the output map_matrix, which (as always) is a 2-by-3 matrix of floating-point numbers).

If we define $\alpha = \text{scale} \cdot \cos(\text{angle})$ and $\beta = \text{scale} \cdot \sin(\text{angle})$ then this function computes the map_matrix to be:

$$\begin{bmatrix} \alpha & \beta & (1-\alpha) \cdot \text{center}_x - \beta \cdot \text{center}_y \\ -\beta & \alpha & \beta \cdot \text{center}_x + (1-\alpha) \cdot \text{center}_y \end{bmatrix}$$

You can combine these methods of setting the map_matrix to obtain, for example, an image that is rotated, scaled, *and* warped.

Sparse affine transformations

We have explained that cvWarpAffine() is the right way to handle dense mappings. For sparse mappings (i.e., mappings of lists of individual points), it is best to use cvTransform():

```
void cvTransform(
    const CvArr* src,
    CvArr*       dst,
    const CvMat* transmat,
    const CvMat* shiftvec = NULL
);
```

In general, src is an N-by-1 array with D_s channels, where N is the number of points to be transformed and D_s is the dimension of those source points. The output array dst must be the same size but may have a different number of channels, D_d. The transformation matrix transmat is a D_s-by-D_d matrix that is then applied to every element of src, after which the results are placed into dst. The optional vector shiftvec, if non-NULL, must be a D_s-by-1 array, which is added to each result before the result is placed in dst.

In our case of an affine transformation, there are two ways to use cvTransform() that depend on how we'd like to represent our transformation. In the first method, we decompose our transformation into the 2-by-2 part (which does rotation, scaling, and warping) and the 2-by-1 part (which does the transformation). Here our input is an N-by-1 array with two channels, transmat is our local homogeneous transformation, and shiftvec contains any needed displacement. The second method is to use our usual 2-by-3 representation of the affine transformation. In this case the input array src is a three-channel array within which we must set all third-channel entries to 1 (i.e., the points must be supplied in homogeneous coordinates). Of course, the output array will still be a two-channel array.

Perspective Transform

To gain the greater flexibility offered by perspective transforms (homographies), we need a new function that will allow us to express this broader class of transformations. First we remark that, even though a perspective projection is specified completely by a single matrix, the projection is not actually a linear transformation. This is because the transformation requires division by the final dimension (usually Z; see Chapter 11) and thus loses a dimension in the process.

As with affine transformations, image operations (dense transformations) are handled by different functions than transformations on point sets (sparse transformations).

Dense perspective transform

The dense perspective transform uses an OpenCV function that is analogous to the one provided for dense affine transformations. Specifically, cvWarpPerspective() has all of the same arguments as cvWarpAffine() but with the small, but crucial, distinction that the map matrix must now be 3-by-3.

```
void cvWarpPerspective(
    const CvArr* src,
    CvArr*       dst,
    const CvMat* map_matrix,
    int          flags    = CV_INTER_LINEAR + CV_WARP_FILL_OUTLIERS,
    CvScalar     fillval   = cvScalarAll(0)
);
```

The flags are the same here as for the affine case.

Computing the perspective map matrix

As with the affine transformation, for filling the map_matrix in the preceding code we have a convenience function that can compute the transformation matrix from a list of point correspondences:

```
CvMat* cvGetPerspectiveTransform(
    const CvPoint2D32f* pts_src,
    const CvPoint2D32f* pts_dst,
    CvMat*              map_matrix
);
```

The pts_src and pts_dst are now arrays of four (not three) points, so we can independently control how the corners of (typically) a rectangle in pts_src are mapped to (generally) some rhombus in pts_dst. Our transformation is completely defined by the specified destinations of the four source points. As mentioned earlier, for perspective transformations we must allocate a 3-by-3 array for map_matrix; see Example 6-3 for sample code. Other than the 3-by-3 matrix and the shift from three to four control points, the perspective transformation is otherwise exactly analogous to the affine transformation we already introduced.

Example 6-3. Code for perspective transformation

```
// Usage: warp <image>
//
#include <cv.h>
#include <highgui.h>

int main(int argc, char** argv) {

  CvPoint2D32f srcQuad[4], dstQuad[4];
  CvMat*       warp_matrix = cvCreateMat(3,3,CV_32FC1);
  IplImage     *src, *dst;
```

Example 6-3. Code for perspective transformation (continued)

```
if( argc == 2 && ((src=cvLoadImage(argv[1],1)) != 0 )) {

    dst = cvCloneImage(src);
    dst->origin = src->origin;
    cvZero(dst);

    srcQuad[0].x = 0;                        //src Top left
    srcQuad[0].y = 0;
    srcQuad[1].x = src->width - 1;   //src Top right
    srcQuad[1].y = 0;
    srcQuad[2].x = 0;                        //src Bottom left
    srcQuad[2].y = src->height - 1;
    srcQuad[3].x = src->width - 1;   //src Bot right
    srcQuad[3].y = src->height - 1;

    dstQuad[0].x = src->width*0.05; //dst Top left
    dstQuad[0].y = src->height*0.33;
    dstQuad[1].x = src->width*0.9;   //dst Top right
    dstQuad[1].y = src->height*0.25;
    dstQuad[2].x = src->width*0.2;   //dst Bottom left
    dstQuad[2].y = src->height*0.7;
    dstQuad[3].x = src->width*0.8;   //dst Bot right
    dstQuad[3].y = src->height*0.9;

    cvGetPerspectiveTransform(
      srcQuad,
      dstQuad,
      warp_matrix
    );
    cvWarpPerspective( src, dst, warp_matrix );
    cvNamedWindow( "Perspective_Warp", 1 );
      cvShowImage( "Perspective_Warp", dst );
      cvWaitKey();
    }
    cvReleaseImage(&dst);
    cvReleaseMat(&warp_matrix);
    return 0;
  }
}
```

Sparse perspective transformations

There is a special function, cvPerspectiveTransform(), that performs perspective transformations on lists of points; we cannot use cvTransform(), which is limited to linear operations. As such, it cannot handle perspective transforms because they require division
by the third coordinate of the homogeneous representation ($x = f * X/Z$, $y = f * Y/Z$). The
special function cvPerspectiveTransform() takes care of this for us.

```
void cvPerspectiveTransform(
    const CvArr* src,
    CvArr*        dst,
    const CvMat* mat
);
```

As usual, the src and dst arguments are (respectively) the array of source points to be transformed and the array of destination points; these arrays should be of three-channel, floating-point type. The matrix mat can be either a 3-by-3 or a 4-by-4 matrix. If it is 3-by-3 then the projection is from two dimensions to two; if the matrix is 4-by-4, then the projection is from four dimensions to three.

In the current context we are transforming a set of points in an image to another set of points in an image, which sounds like a mapping from two dimensions to two dimensions. But this is not exactly correct, because the perspective transformation is actually mapping points on a two-dimensional plane embedded in a three-dimensional space back down to a (different) two-dimensional subspace. Think of this as being just what a camera does (we will return to this topic in greater detail when discussing cameras in later chapters). The camera takes points in three dimensions and maps them to the two dimensions of the camera imager. This is essentially what is meant when the source points are taken to be in "homogeneous coordinates". We are adding an additional dimension to those points by introducing the Z dimension and then setting all of the Z values to 1. The projective transformation is then projecting back out of that space onto the two-dimensional space of our output. This is a rather long-winded way of explaining why, when mapping points in one image to points in another, you will need a 3-by-3 matrix.

Output of the code in Example 6-3 is shown in Figure 6-14 for affine and perspective transformations. Compare this with the diagrams of Figure 6-13 to see how this works with real images. In Figure 6-14, we transformed the whole image. This isn't necessary; we could have used the src_pts to define a smaller (or larger!) region in the source image to be transformed. We could also have used ROIs in the source or destination image in order to limit the transformation.

CartToPolar and PolarToCart

The functions cvCartToPolar() and cvPolarToCart() are employed by more complex routines such as cvLogPolar() (described later) but are also useful in their own right. These functions map numbers back and forth between a Cartesian (x, y) space and a polar or radial (r, θ) space (i.e., from *Cartesian* to *polar* coordinates and vice versa). The function formats are as follows:

```
void cvCartToPolar(
  const CvArr* x,
  const CvArr* y,
  CvArr*       magnitude,
  CvArr*       angle            = NULL,
  int          angle_in_degrees = 0
);
void cvPolarToCart(
  const CvArr* magnitude,
  const CvArr* angle,
  CvArr*       x,
  CvArr*       y,
```

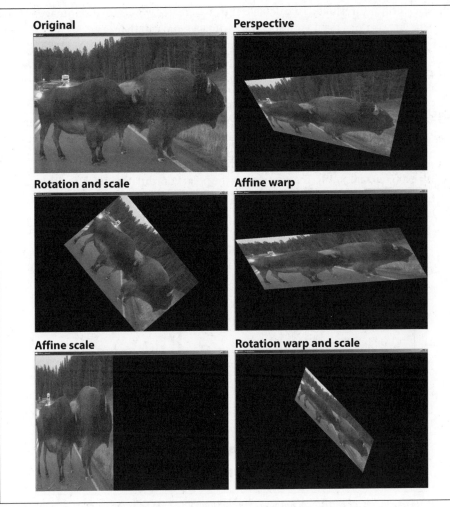

Original

Perspective

Rotation and scale

Affine warp

Affine scale

Rotation warp and scale

Figure 6-14. Perspective and affine mapping of an image

```
    int         angle_in_degrees = 0
);
```

In each of these functions, the first two two-dimensional arrays or images are the input and the second two are the outputs. If an output pointer is set to NULL then it will not be computed. The requirements on these arrays are that they be float or doubles and matching (size, number of channels, and type). The last parameter specifies whether we are working with angles in degrees (0, 360) or in radians (0, 2π).

For an example of where you might use this function, suppose you have already taken the *x*- and *y*-derivatives of an image, either by using cvSobel() or by using convolution functions via cvDFT() or cvFilter2D(). If you stored the *x*-derivatives in an image dx_img and the *y*-derivatives in dy_img, you could now create an edge-angle recognition histogram. That is, you can collect all the angles provided the magnitude or strength of the edge pixel is above a certain threshold. To calculate this, we create two destination images of the same type (integer or float) as the derivative images and call them img_mag and img_angle. If you want the result to be given in degrees, then you can use the function cvCartToPolar(dx_img, dy_img, img_mag, img_angle, 1). We would then fill the histogram from img_angle as long as the corresponding "pixel" in img_mag is above the threshold.

LogPolar

For two-dimensional images, the log-polar transform [Schwartz80] is a change from Cartesian to polar coordinates: $(x,y) \leftrightarrow re^{i\theta}$, where $r = \sqrt{x^2 + y^2}$ and $\exp(i\theta) = \exp(i \cdot \arctan(y/x))$. To separate out the polar coordinates into a (ρ, θ) space that is relative to some center point (x_c, y_c), we take the log so that $\rho = \log(\sqrt{(x-x_c)^2 + (y-y_c)^2})$ and $\theta = \arctan((y-y_c)/(x-x_c))$. For image purposes—when we need to "fit" the interesting stuff into the available image memory—we typically apply a scaling factor m to ρ. Figure 6-15 shows a square object on the left and its encoding in log-polar space.

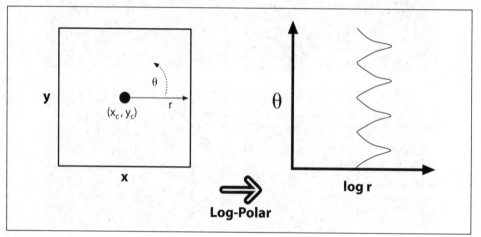

Figure 6-15. The log-polar transform maps (x, y) into $(\log(r),\theta)$; here, a square is displayed in the log-polar coordinate system

The next question is, of course, "Why bother?" The log-polar transform takes its inspiration from the human visual system. Your eye has a small but dense center of photoreceptors in its center (the *fovea*), and the density of receptors fall off rapidly (exponentially) from there. Try staring at a spot on the wall and holding your finger at arm's length in your line of sight. Then, keep staring at the spot and move your finger slowly away; note how the detail rapidly decreases as the image of your finger moves away from your fovea. This structure also has certain nice mathematical properties (beyond the scope of this book) that concern preserving the angles of line intersections.

More important for us is that the log-polar transform can be used to create two-dimensional invariant representations of object views by shifting the transformed image's center of mass to a fixed point in the log-polar plane; see Figure 6-16. On the left are three shapes that we want to recognize as "square". The problem is, they look very different. One is much larger than the others and another is rotated. The log-polar transform appears on the right in Figure 6-16. Observe that size differences in the (x, y) plane are converted to shifts along the $\log(r)$ axis of the log-polar plane and that the rotation differences are converted to shifts along the θ-axis in the log-polar plane. If we take the transformed center of each transformed square in the log-polar plane and then recenter that point to a certain fixed position, then all the squares will show up identically in the log-polar plane. This yields a type of invariance to two-dimensional rotation and scaling.*

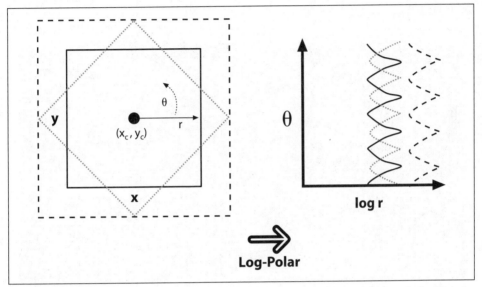

Figure 6-16. Log-polar transform of rotated and scaled squares: size goes to a shift on the $\log(r)$ axis and rotation to a shift on the θ-axis

The OpenCV function for a log-polar transform is `cvLogPolar()`:

```
void cvLogPolar(
  const CvArr*  src,
  CvArr*        dst,
  CvPoint2D32f  center,
  double        m,
  int           flags  = CV_INTER_LINEAR | CV_WARP_FILL_OUTLIERS
);
```

The `src` and `dst` are one- or three-channel color or grayscale images. The parameter center is the center point (x_c, y_c) of the log-polar transform; m is the scale factor, which

---

* In Chapter 13 we'll learn about recognition. For now simply note that it wouldn't be a good idea to derive a log-polar transform for a whole object because such transforms are quite sensitive to the exact location of their center points. What is more likely to work for object recognition is to detect a collection of key points (such as corners or blob locations) around an object, truncate the extent of such views, and then use the centers of those key points as log-polar centers. These local log-polar transforms could then be used to create local features that are (partially) scale- and rotation-invariant and that can be associated with a visual object.

should be set so that the features of interest dominate the available image area. The `flags` parameter allows for different interpolation methods. The interpolation methods are the same set of standard interpolations available in OpenCV (Table 6-1). The interpolation methods can be combined with either or both of the flags `CV_WARP_FILL_OUTLIERS` (to fill points that would otherwise be undefined) or `CV_WARP_INVERSE_MAP` (to compute the reverse mapping from log-polar to Cartesian coordinates).

Sample log-polar coding is given in Example 6-4, which demonstrates the forward and backward (inverse) log-polar transform. The results on a photographic image are shown in Figure 6-17.

Figure 6-17. Log-polar example on an elk with transform centered at the white circle on the left; the output is on the right

Example 6-4. Log-polar transform example

```
// logPolar.cpp : Defines the entry point for the console application.
//
#include <cv.h>
#include <highgui.h>

int main(int argc, char** argv) {
  IplImage* src;
  double    M;
  if( argc == 3 && ((src=cvLoadImage(argv[1],1)) != 0 )) {
    M = atof(argv[2]);
    IplImage* dst  = cvCreateImage( cvGetSize(src), 8, 3 );
    IplImage* src2 = cvCreateImage( cvGetSize(src), 8, 3 );
    cvLogPolar(
      src,
      dst,
      cvPoint2D32f(src->width/4,src->height/2),
      M,
      CV_INTER_LINEAR+CV_WARP_FILL_OUTLIERS
```

Example 6-4. Log-polar transform example (continued)

```
  );
  cvLogPolar(
    dst,
    src2,
    cvPoint2D32f(src->width/4, src->height/2),
    M,
    CV_INTER_LINEAR | CV_WARP_INVERSE_MAP
  );
  cvNamedWindow( "log-polar", 1 );
  cvShowImage( "log-polar", dst );
  cvNamedWindow( "inverse log-polar", 1 );
  cvShowImage( "inverse log-polar", src2 );
  cvWaitKey();
}
return 0;
}
```

Discrete Fourier Transform (DFT)

For any set of values that are indexed by a discrete (integer) parameter, is it possible to define a *discrete Fourier transform* (DFT)* in a manner analogous to the Fourier transform of a continuous function. For N complex numbers $x_0, ..., x_{N-1}$, the one-dimensional DFT is defined by the following formula (where $i = \sqrt{-1}$):

$$f_k = \sum_{n=0}^{N-1} x_n \exp\left(-\frac{2\pi i}{N} kn\right), \quad k = 0, ..., N-1$$

A similar transform can be defined for a two-dimensional array of numbers (of course higher-dimensional analogues exist also):

$$f_{k_x k_y} = \sum_{n_x=0}^{N_x-1} \sum_{n_y=0}^{N_y-1} x_{n_x n_y} \exp\left(-\frac{2\pi i}{N_x} k_x n_x\right) \exp\left(-\frac{2\pi i}{N_y} k_y n_y\right)$$

In general, one might expect that the computation of the N different terms f_k would require $O(N^2)$ operations. In fact, there are a number of *fast Fourier transform* (FFT) algorithms capable of computing these values in $O(N \log N)$ time. The OpenCV function cvDFT() implements one such FFT algorithm. The function cvDFT() can compute FFTs for one- and two-dimensional arrays of inputs. In the latter case, the two-dimensional transform can be computed or, if desired, only the one-dimensional transforms of each individual row can be computed (this operation is much faster than calling cvDFT() many separate times).

---

* Joseph Fourier [Fourier] was the first to find that some functions can be decomposed into an infinite series of other functions, and doing so became a field known as Fourier analysis. Some key text on methods of decomposing functions into their Fourier series are Morse for physics [Morse53] and Papoulis in general [Papoulis62]. The fast Fourier transform was invented by Cooley and Tukeye in 1965 [Cooley65] though Carl Gauss worked out the key steps as early as 1805 [Johnson84]. Early use in computer vision is described by Ballard and Brown [Ballard82].

```
void cvDFT(
    const CvArr* src,
    CvArr*       dst,
    int          flags,
    int          nonzero_rows = 0
);
```

The input and the output arrays must be floating-point types and may be single- or double-channel arrays. In the single-channel case, the entries are assumed to be real numbers and the output will be packed in a special space-saving format (inherited from the same older IPL library as the IplImage structure). If the source and channel are two-channel matrices or images, then the two channels will be interpreted as the real and imaginary components of the input data. In this case, there will be no special packing of the results, and some space will be wasted with a lot of 0s in both the input and output arrays.*

The special packing of result values that is used with single-channel output is as follows.

For a one-dimensional array:

Re Y_0	Re Y_1	Im Y_1	Re Y_2	Im Y_2	$\cdots$	Re $Y_{(N/2-1)}$	Im $Y_{(N/2-1)}$	Re $Y_{(N/2)}$

For a two-dimensional array:

Re Y_{00}	Re Y_{01}	Im Y_{01}	Re Y_{02}	Im Y_{02}	$\cdots$	Re $Y_{0(Nx/2-1)}$	Im $Y_{0(Nx/2-1)}$	Re $Y_{0(Nx/2)}$
Re Y_{10}	Re Y_{11}	Im Y_{11}	Re Y_{12}	Im Y_{12}	$\cdots$	Re $Y_{1(Nx/2-1)}$	Im $Y_{1(Nx/2-1)}$	Re $Y_{1(Nx/2)}$
Re Y_{20}	Re Y_{21}	Im Y_{21}	Re Y_{22}	Im Y_{22}	$\cdots$	Re $Y_{2(Nx/2-1)}$	Im $Y_{2(Nx/2-1)}$	Re $Y_{2(Nx/2)}$
$\vdots$	$\vdots$	$\vdots$	$\vdots$	$\vdots$	$\vdots$	$\vdots$	$\vdots$	$\vdots$
Re $Y_{(Ny/2-1)0}$	Re $Y_{(Ny-3)1}$	Im $Y_{(Ny-3)1}$	Re $Y_{(Ny-3)2}$	Im $Y_{(Ny-3)2}$	$\cdots$	Re $Y_{(Ny-3)(Nx/2-1)}$	Im $Y_{(Ny-3)(Nx/2-1)}$	Re $Y_{(Ny-3)(Nx/2)}$
Im $Y_{(Ny/2-1)0}$	Re $Y_{(Ny-2)1}$	Im $Y_{(Ny-2)1}$	Re $Y_{(Ny-2)2}$	Im $Y_{(Ny-2)2}$	$\cdots$	Re $Y_{(Ny-2)(Nx/2-1)}$	Im $Y_{(Ny-2)(Nx/2-1)}$	Re $Y_{(Ny-2)(Nx/2)}$
Re $Y_{(Ny/2)0}$	Re $Y_{(Ny-1)1}$	Im $Y_{(Ny-1)1}$	Re $Y_{(Ny-1)2}$	Im $Y_{(Ny-1)2}$	$\cdots$	Re $Y_{(Ny-1)(Nx/2-1)}$	Im $Y_{(Ny-1)(Nx/2-1)}$	Re $Y_{(Ny-1)(Nx/2)}$

It is worth taking a moment to look closely at the indices on these arrays. The issue here is that certain values are guaranteed to be 0 (more accurately, certain values of f_k are guaranteed to be real). It should also be noted that the last row listed in the table will be present only if N_y is even and that the last column will be present only if N_x is even. (In the case of the 2D array being treated as N_y 1D arrays rather than a full 2D transform, all of the result rows will be analogous to the single row listed for the output of the 1D array).

---

* When using this method, you must be sure to explicitly set the imaginary components to 0 in the two-channel representation. An easy way to do this is to create a matrix full of 0s using cvZero() for the imaginary part and then to call cvMerge() with a real-valued matrix to form a temporary complex array on which to run cvDFT() (possibly in-place). This procedure will result in full-size, unpacked, complex matrix of the spectrum.

The third argument, called flags, indicates exactly what operation is to be done. The transformation we started with is known as a *forward transform* and is selected with the flag CV_DXT_FORWARD. The inverse transform* is defined in exactly the same way except for a change of sign in the exponential and a scale factor. To perform the inverse transform without the scale factor, use the flag CV_DXT_INVERSE. The flag for the scale factor is CV_DXT_SCALE, and this results in all of the output being scaled by a factor of $1/N$ (or $1/N_x N_y$ for a 2D transform). This scaling is necessary if the sequential application of the forward transform and the inverse transform is to bring us back to where we started. Because one often wants to combine CV_DXT_INVERSE with CV_DXT_SCALE, there are several shorthand notations for this kind of operation. In addition to just combining the two operations with OR, you can use CV_DXT_INV_SCALE (or CV_DXT_INVERSE_SCALE if you're not into that brevity thing). The last flag you may want to have handy is CV_DXT_ROWS, which allows you to tell cvDFT() to treat a two-dimensional array as a collection of one-dimensional arrays that should each be transformed separately as if they were N_y distinct vectors of length N_x. This significantly reduces overhead when doing many transformations at a time (especially when using Intel's optimized IPP libraries). By using CV_DXT_ROWS it is also possible to implement three-dimensional (and higher) DFT.

In order to understand the last argument, nonzero_rows, we must digress for a moment. In general, DFT algorithms will strongly prefer vectors of some lengths over others or arrays of some sizes over others. In most DFT algorithms, the preferred sizes are powers of 2 (i.e., 2^n for some integer n). In the case of the algorithm used by OpenCV, the preference is that the vector lengths, or array dimensions, be $2^p 3^q 5^r$, for some integers p, q, and r. Hence the usual procedure is to create a somewhat larger array (for which purpose there is a handy utility function, cvGetOptimalDFTSize(), which takes the length of your vector and returns the first equal or larger appropriate number size) and then use cvGetSubRect() to copy your array into the somewhat roomier zero-padded array. Despite the need for this padding, it is possible to indicate to cvDFT() that you really do not care about the transform of those rows that you had to add down below your actual data (or, if you are doing an inverse transform, which rows in the result you do not care about). In either case, you can use nonzero_rows to indicate how many rows can be safely ignored. This will provide some savings in computation time.

Spectrum Multiplication

In many applications that involve computing DFTs, one must also compute the per-element multiplication of two spectra. Because such results are typically packed in their special high-density format and are usually complex numbers, it would be tedious to unpack them and handle the multiplication via the "usual" matrix operations. Fortunately, OpenCV provides the handy cvMulSpectrums() routine, which performs exactly this function as well as a few other handy things.

---

* With the inverse transform, the input is packed in the special format described previously. This makes sense because, if we first called the forward DFT and then ran the inverse DFT on the results, we would expect to wind up with the original data—that is, of course, if we remember to use the CV_DXT_SCALE flag!

```
void cvMulSpectrums(
    const CvArr*    src1,
    const CvArr*    src2,
    CvArr*          dst,
    int             flags
);
```

Note that the first two arguments are the usual input arrays, though in this case they are spectra from calls to cvDFT(). The third argument must be a pointer to an array—of the same type and size as the first two—that will be used for the results. The final argument, flags, tells cvMulSpectrums() exactly what you want done. In particular, it may be set to 0 (CV_DXT_FORWARD) for implementing the above pair multiplication or set to CV_DXT_MUL_CONJ if the element from the first array is to be multiplied by the complex conjugate of the corresponding element of the second array. The flags may also be combined with CV_DXT_ROWS in the two-dimensional case if each array row 0 is to be treated as a separate spectrum (remember, if you created the spectrum arrays with CV_DXT_ROWS then the data packing is slightly different than if you created them without that function, so you must be consistent in the way you call cvMulSpectrums).

Convolution and DFT

It is possible to greatly increase the speed of a convolution by using DFT via the convolution theorem [Titchmarsh26] that relates convolution in the spatial domain to multiplication in the Fourier domain [Morse53; Bracewell65; Arfken85].* To accomplish this, one first computes the Fourier transform of the image and then the Fourier transform of the convolution filter. Once this is done, the convolution can be performed in the transform space in linear time with respect to the number of pixels in the image. It is worthwhile to look at the source code for computing such a convolution, as it also will provide us with many good examples of using cvDFT(). The code is shown in Example 6-5, which is taken directly from the OpenCV reference.

Example 6-5. Use of cvDFT() to accelerate the computation of convolutions

```
// Use DFT to accelerate the convolution of array A by kernel B.
// Place the result in array V.
//
void speedy_conv olution(
  const CvMat* A,  // Size: M1xN1
  const CvMat* B,  // Size: M2xN2
  CvMat*       C   // Size:(A->rows+B->rows-1)x(A->cols+B->cols-1)
) {

  int dft_M = cvGetOptimalDFTSize( A->rows+B->rows-1 );
  int dft_N = cvGetOptimalDFTSize( A->cols+B->cols-1 );

  CvMat* dft_A = cvCreateMat( dft_M, dft_N, A->type );
  CvMat* dft_B = cvCreateMat( dft_M, dft_N, B->type );
  CvMat tmp;
```

---

* Recall that OpenCV's DFT algorithm implements the FFT whenever the data size makes the FFT faster.

Example 6-5. Use of cvDFT() to accelerate the computation of convolutions (continued)

```
  // copy A to dft_A and pad dft_A with zeros
  //
  cvGetSubRect( dft_A, &tmp, cvRect(0,0,A->cols,A->rows));
  cvCopy( A, &tmp );
  cvGetSubRect(
    dft_A,
    &tmp,
    cvRect( A->cols, 0, dft_A->cols-A->cols, A->rows )
  );
  cvZero( &tmp );

  // no need to pad bottom part of dft_A with zeros because of
  // use nonzero_rows parameter in cvDFT() call below
  //
  cvDFT( dft_A, dft_A, CV_DXT_FORWARD, A->rows );

  // repeat the same with the second array
  //
  cvGetSubRect( dft_B, &tmp, cvRect(0,0,B->cols,B->rows) );
  cvCopy( B, &tmp );
  cvGetSubRect(
    dft_B,
    &tmp,
    cvRect( B->cols, 0, dft_B->cols-B->cols, B->rows )
  );
  cvZero( &tmp );

  // no need to pad bottom part of dft_B with zeros because of
  // use nonzero_rows parameter in cvDFT() call below
  //
  cvDFT( dft_B, dft_B, CV_DXT_FORWARD, B->rows );

  // or CV_DXT_MUL_CONJ to get correlation rather than convolution
  //
  cvMulSpectrums( dft_A, dft_B, dft_A, 0 );

  // calculate only the top part
  //
  cvDFT( dft_A, dft_A, CV_DXT_INV_SCALE, C->rows );
  cvGetSubRect( dft_A, &tmp, cvRect(0,0,conv->cols,C->rows) );

  cvCopy( &tmp, C );

  cvReleaseMat( dft_A );
  cvReleaseMat( dft_B );
}
```

In Example 6-5 we can see that the input arrays are first created and then initialized. Next, two new arrays are created whose dimensions are optimal for the DFT algorithm. The original arrays are copied into these new arrays and then the transforms are computed. Finally, the spectra are multiplied together and the inverse transform is applied to the product. The transforms are the slowest* part of this operation; an N-by-N image takes $O(N^2 \log N)$ time and so the entire process is also completed in that time (assuming that $N > M$ for an M-by-M convolution kernel). This time is much faster than $O(N2M^2)$, the non-DFT convolution time required by the more naïve method.

Discrete Cosine Transform (DCT)

For real-valued data it is often sufficient to compute what is, in effect, only half of the discrete Fourier transform. The *discrete cosine transform* (DCT) [Ahmed74; Jain77] is defined analogously to the full DFT by the following formula:

$$c_k = \sum_{n=0}^{N-1} \sqrt{n = \begin{cases} \dfrac{1}{N} & \text{if } n=0 \\ \dfrac{2}{N} & \text{else} \end{cases}} \cdot x_n \cdot \cos\left(-\pi \frac{(2k+1)n}{N}\right)$$

Observe that, by convention, the normalization factor is applied to both the cosine transform and its inverse. Of course, there is a similar transform for higher dimensions.

The basic ideas of the DFT apply also to the DCT, but now all the coefficients are real-valued. Astute readers might object that the cosine transform is being applied to a vector that is not a manifestly even function. However, with cvDCT() the algorithm simply treats the vector as if it were extended to negative indices in a mirrored manner.

The actual OpenCV call is:

```
void cvDCT(
  const CvArr* src,
  CvArr*       dst,
  int          flags
);
```

The cvDCT() function expects arguments like those for cvDFT() except that, because the results are real-valued, there is no need for any special packing of the result array (or of the input array in the case of an inverse transform). The flags argument can be set to CV_DXT_FORWARD or CV_DXT_INVERSE, and either may be combined with CV_DXT_ROWS with the same effect as with cvDFT(). Because of the different normalization convention, both the forward and inverse cosine transforms always contain their respective contribution to the overall normalization of the transform; hence CV_DXT_SCALE plays no role in cvDCT.

Integral Images

OpenCV allows you to calculate an integral image easily with the appropriately named cvIntegral() function. An *integral image* [Viola04] is a data structure that allows rapid

---

* By "slowest" we mean "asymptotically slowest"—in other words, that this portion of the algorithm takes the most time for very large N. This is an important distinction. In practice, as we saw in the earlier section on convolutions, it is not always optimal to pay the overhead for conversion to Fourier space. In general, when convolving with a small kernel it will not be worth the trouble to make this transformation.

summing of subregions. Such summations are useful in many applications; a notable one is the computation of *Haar wavelets*, which are used in some face recognition and similar algorithms.

```
void cvIntegral(
    const CvArr*  image,
    CvArr*        sum,
    CvArr*        sqsum      = NULL,
    CvArr*        tilted_sum = NULL
);
```

The arguments to cvIntegral() are the original image as well as pointers to destination images for the results. The argument sum is required; the others, sqsum and tilted_sum, may be provided if desired. (Actually, the arguments need not be images; they could be matrices, but in practice, they are usually images.) When the input image is 8-bit unsigned, the sum or tilted_sum may be 32-bit integer or floating-point arrays. For all other cases, the sum or tilted_sum must be floating-point valued (either 32- or 64-bit). The result "images" must always be floating-point. If the input image is of size *W*-by-*H*, then the output images must be of size $(W + 1)$-by-$(H + 1)$.*

An integral image sum has the form:

$$\text{sum}(X,Y) = \sum_{x \leq X} \sum_{y \leq Y} \text{image}(x,y)$$

The optional sqsum image is the sum of squares:

$$\text{sum}(X,Y) = \sum_{x \leq X} \sum_{y \leq Y} (\text{image}(x,y))^2$$

and the tilted_sum is like the sum except that it is for the image rotated by 45 degrees:

$$\text{tilt\_sum}(X,Y) = \sum_{y \leq Y} \sum_{\text{abs}(x-X) \leq y} \text{image}(x,y)$$

Using these integral images, one may calculate sums, means, and standard deviations over arbitrary upright or "tilted" rectangular regions of the image. As a simple example, to sum over a simple rectangular region described by the corner points $(x1, y1)$ and $(x2, y2)$, where $x2 > x1$ and $y2 > y1$, we'd compute:

$$\sum_{x1 \leq x \leq x2} \sum_{y1 \leq y \leq y2} [\text{image}(x,y)]$$
$$= [\text{sum}(x2,y2) - \text{sum}(x1-1,y2) - \text{sum}(x2,y1-1) + \text{sum}(x1-1,y1-1)]$$

In this way, it is possible to do fast blurring, approximate gradients, compute means and standard deviations, and perform fast block correlations even for variable window sizes.

---

* This is because we need to put in a buffer of zero values along the *x*-axis and *y*-axis in order to make computation efficient.

To make this all a little more clear, consider the 7-by-5 image shown in Figure 6-18; the region is shown as a bar chart in which the height associated with the pixels represents the brightness of those pixel values. The same information is shown in Figure 6-19, numerically on the left and in integral form on the right. Integral images (I') are computed by going across rows, proceeding row by row using the previously computed integral image values together with the current raw image (I) pixel value $I(x, y)$ to calculate the next integral image value as follows:

$$I'(x,y) = I(x,y) + I'(x-1,y) + I'(x,y-1) - I'(x-1,y-1)$$

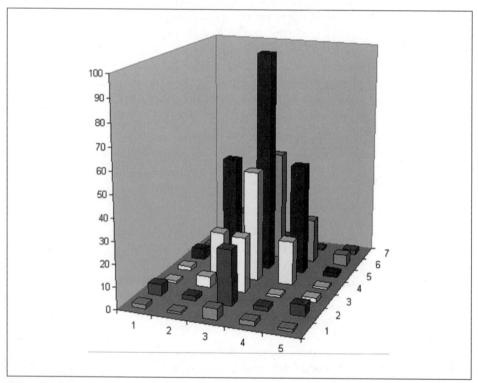

Figure 6-18. Simple 7-by-5 image shown as a bar chart with x, y, and height equal to pixel value

The last term is subtracted off because this value is double-counted when adding the second and third terms. You can verify that this works by testing some values in Figure 6-19.

When using the integral image to compute a region, we can see by Figure 6-19 that, in order to compute the central rectangular area bounded by the 20s in the original image, we'd calculate 398 – 9 – 10 + 1 = 380. Thus, a rectangle of any size can be computed using four measurements (resulting in $O(1)$ computational complexity).

1	2	5	1	2
2	20	50	20	5
5	50	100	50	2
2	20	50	20	1
1	5	25	1	2
5	2	25	2	5
2	1	5	2	1

1	3	8	9	11
3	25	80	101	108
8	80	235	306	315
10	102	307	398	408
11	108	338	430	442
16	115	370	464	481
18	118	378	474	492

Figure 6-19. The 7-by-5 image of Figure 6-18 shown numerically at left (with the origin assumed to be the upper-left) and converted to an integral image at right

Distance Transform

The *distance transform* of an image is defined as a new image in which every output pixel is set to a value equal to the distance to the nearest zero pixel in the input image. It should be immediately obvious that the typical input to a distance transform should be some kind of edge image. In most applications the input to the distance transform is an output of an edge detector such as the Canny edge detector that has been inverted (so that the edges have value zero and the non-edges are nonzero).

In practice, the distance transform is carried out by using a mask that is typically a 3-by-3 or 5-by-5 array. Each point in the array defines the "distance" to be associated with a point in that particular position relative to the center of the mask. Larger distances are built up (and thus approximated) as sequences of "moves" defined by the entries in the mask. This means that using a larger mask will yield more accurate distances.

Depending on the desired distance metric, the appropriate mask is automatically selected from a set known to OpenCV. It is also possible to tell OpenCV to compute "exact" distances according to some formula appropriate to the selected metric, but of course this is much slower.

The distance metric can be any of several different types, including the classic L2 (Cartesian) distance metric; see Table 6-2 for a listing. In addition to these you may define a custom metric and associate it with your own custom mask.

Table 6-2. Possible values for distance_type argument to cvDistTransform()

Value of distance_type	Metric
CV_DIST_L2	$\rho(r) = \dfrac{r^2}{2}$
CV_DIST_L1	$\rho(r) = r$
CV_DIST_L12	$\rho(r) = 2\left[\sqrt{1 + \dfrac{r^2}{2}} - 1\right]$
CV_DIST_FAIR	$\rho(r) = C^2\left[\dfrac{r}{C} - \log\left(1 + \dfrac{r}{C}\right)\right],\ C = 1.3998$

Table 6-2. Possible values for distance_type argument to cvDistTransform() (continued)

Value of distance_type	Metric
CV_DIST_WELSCH	$\rho(r) = \dfrac{C^2}{2}\left[1 - \exp\left(-\left(\dfrac{r}{C}\right)^2\right)\right],\ C = 2.9846$
CV_DIST_USER	User-defined distance

When calling the OpenCV distance transform function, the output image should be a 32-bit floating-point image (i.e., IPL_DEPTH_32F).

```
Void cvDistTransform(
    const CvArr* src,
    CvArr*       dst,
    int          distance_type = CV_DIST_L2,
    int          mask_size      = 3,
    const float* kernel         = NULL,
    CvArr*       labels         = NULL
);
```

There are several optional parameters when calling cvDistTransform(). The first is distance_type, which indicates the distance metric to be used. The available values for this argument are defined in Borgefors (1986) [Borgefors86].

After the distance type is the mask_size, which may be 3 (choose CV_DIST_MASK_3) or 5 (choose CV_DIST_MASK_5); alternatively, distance computations can be made without a kernel* (choose CV_DIST_MASK_PRECISE). The kernel argument to cvDistanceTransform() is the distance mask to be used in the case of custom metric. These kernels are constructed according to the method of Gunilla Borgefors, two examples of which are shown in Figure 6-20. The last argument, labels, indicates that associations should be made between individual points and the nearest connected component consisting of zero pixels. When labels is non-NULL, it must be a pointer to an array of integer values the same size as the input and output images. When the function returns, this image can be read to determine which object was closest to the particular pixel under consideration. Figure 6-21 shows the outputs of distance transforms on a test pattern and a photographic image.

Histogram Equalization

Cameras and image sensors must usually deal not only with the contrast in a scene but also with the image sensors' exposure to the resulting light in that scene. In a standard camera, the shutter and lens aperture settings juggle between exposing the sensors to too much or too little light. Often the range of contrasts is too much for the sensors to deal with; hence there is a trade-off between capturing the dark areas (e.g., shadows), which requires a longer exposure time, and the bright areas, which require shorter exposure to avoid saturating "whiteouts."

---

* The exact method comes from Pedro F. Felzenszwalb and Daniel P. Huttenlocher [Felzenszwalb63].

| **User defined 3x3 mask** | | | | | | |
(a=1, b=1.5)						
4.5	4	3.5	3	3.5	4	4.5
4	3	2.5	2	2.5	3	4
3.5	2.5	1.5	1	1.5	2.5	3.5
3	2	1	0	1	2	3
3.5	2.5	1.5	1	1.5	2.5	3.5
4	3	2.5	2	2.5	3	4
4.5	4	3.5	3	3.5	4	4.5

| **User defined 5x5 mask** | | | | | | |
(a=1, b=1.5, c=2)						
4.5	3.5	3	3	3	3.5	4.5
3.5	3	2	2	2	3	3.5
3	2	1.5	1	1.5	2	3
3	2	1	0	1	2	3
3	2	1.5	1	1.5	2	3
3.5	3	2	2	2	3	3.5
4	3.5	3	3	3	3.5	4

Figure 6-20. Two custom distance transform masks

Figure 6-21. First a Canny edge detector was run with param1=100 and param2=200; then the distance transform was run with the output scaled by a factor of 5 to increase visibility

After the picture has been taken, there's nothing we can do about what the sensor recorded; however, we can still take what's there and try to expand the dynamic range of the image. The most commonly used technique for this is histogram equalization.[*][†] In Figure 6-22 we can see that the image on the left is poor because there's not much variation of the range of values. This is evident from the histogram of its intensity values on the right. Because we are dealing with an 8-bit image, its intensity values can range from 0 to 255, but the histogram shows that the actual intensity values are all clustered near the middle of the available range. Histogram equalization is a method for stretching this range out.

Figure 6-22. The image on the left has poor contrast, as is confirmed by the histogram of its intensity values on the right

The underlying math behind histogram equalization involves mapping one distribution (the given histogram of intensity values) to another distribution (a wider and, ideally, uniform distribution of intensity values). That is, we want to spread out the y-values of the original distribution as evenly as possible in the new distribution. It turns out that there is a good answer to the problem of spreading out distribution values: the remapping function should be the *cumulative distribution function*. An example of the cumulative density function is shown in Figure 6-23 for the somewhat idealized case of a distribution that was originally pure Gaussian. However, cumulative density can be applied to any distribution; it is just the running sum of the original distribution from its negative to its positive bounds.

We may use the cumulative distribution function to remap the original distribution as an equally spread distribution (see Figure 6-24) simply by looking up each y-value in the original distribution and seeing where it should go in the equalized distribution.

---

* If you are wondering why histogram equalization is not in the chapter on histograms (Chapter 7), the reason is that histogram equalization makes no explicit use of any histogram data types. Although histograms are used internally, the function (from the user's perspective) requires no histograms at all.

† Histogram equalization is an old mathematical technique; its use in image processing is described in various textbooks [Jain86; Russ02; Acharya05], conference papers [Schwarz78], and even in biological vision [Laughlin81].

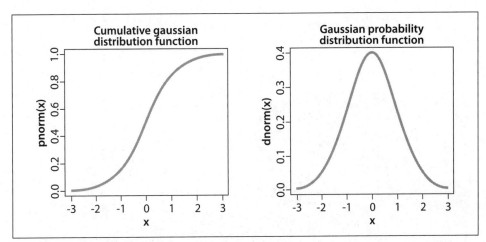

Figure 6-23. *Result of cumulative distribution function (left) on a Gaussian distribution (right)*

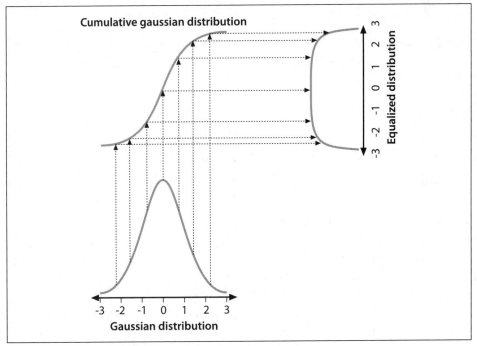

Figure 6-24. *Using the cumulative density function to equalize a Gaussian distribution*

For continuous distributions the result will be an exact equalization, but for digitized/ discrete distributions the results may be far from uniform.

Applying this equalization process to Figure 6-22 yields the equalized intensity distribution histogram and resulting image in Figure 6-25. This whole process is wrapped up in one neat function:

```
void  cvEqualizeHist(
  const CvArr* src,
  CvArr*       dst
);
```

Figure 6-25. Histogram equalized results: the spectrum has been spread out

In cvEqualizeHist(), the source and destination must be single-channel, 8-bit images of the same size. For color images you will have to separate the channels and process them one by one.

Exercises

1. Use cvFilter2D() to create a filter that detects only 60 degree lines in an image. Display the results on a sufficiently interesting image scene.

2. Separable kernels. Create a 3-by-3 Gaussian kernel using rows [(1/16, 2/16, 1/16), (2/16, 4/16, 2/16), (1/16, 2/16, 1/16)] and with anchor point in the middle.

   a. Run this kernel on an image and display the results.

   b. Now create two one-dimensional kernels with anchors in the center: one going "across" (1/4, 2/4, 1/4), and one going down (1/4, 2/4, 1/4). Load the same original image and use cvFilter2D() to convolve the image twice, once with the first 1D kernel and once with the second 1D kernel. Describe the results.

   c. Describe the order of complexity (number of operations) for the kernel in part a and for the kernels in part b. The difference is the advantage of being able to use separable kernels and the entire Gaussian class of filters—or any linearly decomposable filter that is separable, since convolution is a linear operation.

3. Can you make a separable kernel from the filter shown in Figure 6-5? If so, show what it looks like.

4. In a drawing program such as PowerPoint, draw a series of concentric circles forming a bull's-eye.

a. Make a series of lines going into the bull's-eye. Save the image.

b. Using a 3-by-3 aperture size, take and display the first-order x- and y-derivatives of your picture. Then increase the aperture size to 5-by-5, 9-by-9, and 13-by-13. Describe the results.

5. Create a new image that is just a 45 degree line, white on black. For a given series of aperture sizes, we will take the image's first-order x-derivative (dx) and first-order y-derivative (dy). We will then take measurements of this line as follows. The (dx) and (dy) images constitute the gradient of the input image. The magnitude at location (i, j) is $mag(i, j) = \sqrt{dx^2(i, j) + dy^2(i, j)}$ and the angle is $\theta(i, j) = \arctan(dy(i, j)dx(i, j))$. Scan across the image and find places where the magnitude is at or near maximum. Record the angle at these places. Average the angles and report that as the measured line angle.

a. Do this for a 3-by-3 aperture Sobel filter.

b. Do this for a 5-by-5 filter.

c. Do this for a 9-by-9 filter.

d. Do the results change? If so, why?

6. Find and load a picture of a face where the face is frontal, has eyes open, and takes up most or all of the image area. Write code to find the pupils of the eyes.

> A Laplacian "likes" a bright central point surrounded by dark. Pupils are just the opposite. Invert and convolve with a sufficiently large Laplacian.

7. In this exercise we learn to experiment with parameters by setting good lowThresh and highThresh values in cvCanny(). Load an image with suitably interesting line structures. We'll use three different high:low threshold settings of 1.5:1, 2.75:1, and 4:1.

a. Report what you see with a high setting of less than 50.

b. Report what you see with high settings between 50 and 100.

c. Report what you see with high settings between 100 and 150.

d. Report what you see with high settings between 150 and 200.

e. Report what you see with high settings between 200 and 250.

f. Summarize your results and explain what happens as best you can.

8. Load an image containing clear lines and circles such as a side view of a bicycle. Use the Hough line and Hough circle calls and see how they respond to your image.

9. Can you think of a way to use the Hough transform to identify any kind of shape with a distinct perimeter? Explain how.

10. Look at the diagrams of how the log-polar function transforms a square into a wavy line.

a. Draw the log-polar results if the log-polar center point were sitting on one of the corners of the square.

b. What would a circle look like in a log-polar transform if the center point were inside the circle and close to the edge?

c. Draw what the transform would look like if the center point were sitting just outside of the circle.

11. A log-polar transform takes shapes of different rotations and sizes into a space where these correspond to shifts in the θ-axis and $\log(r)$ axis. The Fourier transform is translation invariant. How can we use these facts to force shapes of different sizes and rotations to automatically give equivalent representations in the log-polar domain?

12. Draw separate pictures of large, small, large rotated, and small rotated squares. Take the log-polar transform of these each separately. Code up a two-dimensional shifter that takes the center point in the resulting log-polar domain and shifts the shapes to be as identical as possible.

13. Take the Fourier transform of a small Gaussian distribution and the Fourier transform of an image. Multiply them and take the inverse Fourier transform of the results. What have you achieved? As the filters get bigger, you will find that working in the Fourier space is much faster than in the normal space.

14. Load an interesting image, convert it to grayscale, and then take an integral image of it. Now find vertical and horizontal edges in the image by using the properties of an integral image.

> Use long skinny rectangles; subtract and add them in place.

15. Explain how you could use the distance transform to automatically align a known shape with a test shape when the scale is known and held fixed. How would this be done over multiple scales?

16. Practice histogram equalization on images that you load in, and report the results.

17. Load an image, take a perspective transform, and then rotate it. Can this transform be done in one step?

CHAPTER 7

Histograms and Matching

In the course of analyzing images, objects, and video information, we frequently want to represent what we are looking at as a *histogram*. Histograms can be used to represent such diverse things as the color distribution of an object, an edge gradient template of an object [Freeman95], and the distribution of probabilities representing our current hypothesis about an object's location. Figure 7-1 shows the use of histograms for rapid gesture recognition. Edge gradients were collected from "up", "right", "left", "stop" and "OK" hand gestures. A webcam was then set up to watch a person who used these gestures to control web videos. In each frame, color interest regions were detected from the incoming video; then edge gradient directions were computed around these interest regions, and these directions were collected into orientation bins within a histogram. The histograms were then matched against the gesture models to recognize the gesture. The vertical bars in Figure 7-1 show the match levels of the different gestures. The gray horizontal line represents the threshold for acceptance of the "winning" vertical bar corresponding to a gesture model.

Histograms find uses in many computer vision applications. Histograms are used to detect scene transitions in videos by marking when the edge and color statistics markedly change from frame to frame. They are used to identify interest points in images by assigning each interest point a "tag" consisting of histograms of nearby features. Histograms of edges, colors, corners, and so on form a general feature type that is passed to classifiers for object recognition. Sequences of color or edge histograms are used to identify whether videos have been copied on the web, and the list goes on. Histograms are one of the classic tools of computer vision.

Histograms are simply collected *counts* of the underlying data organized into a set of predefined *bins*. They can be populated by counts of features computed from the data, such as gradient magnitudes and directions, color, or just about any other characteristic. In any case, they are used to obtain a statistical picture of the underlying distribution of data. The histogram usually has fewer dimensions than the source data. Figure 7-2 depicts a typical situation. The figure shows a two-dimensional distribution of points (upper left); we impose a grid (upper right) and count the data points in each *grid cell*, yielding a one-dimensional histogram (lower right). Because the raw data points can

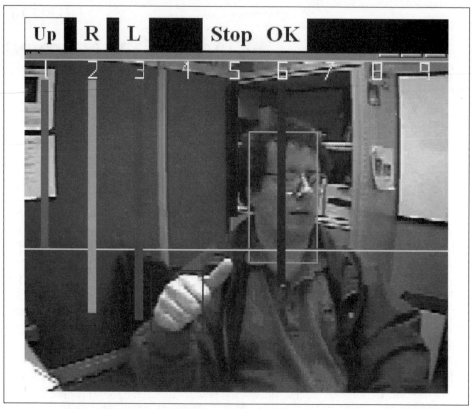

Figure 7-1. Local histograms of gradient orientations are used to find the hand and its gesture; here the "winning" gesture (longest vertical bar) is a correct recognition of "L" (move left)

represent just about anything, the histogram is a handy way of representing whatever it is that you have learned from your image.

Histograms that represent continuous distributions do so by implicitly averaging the number of points in each grid cell.* This is where problems can arise, as shown in Figure 7-3. If the grid is too wide (upper left), then there is too much averaging and we lose the structure of the distribution. If the grid is too narrow (upper right), then there is not enough averaging to represent the distribution accurately and we get small, "spiky" cells.

OpenCV has a data type for representing histograms. The histogram data structure is capable of representing histograms in one or many dimensions, and it contains all the data necessary to track bins of both uniform and nonuniform sizes. And, as you might expect, it comes equipped with a variety of useful functions which will allow us to easily perform common operations on our histograms.

---

* This is also true of histograms representing information that falls naturally into discrete groups when the histogram uses fewer bins than the natural description would suggest or require. An example of this is representing 8-bit intensity values in a 10-bin histogram: each bin would then combine the points associated with approximately 25 different intensities, (erroneously) treating them all as equivalent.

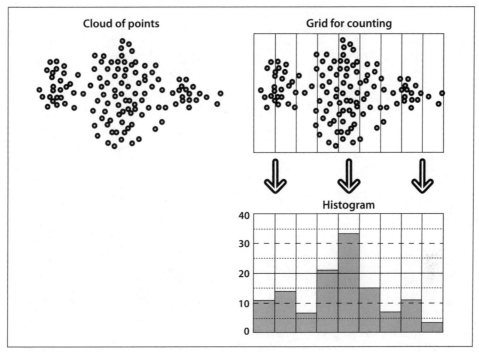

Figure 7-2. Typical histogram example: starting with a cloud of points (upper left), a counting grid is imposed (upper right) that yields a one-dimensional histogram of point counts (lower right)

Basic Histogram Data Structure

Let's start out by looking directly at the CvHistogram data structure.

```
typedef struct CvHistogram
{
    int     type;
    CvArr*  bins;
    float   thresh[CV_MAX_DIM][2]; // for uniform histograms
    float** thresh2;               // for nonuniform histograms
    CvMatND mat;                   // embedded matrix header
                                   // for array histograms
}
CvHistogram;
```

This definition is deceptively simple, because much of the internal data of the histogram is stored inside of the CvMatND structure. We create new histograms with the following routine:

```
CvHistogram* cvCreateHist(
    int     dims,
    int*    sizes,
    int     type,
    float** ranges  = NULL,
    int     uniform = 1
);
```

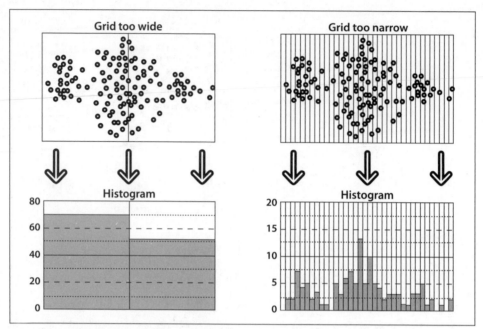

Figure 7-3. A histogram's accuracy depends on its grid size: a grid that is too wide yields too much spatial averaging in the histogram counts (left); a grid that is too small yields "spiky" and singleton results from too little averaging (right)

The argument dims indicates how many dimensions we want the histogram to have. The sizes argument must be an array of integers whose length is equal to dims. Each integer in this array indicates how many bins are to be assigned to the corresponding dimension. The type can be either CV_HIST_ARRAY, which is used for multidimensional histograms to be stored using the dense multidimensional matrix structure (i.e., CvMatND), or CV_HIST_SPARSE* if the data is to be stored using the sparse matrix representation (CvSparseMat). The argument ranges can have one of two forms. For a uniform histogram, ranges is an array of floating-point value pairs,[†] where the number of value pairs is equal to the number of dimensions. For a nonuniform histogram, the pairs used by the uniform histogram are replaced by arrays containing the values by which the nonuniform bins are separated. If there are N bins, then there will be $N + 1$ entries in each of these subarrays. Each array of values starts with the bottom edge of the lowest bin and ends with the top edge of the highest bin.[‡] The Boolean argument uniform indicates if the histogram is to have

---

* For you old timers, the value CV_HIST_TREE is still supported, but it is identical to CV_HIST_SPARSE.

† These "pairs" are just C-arrays with only two entries.

‡ To clarify: in the case of a uniform histogram, if the lower and upper ranges are set to 0 and 10, respectively, and if there are two bins, then the bins will be assigned to the respective intervals [0, 5) and [5, 10]. In the case of a nonuniform histogram, if the size dimension i is 4 and if the corresponding ranges are set to (0, 2, 4, 9, 10), then the resulting bins will be assigned to the following (nonuniform) intervals: [0, 2), [2,4), [4, 9), and [9, 10].

uniform bins and thus how the ranges value is interpreted;* if set to a nonzero value, the bins are uniform. It is possible to set ranges to NULL, in which case the ranges are simply "unknown" (they can be set later using the specialized function cvSetHistBinRanges()). Clearly, you had better set the value of ranges before you start using the histogram.

```
void cvSetHistBinRanges(
    CvHistogram* hist,
    float**      ranges,
    int          uniform = 1
);
```

The arguments to cvSetHistRanges() are exactly the same as the corresponding arguments for cvCreateHist(). Once you are done with a histogram, you can *clear* it (i.e., reset all of the bins to 0) if you plan to reuse it or you can de-allocate it with the usual *release*-type function.

```
void cvClearHist(
    CvHistogram*  hist
);
void cvReleaseHist(
    CvHistogram** hist
);
```

As usual, the release function is called with a pointer to the histogram pointer you obtained from the create function. The histogram pointer is set to NULL once the histogram is de-allocated.

Another useful function helps create a histogram from data we already have lying around:

```
CvHistogram*  cvMakeHistHeaderForArray(
    int           dims,
    int*          sizes,
    CvHistogram*  hist,
    float*        data,
    float**       ranges  = NULL,
    int           uniform = 1
);
```

In this case, hist is a pointer to a CvHistogram data structure and data is a pointer to an area of size sizes[0]*sizes[1]*...*sizes[dims-1] for storing the histogram bins. Notice that data is a pointer to float because the internal data representation for the histogram is always of type float. The return value is just the same as the hist value we passed in. Unlike the cvCreateHist() routine, there is no type argument. All histograms created by cvMakeHistHeaderForArray() are dense histograms. One last point before we move on: since you (presumably) allocated the data storage area for the histogram bins yourself, there is no reason to call cvReleaseHist() on your CvHistogram structure. You will have to clean up the header structure (if you did not allocate it on the stack) and, of course, clean up your data as well; but since these are "your" variables, you are assumed to be taking care of this in your own way.

---

* Have no fear that this argument is type int, because the only meaningful distinction is between zero and nonzero.

Accessing Histograms

There are several ways to access a histogram's data. The most straightforward method is to use OpenCV's accessor functions.

```
double cvQueryHistValue_1D(
    CvHistogram* hist,
    int          idx0
);
double cvQueryHistValue_2D(
    CvHistogram* hist,
    int          idx0,
    int          idx1
);
double cvQueryHistValue_3D(
    CvHistogram* hist,
    int          idx0,
    int          idx1,
    int          idx2
);
double cvQueryHistValue_nD(
    CvHistogram* hist,
    int*         idxN
);
```

Each of these functions returns a floating-point number for the value in the appropriate bin. Similarly, you can set (or get) histogram bin values with the functions that return a pointer to a bin (*not* to a bin's value):

```
float* cvGetHistValue_1D(
    CvHistogram* hist,
    int          idx0
);
float* cvGetHistValue_2D(
    CvHistogram* hist,
    int          idx0,
    int          idx1
);
float* cvGetHistValue_3D(
    CvHistogram* hist,
    int          idx0,
    int          idx1,
    int          idx2
);
float* cvGetHistValue_nD(
    CvHistogram* hist,
    int*         idxN
);
```

These functions look a lot like the cvGetReal*D and cvPtr*D families of functions, and in fact they are pretty much the same thing. Inside of these calls are essentially those same matrix accessors called with the matrix hist->bins passed on to them. Similarly, the functions for sparse histograms inherit the behavior of the corresponding sparse matrix functions. If you attempt to access a nonexistent bin using a GetHist*() function in a sparse histogram, then that bin is automatically created and its value set to 0. Note that QueryHist*() functions do not create missing bins.

This leads us to the more general topic of accessing the histogram. In many cases, for *dense histograms* we will want to access the bins member of the histogram directly. Of course, we might do this just as part of data access. For example, we might want to access all of the elements in a dense histogram sequentially, or we might want to access bins directly for performance reasons, in which case we might use hist->mat.data.fl (again, for dense histograms). Other reasons for accessing histograms include finding how many dimensions it has or what regions are represented by its individual bins. For this information we can use the following tricks to access either the actual data in the CvHistogram structure or the information imbedded in the CvMatND structure known as mat.

```
int n_dimension            = histogram->mat.dims;
int dim_i_nbins            = histogram->mat.dim[ i ].size;

// uniform histograms
int dim_i_bin_lower_bound  = histogram->thresh[ i ][ 0 ];
int dim_i_bin_upper_bound  = histogram->thresh[ i ][ 1 ];

// nonuniform histograms
int dim_i_bin_j_lower_bound = histogram->thresh2[ i ][ j ];
int dim_j_bin_j_upper_bound = histogram->thresh2[ i ][ j+1 ];
```

As you can see, there's a lot going on inside the histogram data structure.

Basic Manipulations with Histograms

Now that we have this great data structure, we will naturally want to do some fun stuff with it. First let's hit some of the basics that will be used over and over; then we'll move on to some more complicated features that are used for more specialized tasks.

When dealing with a histogram, we typically just want to accumulate information into its various bins. Once we have done this, however, it is often desirable to work with the histogram in *normalized form*, so that individual bins will then represent the fraction of the total number of events assigned to the entire histogram:

```
cvNormalizeHist( CvHistogram* hist, double factor );
```

Here hist is your histogram and factor is the number to which you would like to normalize the histogram (which will usually be 1). If you are following closely then you may have noticed that the argument factor is a double although the internal data type of CvHistogram() is always float—further evidence that OpenCV is a work in progress!

The next handy function is the threshold function:

```
cvThreshHist( CvHistogram* hist, double factor );
```

The argument factor is the cutoff for the threshold. The result of thresholding a histogram is that all bins whose value is below the threshold factor are set to 0. Recalling the image thresholding function cvThreshold(), we might say that the histogram thresholding function is analogous to calling the image threshold function with the argument threshold_type set to CV_THRESH_TOZERO. Unfortunately, there are no convenient histogram thresholding functions that provide operations analogous to the other threshold types. In practice, however, cvThreshHist() is the one you'll probably want because with real data we often end up with some bins that contain just a few data points. Such bins are mostly noise and thus should usually be zeroed out.

Another useful function is cvCopyHist(), which (as you might guess) copies the information from one histogram into another.

```
void cvCopyHist(const CvHistogram* src, CvHistogram** dst );
```

This function can be used in two ways. If the destination histogram *dst is a histogram of the same size as src, then both the data and the bin ranges from src will be copied into *dst. The other way of using cvCopyHist() is to set *dst to NULL. In this case, a new histogram will be allocated that has the same size as src and then the data and bin ranges will be copied (this is analogous to the image function cvCloneImage()). It is to allow this kind of cloning that the second argument dst is a pointer to a pointer to a histogram—unlike the src, which is just a pointer to a histogram. If *dst is NULL when cvCopyHist() is called, then *dst will be set to the pointer to the newly allocated histogram when the function returns.

Proceeding on our tour of useful histogram functions, our next new friend is cvGetMinMax HistValue(), which reports the minimal and maximal values found in the histogram.

```
void cvGetMinMaxHistValue(
    const CvHistogram* hist,
    float*             min_value,
    float*             max_value,
    int*               min_idx  = NULL,
    int*               max_idx  = NULL
);
```

Thus, given a histogram hist, cvGetMinMaxHistValue() will compute its largest and smallest values. When the function returns, *min_value and *max_value will be set to those respective values. If you don't need one (or both) of these results, then you may set the corresponding argument to NULL. The next two arguments are optional; if you leave them set to their default value (NULL), they will do nothing. However, if they are non-NULL pointers to int then the integer values indicated will be filled with the location index of the minimal and maximal values. In the case of multi-dimensional histograms, the arguments min_idx and max_idx (if not NULL) are assumed to point to an array of integers whose length is equal to the dimensionality of the histogram. If more than one bin in the histogram has the same minimal (or maximal) value, then the bin that will be returned is the one with the smallest index (in lexicographic order for multidimensional histograms).

After collecting data in a histogram, we often use cvGetMinMaxHistValue() to find the minimum value and then "threshold away" bins with values near this minimum using cvThreshHist() before finally normalizing the histogram via cvNormalizeHist().

Last, but certainly not least, is the automatic computation of histograms from images. The function cvCalcHist() performs this crucial task:

```
void cvCalcHist(
    IplImage**   image,
```

```
        CvHistogram* hist,
        int            accumulate = 0,
        const CvArr* mask       = NULL
    );
```

The first argument, image, is a pointer to an array of IplImage* pointers.* This allows us to pass in many image planes. In the case of a multi-channel image (e.g., HSV or RGB) we will have to cvSplit() (see Chapter 3 or Chapter 5) that image into planes before calling cvCalcHist(). Admittedly that's a bit of a pain, but consider that frequently you'll also want to pass in multiple image planes that contain different filtered versions of an image—for example, a plane of gradients or the *U*- and *V*-planes of *YUV*. Then what a mess it would be when you tried to pass in several images with various numbers of channels (and you can be sure that someone, somewhere, would want just some of those channels in those images!). To avoid this confusion, all images passed to cvCalcHist() are assumed (read "required") to be single-channel images. When the histogram is populated, the bins will be identified by the tuples formed across these multiple images. The argument hist must be a histogram of the appropriate dimensionality (i.e., of dimension equal to the number of image planes passed in through image). The last two arguments are optional. The accumulate argument, if nonzero, indicates that the histogram hist should not be cleared before the images are read; note that accumulation allows cvCalcHist() to be called multiple times in a data collection loop. The final argument, mask, is the usual optional Boolean mask; if non-NULL, only pixels corresponding to nonzero entries in the mask image will be included in the computed histogram.

Comparing Two Histograms

Yet another indispensable tool for working with histograms, first introduced by Swain and Ballard [Swain91] and further generalized by Schiele and Crowley [Schiele96], is the ability to compare two histograms in terms of some specific criteria for similarity. The function cvCompareHist() does just this.

```
double cvCompareHist(
    const CvHistogram* hist1,
    const CvHistogram* hist2,
    int                method
);
```

The first two arguments are the histograms to be compared, which should be of the same size. The third argument is where we select our desired distance metric. The four available options are as follows.

Correlation (method = CV_COMP_CORREL)

$$d_{\text{correl}}(H_1, H_2) = \frac{\sum_i H_1'(i) \cdot H_2'(i)}{\sqrt{\sum_i H_1'^2(i) \cdot H_2'^2(i)}}$$

---

* Actually, you could also use CvMat* matrix pointers here.

where $H'_k(i) = H_k(i) - (1/N)\left(\sum_j H_k(j)\right)$ and N equals the number of bins in the histogram.

For *correlation*, a high score represents a better match than a low score. A perfect match is 1 and a maximal mismatch is –1; a value of 0 indicates no correlation (random association).

Chi-square (method = CV_COMP_CHISQR)

$$d_{\text{chi-square}}(H_1, H_2) = \sum_i \frac{(H_1(i) - H_2(i))^2}{H_1(i) + H_2(i)}$$

For *chi-square*,* a low score represents a better match than a high score. A perfect match is 0 and a total mismatch is unbounded (depending on the size of the histogram).

Intersection (method = CV_COMP_INTERSECT)

$$d_{\text{intersection}}(H_1, H_2) = \sum_i \min(H_1(i), H_2(i))$$

For *histogram intersection*, high scores indicate good matches and low scores indicate bad matches. If both histograms are normalized to 1, then a perfect match is 1 and a total mismatch is 0.

Bhattacharyya distance (method = CV_COMP_BHATTACHARYYA)

$$d_{\text{Bhattacharyya}}(H_1, H_2) = \sqrt{1 - \sum_i \frac{\sqrt{H_1(i) \cdot H_2(i)}}{\sqrt{\sum_i H_1(i) \cdot \sum_i H_2(i)}}}$$

For *Bhattacharyya matching* [Bhattacharyya43], low scores indicate good matches and high scores indicate bad matches. A perfect match is 0 and a total mismatch is a 1.

With CV_COMP_BHATTACHARYYA, a special factor in the code is used to normalize the input histograms. In general, however, you should normalize histograms *before* comparing them because concepts like histogram intersection make little sense (even if allowed) without normalization.

The simple case depicted in Figure 7-4 should clarify matters. In fact, this is about the simplest case that could be imagined: a one-dimensional histogram with only two bins. The model histogram has a 1.0 value in the left bin and a 0.0 value in the right bin. The last three rows show the comparison histograms and the values generated for them by the various metrics (the EMD metric will be explained shortly).

---

* The chi-square test was invented by Karl Pearson [Pearson] who founded the field of mathematical statistics.

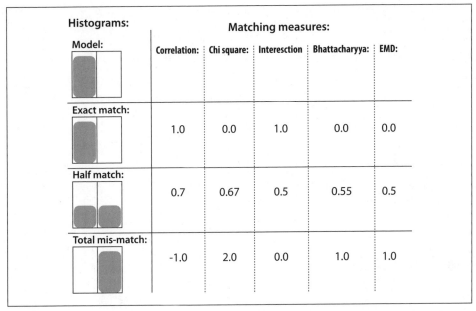

Histograms:	Matching measures:				
Model:	Correlation:	Chi square:	Interesction	Bhattacharyya:	EMD:
Exact match:	1.0	0.0	1.0	0.0	0.0
Half match:	0.7	0.67	0.5	0.55	0.5
Total mis-match:	-1.0	2.0	0.0	1.0	1.0

Figure 7-4. Histogram matching measures

Figure 7-4 provides a quick reference for the behavior of different matching types, but there is something disconcerting here, too. If histogram bins shift by just one slot—as with the chart's first and third comparison histograms—then all these matching methods (except EMD) yield a maximal mismatch even though these two histograms have a similar "shape". The rightmost column in Figure 7-4 reports values returned by EMD, a type of distance measure. In comparing the third to the model histogram, the EMD measure quantifies the situation precisely: the third histogram has moved to the right by one unit. We shall explore this measure further in the "Earth Mover's Distance" section to follow.

In the authors' experience, intersection works well for quick-and-dirty matching and chi-square or Bhattacharyya work best for slower but more accurate matches. The EMD measure gives the most intuitive matches but is much slower.

Histogram Usage Examples

It's probably time for some helpful examples. The program in Example 7-1 (adapted from the OpenCV code bundle) shows how we can use some of the functions just discussed. This program computes a hue-saturation histogram from an incoming image and then draws that histogram as an illuminated grid.

Example 7-1. Histogram computation and display

```
#include <cv.h>
#include <highgui.h>

int main( int argc, char** argv ) {
```

Example 7-1. Histogram computation and display (continued)

```
IplImage* src;

if( argc == 2 && (src=cvLoadImage(argv[1], 1))!= 0) {

    // Compute the HSV image and decompose it into separate planes.
    //
    IplImage* hsv = cvCreateImage( cvGetSize(src), 8, 3 );
    cvCvtColor( src, hsv, CV_BGR2HSV );

    IplImage* h_plane  = cvCreateImage( cvGetSize(src), 8, 1 );
    IplImage* s_plane  = cvCreateImage( cvGetSize(src), 8, 1 );
    IplImage* v_plane  = cvCreateImage( cvGetSize(src), 8, 1 );
    IplImage* planes[] = { h_plane, s_plane };
    cvCvtPixToPlane( hsv, h_plane, s_plane, v_plane, 0 );

    // Build the histogram and compute its contents.
    //
    int h_bins = 30, s_bins = 32;
    CvHistogram* hist;
    {
      int    hist_size[] = { h_bins, s_bins };
      float  h_ranges[]  = { 0, 180 };            // hue is [0,180]
      float  s_ranges[]  = { 0, 255 };
      float* ranges[]    = { h_ranges, s_ranges };
      hist = cvCreateHist(
        2,
        hist_size,
        CV_HIST_ARRAY,
        ranges,
        1
      );
    }
    cvCalcHist( planes, hist, 0, 0 ); //Compute histogram
    cvNormalizeHist( hist[i], 1.0 );   //Normalize it

    // Create an image to use to visualize our histogram.
    //
    int scale = 10;
    IplImage* hist_img = cvCreateImage(
      cvSize( h_bins * scale, s_bins * scale ),
      8,
      3
    );
    cvZero( hist_img );

    // populate our visualization with little gray squares.
    //
    float max_value = 0;
    cvGetMinMaxHistValue( hist, 0, &max_value, 0, 0 );

    for( int h = 0; h < h_bins; h++ ) {
        for( int s = 0; s < s_bins; s++ ) {
```

Example 7-1. Histogram computation and display (continued)

```
        float bin_val = cvQueryHistValue_2D( hist, h, s );
        int intensity = cvRound( bin_val * 255 / max_value );
        cvRectangle(
          hist_img,
          cvPoint( h*scale, s*scale ),
          cvPoint( (h+1)*scale - 1, (s+1)*scale - 1),
          CV_RGB(intensity,intensity,intensity),
          CV_FILLED
        );
      }
    }

    cvNamedWindow( "Source", 1 );
    cvShowImage(  "Source", src );

    cvNamedWindow( "H-S Histogram", 1 );
    cvShowImage(  "H-S Histogram", hist_img );

    cvWaitKey(0);
  }
}
```

In this example we have spent a fair amount of time preparing the arguments for cvCalcHist(), which is not uncommon. We also chose to normalize the colors in the visualization rather than normalizing the histogram itself, although the reverse order might be better for some applications. In this case it gave us an excuse to call cvGetMinMaxHistValue(), which was reason enough not to reverse the order.

Let's look at a more practical example: color histograms taken from a human hand under various lighting conditions. The left column of Figure 7-5 shows images of a hand in an indoor environment, a shaded outdoor environment, and a sunlit outdoor environment. In the middle column are the blue, green, and red (BGR) histograms corresponding to the observed flesh tone of the hand. In the right column are the corresponding HSV histograms, where the vertical axis is V (value), the radius is S (saturation) and the angle is H (hue). Notice that indoors is darkest, outdoors in the shade brighter, and outdoors in the sun brightest. Observe also that the colors shift around somewhat as a result of the changing color of the illuminating light.

As a test of histogram comparison, we could take a portion of one palm (e.g., the top half of the indoor palm), and compare the histogram representation of the colors in that image either with the histogram representation of the colors in the remainder of that image or with the histogram representations of the other two hand images. Flesh tones are often easier to pick out after conversion to an HSV color space. It turns out that restricting ourselves to the hue and saturation planes is not only sufficient but also helps with recognition of flesh tones across ethnic groups. The matching results for our experiment are shown in Table 7-1, which confirms that lighting can cause severe mismatches in color. Sometimes normalized BGR works better than HSV in the context of lighting changes.

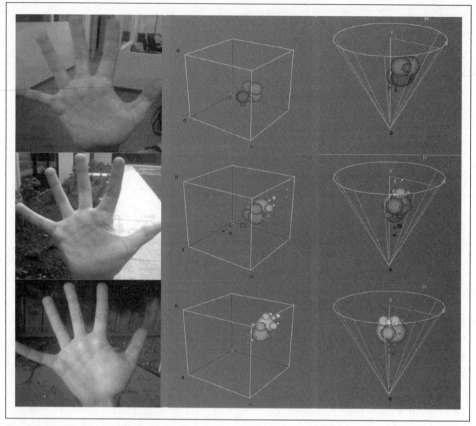

Figure 7-5. Histogram of flesh colors under indoor (upper left), shaded outdoor (middle left), and outdoor (lower left) lighting conditions; the middle and right-hand columns display the associated BGR and HSV histograms, respectively

Table 7-1. Histogram comparison, via four matching methods, of palm-flesh colors in upper half of indoor palm with listed variant palm-flesh color

Comparison	CORREL	CHISQR	INTERSECT	BHATTACHARYYA
Indoor lower half	0.96	0.14	0.82	0.2
Outdoor shade	0.09	1.57	0.13	0.8
Outdoor sun	−0.0	1.98	0.01	0.99

Some More Complicated Stuff

Everything we've discussed so far was reasonably basic. Each of the functions provided for a relatively obvious need. Collectively, they form a good foundation for much of what you might want to do with histograms in the context of computer vision (and probably in other contexts as well). At this point we want to look at some more complicated routines available within OpenCV that are extremely useful in certain applications. These routines include a more sophisticated method of comparing two histograms as well as tools for computing and/or visualizing which portions of an image contribute to a given portion of a histogram.

Earth Mover's Distance

Lighting changes can cause shifts in color values (see Figure 7-5), although such shifts tend not to change the shape of the histogram of color values, but shift the color value locations and thus cause the histogram-matching schemes we've learned about to fail. If instead of a histogram *match* measure we used a histogram *distance* measure, then we could still match like histograms to like histograms even when the second histogram has shifted its been by looking for small distance measures. Earth mover's distance (EMD) [Rubner00] is such a metric; it essentially measures how much work it would take to "shovel" one histogram shape into another, including moving part (or all) of the histogram to a new location. It works in any number of dimensions.

Return again to Figure 7-4; we see the "earthshoveling" nature of EMD's distance measure in the rightmost column. An exact match is a distance of 0. Half a match is half a "shovel full", the amount it would take to spread half of the left histogram into the next slot. Finally, moving the entire histogram one step to the right would require an entire unit of distance (i.e., to change the model histogram into the "totally mismatched" histogram).

The EMD algorithm itself is quite general; it allows users to set their own distance metric or their own cost-of-moving matrix. One can record where the histogram "material" flowed from one histogram to another, and one can employ nonlinear distance metrics derived from prior information about the data. The EMD function in OpenCV is cvCalcEMD2():

```
float cvCalcEMD2(
    const CvArr*       signature1,
    const CvArr*       signature2,
    int                distance_type,
    CvDistanceFunction distance_func = NULL,
    const CvArr*       cost_matrix   = NULL,
    CvArr*             flow          = NULL,
    float*             lower_bound   = NULL,
    void*              userdata      = NULL
);
```

The cvCalcEMD2() function has enough parameters to make one dizzy. This may seem rather complex for such an intuitive function, but the complexity stems from all the subtle configurable dimensions of the algorithm.* Fortunately, the function can be used in its more basic and intuitive form and without most of the arguments (note all the "=NULL" defaults in the preceding code). Example 7-2 shows the simplified version.

---

* If you want all of the gory details, we recommend that you read the 1989 paper by S. Peleg, M. Werman, and H. Rom, "A Unified Approach to the Change of Resolution: Space and Gray-Level," and then take a look at the relevant entries in the OpenCV user manual that are included in the release ...\opencv\docs\ref\ opencvref_cv.htm.

Example 7-2. Simple EMD interface

```
float cvCalcEMD2(
    const CvArr* signature1,
    const CvArr* signature2,
    int          distance_type
);
```

The parameter distance_type for the simpler version of cvCalcEMD2() is either *Manhattan distance* (CV_DIST_L1) or *Euclidean distance* (CV_DIST_L2). Although we're applying the EMD to histograms, the interface prefers that we talk to it in terms of *signatures* for the first two array parameters.

These signature arrays are always of type float and consist of rows containing the histogram bin count followed by its coordinates. For the one-dimensional histogram of Figure 7-4, the signatures (listed array rows) for the lefthand column of histograms (skipping the model) would be as follows: top, [1, 0; 0, 1]; middle, [0.5, 0; 0.5, 1]; bottom, [0, 0; 1, 1]. If we had a bin in a three-dimensional histogram with a bin count of 537 at (x, y, z) index (7, 43, 11), then the signature row for that bin would be [537, 7; 43, 11]. This is how we perform the necessary step of converting histograms into signatures.

As an example, suppose we have two histograms, hist1 and hist2, that we want to convert to two signatures, sig1 and sig2. Just to make things more difficult, let's suppose that these are two-dimensional histograms (as in the preceding code examples) of dimension h_bins by s_bins. Example 7-3 shows how to convert these two histograms into two signatures.

Example 7-3. Creating signatures from histograms for EMD

```
//Convert histograms into signatures for EMD matching
//assume we already have 2D histograms hist1 and hist2
//that are both of dimension h_bins by s_bins (though for EMD,
// histograms don't have to match in size).
//
CvMat* sig1,sig2;
int numrows = h_bins*s_bins;

//Create matrices to store the signature in
//
sig1 = cvCreateMat(numrows, 3, CV_32FC1); //1 count + 2 coords = 3
sig2 = cvCreateMat(numrows, 3, CV_32FC1); //sigs are of type float.

//Fill signatures for the two histograms
//
for( int h = 0; h < h_bins; h++ ) {
    for( int s = 0; s < s_bins; s++ ) {
        float bin_val = cvQueryHistValue_2D( hist1, h, s );
        cvSet2D(sig1,h*s_bins + s,0,cvScalar(bin_val)); //bin value
        cvSet2D(sig1,h*s_bins + s,1,cvScalar(h));        //Coord 1
        cvSet2D(sig1,h*s_bins + s,2,cvScalar(s));        //Coord 2
```

Example 7-3. Creating signatures from histograms for EMD (continued)

```
        bin_val = cvQueryHistValue_2D( hist2, h, s );
        cvSet2D(sig2,h*s_bins + s,0,cvScalar(bin_val)); //bin value
        cvSet2D(sig2,h*s_bins + s,1,cvScalar(h));        //Coord 1
        cvSet2D(sig2,h*s_bins + s,2,cvScalar(s));        //Coord 2
    }
}
```

Notice in this example* that the function cvSet2D() takes a CvScalar() array to set its value even though each entry in this particular matrix is a single float. We use the inline convenience macro cvScalar() to accomplish this task. Once we have our histograms converted into signatures, we are ready to get the distance measure. Choosing to measure by Euclidean distance, we now add the code of Example 7-4.

Example 7-4. Using EMD to measure the similarity between distributions

```
// Do EMD AND REPORT
//
float emd = cvCalcEMD2(sig1,sig2,CV_DIST_L2);
printf("%f; ",emd);
```

Back Projection

Back projection is a way of recording how well the pixels (for cvCalcBackProject()) or patches of pixels (for cvCalcBackProjectPatch()) fit the distribution of pixels in a histogram model. For example, if we have a histogram of flesh color then we can use back projection to find flesh color areas in an image. The function call for doing this kind of lookup is:

```
    void cvCalcBackProject(
        IplImage**         image,
        CvArr*             back_project,
        const CvHistogram* hist
    );
```

We have already seen the array of single channel images IplImage** image in the function cvCalcHist() (see the section "Basic Manipulations with Histograms"). The number of images in this array is exactly the same—and in the same order—as used to construct the histogram model hist. Example 7-1 showed how to convert an image into single-channel planes and then make an array of them. The image or array back_project is a single-channel 8-bit or floating-point image of the same size as the input images in the array. The values in back_project are set to the values in the associated bin in hist. If the histogram is normalized, then this value can be associated with a conditional probability value (i.e., the probability that a pixel in image is a member of the type characterized

---

* Using cvSetReal2D() or cvmSet() would have been more compact and efficient here, but the example is clearer this way and the extra overhead is small compared to the actual distance calculation in EMD.

by the histogram in hist).* In Figure 7-6, we use a flesh-color histogram to derive a probability of flesh image.

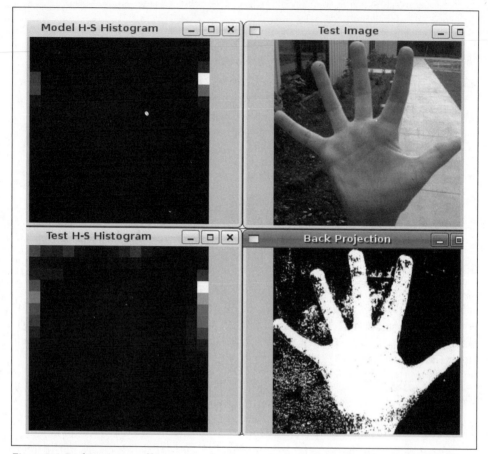

Figure 7-6. Back projection of histogram values onto each pixel based on its color: the HSV flesh-color histogram (upper left) is used to convert the hand image (upper right) into the flesh-color probability image (lower right); the lower left panel is the histogram of the hand image

---

* Specifically, in the case of our flesh-tone H-S histogram, if C is the color of the pixel and F is the probability that a pixel is flesh, then this probability map gives us $p(C|F)$, the probability of drawing that color if the pixel actually is flesh. This is not quite the same as $p(F|C)$, the probability that the pixel is flesh given its color. However, these two probabilities are related by Bayes' theorem [Bayes1763] and so, if we know the overall probability of encountering a flesh-colored object in a scene as well as the total probability of encountering of the range of flesh colors, then we can compute $p(F|C)$ from $p(C|F)$. Specifically, Bayes' theorem establishes the following relation:

$$p(F|C) = \frac{p(F)}{p(C)} p(C|F)$$

When `back_project` is a byte image rather than a float image, you should either not normalize the histogram or else scale it up before use. The reason is that the highest possible value in a normalized histogram is 1, so anything less than that will be rounded down to 0 in the 8-bit image. You might also need to scale `back_project` in order to see the values with your eyes, depending on how high the values are in your histogram.

Patch-based back projection

We can use the basic back-projection method to model whether or not a particular pixel is likely to be a member of a particular object type (when that object type was modeled by a histogram). This is not exactly the same as computing the probability of the presence of a particular object. An alternative method would be to consider subregions of an image and the feature (e.g., color) histogram of that subregion and to ask whether the histogram of features for the subregion matches the model histogram; we could then associate with each such subregion a probability that the modeled object is, in fact, present in that subregion.

Thus, just as `cvCalcBackProject()` allows us to compute if a pixel might be part of a known object, `cvCalcBackProjectPatch()` allows us to compute if a patch might contain a known object. The `cvCalcBackProjectPatch()` function uses a sliding window over the entire input image, as shown in Figure 7-7. At each location in the input array of images, all the pixels in the patch are used to set one pixel in the destination image corresponding to the center of the patch. This is important because many properties of images such as textures cannot be determined at the level of individual pixels, but instead arise from groups of pixels.

For simplicity in these examples, we've been sampling color to create our histogram models. Thus in Figure 7-6 the whole hand "lights up" because pixels there match the flesh color histogram model well. Using patches, we can detect statistical properties that occur over local regions, such as the variations in local intensity that make up a texture on up to the configuration of properties that make up a whole object. Using local patches, there are two ways one might consider applying `cvCalcBackProjectPatch()`: as a region detector when the sampling window is smaller than the object and as an object detector when the sampling window is the size of the object. Figure 7-8 shows the use of `cvCalcBackProjectPatch()` as a region detector. We start with a histogram model of palm-flesh color and a small window is moved over the image such that each pixel in the back projection image records the probability of palm-flesh at that pixel given all the pixels in the surrounding window in the original image. In Figure 7-8 the hand is much larger than the scanning window and the palm region is preferentially detected. Figure 7-9 starts with a histogram model collected from blue mugs. In contrast to Figure 7-8 where regions were detected, Figure 7-9 shows how `cvCalcBackProjectPatch()` can be used as an object detector. When the window size is roughly the same size as the objects we are hoping to find in an image, the whole object "lights up" in the back projection

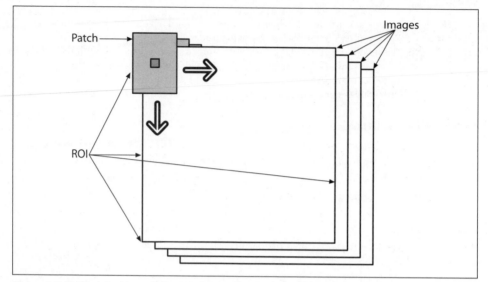

Figure 7-7. Back projection: a sliding patch over the input image planes is used to set the corresponding pixel (at the center of the patch) in the destination image; for normalized histogram models, the resulting image can be interpreted as a probability map indicating the possible presence of the object (this figure is taken from the OpenCV reference manual)

image. Finding peaks in the back projection image then corresponds to finding the location of objects (in Figure 7-9, a mug) that we are looking for.

The function provided by OpenCV for back projection by patches is:

```
void cvCalcBackProjectPatch(
    IplImage**    images,
    CvArr*        dst,
    CvSize        patch_size,
    CvHistogram*  hist,
    int           method,
    float         factor
);
```

Here we have the same array of single-channel images that was used to create the histogram using cvCalcHist(). However, the destination image dst is different: it can only be a single-channel, floating-point image with size (images[0][0].width - patch_size.x + 1, images[0][0].height - patch_size.y + 1). The explanation for this size (see Figure 7-7) is that the center pixel in the patch is used to set the corresponding location in dst, so we lose half a patch dimension along the edges of the image on every side. The parameter patch_size is exactly what you would expect (the size of the patch) and may be set using the convenience macro cvSize(width, height). We are already familiar with the histogram parameter; as with cvCalcBackProject(), this is the model histogram to which individual windows will be compared. The parameter for comparison method takes as arguments exactly the same method types as used in cvCompareHist() (see the

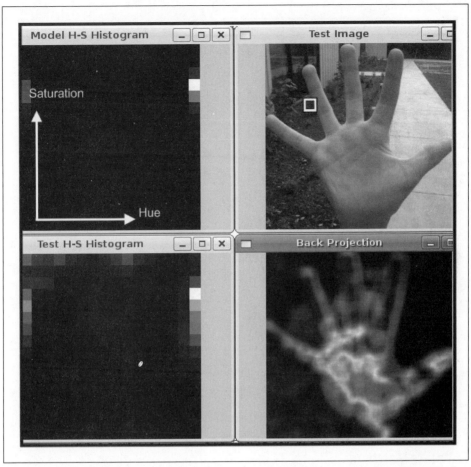

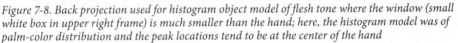

Figure 7-8. Back projection used for histogram object model of flesh tone where the window (small white box in upper right frame) is much smaller than the hand; here, the histogram model was of palm-color distribution and the peak locations tend to be at the center of the hand

"Comparing Two Histograms" section).* The final parameter, factor, is the normalization level; this parameter is the same as discussed previously in connection with cvNormalizeHist(). You can set it to 1 or, as a visualization aid, to some larger number. Because of this flexibility, you are always free to normalize your hist model before using cvCalcBackProjectPatch().

A final question comes up: Once we have a probability of object image, how do we use that image to find the object that we are searching for? For search, we can use the cvMinMaxLoc() discussed in Chapter 3. The maximum location (assuming you smooth a bit first) is the most likely location of the object in an image. This leads us to a slight digression, template matching.

---

* You must be careful when choosing a method, because some indicate best match with a return value of 1 and others with a value of 0.

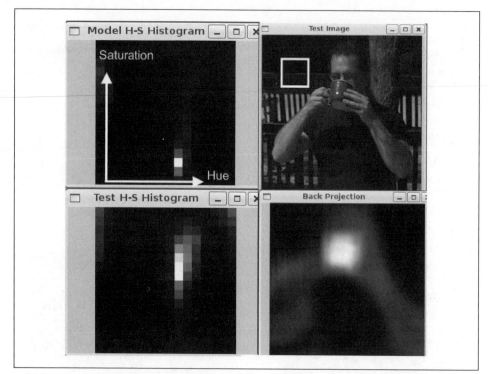

Figure 7-9. Using cvCalcBackProjectPatch() to locate an object (here, a coffee cup) whose size approximately matches the patch size (white box in upper right panel): the sought object is modeled by a hue-saturation histogram (upper left), which can be compared with an HS histogram for the image as a whole (lower left); the result of cvCalcBackProjectPatch() (lower right) is that the object is easily picked out from the scene by virtue of its color

Template Matching

Template matching via cvMatchTemplate() is not based on histograms; rather, the function matches an actual image patch against an input image by "sliding" the patch over the input image using one of the matching methods described in this section.

If, as in Figure 7-10, we have an image patch containing a face, then we can slide that face over an input image looking for strong matches that would indicate another face is present. The function call is similar to that of cvCalcBackProjectPatch():

```
void cvMatchTemplate(
    const CvArr* image,
    const CvArr* templ,
    CvArr*       result,
    int          method
);
```

Instead of the array of input image planes that we saw in cvCalcBackProjectPatch(), here we have a single 8-bit or floating-point plane or color image as input. The matching model in templ is just a patch from a similar image containing the object for which

Figure 7-10. cvMatchTemplate() sweeps a template image patch across another image looking for matches

you are searching. The output object image will be put in the result image, which is a single-channel byte or floating-point image of size (images->width - patch_size.x + 1, rimages->height - patch_size.y + 1), as we saw previously in cvCalcBackProjectPatch(). The matching method is somewhat more complex, as we now explain. We use I to denote the input image, T the template, and R the result.

Square difference matching method (method = CV_TM_SQDIFF)

These methods match the squared difference, so a perfect match will be 0 and bad matches will be large:

$$R_{\text{sq\_diff}}(x,y) = \sum_{x',y'} [T(x',y') - I(x+x',y+y')]^2$$

Correlation matching methods (method = CV_TM_CCORR)

These methods multiplicatively match the template against the image, so a perfect match will be large and bad matches will be small or 0.

$$R_{\text{ccorr}}(x,y) = \sum_{x',y'} [T(x',y') \cdot I(x+x',y+y')]^2$$

Correlation coefficient matching methods (method = CV_TM_CCOEFF)

These methods match a template relative to its mean against the image relative to its mean, so a perfect match will be 1 and a perfect mismatch will be –1; a value of 0 simply means that there is no correlation (random alignments).

$$R_{ccoeff}(x,y) = \sum_{x',y'} [T'(x',y') \cdot I'(x+x',y+y')]^2$$

$$T'(x',y') = T(x',y') - \frac{1}{(w \cdot h) \sum_{x'',y''} T(x'',y'')}$$

$$I'(x+x',y+y') = I(x+x',y+y') - \frac{1}{(w \cdot h) \sum_{x'',y''} I(x+x'',y+y'')}$$

Normalized methods

For each of the three methods just described, there are also normalized versions first developed by Galton [Galton] as described by Rodgers [Rodgers88]. The normalized methods are useful because, as mentioned previously, they can help reduce the effects of lighting differences between the template and the image. In each case, the normalization coefficient is the same:

$$Z(x,y) = \sqrt{\sum_{x',y'} T(x',y')^2 \cdot \sum_{x',y'} I(x+x',y+x')^2}$$

The values for method that give the normalized computations are listed in Table 7-2.

Table 7-2. Values of the method parameter for normalized template matching

Value of method parameter	Computed result
CV_TM_SQDIFF_NORMED	$R_{sq\_diff\_normed}(x,y) = \dfrac{R_{sq\_diff}(x,y)}{Z(x,y)}$
CV_TM_CCORR_NORMED	$R_{ccor\_normed}(x,y) = \dfrac{R_{ccor}(x,y)}{Z(x,y)}$
CV_TM_CCOEFF_NORMED	$R_{ccoeff\_normed}(x,y) = \dfrac{R_{ccoeff}(x,y)}{Z(x,y)}$

As usual, we obtain more accurate matches (at the cost of more computations) as we move from simpler measures (square difference) to the more sophisticated ones (correlation coefficient). It's best to do some test trials of all these settings and then choose the one that best trades off accuracy for speed in your application.

Again, be careful when interpreting your results. The square-difference methods show best matches with a minimum, whereas the correlation and correlation-coefficient methods show best matches at maximum points.

As in the case of cvCalcBackProjectPatch(), once we use cvMatchTemplate() to obtain a matching result image we can then use cvMinMaxLoc() to find the location of the best match. Again, we want to ensure there's an area of good match around that point in order to avoid random template alignments that just happen to work well. A good match should have good matches nearby, because slight misalignments of the template shouldn't vary the results too much for real matches. Looking for the best matching "hill" can be done by slightly smoothing the result image before seeking the maximum (for correlation or correlation-coefficient) or minimum (for square-difference) matching methods. The morphological operators can also be helpful in this context.

Example 7-5 should give you a good idea of how the different template matching techniques behave. This program first reads in a template and image to be matched and then performs the matching via the methods we've discussed here.

Example 7-5. Template matching

```
// Template matching.
//    Usage: matchTemplate image template
//
#include <cv.h>
#include <cxcore.h>
#include <highgui.h>
#include <stdio.h>
int main( int argc, char** argv ) {
    IplImage *src, *templ,*ftmp[6]; //ftmp will hold results
    int i;
    if( argc == 3){
        //Read in the source image to be searched:
        if((src=cvLoadImage(argv[1], 1))== 0) {
            printf("Error on reading src image %s\n",argv[i]);
            return(-1);
        }
        //Read in the template to be used for matching:
        if((templ=cvLoadImage(argv[2], 1))== 0) {
            printf("Error on reading template %s\n",argv[2]);
                return(-1);
        }
        //ALLOCATE OUTPUT IMAGES:
        int iwidth = src->width - templ->width + 1;
        int iheight = src->height - templ->height + 1;
        for(i=0; i<6; ++i){
            ftmp[i] = cvCreateImage(
                            cvSize(iwidth,iheight),32,1);
        }
        //DO THE MATCHING OF THE TEMPLATE WITH THE IMAGE:
```

Example 7-5. Template matching (continued)

```
for(i=0; i<6; ++i){
    cvMatchTemplate( src, templ, ftmp[i], i);
    cvNormalize(ftmp[i],ftmp[i],1,0,CV_MINMAX)*;
}
//DISPLAY
cvNamedWindow( "Template", 0 );
cvShowImage(   "Template", templ );
cvNamedWindow( "Image", 0 );
cvShowImage(   "Image", src );
cvNamedWindow( "SQDIFF", 0 );
cvShowImage(   "SQDIFF", ftmp[0] );
cvNamedWindow( "SQDIFF_NORMED", 0 );
cvShowImage(   "SQDIFF_NORMED", ftmp[1] );
cvNamedWindow( "CCORR", 0 );
cvShowImage(   "CCORR", ftmp[2] );
cvNamedWindow( "CCORR_NORMED", 0 );
cvShowImage(   "CCORR_NORMED", ftmp[3] );
cvNamedWindow( "CCOEFF", 0 );
cvShowImage(   "CCOEFF", ftmp[4] );
cvNamedWindow( "CCOEFF_NORMED", 0 );
cvShowImage(   "CCOEFF_NORMED", ftmp[5] );
//LET USER VIEW RESULTS:
cvWaitKey(0);
}
else { printf("Call should be: "
              "matchTemplate image template \n");}
}
```

Note the use of cvNormalize() in this code, which allows us to display the results in a consistent way (recall that some of the matching methods can return negative-valued results. We use the CV_MINMAX flag when normalizing; this tells the function to shift and scale the floating-point images so that all returned values are between 0 and 1. Figure 7-11 shows the results of sweeping the face template over the source image (shown in Figure 7-10) using each of cvMatchTemplate()'s available matching methods. In outdoor imagery especially, it's almost always better to use one of the normalized methods. Among those, correlation coefficient gives the most clearly delineated match—but, as expected, at a greater computational cost. For a specific application, such as automatic parts inspection or tracking features in a video, you should try all the methods and find the speed and accuracy trade-off that best serves your needs.

---

* You can often get more pronounced match results by raising the matches to a power (e.g., cvPow(ftmp[i], ftmp[i], 5);). In the case of a result which is normalized between 0.0 and 1.0, then you can immediately see that a good match of 0.99 taken to the fifth power is not much reduced (0.99^5=0.95) while a poorer score of 0.20 is reduced substantially (0.50^5=0.03).

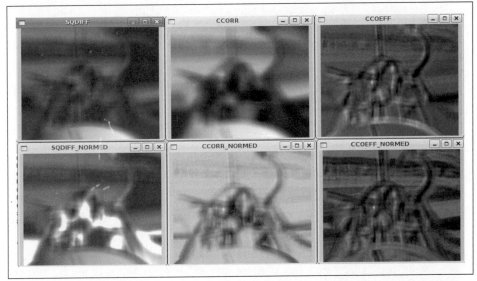

Figure 7-11. Match results of six matching methods for the template search depicted in Figure 7-10: the best match for square difference is 0 and for the other methods it's the maximum point; thus, matches are indicated by dark areas in the left column and by bright spots in the other two columns

Exercises

1. Generate 1,000 random numbers r_i between 0 and 1. Decide on a bin size and then take a histogram of $1/r_i$.

    a. Are there similar numbers of entries (i.e., within a factor of ±10) in each histogram bin?

    b. Propose a way of dealing with distributions that are highly nonlinear so that each bin has, within a factor of 10, the same amount of data.

2. Take three images of a hand in each of the three lighting conditions discussed in the text. Use cvCalcHist() to make an RGB histogram of the flesh color of one of the hands photographed indoors.

    a. Try using just a few large bins (e.g., 2 per dimension), a medium number of bins (16 per dimension) and many bins (256 per dimension). Then run a matching routine (using all histogram matching methods) against the other indoor lighting images of hands. Describe what you find.

    b. Now add 8 and then 32 bins per dimension and try matching across lighting conditions (train on indoor, test on outdoor). Describe the results.

3. As in exercise 2, gather RGB histograms of hand flesh color. Take one of the indoor histogram samples as your model and measure EMD (earth mover's distance) against the second indoor histogram and against the first outdoor shaded and first outdoor sunlit histograms. Use these measurements to set a distance threshold.

a. Using this EMD threshold, see how well you detect the flesh histogram of the third indoor histogram, the second outdoor shaded, and the second outdoor sunlit histograms. Report your results.

b. Take histograms of randomly chosen nonflesh background patches to see how well your EMD discriminates. Can it reject the background while matching the true flesh histograms?

4. Using your collection of hand images, design a histogram that can determine under which of the three lighting conditions a given image was captured. Toward this end, you should create features—perhaps sampling from parts of the whole scene, sampling brightness values, and/or sampling relative brightness (e.g., from top to bottom patches in the frame) or gradients from center to edges.

5. Assemble three histograms of flesh models from each of our three lighting conditions.

a. Use the first histograms from indoor, outdoor shaded, and outdoor sunlit as your models. Test each one of these against the second images in each respective class to see how well the flesh-matching score works. Report matches.

b. Use the "scene detector" you devised in part a, to create a "switching histogram" model. First use the scene detector to determine which histogram model to use: indoor, outdoor shaded, or outdoor sunlit. Then use the corresponding flesh model to accept or reject the second flesh patch under all three conditions. How well does this switching model work?

6. Create a flesh-region interest (or "attention") detector.

a. Just indoors for now, use several samples of hand and face flesh to create an RGB histogram.

b. Use cvCalcBackProject() to find areas of flesh.

c. Use cvErode() from Chapter 5 to clean up noise and then cvFloodFill() (from the same chapter) to find large areas of flesh in an image. These are your "attention" regions.

7. Try some hand-gesture recognition. Photograph a hand about 2 feet from the camera, create some (nonmoving) hand gestures: thumb up, thumb left, thumb right.

a. Using your attention detector from exercise 6, take image gradients in the area of detected flesh around the hand and create a histogram model for each of the three gestures. Also create a histogram of the face (if there's a face in the image) so that you'll have a (nongesture) model of that large flesh region. You might also take histograms of some similar but nongesture hand positions, just so they won't be confused with the actual gestures.

b. Test for recognition using a webcam: use the flesh interest regions to find "potential hands"; take gradients in each flesh region; use histogram matching above a threshold to detect the gesture. If two models are above threshold, take the better match as the winner.

   c. Move your hand 1–2 feet further back and see if the gradient histogram can still recognize the gestures. Report.

8. Repeat exercise 7 but with EMD for the matching. What happens to EMD as you move your hand back?

9. With the same images as before but with captured image patches instead of histograms of the flesh around the hand, use cvMatchTemplate() instead of histogram matching. What happens to template matching when you move your hand backwards so that its size is smaller in the image?

CHAPTER 8

Contours

Although algorithms like the Canny edge detector can be used to find the edge pixels that separate different segments in an image, they do not tell you anything about those edges as entities in themselves. The next step is to be able to assemble those edge pixels into contours. By now you have probably come to expect that there is a convenient function in OpenCV that will do exactly this for you, and indeed there is: cvFindContours(). We will start out this chapter with some basics that we will need in order to use this function. Specifically, we will introduce memory storages, which are how OpenCV functions gain access to memory when they need to construct new objects dynamically; then we will learn some basics about sequences, which are the objects used to represent contours generally. With those concepts in hand, we will get into contour finding in some detail. Thereafter we will move on to the many things we can do with contours after they've been computed.

Memory Storage

OpenCV uses an entity called a *memory storage* as its method of handling memory allocation for dynamic objects. Memory storages are linked lists of memory blocks that allow for fast allocation and de-allocation of continuous sets of blocks. OpenCV functions that require the ability to allocate memory as part of their normal functionality will require access to a memory storage from which to get the memory they require (typically this includes any function whose output is of variable size).

Memory storages are handled with the following four routines:

```
CvMemStorage* cvCreateMemStorage(
    int           block_size = 0
);
void cvReleaseMemStorage(
    CvMemStorage** storage
);
void cvClearMemStorage(
    CvMemStorage*  storage
);
void* cvMemStorageAlloc(
    CvMemStorage*  storage,
```

```
  size_t        size
);
```

To create a memory storage, the function `cvCreateMemStorage()` is used. This function takes as an argument a block size, which gives the size of memory blocks inside the store. If this argument is set to 0 then the default block size (64kB) will be used. The function returns a pointer to a new memory store.

The `cvReleaseMemStorage()` function takes a pointer to a valid memory storage and then de-allocates the storage. This is essentially equivalent to the OpenCV de-allocations of images, matrices, and other structures.

You can empty a memory storage by calling `cvClearMemStorage()`, which also takes a pointer to a valid storage. You must be aware of an important feature of this function: it is the only way to release (and thereafter reuse) memory allocated to a memory storage. This might not seem like much, but there will be other routines that delete objects inside of memory storages (we will introduce one of these momentarily) but do *not* return the memory they were using. In short, only `cvClearMemStorage()` (and, of course, `cvReleaseMemStorage()`) recycle the storage memory.* Deletion of any dynamic structure (CvSeq, CvSet, etc.) never returns any memory back to storage (although the structures are able to reuse some memory once taken from the storage for their own data).

You can also allocate your own continuous blocks from a memory store—in a manner analogous to the way `malloc()` allocates memory from the heap—with the function `cvMemStorageAlloc()`. In this case you simply provide a pointer to the storage and the number of bytes you need. The return is a pointer of type `void*` (again, similar to `malloc()`).

Sequences

One kind of object that can be stored inside a memory storage is a *sequence*. Sequences are themselves linked lists of other structures. OpenCV can make sequences out of many different kinds of objects. In this sense you can think of the sequence as something similar to the generic container classes (or container class templates) that exist in various other programming languages. The sequence construct in OpenCV is actually a *deque*, so it is very fast for random access and for additions and deletions from either end but a little slow for adding and deleting objects in the middle.

The sequence structure itself (see Example 8-1) has some important elements that you should be aware of. The first, and one you will use often, is total. This is the total number of points or objects in the sequence. The next four important elements are pointers to other sequences: h_prev, h_next, v_prev, and v_next. These four pointers are part of what are called CV_TREE_NODE_FIELDS; they are used not to indicate elements inside of the sequence but rather to connect different sequences to one another. Other objects in the OpenCV universe also contain these tree node fields. Any such objects can be

---

* Actually, one other function, called `cvRestoreMemStoragePos()`, can restore memory to the storage. But this function is primarily for the library's internal use and is beyond the scope of this book.

assembled, by means of these pointers, into more complicated superstructures such as lists, trees, or other graphs. The variables h_prev and h_next can be used alone to create a simple linked list. The other two, v_prev and v_next, can be used to create more complex topologies that relate nodes to one another. It is by means of these four pointers that cvFindContours() will be able to represent all of the contours it finds in the form of rich structures such as contour trees.

Example 8-1. Internal organization of CvSeq sequence structure

```
typedef struct CvSeq {
    int          flags;            // miscellaneous flags
    int          header_size;      // size of sequence header
    CvSeq*       h_prev;           // previous sequence
    CvSeq*       h_next;           // next sequence
    CvSeq*       v_prev;           // 2nd previous sequence
    CvSeq*       v_next;           // 2nd next sequence
    int          total;            // total number of elements
    int          elem_size;        // size of sequence element in byte
    char*        block_max;        // maximal bound of the last block
    char*        ptr;              // current write pointer
    int          delta_elems;      // how many elements allocated
                                   // when the sequence grows
    CvMemStorage* storage;         // where the sequence is stored
    CvSeqBlock* free_blocks;       // free blocks list
    CvSeqBlock* first;             // pointer to the first sequence block
}
```

Creating a Sequence

As we have alluded to already, sequences can be returned from various OpenCV functions. In addition to this, you can, of course, create sequences yourself. Like many objects in OpenCV, there is an allocator function that will create a sequence for you and return a pointer to the resulting data structure. This function is called cvCreateSeq().

```
    CvSeq* cvCreateSeq(
        int          seq_flags,
        int          header_size,
        int          elem_size,
        CvMemStorage* storage
    );
```

This function requires some additional flags, which will further specify exactly what sort of sequence we are creating. In addition it needs to be told the size of the sequence header itself (which will always be sizeof(CvSeq)*) and the size of the objects that the sequence will contain. Finally, a memory storage is needed from which the sequence can allocate memory when new elements are added to the sequence.

---

* Obviously, there must be some other value to which you can set this argument or it would not exist. This argument is needed because sometimes we want to extend the CvSeq "class". To extend CvSeq, you create your own struct using the CV_SEQUENCE_FIELDS() macro in the structure definition of the new type; note that, when using an extended structure, the size of that structure must be passed. This is a pretty esoteric activity in which only serious gurus are likely to participate.

These flags are of three different categories and can be combined using the bitwise OR operator. The first category determines the type of objects* from which the sequence is to be constructed. Many of these types might look a bit alien to you, and some are primarily for internal use by other OpenCV functions. Also, some of the flags are meaningful only for certain kinds of sequences (e.g., CV_SEQ_FLAG_CLOSED is meaningful only for sequences that in some way represent a polygon).

CV_SEQ_ELTYPE_POINT

> (x,y)

CV_SEQ_ELTYPE_CODE

> Freeman code: 0..7

CV_SEQ_ELTYPE_POINT

> Pointer to a point: &(x,y)

CV_SEQ_ELTYPE_INDEX

> Integer index of a point: #(x,y)

CV_SEQ_ELTYPE_GRAPH_EDGE

> &next_o,&next_d,&vtx_o,&vtx_d

CV_SEQ_ELTYPE_GRAPH_VERTEX

> first_edge, &(x,y)

CV_SEQ_ELTYPE_TRIAN_ATR

> Vertex of the binary tree

CV_SEQ_ELTYPE_CONNECTED_COMP

> Connected component

CV_SEQ_ELTYPE_POINT3D

> (x,y,z)

The second category indicates the nature of the sequence, which can be any of the following.

CV_SEQ_KIND_SET

> A set of objects

CV_SEQ_KIND_CURVE

> A curve defined by the objects

CV_SEQ_KIND_BIN_TREE

> A binary tree of the objects

---

* The types in this first listing are used only rarely. To create a sequence whose elements are tuples of numbers, use CV_32SC2, CV_32FC4, etc. To create a sequence of elements of your own type, simply pass 0 and specify the correct elem_size.

CV_SEQ_KIND_GRAPH

    A graph with the objects as nodes

The third category consists of additional feature flags that indicate some other property of the sequence.

CV_SEQ_FLAG_CLOSED

    Sequence is closed (polygons)

CV_SEQ_FLAG_SIMPLE

    Sequence is simple (polygons)

CV_SEQ_FLAG_CONVEX

    Sequence is convex (polygons)

CV_SEQ_FLAG_HOLE

    Sequence is a hole (polygons)

Deleting a Sequence

```
void cvClearSeq(
  CvSeq* seq
);
```

When you want to delete a sequence, you can use cvClearSeq(), a routine that clears all elements of the sequence. However, this function does not return allocated blocks in the memory store either to the store or to the system; the memory allocated by the sequence can be reused only by the same sequence. If you want to retrieve that memory for some other purpose, you must clear the memory store via cvClearMemStore().

Direct Access to Sequence Elements

Often you will find yourself wanting to directly access a particular member of a sequence. Though there are several ways to do this, the most direct way—and the correct way to access a randomly chosen element (as opposed to one that you happen to know is at the ends)—is to use cvGetSeqElem().

```
char* cvGetSeqElem( seq, index )
```

More often than not, you will have to cast the return pointer to whatever type you know the sequence to be. Here is an example usage of cvGetSeqElem() to print the elements in a sequence of points (such as might be returned by cvFindContours(), which we will get to shortly):

```
for( int i=0; i<seq->total; ++i ) {
  CvPoint* p = (CvPoint*)cvGetSeqElem ( seq, i );
  printf("(%d,%d)\n", p->x, p->y );
}
```

You can also check to see where a particular element is located in a sequence. The function cvSeqElemIdx() does this for you:

```
int cvSeqElemIdx(
  const CvSeq*  seq,
  const void*   element,
  CvSeqBlock** block   = NULL
);
```

This check takes a bit of time, so it is not a particularly efficient thing to do (the time for the search is proportional to the size of the sequence). Note that cvSeqElemIdx() takes as arguments a pointer to your sequence and a pointer to the element for which you are searching.* Optionally, you may also supply a pointer to a sequence memory block pointer. If this is non-NULL, then the location of the block in which the sequence element was found will be returned.

Slices, Copying, and Moving Data

Sequences are copied with cvCloneSeq(), which does a deep copy of a sequence and creates another entirely separate sequence structure.

```
CvSeq* cvCloneSeq(
  const CvSeq*  seq,
  CvMemStorage* storage  = NULL
)
```

This routine is actually just a wrapper for the somewhat more general routine cvSeq Slice(). This latter routine can pull out just a subsection of an array; it can also do either a deep copy or just build a new header to create an alternate "view" on the same data elements.

```
CvSeq* cvSeqSlice(
  const CvSeq*  seq,
  CvSlice       slice,
  CvMemStorage* storage   = NULL,
  int           copy_data = 0
);
```

You will notice that the argument slice to cvSeqSlice() is of type CvSlice. A slice can be defined using either the convenience function cvSlice(a,b) or the macro CV_WHOLE_SEQ. In the former case, only those elements starting at a and continuing through b are included in the copy (b may also be set to CV_WHOLE_SEQ_END_INDEX to indicate the end of the array). The argument copy_data is how we decide if we want a "deep" copy (i.e., if we want the data elements themselves to be copied and for those new copies to be the elements of the new sequence).

Slices can be used to specify elements to remove from a sequence using cvSeqRemoveSlice() or to insert into a sequence using cvSeqInsertSlice().

```
void cvSeqRemoveSlice(
  CvSeq*        seq,
  CvSlice       slice
);
```

---

* Actually, it would be more accurate to say that cvSeqElemIdx() takes the *pointer* being searched for. This is because cvSeqElemIdx() is not searching for an element in the sequence that is equal to *element; rather, it is searching for the element that is at the location given by element.

```
void cvSeqInsertSlice(
    CvSeq*          seq,
    int             before_index,
    const CvArr*    from_arr
);
```

With the introduction of a comparison function, it is also possible to sort or search a (sorted) sequence. The comparison function must have the following prototype:

```
typedef int (*CvCmpFunc)(const void* a, const void* b, void* userdata );
```

Here a and b are pointers to elements of the type being sorted, and userdata is just a pointer to any additional data structure that the caller doing the sorting or searching can provide at the time of execution. The comparison function should return -1 if a is greater than b, +1 if a is less than b, and 0 if a and b are equal.

With such a comparison function defined, a sequence can be sorted by cvSeqSort(). The sequence can also be searched for an element (or for a pointer to an element) elem using cvSeqSearch(). This searching is done in order $O(\log n)$ time if the sequence is already sorted (is_sorted=1). If the sequence is unsorted, then the comparison function is not needed and the search will take $O(n)$ time. On completion, the search will set *elem_idx to the index of the found element (if it was found at all) and return a pointer to that element. If the element was not found, then NULL is returned.

```
void cvSeqSort(
    CvSeq*          seq,
    CvCmpFunc       func,
    void*           userdata    = NULL
);
char* cvSeqSearch(
    CvSeq*          seq,
    const void*     elem,
    CvCmpFunc       func,
    int             is_sorted,
    int*            elem_idx,
    void*           userdata    = NULL
);
```

A sequence can be inverted (reversed) in a single call with the function cvSeqInvert(). This function does not change the data in any way, but it reorganizes the sequence so that the elements appear in the opposite order.

```
void cvSeqInvert(
    CvSeq*          seq
);
```

OpenCV also supports a method of partitioning a sequence* based on a user-supplied criterion via the function cvSeqPartition(). This partitioning uses the same sort of comparison function as described previously but with the expectation that the function will return a nonzero value if the two arguments are equal and zero if they are not (i.e., the opposite convention as is used for searching and sorting).

---

* For more on partitioning, see Hastie, Tibshirani, and Friedman [Hastie01].

```
int cvSeqPartition(
  const CvSeq*   seq,
  CvMemStorage*  storage,
  CvSeq**        labels,
  CvCmpFunc      is_equal,
  void*          userdata
);
```

The partitioning requires a memory storage so that it can allocate memory to express the output of the partitioning. The argument labels should be a pointer to a sequence pointer. When cvSeqPartition() returns, the result will be that labels will now indicate a sequence of integers that have a one-to-one correspondence with the elements of the partitioned sequence seq. The values of these integers will be, starting at 0 and incrementing from there, the "names" of the partitions that the points in seq were to be assigned. The pointer userdata is the usual pointer that is just transparently passed to the comparison function.

In Figure 8-1, a group of 100 points are randomly distributed on 100-by-100 canvas. Then cvSeqPartition() is called on these points, where the comparison function is based on Euclidean distance. The comparison function is set to return true (1) if the distance is less than or equal to 5 and to return false (0) otherwise. The resulting clusters are labeled with their integer ordinal from labels.

Using a Sequence As a Stack

As stated earlier, a sequence in OpenCV is really a linked list. This means, among other things, that it can be accessed efficiently from either end. As a result, it is natural to use a sequence of this kind as a stack when circumstances call for one. The following six functions, when used in conjunction with the CvSeq structure, implement the behavior required to use the sequence as a stack (more properly, a deque, because these functions allow access to both ends of the list).

```
char*  cvSeqPush(
  CvSeq* seq,
  void*  element  = NULL
);
char*  cvSeqPushFront(
  CvSeq* seq,
  void*  element = NULL
);
void   cvSeqPop(
  CvSeq* seq,
  void*  element  = NULL
);
void   cvSeqPopFront(
  CvSeq* seq,
  void*  element  = NULL
);
void cvSeqPushMulti(
  CvSeq* seq,
  void*  elements,
  int    count,
```

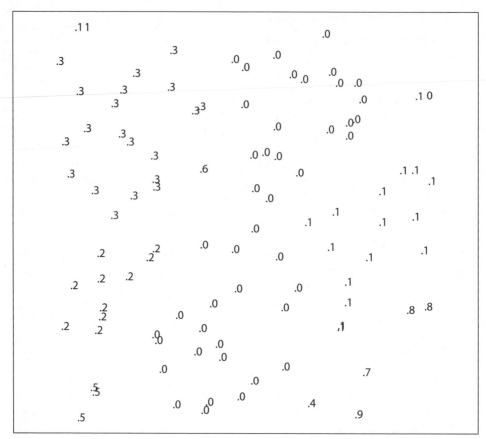

Figure 8-1. A sequence of 100 points on a 100-by-100 canvas, partitioned by distance D ≤ 5

```
    int    in_front = 0
);
void cvSeqPopMulti(
  CvSeq* seq,
  void*  elements,
  int    count,
  int    in_front = 0
);
```

The primary modes of accessing the sequence are cvSeqPush(), cvSeqPushFront(), cvSeqPop(), and cvSeqPopFront(). Because these routines act on the ends of the sequence, all of them operate in $O(l)$ time (i.e., independent of the size of the sequence). The Push functions return an argument to the element pushed into the sequence, and the Pop functions will optionally save the popped element if a pointer is provided to a location where the object can be copied. The cvSeqPushMulti() and cvSeqPopMulti() variants will push or pop several items at a time. Both take a separate argument to distinguish the front from the back; you can set in_front to either CV_FRONT (1) or to CV_BACK (0) and so determine from where you'll be pushing or popping.

Inserting and Removing Elements

```
char* cvSeqInsert(
  CvSeq* seq,
  int    before_index,
  void*  element  = NULL
);
void cvSeqRemove(
  CvSeq* seq,
  int    index
);
```

Objects can be inserted into and removed from the middle of a sequence by using cvSeqInsert() and cvSeqRemove(), respectively, but remember that these are not very fast. On average, they take time proportional to the total size of the sequence.

Sequence Block Size

One function whose purpose may not be obvious at first glance is cvSetSeqBlockSize(). This routine takes as arguments a sequence and a new block size, which is the size of blocks that will be allocated out of the memory store when new elements are needed in the sequence. By making this size big you are less likely to fragment your sequence across disconnected memory blocks; by making it small you are less likely to waste memory. The default value is 1,000 bytes, but this can be changed at any time.*

```
void    cvSetSeqBlockSize(
  CvSeq* seq,
  Int     delta_elems
);
```

Sequence Readers and Sequence Writers

When you are working with sequences and you want the highest performance, there are some special methods for accessing and modifying them that (although they require a bit of special care to use) will let you do what you want to do with a minimum of overhead. These functions make use of special structures to keep track of the state of what they are doing; this allows many actions to be done in sequence and the necessary final bookkeeping to be done only after the last action.

For writing, this control structure is called CvSeqWriter. The writer is initialized with the function cvStartWriteSeq() and is "closed" with cvEndWriteSeq(). While the sequence writing is "open", new elements can be added to the sequence with the macro CV_WRITE_ SEQ(). Notice that the writing is done with a macro and not a function call, which saves even the overhead of entering and exiting that code. Using the writer is faster than using cvSeqPush(); however, not all the sequence headers are updated immediately by this macro, so the added element will be essentially invisible until you are done writing. It will become visible when the structure is completely updated by cvEndWriteSeq().

---

* Effective with the beta 5 version of OpenCV, this size is automatically increased if the sequence becomes big; hence you'll not need to worry about it under normal circumstances.

If necessary, the structure can be brought up-to-date (without actually closing the writer) by calling cvFlushSeqWriter().

```
void    cvStartWriteSeq(
  int            seq_flags,
  int            header_size,
  int            elem_size,
  CvMemStorage*  storage,
  CvSeqWriter*   writer
);
void    cvStartAppendToSeq(
  CvSeq*         seq,
  CvSeqWriter*   writer
);
CvSeq*  cvEndWriteSeq(
  CvSeqWriter*   writer
);
void    cvFlushSeqWriter(
  CvSeqWriter*   writer
);

CV_WRITE_SEQ_ELEM( elem, writer )
CV_WRITE_SEQ_ELEM_VAR( elem_ptr, writer )
```

The arguments to these functions are largely self-explanatory. The seq_flags, header_size, and elem_size arguments to cvStartWriteSeq() are identical to the corresponding arguments to cvCreateSeq(). The function cvStartAppendToSeq() initializes the writer to begin adding new elements to the end of the existing sequence seq. The macro CV_WRITE_SEQ_ELEM() requires the element to be written (e.g., a CvPoint) and a pointer to the writer; a new element is added to the sequence and the element elem is copied into that new element.

Putting these all together into a simple example, we will create a writer and append a hundred random points drawn from a 320-by-240 rectangle to the new sequence.

```
CvSeqWriter writer;
cvStartWriteSeq( CV_32SC2, sizeof(CvSeq), sizeof(CvPoint), storage, &writer );
for( i = 0; i < 100; i++ )
{
    CvPoint pt; pt.x = rand()%320; pt.y = rand()%240;
    CV_WRITE_SEQ_ELEM( pt, writer );
}
CvSeq* seq = cvEndWriteSeq( &writer );
```

For reading, there is a similar set of functions and a few more associated macros.

```
void  cvStartReadSeq(
  const CvSeq* seq,
  CvSeqReader* reader,
  int          reverse   = 0
);
int   cvGetSeqReaderPos(
  CvSeqReader* reader
);
void  cvSetSeqReaderPos(
  CvSeqReader* reader,
```

```
    int         index,
    int         is_relative = 0
);

CV_NEXT_SEQ_ELEM( elem_size, reader )
CV_PREV_SEQ_ELEM( elem_size, reader )
CV_READ_SEQ_ELEM( elem, reader )
CV_REV_READ_SEQ_ELEM( elem, reader )
```

The structure CvSeqReader, which is analogous to CvSeqWriter, is initialized with the function cvStartReadSeq(). The argument reverse allows for the sequence to be read either in "normal" order (reverse=0) or backwards (reverse=1). The function cvGetSeqReaderPos() returns an integer indicating the current location of the reader in the sequence. Finally, cvSetSeqReaderPos() allows the reader to "seek" to an arbitrary location in the sequence. If the argument is_relative is nonzero, then the index will be interpreted as a relative offset to the current reader position. In this case, the index may be positive or negative.

The two macros CV_NEXT_SEQ_ELEM() and CV_PREV_SEQ_ELEM() simply move the reader forward or backward one step in the sequence. They do no error checking and thus cannot help you if you unintentionally step off the end of the sequence. The macros CV_READ_SEQ_ELEM() and CV_REV_READ_SEQ_ELEM() are used to read from the sequence. They will both copy the "current" element at which the reader is pointed onto the variable elem and then step the reader one step (forward or backward, respectively). These latter two macros expect just the name of the variable to be copied to; the address of that variable will be computed inside of the macro.

Sequences and Arrays

You may often find yourself wanting to convert a sequence, usually full of points, into an array.

```
void*  cvCvtSeqToArray(
  const CvSeq* seq,
  void*        elements,
  CvSlice      slice    = CV_WHOLE_SEQ
);
CvSeq* cvMakeSeqHeaderForArray(
  int          seq_type,
  int          header_size,
  int          elem_size,
  void*        elements,
  int          total,
  CvSeq*       seq,
  CvSeqBlock*  block
);
```

The function cvCvtSeqToArray() copies the content of the sequence into a continuous memory array. This means that if you have a sequence of 20 elements of type CvPoint then the function will require a pointer, elements, to enough space for 40 integers. The third (optional) argument is slice, which can be either an object of type CvSlice or the macro CV_WHOLE_SEQ (the latter is the default value). If CV_WHOLE_SEQ is selected, then the entire sequence is copied.

The opposite functionality to cvCvtSeqToArray() is implemented by cvMakeSeqHeaderFor Array(). In this case, you can build a sequence from an existing array of data. The function's first few arguments are identical to those of cvCreateSeq(). In addition to requiring the data (elements) to copy in and the number (total) of data items, you must provide a sequence header (seq) and a sequence memory block structure (block). Sequences created in this way are not exactly the same as sequences created by other methods. In particular, you will not be able to subsequently alter the data in the created sequence.

Contour Finding

We are finally ready to start talking about *contours*. To start with, we should define exactly what a contour is. A contour is a list of points that represent, in one way or another, a curve in an image. This representation can be different depending on the circumstance at hand. There are many ways to represent a curve. Contours are represented in OpenCV by sequences in which every entry in the sequence encodes information about the location of the next point on the curve. We will dig into the details of such sequences in a moment, but for now just keep in mind that a contour is represented in OpenCV by a CvSeq sequence that is, one way or another, a sequence of points.

The function cvFindContours() computes contours from binary images. It can take images created by cvCanny(), which have edge pixels in them, or images created by functions like cvThreshold() or cvAdaptiveThreshold(), in which the edges are implicit as boundaries between positive and negative regions.*

Before getting to the function prototype, it is worth taking a moment to understand exactly what a contour is. Along the way, we will encounter the concept of a contour tree, which is important for understanding how cvFindContours() (retrieval methods derive from Suzuki [Suzuki85]) will communicate its results to us.

Take a moment to look at Figure 8-2, which depicts the functionality of cvFindContours(). The upper part of the figure shows a test image containing a number of white regions (labeled A through E) on a dark background.† The lower portion of the figure depicts the same image along with the contours that will be located by cvFindContours(). Those contours are labeled cX or hX, where "c" stands for "contour", "h" stands for "hole", and "X" is some number. Some of those contours are dashed lines; they represent *exterior boundaries* of the white regions (i.e., nonzero regions). OpenCV and cvFindContours() distinguish between these exterior boundaries and the dotted lines, which you may think of either as *interior boundaries* or as the exterior boundaries of *holes* (i.e., zero regions).

---

* There are some subtle differences between passing edge images and binary images to cvFindContours(); we will discuss those shortly.

† For clarity, the dark areas are depicted as gray in the figure, so simply imagine that this image is thresholded such that the gray areas are set to black before passing to cvFindContours().

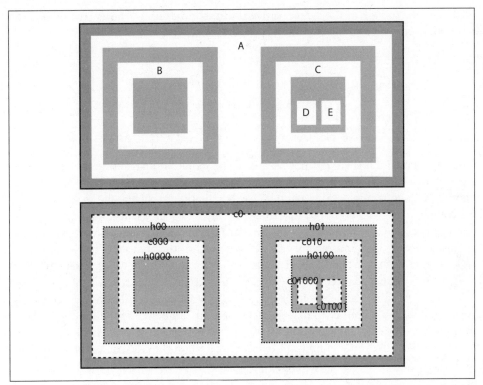

Figure 8-2. A test image (above) passed to cvFindContours() (below): the found contours may be either of two types, exterior "contours" (dashed lines) or "holes" (dotted lines)

The concept of containment here is important in many applications. For this reason, OpenCV can be asked to assemble the found contours into a *contour tree** that encodes the containment relationships in its structure. A contour tree corresponding to this test image would have the contour called c0 at the root node, with the holes h00 and h01 as its children. Those would in turn have as children the contours that they directly contain, and so on.

 It is interesting to note the consequences of using cvFindContours() on an image generated by cvCanny() or a similar edge detector relative to what happens with a binary image such as the test image shown in Figure 8-1. Deep down, cvFindContours() does not really know anything about edge images. This means that, to cvFindContours(), an "edge" is just a very thin "white" area. As a result, for every exterior contour there will be a hole contour that almost exactly coincides with it. This hole is actually just inside of the exterior boundary. You can think of it as the white-to-black transition that marks the interior edge of the edge.

---

* Contour trees first appeared in Reeb [Reeb46] and were further developed by [Bajaj97], [Kreveld97], [Pascucci02], and [Carr04].

Now it's time to look at the cvFindContours() function itself: to clarify exactly how we tell it what we want and how we interpret its response.

```
int cvFindContours(
    IplImage*              img,
    CvMemStorage*          storage,
    CvSeq**                firstContour,
    int                    headerSize = sizeof(CvContour),
    CvContourRetrievalMode mode       = CV_RETR_LIST,
    CvChainApproxMethod    method     = CV_CHAIN_APPROX_SIMPLE
);
```

The first argument is the input image; this image should be an 8-bit single-channel image and will be interpreted as binary (i.e., as if all nonzero pixels are equivalent to one another). When it runs, cvFindContours() will actually use this image as scratch space for computation, so if you need that image for anything later you should make a copy and pass that to cvFindContours(). The next argument, storage, indicates a place where cvFindContours() can find memory in which to record the contours. This storage area should have been allocated with cvCreateMemStorage(), which we covered earlier in the chapter. Next is firstContour, which is a pointer to a CvSeq*. The function cvFind Contours() will allocate this pointer for you, so you shouldn't allocate it yourself. Instead, just pass in a pointer to that pointer so that it can be set by the function. No allocation/de-allocation (new/delete or malloc/free) is needed. It is at this location (i.e., *firstContour) that you will find a pointer to the head of the constructed contour tree.*
The return value of cvFindContours() is the total number of contours found.

```
CvSeq* firstContour = NULL;
cvFindContours( …, &firstContour, … );
```

The headerSize is just telling cvFindContours() more about the objects that it will be allocating; it can be set to sizeof(CvContour) or to sizeof(CvChain) (the latter is used when the approximation method is set to CV_CHAIN_CODE).† Finally, we have the mode and method, which (respectively) further clarify exactly *what* is to be computed and *how* it is to be computed.

The mode variable can be set to any of four options: CV_RETR_EXTERNAL, CV_RETR_LIST, CV_ RETR_CCOMP, or CV_RETR_TREE. The value of mode indicates to cvFindContours() exactly what contours we would like found and how we would like the result presented to us. In particular, the manner in which the tree node variables (h_prev, h_next, v_prev, and v_next) are used to "hook up" the found contours is determined by the value of mode. In Figure 8-3, the resulting topologies are shown for all four possible values of mode. In every case, the structures can be thought of as "levels" which are related by the "horizontal" links (h_next and h_prev), and those levels are separated from one another by the "vertical" links (v_next and v_prev).

---

* As we will see momentarily, contour trees are just one way that cvFindContours() can organize the contours it finds. In any case, they will be organized using the CV_TREE_NODE_FIELDS elements of the contours that we introduced when we first started talking about sequences.

† In fact, headerSize can be an arbitrary number equal to or greater than the values listed.

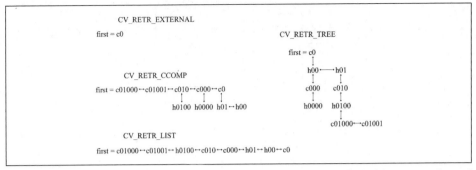

Figure 8-3. The way in which the tree node variables are used to "hook up" all of the contours located by cvFindContours()

CV_RETR_EXTERNAL

> Retrieves only the extreme outer contours. In Figure 8-2, there is only one exterior contour, so Figure 8-3 indicates the first contour points to that outermost sequence and that there are no further connections.

CV_RETR_LIST

> Retrieves all the contours and puts them in the list. Figure 8-3 depicts the list resulting from the test image in Figure 8-2. In this case, eight contours are found and they are all connected to one another by h_prev and h_next (v_prev and v_next are not used here.)

CV_RETR_CCOMP

> Retrieves all the contours and organizes them into a two-level hierarchy, where the top-level boundaries are external boundaries of the components and the second-level boundaries are boundaries of the holes. Referring to Figure 8-3, we can see that there are five exterior boundaries, of which three contain holes. The holes are connected to their corresponding exterior boundaries by v_next and v_prev. The outermost boundary c0 contains two holes. Because v_next can contain only one value, the node can only have one child. All of the holes inside of c0 are connected to one another by the h_prev and h_next pointers.

CV_RETR_TREE

> Retrieves all the contours and reconstructs the full hierarchy of nested contours. In our example (Figures 8-2 and 8-3), this means that the root node is the outermost contour c0. Below c0 is the hole h00, which is connected to the other hole h01 at the same level. Each of those holes in turn has children (the contours c000 and c010, respectively), which are connected to their parents by vertical links. This continues down to the most-interior contours in the image, which become the leaf nodes in the tree.

The next five values pertain to the method (i.e., how the contours are approximated).

CV_CHAIN_CODE

Outputs contours in the Freeman chain code;* all other methods output polygons (sequences of vertices).[†]

CV_CHAIN_APPROX_NONE

Translates all the points from the chain code into points.

CV_CHAIN_APPROX_SIMPLE

Compresses horizontal, vertical, and diagonal segments, leaving only their ending points.

CV_CHAIN_APPROX_TC89_L1 or CV_CHAIN_APPROX_TC89_KCOS

Applies one of the flavors of the Teh-Chin chain approximation algorithm.

CV_LINK_RUNS

Completely different algorithm (from those listed above) that links horizontal segments of 1s; the only retrieval mode allowed by this method is CV_RETR_LIST.

Contours Are Sequences

As you can see, there is a lot to sequences and contours. The good news is that, for our current purpose, we need only a small amount of what's available. When cvFindContours() is called, it will give us a bunch of sequences. These sequences are all of one specific type; as we saw, which particular type depends on the arguments passed to cvFindContours(). Recall that the default mode is CV_RETR_LIST and the default method is CV_CHAIN_APPROX_SIMPLE.

These sequences are sequences of points; more precisely, they are contours—the actual topic of this chapter. The key thing to remember about contours is that they are just a special case of sequences.[‡] In particular, they are sequences of points representing some kind of curve in (image) space. Such a chain of points comes up often enough that we might expect special functions to help us manipulate them. Here is a list of these functions.

```
int cvFindContours(
    CvArr*          image,
    CvMemStorage*   storage,
    CvSeq**         first_contour,
    int             header_size  = sizeof(CvContour),
    int             mode         = CV_RETR_LIST,
    int             method       = CV_CHAIN_APPROX_SIMPLE,
```

---

* Freeman chain codes will be discussed in the section entitled "Contours Are Sequences".

† Here "vertices" means points of type CvPoint. The sequences created by cvFindContours() are the same as those created with cvCreateSeq() with the flag CV_SEQ_ELTYPE_POINT. (That function and flag will be described in detail later in this chapter.)

‡ OK, there's a little more to it than this, but we did not want to be sidetracked by technicalities and so will clarify in this footnote. The type CvContour is not identical to CvSeq. In the way such things are handled in OpenCV, CvContour is, in effect, derived from CvSeq. The CvContour type has a few extra data members, including a color and a CvRect for stashing its bounding box.

```
    CvPoint        offset      = cvPoint(0,0)
);
CvContourScanner cvStartFindContours(
    CvArr*         image,
    CvMemStorage*  storage,
    int            header_size = sizeof(CvContour),
    int            mode        = CV_RETR_LIST,
    int            method      = CV_CHAIN_APPROX_SIMPLE,
    CvPoint        offset      = cvPoint(0,0)
);
CvSeq* cvFindNextContour(
    CvContourScanner scanner
);
void   cvSubstituteContour(
    CvContourScanner scanner,
    CvSeq*             new_contour
);
CvSeq* cvEndFindContour(
    CvContourScanner* scanner
);
CvSeq* cvApproxChains(
    CvSeq*         src_seq,
    CvMemStorage*  storage,
    int            method            = CV_CHAIN_APPROX_SIMPLE,
    double         parameter         = 0,
    int            minimal_perimeter = 0,
    int            recursive         = 0
);
```

First is the cvFindContours() function, which we encountered earlier. The second function, cvStartFindContours(), is closely related to cvFindContours() except that it is used when you want the contours one at a time rather than all packed up into a higher-level structure (in the manner of cvFindContours()). A call to cvStartFindContours() returns a CvSequenceScanner. The scanner contains some simple state information about what has and what has not been read out.* You can then call cvFindNextContour() on the scanner to successively retrieve all of the contours found. A NULL return means that no more contours are left.

cvSubstituteContour() allows the contour to which a scanner is currently pointing to be replaced by some other contour. A useful characteristic of this function is that, if the new_contour argument is set to NULL, then the current contour will be deleted from the chain or tree to which the scanner is pointing (and the appropriate updates will be made to the internals of the affected sequence, so there will be no pointers to nonexistent objects).

Finally, cvEndFindContour() ends the scanning and sets the scanner to a "done" state. Note that the sequence the scanner was scanning is not deleted; in fact, the return value of cvEndFindContour() is a pointer to the first element in the sequence.

---

* It is important not to confuse a CvSequenceScanner with the similarly named CvSeqReader. The latter is for reading the elements in a sequence, whereas the former is used to read from what is, in effect, a list of sequences.

The final function is cvApproxChains(). This function converts Freeman chains to polygonal representations (precisely or with some approximation). We will discuss cvApproxPoly() in detail later in this chapter (see the section "Polygon Approximations").

Freeman Chain Codes

Normally, the contours created by cvFindContours() are sequences of vertices (i.e., points). An alternative representation can be generated by setting the method to CV_CHAIN_CODE. In this case, the resulting contours are stored internally as *Freeman chains* [Freeman67] (Figure 8-4). With a Freeman chain, a polygon is represented as a sequence of steps in one of eight directions; each step is designated by an integer from 0 to 7. Freeman chains have useful applications in recognition and other contexts. When working with Freeman chains, you can read out their contents via two "helper" functions:

```
void cvStartReadChainPoints(
    CvChain*         chain,
    CvChainPtReader* reader
);
CvPoint cvReadChainPoint(
    CvChainPtReader* reader
);
```

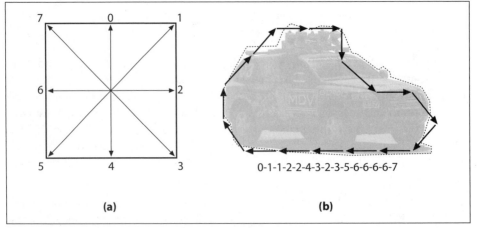

Figure 8-4. Panel a, Freeman chain moves are numbered 0–7; panel b, contour converted to a Freeman chain-code representation starting from the back bumper

The first function takes a chain as its argument and the second function is a chain reader. The CvChain structure is a form of CvSeq.* Just as CvContourScanner iterates through different contours, CvChainPtReader iterates through a single contour represented by a chain. In this respect, CvChainPtReader is similar to the more general CvSeqReader, and

---

* You may recall a previous mention of "extensions" of the CvSeq structure; CvChain is such an extension. It is defined using the CV_SEQUENCE_FIELDS() macro and has one extra element in it, a CvPoint representing the origin. You can think of CvChain as being "derived from" CvSeq. In this sense, even though the return type of cvApproxChains() is indicated as CvSeq*, it is really a pointer to a chain and is not a normal sequence.

cvStartReadChainPoints plays the role of cvStartReadSeq. As you might expect, CvChain-PtReader returns NULL when there's nothing left to read.

Drawing Contours

One of our most basic tasks is drawing a contour on the screen. For this we have cvDrawContours():

```
void  cvDrawContours(
    CvArr*    img,
    CvSeq*    contour,
    CvScalar  external_color,
    CvScalar  hole_color,
    int       max_level,
    int       thickness    = 1,
    int       line_type    = 8,
    CvPoint   offset        = cvPoint(0,0)
);
```

The first argument is simple: it is the image on which to draw the contours. The next argument, contour, is not quite as simple as it looks. In particular, it is really treated as the root node of a contour tree. Other arguments (primarily max_level) will determine what is to be done with the rest of the tree. The next argument is pretty straightforward: the color with which to draw the contour. But what about hole_color? Recall that OpenCV distinguishes between contours that are exterior contours and those that are hole contours (the dashed and dotted lines, respectively, in Figure 8-2). When drawing either a single contour or all contours in a tree, any contour that is marked as a "hole" will be drawn in this alternative color.

The max_level tells cvDrawContours() how to handle any contours that might be attached to contour by means of the node tree variables. This argument can be set to indicate the maximum depth to traverse in the drawing. Thus, max_level=0 means that all the contours on the same level as the input level (more exactly, the input contour and the contours next to it) are drawn, max_level=1 means that all the contours on the same level as the input and their children are drawn, and so forth. If the contours in question were produced by cvFindContours() using either CV_RETR_CCOMP or CV_RETR_TREE mode, then the additional idiom of negative values for max_level is also supported. In this case, max_level=-1 is interpreted to mean that only the input contour will be drawn, max_level=-2 means that the input contour and its *direct* children will the drawn, and so on. The sample code in .../opencv/samples/c/contours.c illustrates this point.

The parameters thickness and line_type have their usual meanings.[*] Finally, we can give an offset to the draw routine so that the contour will be drawn elsewhere than at the absolute coordinates by which it was defined. This feature is particularly useful when the contour has already been converted to center-of-mass or other local coordinates.

---

[*] In particular, thickness=-1 (aka CV_FILLED) is useful for converting the contour tree (or an individual contour) back to the black-and-white image from which it was extracted. This feature, together with the offset parameter, can be used to do some quite complex things with contours: intersect and merge contours, test points quickly against the contours, perform morphological operations (erode/dilate), etc.

More specifically, offset would be helpful if we ran cvFindContours() one or more times in different image subregions (ROIs) and thereafter wanted to display all the results within the original large image. Conversely, we could use offset if we'd extracted a contour from a large image and then wanted to form a small mask for this contour.

A Contour Example

Our Example 8-2 is drawn from the OpenCV package. Here we create a window with an image in it. A trackbar sets a simple threshold, and the contours in the thresholded image are drawn. The image is updated whenever the trackbar is adjusted.

Example 8-2. Finding contours based on a trackbar's location; the contours are updated whenever the trackbar is moved

```
#include <cv.h>
#include <highgui.h>

IplImage*      g_image    = NULL;
IplImage*      g_gray     = NULL;
int            g_thresh   = 100;
CvMemStorage*  g_storage  = NULL;

void on_trackbar(int) {
  if( g_storage==NULL ) {
    g_gray = cvCreateImage( cvGetSize(g_image), 8, 1 );
    g_storage = cvCreateMemStorage(0);
  } else {
    cvClearMemStorage( g_storage );
  }
  CvSeq* contours = 0;
  cvCvtColor( g_image, g_gray, CV_BGR2GRAY );
  cvThreshold( g_gray, g_gray, g_thresh, 255, CV_THRESH_BINARY );
  cvFindContours( g_gray, g_storage, &contours );
  cvZero( g_gray );
  if( contours )
    cvDrawContours(
      g_gray,
      contours,
      cvScalarAll(255),
      cvScalarAll(255),
      100
    );
  cvShowImage( "Contours", g_gray );
}

int main( int argc, char** argv )
{
  if( argc != 2 || !(g_image = cvLoadImage(argv[1])) )
  return -1;
  cvNamedWindow( "Contours", 1 );
  cvCreateTrackbar(
    "Threshold",
    "Contours",
    &g_thresh,
```

```
    255,
    on_trackbar
  );
  on_trackbar(0);
  cvWaitKey();
  return 0;
}
```

Here, everything of interest to us is happening inside of the function on_trackbar(). If the global variable g_storage is still at its (NULL) initial value, then cvCreateMemStorage(0) creates the memory storage and g_gray is initialized to a blank image the same size as g_image but with only a single channel. If g_storage is non-NULL, then we've been here before and thus need only empty the storage so it can be reused. On the next line, a CvSeq* pointer is created; it is used to point to the sequence that we will create via cvFindContours().

Next, the image g_image is converted to grayscale and thresholded such that only those pixels brighter than g_thresh are retained as nonzero. The cvFindContours() function is then called on this thresholded image. If any contours were found (i.e., if contours is non-NULL), then cvDrawContours() is called and the contours are drawn (in white) onto the grayscale image. Finally, that image is displayed and the structures we allocated at the beginning of the callback are released.

Another Contour Example

In this example, we find contours on an input image and then proceed to draw them one by one. This is a good example to play with yourself and see what effects result from changing either the contour finding mode (CV_RETR_LIST in the code) or the max_depth that is used to draw the contours (0 in the code). If you set max_depth to a larger number, notice that the example code steps through the contours returned by cvFindContours() by means of h_next. Thus, for some topologies (CV_RETR_TREE, CV_RETR_CCOMP, etc.), you may see the same contour more than once as you step through. See Example 8-3.

Example 8-3. Finding and drawing contours on an input image

```
int main(int argc, char* argv[]) {

  cvNamedWindow( argv[0], 1 );

  IplImage* img_8uc1 = cvLoadImage( argv[1], CV_LOAD_IMAGE_GRAYSCALE );
  IplImage* img_edge = cvCreateImage( cvGetSize(img_8uc1), 8, 1 );
  IplImage* img_8uc3 = cvCreateImage( cvGetSize(img_8uc1), 8, 3 );

  cvThreshold( img_8uc1, img_edge, 128, 255, CV_THRESH_BINARY );

  CvMemStorage* storage = cvCreateMemStorage();
  CvSeq* first_contour  = NULL;
```

Example 8-3. Finding and drawing contours on an input image (continued)

```
int Nc = cvFindContours(
  img_edge,
  storage,
  &first_contour,
  sizeof(CvContour),
  CV_RETR_LIST  // Try all four values and see what happens
);

int n=0;
printf( "Total Contours Detected: %d\n", Nc );
for( CvSeq* c=first_contour; c!=NULL; c=c->h_next ) {
  cvCvtColor( img_8uc1, img_8uc3, CV_GRAY2BGR );
  cvDrawContours(
    img_8uc3,
    c,
    CVX_RED,
    CVX_BLUE,
    0,          // Try different values of max_level, and see what happens
    2,
    8
  );
  printf("Contour #%d\n", n );
  cvShowImage( argv[0], img_8uc3 );
  printf(" %d elements:\n", c->total );
  for( int i=0; i<c->total; ++i ) {
  CvPoint* p = CV_GET_SEQ_ELEM( CvPoint, c, i );
    printf("    (%d,%d)\n", p->x, p->y );
  }
  cvWaitKey(0);
  n++;
}

printf("Finished all contours.\n");
cvCvtColor( img_8uc1, img_8uc3, CV_GRAY2BGR );
cvShowImage( argv[0], img_8uc3 );
cvWaitKey(0);

cvDestroyWindow( argv[0] );

cvReleaseImage( &img_8uc1 );
cvReleaseImage( &img_8uc3 );
cvReleaseImage( &img_edge );

return 0;
}
```

More to Do with Contours

When analyzing an image, there are many different things we might want to do with contours. After all, most contours are—or are candidates to be—things that we are interested in identifying or manipulating. The various relevant tasks include characterizing the contours in various ways, simplifying or approximating them, matching them to templates, and so on.

In this section we will examine some of these common tasks and visit the various functions built into OpenCV that will either do these things for us or at least make it easier for us to perform our own tasks.

Polygon Approximations

If we are drawing a contour or are engaged in shape analysis, it is common to approximate a contour representing a polygon with another contour having fewer vertices. There are many different ways to do this; OpenCV offers an implementation of one of them.* The routine cvApproxPoly() is an implementation of this algorithm that will act on a sequence of contours:

```
CvSeq*  cvApproxPoly(
    const void*     src_seq,
    int             header_size,
    CvMemStorage*   storage,
    int             method,
    double          parameter,
    int             recursive = 0
);
```

We can pass a list or a tree sequence containing contours to cvApproxPoly(), which will then act on all of the contained contours. The return value of cvApproxPoly() is actually just the first contour, but you can move to the others by using the h_next (and v_next, as appropriate) elements of the returned sequence.

Because cvApproxPoly() needs to create the objects that it will return a pointer to, it requires the usual CvMemStorage* pointer and header size (which, as usual, is set to sizeof(CvContour)).

The method argument is always set to CV_POLY_APPROX_DP (though other algorithms could be selected if they become available). The next two arguments are specific to the method (of which, for now, there is but one). The parameter argument is the precision parameter for the algorithm. To understand how this parameter works, we must take a moment to review the actual algorithm.† The last argument indicates whether the algorithm should (as mentioned previously) be applied to every contour that can be reached via the h_next and v_next pointers. If this argument is 0, then only the contour directly pointed to by src_seq will be approximated.

So here is the promised explanation of how the algorithm works. In Figure 8-5, starting with a contour (panel b), the algorithm begins by picking two extremal points and connecting them with a line (panel c). Then the original polygon is searched to find the point farthest from the line just drawn, and that point is added to the approximation.

---

* For aficionados, the method used by OpenCV is the Douglas-Peucker (DP) approximation [Douglas73]. Other popular methods are the Rosenfeld-Johnson [Rosenfeld73] and Teh-Chin [Teh89] algorithms.

† If that's too much trouble, then just set this parameter to a small fraction of the total curve length.

The process is iterated (panel d), adding the next most distant point to the accumulated approximation, until all of the points are less than the distance indicated by the precision parameter (panel f). This means that good candidates for the parameter are some fraction of the contour's length, or of the length of its bounding box, or a similar measure of the contour's overall size.

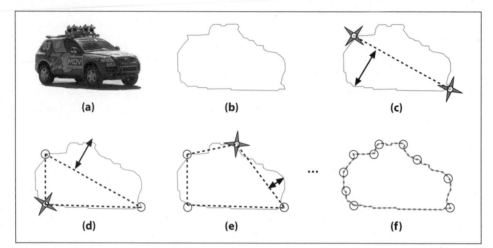

Figure 8-5. Visualization of the DP algorithm used by cvApproxPoly(): the original image (a) is approximated by a contour (b) and then, starting from the first two maximally separated vertices (c), the additional vertices are iteratively selected from that contour (d)–(f)

Closely related to the approximation just described is the process of finding dominant points. A *dominant point* is defined as a point that has more information about the curve than do other points. Dominant points are used in many of the same contexts as polygon approximations. The routine cvFindDominantPoints() implements what is known as the IPAN* [Chetverikov99] algorithm.

```
CvSeq* cvFindDominantPoints(
    CvSeq*          contour,
    CvMemStorage*   storage,
    int             method      = CV_DOMINANT_IPAN,
    double          parameter1  = 0,
    double          parameter2  = 0,
    double          parameter3  = 0,
    double          parameter4  = 0
);
```

In essence, the IPAN algorithm works by scanning along the contour and trying to construct triangles on the interior of the curve using the available vertices. That triangle is characterized by its size and the opening angle (see Figure 8-6). The points with large opening angles are retained provided that their angles are smaller than a specified global threshold *and* smaller than their neighbors.

---

* For "Image and Pattern Analysis Group," Hungarian Academy of Sciences. The algorithm is often referred to as "IPAN99" because it was first published in 1999.

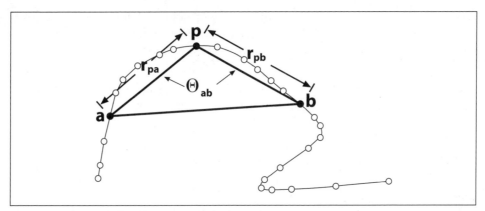

Figure 8-6. The IPAN algorithm uses triangle abp to characterize point p

The routine cvFindDominantPoints() takes the usual CvSeq* and CvMemStorage* arguments. It also requires a method, which (as with cvApproxPoly()) can take only one argument at this time: CV_DOMINANT_IPAN.

The next four arguments are: a minimal distance d_{min}, a maximal distance d_{max}, a *neighborhood distance* d_n, and a maximum angle θ_{max}. As shown in Figure 8-6, the algorithm first constructs all triangles for which r_{pa} and r_{pb} fall between d_{min} and d_{max} and for which $\theta_{ab} < \theta_{max}$. This is followed by a second pass in which only those points p with the smallest associated value of θ_{ab} in the neighborhood d_n are retained (the value of d_n should never exceed d_{max}). Typical values for d_{min}, d_{max}, d_n, and θ_{max} are 7, 9, 9, and 150 (the last argument is an angle and is measured in degrees).

Summary Characteristics

Another task that one often faces with contours is computing their various *summary characteristics*. These might include length or some other form of size measure of the overall contour. Other useful characteristics are the *contour moments*, which can be used to summarize the gross shape characteristics of a contour (we will address these in the next section).

Length

The subroutine cvContourPerimeter() will take a contour and return its length. In fact, this function is actually a macro for the somewhat more general cvArcLength().

```
double  cvArcLength(
   const void* curve,
   CvSlice     slice    = CV_WHOLE_SEQ,
   int         is_closed = -1
);
#define cvContourPerimeter( contour )            \
   cvArcLength( contour, CV_WHOLE_SEQ, 1 )
```

The first argument of cvArcLength() is the contour itself, whose form may be either a sequence of points (CvContour* or CvSeq*) or an *n*-by-2 array of points. Next are the slice argument and a Boolean indicating whether the contour should be treated as closed (i.e., whether the last point should be treated as connected to the first). The slice argument allows us to select only some subset of the points in the curve.*

Closely related to cvArcLegth() is cvContourArea(), which (as its name suggests) computes the area of a contour. It takes the contour as an argument and the same slice argument as cvArcLength().

```
double  cvContourArea(
    const CvArr* contour,
    CvSlice      slice = CV_WHOLE_SEQ
);
```

Bounding boxes

Of course the length and area are simple characterizations of a contour. The next level of detail might be to summarize them with a bounding box or bounding circle or ellipse. There are two ways to do the former, and there is a single method for doing each of the latter.

```
CvRect  cvBoundingRect(
    CvArr* points,
    int    update         = 0
);
CvBox2D  cvMinAreaRect2(
    const CvArr*  points,
    CvMemStorage* storage = NULL
);
```

The simplest technique is to call cvBoundingRect(); it will return a CvRect that bounds the contour. The points used for the first argument can be either a contour (CvContour*) or an *n*-by-1, two-channel matrix (CvMat*) containing the points in the sequence. To understand the second argument, update, we must harken back to footnote 8. Remember that CvContour is not exactly the same as CvSeq; it does everything CvSeq does but also a little bit more. One of those CvContour extras is a CvRect member for referring to its own bounding box. If you call cvBoundingRect() with update set to 0 then you will just get the contents of that data member; but if you call with update set to 1, the bounding box will be computed (and the associated data member will also be updated).

One problem with the bounding rectangle from cvBoundingRect() is that it is a CvRect and so can only represent a rectangle whose sides are oriented horizontally and vertically. In contrast, the routine cvMinAreaRect2() returns the minimal rectangle that will bound your contour, and this rectangle may be inclined relative to the vertical; see Figure 8-7. The arguments are otherwise similar to cvBoundingRect(). The OpenCV data type CvBox2D is just what is needed to represent such a rectangle.

---

* Almost always, the default value CV_WHOLE_SEQ is used. The structure CvSlice contains only two elements: start_index and end_index. You can create your own slice to put here using the helper constructor function cvSlice(int start, int end). Note that CV_WHOLE_SEQ is just shorthand for a slice starting at 0 and ending at some very large number.

```
typedef struct CvBox2D  {
  CvPoint2D32f center;
  CvSize2D32f  size;
  float        angle;
} CvBox2D;
```

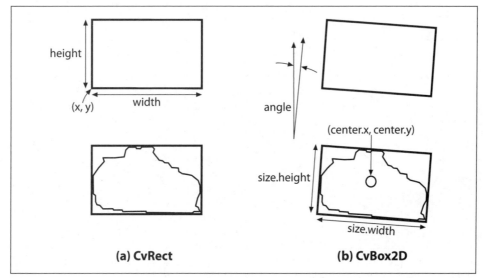

Figure 8-7. CvRect can represent only upright rectangles, but CvBox2D can handle rectangles of any inclination

Enclosing circles and ellipses

Next we have cvMinEnclosingCircle().* This routine works pretty much the same as the bounding box routines, with the same flexibility of being able to set points to be either a sequence or an array of two-dimensional points.

```
int  cvMinEnclosingCircle(
    const CvArr*  points,
    CvPoint2D32f* center,
    float*        radius
);
```

There is no special structure in OpenCV for representing circles, so we need to pass in pointers for a center point and a floating-point variable radius that can be used by cvMinEnclosingCircle() to report the results of its computations.

As with the minimal enclosing circle, OpenCV also provides a method for fitting an ellipse to a set of points:

```
CvBox2D cvFitEllipse2(
    const CvArr* points
);
```

---

* For more information on the inner workings of these fitting techniques, see Fitzgibbon and Fisher [Fitzgibbon95] and Zhang [Zhang96].

The subtle difference between cvMinEnclosingCircle() and cvFitEllipse2() is that the former simply computes the smallest circle that completely encloses the given contour, whereas the latter uses a fitting function and returns the ellipse that is the best approximation to the contour. This means that not all points in the contour will be enclosed in the ellipse returned by cvFitEllipse2(). The fitting is done using a least-squares fitness function.

The results of the fit are returned in a CvBox2D structure. The indicated box exactly encloses the ellipse. See Figure 8-8.

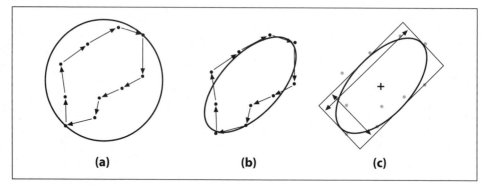

Figure 8-8. Ten-point contour with the minimal enclosing circle superimposed (a) and with the best-fitting ellipsoid (b); a box (c) is used by OpenCV to represent that ellipsoid

Geometry

When dealing with bounding boxes and other summary representations of polygon contours, it is often desirable to perform such simple geometrical checks as polygon overlap or a fast overlap check between bounding boxes. OpenCV provides a small but handy set of routines for this sort of geometrical checking.

```
CvRect cvMaxRect(
    const CvRect* rect1,
    const CvRect* rect2
);
void cvBoxPoints(
    CvBox2D        box,
    CvPoint2D32f   pt[4]
);
CvSeq* cvPointSeqFromMat(
    int            seq_kind,
    const CvArr*   mat,
    CvContour*     contour_header,
    CvSeqBlock*    block
);
double cvPointPolygonTest(
    const CvArr*   contour,
    CvPoint2D32f   pt,
    int            measure_dist
);
```

The first of these functions, cvMaxRect(), computes a new rectangle from two input rectangles. The new rectangle is the smallest rectangle that will bound both inputs.

Next, the utility function cvBoxPoints() simply computes the points at the corners of a CvBox2D structure. You could do this yourself with a bit of trigonometry, but you would soon grow tired of that. This function does this simple pencil pushing for you.

The second utility function, cvPointSeqFromMat(), generates a sequence structure from a matrix. This is useful when you want to use a contour function that does not also take matrix arguments. The input to cvPointSeqFromMat() first requires you to indicate what sort of sequence you would like. The variable seq_kind may be set to any of the following: zero (0), indicating just a point set; CV_SEQ_KIND_CURVE, indicating that the sequence is a curve; or CV_SEQ_KIND_CURVE | CV_SEQ_FLAG_CLOSED, indicating that the sequence is a closed curve. Next you pass in the array of points, which should be an *n*-by-1 array of points. The points should be of type CV_32SC2 or CV_32FC2 (i.e., they should be single-column, two-channel arrays). The next two arguments are pointers to values that will be computed by cvPointSeqFromMat(), and contour_header is a contour structure that you should already have created but whose internals will be filled by the function call. This is similarly the case for block, which will also be filled for you.* Finally the return value is a CvSeq* pointer, which actually points to the very contour structure you passed in yourself. This is a convenience, because you will generally need the sequence address when calling the sequence-oriented functions that motivated you to perform this conversion in the first place.

The last geometrical tool-kit function to be presented here is cvPointPolygonTest(), a function that allows you to test whether a point is inside a polygon (indicated by a sequence). In particular, if the argument measure_dist is nonzero then the function returns the distance to the nearest contour edge; that distance is 0 if the point is inside the contour and positive if the point is outside. If the measure_dist argument is 0 then the return values are simply + 1, – 1, or 0 depending on whether the point is inside, outside, or on an edge (or vertex), respectively. The contour itself can be either a sequence or an *n*-by-1 two-channel matrix of points.

Matching Contours

Now that we have a pretty good idea of what a contour is and of how to work with contours as objects in OpenCV, we would like to take a moment to understand how to use them for some practical purposes. The most common task associated with contours is matching them in some way with one another. We may have two computed contours that we'd like to compare or a computed contour and some abstract template with which we'd like to compare our contour. We will discuss both of these cases.

---

* You will probably never use block. It exists because no actual memory is copied when you call cvPoint SeqFromMat(); instead, a "virtual" memory block is created that actually points to the matrix you yourself provided. The variable block is used to create a reference to that memory of the kind expected by internal sequence or contour calculations.

Moments

One of the simplest ways to compare two contours is to compute *contour moments*. This is a good time for a short digression into precisely what a moment is. Loosely speaking, a moment is a gross characteristic of the contour computed by integrating (or summing, if you like) over all of the pixels of the contour. In general, we define the (p, q) moment of a contour as

$$m_{p,q} = \sum_{i=1}^{n} I(x,y) x^p y^q$$

Here p is the x-order and q is the y-order, whereby *order* means the power to which the corresponding component is taken in the sum just displayed. The summation is over all of the pixels of the contour boundary (denoted by n in the equation). It then follows immediately that if p and q are both equal to 0, then the m_{00} moment is actually just the length in pixels of the contour.*

The function that computes these moments for us is

```
void cvContoursMoments(
    CvSeq*     contour,
    CvMoments* moments
)
```

The first argument is the contour we are interested in and the second is a pointer to a structure that we must allocate to hold the return data. The CvMoments structure is defined as follows:

```
typedef struct CvMoments {

    // spatial moments
    double  m00, m10, m01, m20, m11, m02, m30, m21, m12, m03;

    // central moments
    double  mu20, mu11, mu02, mu30, mu21, mu12, mu03;

    // m00 != 0 ? 1/sqrt(m00) : 0
    double  inv_sqrt_m00;

} CvMoments;
```

The cvContoursMoments() function uses only the m00, m01, . . ., m03 elements; the elements with names mu00, . . . are used by other routines.

When working with the CvMoments structure, there is a friendly helper function that will return any particular moment out of the structure:

---

* Mathematical purists might object that m_{00} should be not the contour's length but rather its area. But because we are looking here at a contour and not a filled polygon, the length and the area are actually the same in a discrete pixel space (at least for the relevant distance measure in our pixel space). There are also functions for computing moments of IplImage images; in that case, m_{00} would actually be the area of nonzero pixels.

```
double cvGetSpatialMoment(
    CvMoments* moments,
    Int        x_order,
    int        y_order
);
```

A single call to cvContoursMoments() will instigate computation of all the moments through third order (i.e., m_{30} and m_{03} will be computed, as will m_{21} and m_{12}, but m_{22} will not be).

More About Moments

The moment computation just described gives some rudimentary characteristics of a contour that can be used to compare two contours. However, the moments resulting from that computation are not the best parameters for such comparisons in most practical cases. In particular, one would often like to use *normalized moments* (so that objects of the same shape but dissimilar sizes give similar values). Similarly, the simple moments of the previous section depend on the coordinate system chosen, which means that objects are not matched correctly if they are rotated.

OpenCV provides routines to compute normalized moments as well as *Hu invariant moments* [Hu62]. The CvMoments structure can be computed either with cvMoments or with cvContourMoments. Moreover, cvContourMoments is now just an alias for cvMoments.

A useful trick is to use cvDrawContours() to "paint" an image of the contour and then call one of the moment functions on the resulting drawing. This allows you to control whether or not the contour is filled.

Here are the four functions at your disposal:

```
void cvMoments(
    const CvArr* image,
    CvMoments*   moments,
    int          isBinary = 0
)
double cvGetCentralMoment(
    CvMoments* moments,
    int        x_order,
    int        y_order
)
double cvGetNormalizedCentralMoment(
    CvMoments* moments,
    int        x_order,
    int        y_order
);
void cvGetHuMoments(
    CvMoments*   moments,
    CvHuMoments* HuMoments
);
```

The first function is essentially analogous to cvContoursMoments() except that it takes an image (instead of a contour) and has one extra argument. That extra argument, if set to CV_TRUE, tells cvMoments() to treat all pixels as either 1 or 0, where 1 is assigned to any pixel with a nonzero value. When this function is called, all of the moments—including the central moments (see next paragraph)—are computed at once.

A *central moment* is basically the same as the moments just described except that the values of x and y used in the formulas are displaced by the mean values:

$$\mu_{p,q} = \sum_{i=0}^{n} I(x,y)(x - x_{avg})^p (y - y_{avg})^q$$

where $x_{avg} = m_{10}/m_{00}$ and $y_{avg} = m_{01}/m_{00}$.

The *normalized moments* are the same as the central moments except that they are all divided by an appropriate power of m_{00}:*

$$\eta_{p,q} = \frac{\mu_{p,q}}{m_{00}^{(p+q)/2+1}}$$

Finally, the *Hu invariant moments* are linear combinations of the central moments. The idea here is that, by combining the different normalized central moments, it is possible to create invariant functions representing different aspects of the image in a way that is invariant to scale, rotation, and (for all but the one called h_1) reflection.

The cvGetHuMoments() function computes the Hu moments from the central moments. For the sake of completeness, we show here the actual definitions of the Hu moments:

$$h_1 = \eta_{20} + \eta_{02}$$
$$h_2 = (\eta_{20} - \eta_{02})^2 + 4\eta_{11}^2$$
$$h_3 = (\eta_{30} - 3\eta_{12})^2 + (3\eta_{21} - \eta_{03})^2$$
$$h_4 = (\eta_{30} + \eta_{12})^2 + (\eta_{21} + \eta_{03})^2$$
$$h_5 = (\eta_{30} - 3\eta_{12})(\eta_{30} + \eta_{12})((\eta_{30} + \eta_{12})^2 - 3(\eta_{21} + \eta_{03})^2)$$
$$\quad + (3\eta_{21} - \eta_{03})(\eta_{21} + \eta_{03})(3(\eta_{30} + \eta_{12})^2 - (\eta_{21} + \eta_{03})^2)$$
$$h_6 = (\eta_{20} - \eta_{02})((\eta_{30} + \eta_{12})^2 - (\eta_{21} + \eta_{03})^2) + 4\eta_{11}(\eta_{30} + \eta_{12})(\eta_{21} + \eta_{03})$$
$$h_7 = (3\eta_{21} - \eta_{03})(\eta_{21} + \eta_{03})(3(\eta_{30} + \eta_{12})^2 - (\eta_{21} + \eta_{03})^2)$$
$$\quad - (\eta_{30} - 3\eta_{12})(\eta_{21} + \eta_{03})(3(\eta_{30} + \eta_{12})^2 - (\eta_{21} + \eta_{03})^2)$$

Looking at Figure 8-9 and Table 8-1, we can gain a sense of how the Hu moments behave. Observe first that the moments tend to be smaller as we move to higher orders. This should be no surprise in that, by their definition, higher Hu moments have more

---

* Here, "appropriate" means that the moment is scaled by some power of m_{00} such that the resulting normalized moment is independent of the overall scale of the object. In the same sense that an average is the sum of N numbers divided by N, the higher-order moments also require a corresponding normalization factor.

powers of various normalized factors. Since each of those factors is less than 1, the products of more and more of them will tend to be smaller numbers.

Figure 8-9. Images of five simple characters; looking at their Hu moments yields some intuition concerning their behavior

Table 8-1. Values of the Hu moments for the five simple characters of Figure 8-9

	h_1	h_2	h_3	h_4	h_5	h_6	h_7
A	2.837e−1	1.961e−3	1.484e−2	2.265e−4	−4.152e−7	1.003e−5	−7.941e−9
I	4.578e−1	1.820e−1	0.000	0.000	0.000	0.000	0.000
O	3.791e−1	2.623e−4	4.501e−7	5.858e−7	1.529e−13	7.775e−9	−2.591e−13
M	2.465e−1	4.775e−4	7.263e−5	2.617e−6	−3.607e−11	−5.718e−8	−7.218e−24
F	3.186e−1	2.914e−2	9.397e−3	8.221e−4	3.872e−8	2.019e−5	2.285e−6

Other factors of particular interest are that the "I", which is symmetric under 180 degree rotations and reflection, has a value of exactly 0 for h_3 through h_7; and that the "O", which has similar symmetries, has all nonzero moments. We leave it to the reader to look at the figures, compare the various moments, and so build a basic intuition for what those moments represent.

Matching with Hu Moments

```
double  cvMatchShapes(
  const void* object1,
  const void* object2,
  int         method,
  double      parameter  = 0
);
```

Naturally, with Hu moments we'd like to compare two objects and determine whether they are similar. Of course, there are many possible definitions of "similar". To make this process somewhat easier, the OpenCV function cvMatchShapes() allows us to simply provide two objects and have their moments computed and compared according to a criterion that we provide.

These objects can be either grayscale images or contours. If you provide images, cvMatchShapes() will compute the moments for you before proceeding with the comparison. The method used in cvMatchShapes() is one of the three listed in Table 8-2.

Table 8-2. Matching methods used by cvMatchShapes()

Value of method	cvMatchShapes() return value
CV_CONTOURS_MATCH_I1	$I_1(A,B) = \sum\limits_{i=1}^{7} \left\| \dfrac{1}{m_i^A} - \dfrac{1}{m_i^B} \right\|$
CV_CONTOURS_MATCH_I2	$I_2(A,B) = \sum\limits_{i=1}^{7} \left\| m_i^A - m_i^B \right\|$
CV_CONTOURS_MATCH_I3	$I_3(A,B) = \sum\limits_{i=1}^{7} \dfrac{\left\| m_i^A - m_i^B \right\|}{m_i^A}$

In the table, m_i^A and m_i^B are defined as:

$$m_i^A = \mathrm{sign}(h_i^A) \cdot \log\left|h_i^A\right|$$
$$m_i^B = \mathrm{sign}(h_i^B) \cdot \log\left|h_i^B\right|$$

where h_i^A and h_i^B are the Hu moments of *A* and *B*, respectively.

Each of the three defined constants in Table 8-2 has a different meaning in terms of how the comparison metric is computed. This metric determines the value ultimately returned by cvMatchShapes(). The final parameter argument is not currently used, so we can safely leave it at the default value of 0.

Hierarchical Matching

We'd often like to match two contours and come up with a similarity measure that takes into account the entire structure of the contours being matched. Methods using summary parameters (such as moments) are fairly quick, but there is only so much information they can capture.

For a more accurate measure of similarity, it will be useful first to consider a structure known as a *contour tree*. Contour trees should not be confused with the hierarchical representations of contours that are returned by such functions as cvFindContours(). Instead, they are hierarchical representations of the shape of one particular contour.

Understanding a contour tree will be easier if we first understand how it is constructed. Constructing a contour tree from a contour works from bottom (leaf nodes) to top (the root node). The process begins by searching the perimeter of the shape for triangular protrusions or indentations (every point on the contour that is not exactly collinear with its neighbors). Each such triangle is replaced with the line connecting its two nonadjacent points on the curve;thus, in effect the triangle is either cut off (e.g., triangle D in Figure 8-10), or filled in (triangle C). Each such alteration reduces the contour's number of vertices by 1 and creates a new node in the tree. If such a triangle has original edges on two of its sides, then it is a leaf in the resulting tree; if one of its sides is part of an existing triangle, then it is a parent of that triangle. Iteration of this process ultimately reduces the shape to a quadrangle, which is then cut in half; both resulting triangles are children of the root node.

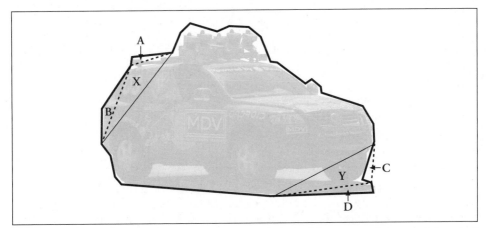

Figure 8-10. Constructing a contour tree: in the first round, the contour around the car produces leaf nodes A, B, C, and D; in the second round, X and Y are produced (X is the parent of A and B, and Y is the parent of C and D)

The resulting binary tree (Figure 8-11) ultimately encodes the shape information about the original contour. Each node is annotated with information about the triangle to which it is associated (information such as the size of the triangle and whether it was created by cutting off or filling in).

Once these trees are constructed, they can be used to effectively compare two contours.* This process begins by attempting to define correspondences between nodes in the two trees and then comparing the characteristics of the corresponding nodes. The end result is a similarity measure between the two trees.

In practice, we need to understand very little about this process. OpenCV provides us with routines to generate contour trees automatically from normal CvContour objects and to convert them back; it also provides the method for comparing the two trees. Unfortunately, the constructed trees are not quite robust (i.e., minor changes in the contour may change the resultant tree significantly). Also, the initial triangle (root of the tree) is chosen somewhat arbitrarily. Thus, to obtain a better representation requires that we first apply cvApproxPoly() and then align the contour (perform a cyclic shift) such that the initial triangle is pretty much rotation-independent.

```
CvContourTree*  cvCreateContourTree(
    const CvSeq*   contour,
    CvMemStorage*  storage,
    double         threshold
```

---

* Some early work in hierarchical matching of contours is described in [Mokhtarian86] and [Neveu86] and to 3D in [Mokhtarian88].

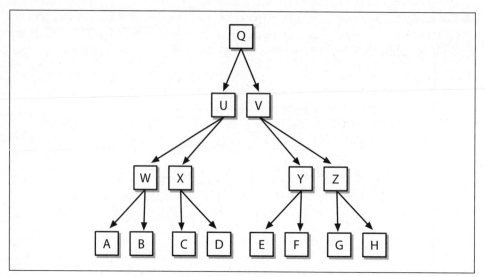

Figure 8-11. A binary tree representation that might correspond to a contour like that of Figure 8-10

```
);
CvSeq*  cvContourFromContourTree(
  const CvContourTree* tree,
  CvMemStorage*        storage,
  CvTermCriteria       criteria
);
double  cvMatchContourTrees(
  const CvContourTree* tree1,
  const CvContourTree* tree2,
  int                  method,
  double               threshold
);
```

This code references CvTermCriteria(), the details of which are given in Chapter 9. For now, you can simply construct a structure using cvTermCriteria() with the following (or similar) defaults:

```
CvTermCriteria termcrit = cvTermCriteria(
  CV_TERMCRIT_ITER | CV_TERMCRIT_EPS, 5, 1 )
);
```

Contour Convexity and Convexity Defects

Another useful way of comprehending the shape of an object or contour is to compute a convex hull for the object and then compute its *convexity defects* [Homma85]. The shapes of many complex objects are well characterized by such defects.

Figure 8-12 illustrates the concept of a convexity defect using an image of a human hand. The convex hull is pictured as a dark line around the hand, and the regions labeled A through H are each "defects" relative to that hull. As you can see, these convexity defects offer a means of characterizing not only the hand itself but also the state of the hand.

```
#define CV_CLOCKWISE        1
#define CV_COUNTER_CLOCKWISE 2
CvSeq* cvConvexHull2(
  const CvArr* input,
  void*        hull_storage  = NULL,
  int          orientation   = CV_CLOCKWISE,
  int          return_points = 0
);
int  cvCheckContourConvexity(
    const CvArr* contour
);
CvSeq*  cvConvexityDefects(
  const CvArr*  contour,
  const CvArr*  convexhull,
  CvMemStorage* storage      = NULL
);
```

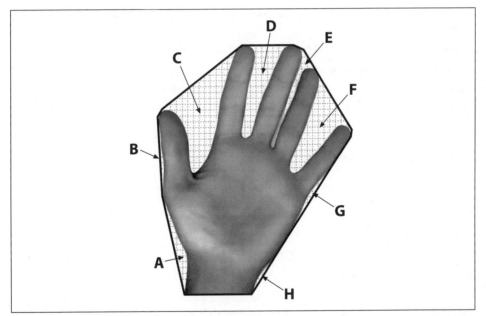

Figure 8-12. Convexity defects: the dark contour line is a convex hull around the hand; the gridded regions (A–H) are convexity defects in the hand contour relative to the convex hull

There are three important OpenCV methods that relate to complex hulls and convexity defects. The first simply computes the hull of a contour that we have already identified, and the second allows us to check whether an identified contour is already convex. The third computes convexity defects in a contour for which the convex hull is known.

The cvConvexHull2() routine takes an array of points as its first argument. This array is typically a matrix with two columns and n rows (i.e., n-by-2), or it can be a contour. The points should be 32-bit integers (CV_32SC1) or floating-point numbers (CV_32FC1). The next argument is the now familiar pointer to a memory storage where space for the result can be allocated. The next argument can be either CV_CLOCKWISE or

CV_COUNTERCLOCKWISE, which will determine the orientation of the points when they are returned by the routine. The final argument, returnPoints, can be either zero (0) or one (1). If set to 1 then the points themselves will be stored in the return array. If it is set to 0, then only indices* will be stored in the return array, indices that refer to the entries in the original array passed to cvConvexHull2().

At this point the astute reader might ask: "If the hull_storage argument is a memory storage, then why is it prototyped as void*?" Good question. The reason is because, in many cases, it is more useful to have the points of the hull returned in the form of an array rather than a sequence. With this in mind, there is another possibility for the hull_storage argument, which is to pass in a CvMat* pointer to a matrix. In this case, the matrix should be one-dimensional and have the same number of entries as there are input points. When cvConvexHull2() is called, it will actually modify the header for the matrix so that the correct number of columns are indicated.†

Sometimes we already have the contour but do not know if it is convex. In this case we can call cvCheckContourConvexity(). This test is simple and fast,‡ but it will not work correctly if the contour passed contains self-intersections.

The third routine, cvConvexityDefects(), actually computes the defects and returns a sequence of the defects. In order to do this, cvConvexityDefects() requires the contour itself, the convex hull, and a memory storage from which to get the memory needed to allocate the result sequence. The first two arguments are CvArr* and are the same form as the input argument to cvConvexHull2().

```
typedef struct CvConvexityDefect {
    // point of the contour where the defect begins
    CvPoint* start;
    // point of the contour where the defect ends
    CvPoint* end;
    // point within the defect farthest from the convex hull
    CvPoint* depth_point;
    // distance between the farthest point and the convex hull
    float depth;
} CvConvexityDefect;
```

The cvConvexityDefects() routine returns a sequence of CvConvexityDefect structures containing some simple parameters that can be used to characterize the defects. The start and end members are points on the hull at which the defect begins and ends. The depth_point indicates the point on the defect that is the farthest from the edge of the hull from which the defect is a deflection. The final parameter, depth, is the distance between the farthest point and the hull edge.

---

* If the input is CvSeq* or CvContour* then what will be stored are pointers to the points.

† You should know that the memory allocated for the data part of the matrix is not re-allocated in any way, so don't expect a rebate on your memory. In any case, since these are C-arrays, the correct memory will be de-allocated when the matrix itself is released.

‡ It actually runs in $O(N)$ time, which is only marginally faster than the $O(N \log N)$ time required to construct a convex hull.

Pairwise Geometrical Histograms

Earlier we briefly visited the Freeman chain codes (FCCs). Recall that a Freeman chain is a representation of a polygon in terms of a sequence of "moves", where each move is of a fixed length and in a particular direction. However, we did not linger on why one might actually want to use such a representation.

There are many uses for Freeman chains, but the most popular one is worth a longer look because the idea underlies the *pairwise geometrical histogram* (PGH).*

The PGH is actually a generalization or extension of what is known as a chain code histogram (CCH). The CCH is a histogram made by counting the number of each kind of step in the Freeman chain code representation of a contour. This histogram has a number of nice properties. Most notably, rotations of the object by 45 degree increments become cyclic transformations on the histogram (see Figure 8-13). This provides a method of shape recognition that is not affected by such rotations.

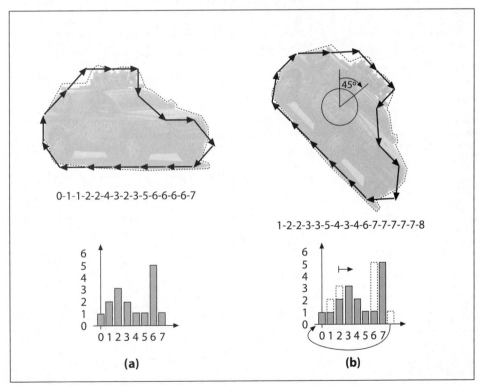

Figure 8-13. Freeman chain code representations of a contour (top) and their associated chain code histograms (bottom); when the original contour (panel a) is rotated 45 degrees clockwise (panel b), the resulting chain code histogram is the same as the original except shifted to the right by one unit

---

* OpenCV implements the method of Iivarinen, Peura, Särelä, and Visa [Iivarinen97].

The PGH is constructed as follows (see Figure 8-14). Each of the edges of the polygon is successively chosen to be the "base edge". Then each of the other edges is considered relative to that base edge and three values are computed: d_{min}, d_{max}, and θ. The d_{min} value is the smallest distance between the two edges, d_{max} is the largest, and θ is the angle between them. The PGH is a two-dimensional histogram whose dimensions are the angle and the distance. In particular: for every edge pair, there is a bin corresponding to (d_{min}, θ) and a bin corresponding to (d_{max}, θ). For each such pair of edges, those two bins are incremented—as are all bins for intermediate values of d (i.e., values between d_{min} and d_{max}).

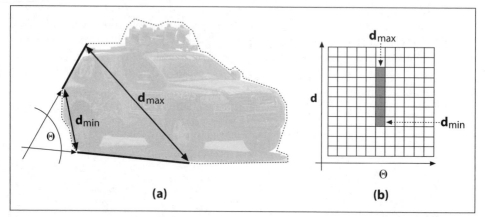

Figure 8-14. Pairwise geometric histogram: every two edge segments of the enclosing polygon have an angle and a minimum and maximum distance (panel a); these numbers are encoded into a two-dimensional histogram (panel b), which is rotation-invariant and can be matched against other objects

The utility of the PGH is similar to that of the FCC. One important difference is that the discriminating power of the PGH is higher, so it is more useful when attempting to solve complex problems involving a greater number of shapes to be recognized and/or a greater variability of background noise. The function used to compute the PGH is

```
void cvCalcPGH(
    const CvSeq* contour,
    CvHistogram* hist
);
```

Here contour can contain integer point coordinates; of course, hist must be two-dimensional.

Exercises

1. Neglecting image noise, does the IPAN algorithm return the same "dominant points" as we zoom in on an object? As we rotate the object?

    a. Give the reasons for your answer.

    b. Try it! Use PowerPoint or a similar program to draw an "interesting" white shape on a black background. Turn it into an image and save. Resize the object several times, saving each time, and reposition it via several different rotations. Read it in to OpenCV, turn it into grayscale, threshold, and find the contour. Then use cvFindDominantPoints() to find the dominant points of the rotated and scaled versions of the object. Are the same points found or not?

2. Finding the extremal points (i.e., the two points that are farthest apart) in a closed contour of N points can be accomplished by comparing the distance of each point to every other point.

   a. What is the complexity of such an algorithm?

   b. Explain how you can do this faster.

3. Create a circular image queue using CvSeq functions.

4. What is the maximal closed contour length that could fit into a 4-by-4 image? What is its contour area?

5. Using PowerPoint or a similar program, draw a white circle of radius 20 on a black background (the circle's circumference will thus be 2 π 20 ≈ 126.7. Save your drawing as an image.

   a. Read the image in, turn it into grayscale, threshold, and find the contour. What is the contour length? Is it the same (within rounding) or different from the calculated length?

   b. Using 126.7 as a base length of the contour, run cvApproxPoly() using as parameters the following fractions of the base length: 90, 66, 33, 10. Find the contour length and draw the results.

6. Using the circle drawn in exercise 5, explore the results of cvFindDominantPoints() as follows.

   a. Vary the d_{min} and d_{max} distances and draw the results.

   b. Then vary the neighborhood distance and describe the resulting changes.

   c. Finally, vary the maximal angle threshold and describe the results.

7. Subpixel corner finding. Create a white-on-black corner in PowerPoint (or similar drawing program) such that the corner sits on exact integer coordinates. Save this as an image and load into OpenCV.

   a. Find and print out the exact coordinates of the corner.

   b. Alter the original image: delete the actual corner by drawing a small black circle over its intersection. Save and load this image, and find the subpixel location of this corner. Is it the same? Why or why not?

8. Suppose we are building a bottle detector and wish to create a "bottle" feature. We have many images of bottles that are easy to segment and find the contours of, but the bottles are rotated and come in various sizes. We can draw the contours and then find the Hu moments to yield an invariant bottle-feature vector. So far, so

good—but should we draw filled-in contours or just line contours? Explain your answer.

9. When using cvMoments() to extract bottle contour moments in exercise 8, how should we set isBinary? Explain your answer.

10. Take the letter shapes used in the discussion of Hu moments. Produce variant images of the shapes by rotating to several different angles, scaling larger and smaller, and combining these transformations. Describe which Hu features respond to rotation, which to scale, and which to both.

11. Make a shape in PowerPoint (or another drawing program) and save it as an image. Make a scaled, a rotated, and a rotated and scaled version of the object and then store these as images. Compare them using cvMatchContourTrees() and cvConvexity Defects(). Which is better for matching the shape? Why?

CHAPTER 9

Image Parts and Segmentation

Parts and Segments

This chapter focuses on how to isolate objects or parts of objects from the rest of the image. The reasons for doing this should be obvious. In video security, for example, the camera mostly looks out on the same boring background, which really isn't of interest. What is of interest is when people or vehicles enter the scene, or when something is left in the scene that wasn't there before. We want to isolate those events and to be able to ignore the endless hours when nothing is changing.

Beyond separating foreground objects from the rest of the image, there are many situations where we want to separate out parts of objects, such as isolating just the face or the hands of a person. We might also want to preprocess an image into meaningful *super pixels*, which are segments of an image that contain things like limbs, hair, face, torso, tree leaves, lake, path, lawn and so on. Using super pixels saves on computation; for example, when running an object classifier over the image, we only need search a box around each super pixel. We might only track the motion of these larger patches and not every point inside.

We saw several image segmentation algorithms when we discussed image processing in Chapter 5. The routines covered in that chapter included image morphology, flood fill, threshold, and pyramid segmentation. This chapter examines other algorithms that deal with finding, filling and isolating objects and object parts in an image. We start with separating foreground objects from learned background scenes. These background modeling functions are not built-in OpenCV functions; rather, they are examples of how we can leverage OpenCV functions to implement more complex algorithms.

Background Subtraction

Because of its simplicity and because camera locations are fixed in many contexts, *background subtraction* (aka *background differencing*) is probably the most fundamental image processing operation for video security applications. Toyama, Krumm, Brumitt, and Meyers give a good overview and comparison of many techniques [Toyama99]. In order to perform background subtraction, we first must "learn" a model of the background.

Once learned, this *background model* is compared against the current image and then the known background parts are subtracted away. The objects left after subtraction are presumably new foreground objects.

Of course "background" is an ill-defined concept that varies by application. For example, if you are watching a highway, perhaps average traffic flow should be considered background. Normally, background is considered to be any static or periodically moving parts of a scene that remain static or periodic over the period of interest. The whole ensemble may have time-varying components, such as trees waving in morning and evening wind but standing still at noon. Two common but substantially distinct environment categories that are likely to be encountered are indoor and outdoor scenes. We are interested in tools that will help us in both of these environments. First we will discuss the weaknesses of typical background models and then will move on to discuss higher-level scene models. Next we present a quick method that is mostly good for indoor static background scenes whose lighting doesn't change much. We will follow this by a "codebook" method that is slightly slower but can work in both outdoor and indoor scenes; it allows for periodic movements (such as trees waving in the wind) and for lighting to change slowly or periodically. This method is also tolerant to learning the background even when there are occasional foreground objects moving by. We'll top this off by another discussion of connected components (first seen in Chapter 5) in the context of cleaning up foreground object detection. Finally, we'll compare the quick background method against the codebook background method.

Weaknesses of Background Subtraction

Although the background modeling methods mentioned here work fairly well for simple scenes, they suffer from an assumption that is often violated: that all the pixels are independent. The methods we describe learn a model for the variations a pixel experiences without considering neighboring pixels. In order to take surrounding pixels into account, we could learn a multipart model, a simple example of which would be an extension of our basic independent pixel model to include a rudimentary sense of the brightness of neighboring pixels. In this case, we use the brightness of neighboring pixels to distinguish when neighboring pixel values are relatively bright or dim. We then learn effectively two models for the individual pixel: one for when the surrounding pixels are bright and one for when the surrounding pixels are dim. In this way, we have a model that takes into account the surrounding *context*. But this comes at the cost of twice as much memory use and more computation, since we now need different values for when the surrounding pixels are bright or dim. We also need twice as much data to fill out this two-state model. We can generalize the idea of "high" and "low" contexts to a multidimensional histogram of single and surrounding pixel intensities as well as make it even more complex by doing all this over a few time steps. Of course, this richer model over space and time would require still more memory, more collected data samples, and more computational resources.

Because of these extra costs, the more complex models are usually avoided. We can often more efficiently invest our resources in cleaning up the *false positive* pixels that result when the independent pixel assumption is violated. The cleanup takes the form of image processing operations (cvErode(), cvDilate(), and cvFloodFill(), mostly) that eliminate stray patches of pixels. We've discussed these routines previously (Chapter 5) in the context of finding large and compact* *connected components* within noisy data. We will employ connected components again in this chapter and so, for now, will restrict our discussion to approaches that assume pixels vary independently.

Scene Modeling

How do we define background and foreground? If we're watching a parking lot and a car comes in to park, then this car is a new foreground object. But should it stay foreground forever? How about a trash can that was moved? It will show up as foreground in two places: the place it was moved to and the "hole" it was moved from. How do we tell the difference? And again, how long should the trash can (and its hole) remain foreground? If we are modeling a dark room and suddenly someone turns on a light, should the whole room become foreground? To answer these questions, we need a higher-level "scene" model, in which we define multiple levels between foreground and background states, and a timing-based method of slowly relegating unmoving foreground patches to background patches. We will also have to detect and create a new model when there is a global change in a scene.

In general, a scene model might contain multiple layers, from "new foreground" to older foreground on down to background. There might also be some motion detection so that, when an object is moved, we can identify both its "positive" aspect (its new location) and its "negative" aspect (its old location, the "hole").

In this way, a new foreground object would be put in the "new foreground" object level and marked as a positive object or a hole. In areas where there was no foreground object, we could continue updating our background model. If a foreground object does not move for a given time, it is demoted to "older foreground," where its pixel statistics are provisionally learned until its learned model joins the learned background model.

For global change detection such as turning on a light in a room, we might use global frame differencing. For example, if many pixels change at once then we could classify it as a global rather than local change and then switch to using a model for the new situation.

A Slice of Pixels

Before we go on to modeling pixel changes, let's get an idea of what pixels in an image can look like over time. Consider a camera looking out a window to a scene of a tree blowing in the wind. Figure 9-1 shows what the pixels in a given line segment of the image look like over 60 frames. We wish to model these kinds of fluctuations. Before doing so, however, we make a small digression to discuss how we sampled this line because it's a generally useful trick for creating features and for debugging.

---

* Here we are using mathematician's definition of "compact," which has nothing to do with size.

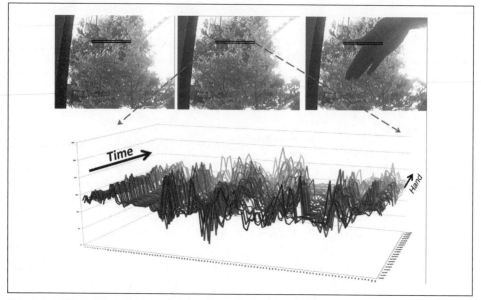

Figure 9-1. Fluctuations of a line of pixels in a scene of a tree moving in the wind over 60 frames: some dark areas (upper left) are quite stable, whereas moving branches (upper center) can vary widely

OpenCV has functions that make it easy to sample an arbitrary line of pixels. The line sampling functions are cvInitLineIterator() and CV_NEXT_LINE_POINT(). The function prototype for cvInitLineIterator() is:

```
int cvInitLineIterator(
    const CvArr*    image,
    CvPoint         pt1,
    CvPoint         pt2,
    CvLineIterator* line_iterator,
    int             connectivity  = 8,
    int             left_to_right = 0
);
```

The input image may be of any type or number of channels. Points pt1 and pt2 are the ends of the line segment. The iterator line_iterator just steps through, pointing to the pixels along the line between the points. In the case of multichannel images, each call to CV_NEXT_LINE_POINT() moves the line_iterator to the next pixel. All the channels are available at once as line_iterator.ptr[0], line_iterator.ptr[1], and so forth. The connectivity can be 4 (the line can step right, left, up, or down) or 8 (the line can additionally step along the diagonals). Finally if left_to_right is set to 0 (false), then line_iterator scans from pt1 to pt2; otherwise, it will go from the leftmost to the rightmost point.* The cvInitLineIterator() function returns the number of points that will be

---

* The left_to_right flag was introduced because a discrete line drawn from pt1 to pt2 does not always match the line from pt2 to pt1. Therefore, setting this flag gives the user a consistent rasterization regardless of the pt1, pt2 order.

iterated over for that line. A companion macro, CV_NEXT_LINE_POINT(line_iterator), steps the iterator from one pixel to another.

Let's take a second to look at how this method can be used to extract some data from a file (Example 9-1). Then we can re-examine Figure 9-1 in terms of the resulting data from that movie file.

Example 9-1. Reading out the RGB values of all pixels in one row of a video and accumulating those values into three separate files

```
// STORE TO DISK A LINE SEGMENT OF BGR PIXELS FROM pt1 to pt2.
//
CvCapture*      capture = cvCreateFileCapture( argv[1] );
int             max_buffer;
IplImage*       rawImage;
int             r[10000],g[10000],b[10000];
CvLineIterator iterator;

FILE *fptrb = fopen("blines.csv","w"); // Store the data here
FILE *fptrg = fopen("glines.csv","w"); // for each color channel
FILE *fptrr = fopen("rlines.csv","w");

// MAIN PROCESSING LOOP:
//
for(;;){
    if( !cvGrabFrame( capture ))
        break;
    rawImage = cvRetrieveFrame( capture );
    max_buffer = cvInitLineIterator(rawImage,pt1,pt2,&iterator,8,0);
    for(int j=0; j<max_buffer; j++){

        fprintf(fptrb,"%d,", iterator.ptr[0]); //Write blue value
        fprintf(fptrg,"%d,", iterator.ptr[1]); //green
        fprintf(fptrr,"%d,", iterator.ptr[2]); //red

        iterator.ptr[2] = 255;   //Mark this sample in red

        CV_NEXT_LINE_POINT(iterator); //Step to the next pixel
    }
    // OUTPUT THE DATA IN ROWS:
    //
    fprintf(fptrb,"/n");fprintf(fptrg,"/n");fprintf(fptrr,"/n");
}
// CLEAN UP:
//
fclose(fptrb); fclose(fptrg); fclose(fptrr);
cvReleaseCapture( &capture );
```

We could have made the line sampling even easier, as follows:

```
    int cvSampleLine(
        const CvArr* image,
        CvPoint      pt1,
        CvPoint      pt2,
```

```
    void*        buffer,
    int          connectivity = 8
);
```

This function simply wraps the function `cvInitLineIterator()` together with the macro `CV_NEXT_LINE_POINT(line_iterator)` from before. It samples from pt1 to pt2; then you pass it a pointer to a buffer of the right type and of length $N_{channels} \times \max(|pt2_x - pt2_x| + 1,$ $|pt2_y - pt2_y| + 1)$. Just like the line iterator, `cvSampleLine()` steps through each channel of each pixel in a multichannel image before moving to the next pixel. The function returns the number of actual elements it filled in the `buffer`.

We are now ready to move on to some methods for modeling the kinds of pixel fluctuations seen in Figure 9-1. As we move from simple to increasingly complex models, we shall restrict our attention to those models that will run in real time and within reasonable memory constraints.

Frame Differencing

The very simplest background subtraction method is to subtract one frame from another (possibly several frames later) and then label any difference that is "big enough" the foreground. This process tends to catch the edges of moving objects. For simplicity, let's say we have three single-channel images: frameTime1, frameTime2, and frameForeground. The image frameTime1 is filled with an older grayscale image, and frameTime2 is filled with the current grayscale image. We could then use the following code to detect the magnitude (absolute value) of foreground differences in frameForeground:

```
cvAbsDiff(
    frameTime1,
    frameTime2,
    frameForeground
);
```

Because pixel values always exhibit noise and fluctuations, we should ignore (set to 0) small differences (say, less than 15), and mark the rest as big differences (set to 255):

```
cvThreshold(
    frameForeground,
    frameForeground,
    15,
    255,
    CV_THRESH_BINARY
);
```

The image frameForeground then marks candidate foreground objects as 255 and background pixels as 0. We need to clean up small noise areas as discussed earlier; we might do this with `cvErode()` or by using connected components. For color images, we could use the same code for each color channel and then combine the channels with `cvOr()`. This method is much too simple for most applications other than merely indicating regions of motion. For a more effective background model we need to keep some statistics about the means and average differences of pixels in the scene. You can look ahead to the section entitled "A quick test" to see examples of frame differencing in Figures 9-5 and 9-6.

Averaging Background Method

The averaging method basically learns the average and standard deviation (or similarly, but computationally faster, the average difference) of each pixel as its model of the background.

Consider the pixel line from Figure 9-1. Instead of plotting one sequence of values for each frame (as we did in that figure), we can represent the variations of each pixel throughout the video in terms of an average and average differences (Figure 9-2). In the same video, a foreground object (which is, in fact, a hand) passes in front of the camera. That foreground object is not nearly as bright as the sky and tree in the background. The brightness of the hand is also shown in the figure.

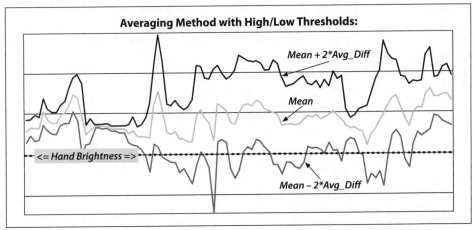

Figure 9-2. Data from Figure 9-1 presented in terms of average differences: an object (a hand) that passes in front of the camera is somewhat darker, and the brightness of that object is reflected in the graph

The averaging method makes use of four OpenCV routines: cvAcc(), to accumulate images over time; cvAbsDiff(), to accumulate frame-to-frame image differences over time; cvInRange(), to segment the image (once a background model has been learned) into foreground and background regions; and cvOr(), to compile segmentations from different color channels into a single mask image. Because this is a rather long code example, we will break it into pieces and discuss each piece in turn.

First, we create pointers for the various scratch and statistics-keeping images we will need along the way. It will prove helpful to sort these pointers according to the type of images they will later hold.

```
//Global storage
//
//Float, 3-channel images
//
IplImage *IavgF,*IdiffF, *IprevF, *IhiF, *IlowF;
```

```
IplImage *Iscratch,*Iscratch2;

//Float, 1-channel images
//
IplImage *Igray1,*Igray2, *Igray3;
IplImage *Ilow1,  *Ilow2, *Ilow3;
IplImage *Ihi1,   *Ihi2,  *Ihi3;

// Byte, 1-channel image
//
IplImage *Imaskt;

//Counts number of images learned for averaging later.
//
float Icount;
```

Next we create a single call to allocate all the necessary intermediate images. For convenience we pass in a single image (from our video) that can be used as a reference for sizing the intermediate images.

```
// I is just a sample image for allocation purposes
// (passed in for sizing)
//
void AllocateImages( IplImage* I ){

   CvSize sz = cvGetSize( I );

   IavgF   = cvCreateImage( sz, IPL_DEPTH_32F, 3 );
   IdiffF  = cvCreateImage( sz, IPL_DEPTH_32F, 3 );
   IprevF  = cvCreateImage( sz, IPL_DEPTH_32F, 3 );
   IhiF    = cvCreateImage( sz, IPL_DEPTH_32F, 3 );
   IlowF   = cvCreateImage( sz, IPL_DEPTH_32F, 3 );
   Ilow1   = cvCreateImage( sz, IPL_DEPTH_32F, 1 );
   Ilow2   = cvCreateImage( sz, IPL_DEPTH_32F, 1 );
   Ilow3   = cvCreateImage( sz, IPL_DEPTH_32F, 1 );
   Ihi1    = cvCreateImage( sz, IPL_DEPTH_32F, 1 );
   Ihi2    = cvCreateImage( sz, IPL_DEPTH_32F, 1 );
   Ihi3    = cvCreateImage( sz, IPL_DEPTH_32F, 1 );
   cvZero( IavgF );
   cvZero( IdiffF );
   cvZero( IprevF );
   cvZero( IhiF );
   cvZero( IlowF );
   Icount  = 0.00001; //Protect against divide by zero

   Iscratch  = cvCreateImage( sz, IPL_DEPTH_32F, 3 );
   Iscratch2 = cvCreateImage( sz, IPL_DEPTH_32F, 3 );
   Igray1    = cvCreateImage( sz, IPL_DEPTH_32F, 1 );
   Igray2    = cvCreateImage( sz, IPL_DEPTH_32F, 1 );
   Igray3    = cvCreateImage( sz, IPL_DEPTH_32F, 1 );
   Imaskt    = cvCreateImage( sz, IPL_DEPTH_8U,  1 );
   cvZero( Iscratch );
   cvZero( Iscratch2 );
}
```

In the next piece of code, we learn the accumulated background image and the accumulated absolute value of frame-to-frame image differences (a computationally quicker proxy* for learning the standard deviation of the image pixels). This is typically called for 30 to 1,000 frames, sometimes taking just a few frames from each second or sometimes taking all available frames. The routine will be called with a three-color channel image of depth 8 bits.

```
// Learn the background statistics for one more frame
// I is a color sample of the background, 3-channel, 8u
//
void accumulateBackground( IplImage *I ){

    static int first = 1;                  // nb. Not thread safe
    cvCvtScale(  I, Iscratch, 1, 0 );    // convert to float
    if( !first ){
       cvAcc( Iscratch, IavgF );
       cvAbsDiff( Iscratch, IprevF, Iscratch2 );
       cvAcc( Iscratch2, IdiffF );
       Icount += 1.0;
    }
    first = 0;
    cvCopy( Iscratch, IprevF );

}
```

We first use cvCvtScale() to turn the raw background 8-bit-per-channel, three-color-channel image into a floating-point three-channel image. We then accumulate the raw floating-point images into IavgF. Next, we calculate the frame-to-frame absolute difference image using cvAbsDiff() and accumulate that into image IdiffF. Each time we accumulate these images, we increment the image count Icount, a global, to use for averaging later.

Once we have accumulated enough frames, we convert them into a statistical model of the background. That is, we compute the means and deviation measures (the average absolute differences) of each pixel:

```
void createModelsfromStats() {

    cvConvertScale( IavgF,   IavgF,( double)(1.0/Icount) );
    cvConvertScale( IdiffF, IdiffF,(double)(1.0/Icount) );

    //Make sure diff is always something
    //
    cvAddS( IdiffF, cvScalar( 1.0, 1.0, 1.0), IdiffF );
    setHighThreshold( 7.0 );
    setLowThreshold( 6.0 );
}
```

---

* Notice our use of the word "proxy." Average difference is not mathematically equivalent to standard deviation, but in this context it is close enough to yield results of similar quality. The advantage of average difference is that it is slightly faster to compute than standard deviation. With only a tiny modification of the code example you can use standard deviations instead and compare the quality of the final results for yourself; we'll discuss this more explicitly later in this section.

In this code, cvConvertScale() calculates the average raw and absolute difference images by dividing by the number of input images accumulated. As a precaution, we ensure that the average difference image is at least 1; we'll need to scale this factor when calculating a foreground-background threshold and would like to avoid the degenerate case in which these two thresholds could become equal.

Both setHighThreshold() and setLowThreshold() are utility functions that set a threshold based on the frame-to-frame average absolute differences. The call setHighThreshold(7.0) fixes a threshold such that any value that is 7 times the average frame-to-frame absolute difference above the average value for that pixel is considered foreground; likewise, setLowThreshold(6.0) sets a threshold bound that is 6 times the average frame-to-frame absolute difference below the average value for that pixel. Within this range around the pixel's average value, objects are considered to be background. These threshold functions are:

```
void setHighThreshold( float scale )
{
    cvConvertScale( IdiffF, Iscratch, scale );
    cvAdd( Iscratch, IavgF, IhiF );
    cvSplit( IhiF, Ihi1, Ihi2, Ihi3, 0 );
}

void setLowThreshold( float scale )
{
    cvConvertScale( IdiffF, Iscratch, scale );
    cvSub( IavgF, Iscratch, IlowF );
    cvSplit( IlowF, Ilow1, Ilow2, Ilow3, 0 );
}
```

Again, in setLowThreshold() and setHighThreshold() we use cvConvertScale() to multiply the values prior to adding or subtracting these ranges relative to IavgF. This action sets the IhiF and IlowF range for each channel in the image via cvSplit().

Once we have our background model, complete with high and low thresholds, we use it to segment the image into foreground (things not "explained" by the background image) and the background (anything that fits within the high and low thresholds of our background model). Segmentation is done by calling:

```
// Create a binary: 0,255 mask where 255 means foreground pixel
// I       Input image, 3-channel, 8u
// Imask   Mask image to be created, 1-channel 8u
//
void backgroundDiff(
  IplImage *I,
  IplImage *Imask
) {
  cvCvtScale(I,Iscratch,1,0); // To float;
  cvSplit( Iscratch, Igray1,Igray2,Igray3, 0 );

  //Channel 1
  //
  cvInRange(Igray1,Ilow1,Ihi1,Imask);
```

```
//Channel 2
//
cvInRange(Igray2,Ilow2,Ihi2,Imaskt);
cvOr(Imask,Imaskt,Imask);

//Channel 3
//
cvInRange(Igray3,Ilow3,Ihi3,Imaskt);
cvOr(Imask,Imaskt,Imask)

//Finally, invert the results
//
cvSubRS( Imask, 255, Imask);
}
```

This function first converts the input image I (the image to be segmented) into a float-ing-point image by calling cvCvtScale(). We then convert the three-channel image into separate one-channel image planes using cvSplit(). These color channel planes are then checked to see if they are within the high and low range of the average background pixel via the cvInRange() function, which sets the grayscale 8-bit depth image Imaskt to max (255) when it's in range and to 0 otherwise. For each color channel we logically OR the segmentation results into a mask image Imask, since strong differences in any color channel are considered evidence of a foreground pixel here. Finally, we invert Imask us-ing cvSubRS(), because foreground should be the values out of range, not in range. The mask image is the output result.

For completeness, we need to release the image memory once we're finished using the background model:

```
void DeallocateImages()
{
    cvReleaseImage( &IavgF);
    cvReleaseImage( &IdiffF );
    cvReleaseImage( &IprevF );
    cvReleaseImage( &IhiF );
    cvReleaseImage( &IlowF );
    cvReleaseImage( &Ilow1 );
    cvReleaseImage( &Ilow2 );
    cvReleaseImage( &Ilow3 );
    cvReleaseImage( &Ihi1 );
    cvReleaseImage( &Ihi2 );
    cvReleaseImage( &Ihi3 );
    cvReleaseImage( &Iscratch );
    cvReleaseImage( &Iscratch2 );
    cvReleaseImage( &Igray1 );
    cvReleaseImage( &Igray2 );
    cvReleaseImage( &Igray3 );
    cvReleaseImage( &Imaskt);
}
```

We've just seen a simple method of learning background scenes and segmenting fore-ground objects. It will work well only with scenes that do not contain moving background components (like a waving curtain or waving trees). It also assumes that the lighting remains fairly constant (as in indoor static scenes). You can look ahead to Figure 9-5 to check the performance of this averaging method.

Accumulating means, variances, and covariances

The averaging background method just described made use of one accumulation function, cvAcc(). It is one of a group of helper functions for accumulating sums of images, squared images, multiplied images, or average images from which we can compute basic statistics (means, variances, covariances) for all or part of a scene. In this section, we'll look at the other functions in this group.

The images in any given function must all have the same width and height. In each function, the input images named image, image1, or image2 can be one- or three-channel byte (8-bit) or floating-point (32F) image arrays. The output accumulation images named sum, sqsum, or acc can be either single-precision (32F) or double-precision (64F) arrays. In the accumulation functions, the mask image (if present) restricts processing to only those locations where the mask pixels are nonzero.

Finding the mean. To compute a mean value for each pixel across a large set of images, the easiest method is to add them all up using cvAcc() and then divide by the total number of images to obtain the mean.

```
void cvAcc(
    const Cvrr*  image,
    CvArr*       sum,
    const CvArr* mask = NULL
);
```

An alternative that is often useful is to use a *running average*.

```
void cvRunningAvg(
    const CvArr* image,
    CvArr*       acc,
    double       alpha,
    const CvArr* mask = NULL
);
```

The running average is given by the following formula:

$$\text{acc}(x,y) = (1-\alpha) \cdot \text{acc}(x,y) + \alpha \cdot \text{image}(x,y) \quad \text{if mask}(x,y) \neq 0$$

For a constant value of α, running averages are not equivalent to the result of summing with cvAcc(). To see this, simply consider adding three numbers (2, 3, and 4) with α set to 0.5. If we were to accumulate them with cvAcc(), then the sum would be 9 and the average 3. If we were to accumulate them with cvRunningAverage(), the first sum would give $0.5 \times 2 + 0.5 \times 3 = 2.5$ and then adding the third term would give $0.5 \times 2.5 + 0.5 \times 4 = 3.25$. The reason the second number is larger is that the most recent contributions are given more weight than those from farther in the past. Such a running average is thus also called a *tracker*. The parameter α essentially sets the amount of time necessary for the influence of a previous frame to fade.

Finding the variance. We can also accumulate squared images, which will allow us to compute quickly the variance of individual pixels.

```
void cvSquareAcc(
    const CvArr* image,
    CvArr*       sqsum,
    const CvArr* mask = NULL
);
```

You may recall from your last class in statistics that the variance of a finite population is defined by the formula:

$$\sigma^2 = \frac{1}{N}\sum_{i=0}^{N-1}(x_i - \bar{x})^2$$

where $\bar{x}$ is the mean of x for all N samples. The problem with this formula is that it entails making one pass through the images to compute $\bar{x}$ and then a second pass to compute σ^2. A little algebra should allow you to convince yourself that the following formula will work just as well:

$$\sigma^2 = \left(\frac{1}{N}\sum_{i=0}^{N-1}x_i^2\right) - \left(\frac{1}{N}\sum_{i=0}^{N-1}x_i\right)^2$$

Using this form, we can accumulate both the pixel values and their squares in a single pass. Then, the variance of a single pixel is just the average of the square minus the square of the average.

Finding the covariance. We can also see how images vary over time by selecting a specific *lag* and then multiplying the current image by the image from the past that corresponds to the given lag. The function cvMultiplyAcc() will perform a pixelwise multiplication of the two images and then add the result to the "running total" in acc:

```
void cvMultiplyAcc(
    const CvArr* image1,
    const CvArr* image2,
    CvArr*       acc,
    const CvArr* mask = NULL
);
```

For covariance, there is a formula analogous to the one we just gave for variance. This formula is also a single-pass formula in that it has been manipulated algebraically from the standard form so as not to require two trips through the list of images:

$$\mathrm{Cov}(x,y) = \left(\frac{1}{N}\sum_{i=0}^{N-1}(x_i y_i)\right) - \left(\frac{1}{N}\sum_{i=0}^{N-1}x_i\right)\left(\frac{1}{N}\sum_{j=0}^{N-1}y_j\right)$$

In our context, x is the image at time t and y is the image at time $t - d$, where d is the lag.

We can use the accumulation functions described here to create a variety of statistics-based background models. The literature is full of variations on the basic model used as our example. You will probably find that, in your own applications, you will tend to extend this simplest model into slightly more specialized versions. A common enhancement, for example, is for the thresholds to be adaptive to some observed global state changes.

Advanced Background Method

Many background scenes contain complicated moving objects such as trees waving in the wind, fans turning, curtains fluttering, et cetera. Often such scenes also contain varying lighting, such as clouds passing by or doors and windows letting in different light.

A nice method to deal with this would be to fit a time-series model to each pixel or group of pixels. This kind of model deals with the temporal fluctuations well, but its disadvantage is the need for a great deal of memory [Toyama99]. If we use 2 seconds of previous input at 30 Hz, this means we need 60 samples for each pixel. The resulting model for each pixel would then encode what it had learned in the form of 60 different adapted *weights*. Often we'd need to gather background statistics for much longer than 2 seconds, which means that such methods are typically impractical on present-day hardware.

To get fairly close to the performance of adaptive filtering, we take inspiration from the techniques of video compression and attempt to form a *codebook** to represent significant states in the background.[†] The simplest way to do this would be to compare a new value observed for a pixel with prior observed values. If the value is close to a prior value, then it is modeled as a perturbation on that color. If it is not close, then it can seed a new group of colors to be associated with that pixel. The result could be envisioned as a bunch of blobs floating in RGB space, each blob representing a separate volume considered likely to be background.

In practice, the choice of RGB is not particularly optimal. It is almost always better to use a color space whose axis is aligned with brightness, such as the YUV color space. (YUV is the most common choice, but spaces such as HSV, where V is essentially brightness, would work as well.) The reason for this is that, empirically, most of the variation in background tends to be along the brightness axis, not the color axis.

The next detail is how to model the "blobs." We have essentially the same choices as before with our simpler model. We could, for example, choose to model the blobs as Gaussian clusters with a mean and a covariance. It turns out that the simplest case, in

---

* The method OpenCV implements is derived from Kim, Chalidabhongse, Harwood, and Davis [Kim05], but rather than learning-oriented tubes in RGB space, for speed, the authors use axis-aligned boxes in YUV space. Fast methods for cleaning up the resulting background image can be found in Martins [Martins99].

† There is a large literature for background modeling and segmentation. OpenCV's implementation is intended to be fast and robust enough that you can use it to collect foreground objects mainly for the purposes of collecting data sets to train classifiers on. Recent work in background subtraction allows arbitrary camera motion [Farin04; Colombari07] and dynamic background models using the mean-shift algorithm [Liu07].

---

which the "blobs" are simply boxes with a learned extent in each of the three axes of our color space, works out quite well. It is the simplest in terms of memory required and in terms of the computational cost of determining whether a newly observed pixel is inside any of the learned boxes.

Let's explain what a codebook is by using a simple example (Figure 9-3). A codebook is made up of boxes that grow to cover the common values seen over time. The upper panel of Figure 9-3 shows a waveform over time. In the lower panel, boxes form to cover a new value and then slowly grow to cover nearby values. If a value is too far away, then a new box forms to cover it and likewise grows slowly toward new values.

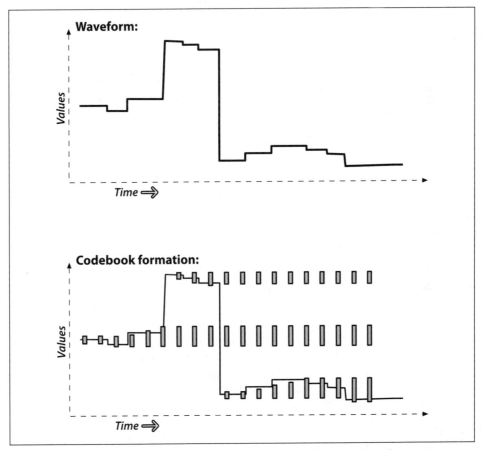

Figure 9-3. Codebooks are just "boxes" delimiting intensity values: a box is formed to cover a new value and slowly grows to cover nearby values; if values are too far away then a new box is formed (see text)

In the case of our background model, we will learn a codebook of boxes that cover three dimensions: the three channels that make up our image at each pixel. Figure 9-4 visualizes the (intensity dimension of the) codebooks for six different pixels learned from the data in Figure 9-1.* This codebook method can deal with pixels that change levels dramatically (e.g., pixels in a windblown tree, which might alternately be one of many colors of leaves, or the blue sky beyond that tree). With this more precise method of modeling, we can detect a foreground object that has values between the pixel values. Compare this with Figure 9-2, where the averaging method cannot distinguish the hand value (shown as a dotted line) from the pixel fluctuations. Peeking ahead to the next section, we see the better performance of the codebook method versus the averaging method shown later in Figure 9-7.

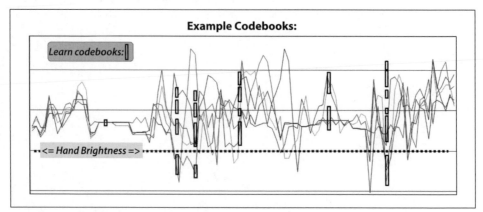

Figure 9-4. Intensity portion of learned codebook entries for fluctuations of six chosen pixels (shown as vertical boxes): codebook boxes accommodate pixels that take on multiple discrete values and so can better model discontinuous distributions; thus they can detect a foreground hand (value at dotted line) whose average value is between the values that background pixels can assume. In this case the codebooks are one dimensional and only represent variations in intensity

In the codebook method of learning a background model, each box is defined by two thresholds (max and min) over each of the three color axes. These box boundary thresholds will expand (max getting larger, min getting smaller) if new background samples fall within a learning threshold (learnHigh and learnLow) above max or below min, respectively. If new background samples fall outside of the box and its learning thresholds, then a new box will be started. In the *background difference* mode there are acceptance thresholds maxMod and minMod; using these threshold values, we say that if a pixel is "close enough" to a max or a min box boundary then we count it as if it were inside the box. A second runtime threshold allows for adjusting the model to specific conditions.

 A situation we will not cover is a pan-tilt camera surveying a large scene. When working with a large scene, it is necessary to stitch together learned models indexed by the pan and tilt angles.

---

* In this case we have chosen several pixels at random from the scan line to avoid excessive clutter. Of course, there is actually a codebook for every pixel.

Structures

It's time to look at all of this in more detail, so let's create an implementation of the codebook algorithm. First, we need our codebook structure, which will simply point to a bunch of boxes in YUV space:

```
typedef struct code_book {
    code_element **cb;
    int numEntries;
    int t;              //count every access
} codeBook;
```

We track how many codebook entries we have in numEntries. The variable t counts the number of points we've accumulated since the start or the last clear operation. Here's how the actual codebook elements are described:

```
#define CHANNELS 3
typedef struct ce {
    uchar learnHigh[CHANNELS]; //High side threshold for learning
    uchar learnLow[CHANNELS];  //Low side threshold for learning
    uchar max[CHANNELS];       //High side of box boundary
    uchar min[CHANNELS];       //Low side of box boundary
    int t_last_update;         //Allow us to kill stale entries
    int stale;                 //max negative run (longest period of inactivity)
} code_element;
```

Each codebook entry consumes four bytes per channel plus two integers, or CHANNELS × 4 + 4 + 4 bytes (20 bytes when we use three channels). We may set CHANNELS to any positive number equal to or less than the number of color channels in an image, but it is usually set to either 1 ("Y", or brightness only) or 3 (YUV, HSV). In this structure, for each channel, max and min are the boundaries of the codebook box. The parameters learnHigh[] and learnLow[] are the thresholds that trigger generation of a new code element. Specifically, a new code element will be generated if a new pixel is encountered whose values do not lie between min - learnLow and max + learnHigh in each of the channels. The time to last update (t_last_update) and stale are used to enable the deletion of seldom-used codebook entries created during learning. Now we can proceed to investigate the functions that use this structure to learn dynamic backgrounds.

Learning the background

We will have one codeBook of code_elements for each pixel. We will need an array of such codebooks that is equal in length to the number of pixels in the images we'll be learning. For each pixel, update_codebook() is called for as many images as are sufficient to capture the relevant changes in the background. Learning may be updated periodically throughout, and clear_stale_entries() can be used to learn the background in the presence of (small numbers of) moving foreground objects. This is possible because the seldom-used "stale" entries induced by a moving foreground will be deleted. The interface to update_codebook() is as follows.

```
/////////////////////////////////////////////////////////////
// int update_codebook(uchar *p, codeBook &c, unsigned cbBounds)
// Updates the codebook entry with a new data point
```

```
//
// p           Pointer to a YUV pixel
// c           Codebook for this pixel
// cbBounds    Learning bounds for codebook (Rule of thumb: 10)
// numChannels Number of color channels we're learning
//
// NOTES:
//      cvBounds must be of length equal to numChannels
//
// RETURN
//    codebook index
//
int update_codebook(
  uchar*    p,
  codeBook& c,
  unsigned* cbBounds,
  int       numChannels
){
   unsigned int high[3],low[3];
   for(n=0; n<numChannels; n++)
   {
      high[n] = *(p+n)+*(cbBounds+n);
      if(high[n] > 255) high[n] = 255;
      low[n] = *(p+n)-*(cbBounds+n);
      if(low[n] < 0) low[n] = 0;
   }
   int matchChannel;

   // SEE IF THIS FITS AN EXISTING CODEWORD
   //
   for(int i=0; i<c.numEntries; i++){
      matchChannel = 0;
      for(n=0; n<numChannels; n++){
         if((c.cb[i]->learnLow[n] <= *(p+n)) &&
         //Found an entry for this channel
         (*(p+n) <= c.cb[i]->learnHigh[n]))
            {
               matchChannel++;
            }
      }
      if(matchChannel == numChannels) //If an entry was found
      {
         c.cb[i]->t_last_update = c.t;
         //adjust this codeword for the first channel
         for(n=0; n<numChannels; n++){
            if(c.cb[i]->max[n] < *(p+n))
            {
               c.cb[i]->max[n] = *(p+n);
            }
            else if(c.cb[i]->min[n] > *(p+n))
            {
               c.cb[i]->min[n] = *(p+n);
            }
         }
         break;
```

```
        }
    }
. . . continued below
```

This function grows or adds a codebook entry when the pixel p falls outside the existing codebook boxes. Boxes grow when the pixel is within cbBounds of an existing box. If a pixel is outside the cbBounds distance from a box, a new codebook box is created. The routine first sets high and low levels to be used later. It then goes through each codebook entry to check whether the pixel value *p is inside the learning bounds of the codebook "box". If the pixel is within the learning bounds for all channels, then the appropriate max or min level is adjusted to include this pixel and the time of last update is set to the current timed count c.t. Next, the update_codebook() routine keeps statistics on how often each codebook entry is hit:

```
. . . continued from above

    // OVERHEAD TO TRACK POTENTIAL STALE ENTRIES
    //
    for(int s=0; s<c.numEntries; s++){

        // Track which codebook entries are going stale:
        //
        int negRun = c.t - c.cb[s]->t_last_update;
        if(c.cb[s]->stale < negRun) c.cb[s]->stale = negRun;

    }

. . . continued below
```

Here, the variable stale contains the largest *negative runtime* (i.e., the longest span of time during which that code was not accessed by the data). Tracking stale entries allows us to delete codebooks that were formed from noise or moving foreground objects and hence tend to become stale over time. In the next stage of learning the background, update_codebook() adds a new codebook if needed:

```
. . . continued from above

    // ENTER A NEW CODEWORD IF NEEDED
    //
    if(i == c.numEntries) //if no existing codeword found, make one
    {
        code_element **foo = new code_element* [c.numEntries+1];
        for(int ii=0; ii<c.numEntries; ii++) {
          foo[ii] = c.cb[ii];
        }
        foo[c.numEntries] = new code_element;
        if(c.numEntries) delete [] c.cb;
        c.cb = foo;
        for(n=0; n<numChannels; n++) {
            c.cb[c.numEntries]->learnHigh[n] = high[n];
            c.cb[c.numEntries]->learnLow[n] = low[n];
            c.cb[c.numEntries]->max[n] = *(p+n);
            c.cb[c.numEntries]->min[n] = *(p+n);
        }
```

```
            c.cb[c.numEntries]->t_last_update = c.t;
            c.cb[c.numEntries]->stale = 0;
            c.numEntries += 1;
    }
```

. . . continued below

Finally, update_codebook() slowly adjusts (by adding 1) the learnHigh and learnLow learning boundaries if pixels were found outside of the box thresholds but still within the high and low bounds:

. . . continued from above

```
    // SLOWLY ADJUST LEARNING BOUNDS
    //
    for(n=0; n<numChannels; n++)
    {
        if(c.cb[i]->learnHigh[n] < high[n]) c.cb[i]->learnHigh[n] += 1;
        if(c.cb[i]->learnLow[n] > low[n]) c.cb[i]->learnLow[n] -= 1;
    }
    return(i);
}
```

The routine concludes by returning the index of the modified codebook. We've now seen how codebooks are learned. In order to learn in the presence of moving foreground objects and to avoid learning codes for spurious noise, we need a way to delete entries that were accessed only rarely during learning.

Learning with moving foreground objects

The following routine, clear_stale_entries(), allows us to learn the background even if there are moving foreground objects.

```
////////////////////////////////////////////////////////////////
//int clear_stale_entries(codeBook &c)
// During learning, after you've learned for some period of time,
// periodically call this to clear out stale codebook entries
//
// c    Codebook to clean up
//
// Return
// number of entries cleared
//
int clear_stale_entries(codeBook &c){
    int staleThresh = c.t>>1;
    int *keep = new int [c.numEntries];
    int keepCnt = 0;
    // SEE WHICH CODEBOOK ENTRIES ARE TOO STALE
    //
    for(int i=0; i<c.numEntries; i++){
        if(c.cb[i]->stale > staleThresh)
            keep[i] = 0; //Mark for destruction
        else
        {
            keep[i] = 1; //Mark to keep
            keepCnt += 1;
```

```
        }
    }
    // KEEP ONLY THE GOOD
    //
    c.t = 0;               //Full reset on stale tracking
    code_element **foo = new code_element* [keepCnt];
    int k=0;
    for(int ii=0; ii<c.numEntries; ii++){
        if(keep[ii])
        {
            foo[k] = c.cb[ii];
            //We have to refresh these entries for next clearStale
            foo[k]->t_last_update = 0;
            k++;
        }
    }
    // CLEAN UP
    //
    delete [] keep;
    delete [] c.cb;
    c.cb = foo;
    int numCleared = c.numEntries - keepCnt;
    c.numEntries = keepCnt;
    return(numCleared);
}
```

The routine begins by defining the parameter staleThresh, which is hardcoded (by a rule of thumb) to be half the total running time count, c.t. This means that, during background learning, if codebook entry i is not accessed for a period of time equal to half the total learning time, then i is marked for deletion (keep[i] = 0). The vector keep[] is allocated so that we can mark each codebook entry; hence it is c.numEntries long. The variable keepCnt counts how many entries we will keep. After recording which codebook entries to keep, we create a new pointer, foo, to a vector of code_element pointers that is keepCnt long, and then the nonstale entries are copied into it. Finally, we delete the old pointer to the codebook vector and replace it with the new, nonstale vector.

Background differencing: Finding foreground objects

We've seen how to create a background codebook model and how to clear it of seldom-used entries. Next we turn to background_diff(), where we use the learned model to segment foreground pixels from the previously learned background:

```
/////////////////////////////////////////////////////////
// uchar background_diff( uchar *p, codeBook &c,
//                        int minMod, int maxMod)
// Given a pixel and a codebook, determine if the pixel is
// covered by the codebook
//
// p           Pixel pointer (YUV interleaved)
// c           Codebook reference
// numChannels Number of channels we are testing
// maxMod      Add this (possibly negative) number onto
```

```
//              max level when determining if new pixel is foreground
// minMod      Subract this (possibly negative) number from
//              min level when determining if new pixel is foreground
//
// NOTES:
// minMod and maxMod must have length numChannels,
// e.g. 3 channels => minMod[3], maxMod[3]. There is one min and
//      one max threshold per channel.
//
// Return
// 0 => background, 255 => foreground
//
uchar background_diff(
  uchar*    p,
  codeBook& c,
  int       numChannels,
  int*      minMod,
  int*      maxMod
) {
  int matchChannel;

  // SEE IF THIS FITS AN EXISTING CODEWORD
  //
  for(int i=0; i<c.numEntries; i++) {
    matchChannel = 0;
    for(int n=0; n<numChannels; n++) {
      if((c.cb[i]->min[n] - minMod[n] <= *(p+n)) &&
         (*(p+n) <= c.cb[i]->max[n] + maxMod[n])) {
        matchChannel++; //Found an entry for this channel
      } else {
        break;
      }
    }
    if(matchChannel == numChannels) {
      break; //Found an entry that matched all channels
    }
  }
  if(i >= c.numEntries) return(255);
  return(0);
}
```

The background differencing function has an inner loop similar to the learning routine update_codebook, except here we look within the learned max and min bounds plus an offset threshold, maxMod and minMod, of each codebook box. If the pixel is within the box plus maxMod on the high side or minus minMod on the low side for each channel, then the matchChannel count is incremented. When matchChannel equals the number of channels, we've searched each dimension and know that we have a match. If the pixel is within a learned box, 255 is returned (a positive detection of foreground); otherwise, 0 is returned (background).

The three functions update_codebook(), clear_stale_entries(), and background_diff() constitute a codebook method of segmenting foreground from learned background.

Using the codebook background model

To use the codebook background segmentation technique, typically we take the following steps.

1. Learn a basic model of the background over a few seconds or minutes using `update_codebook()`.

2. Clean out stale entries with `clear_stale_entries()`.

3. Adjust the thresholds `minMod` and `maxMod` to best segment the known foreground.

4. Maintain a higher-level scene model (as discussed previously).

5. Use the learned model to segment the foreground from the background via `background_diff()`.

6. Periodically update the learned background pixels.

7. At a much slower frequency, periodically clean out stale codebook entries with `clear_stale_entries()`.

A few more thoughts on codebook models

In general, the codebook method works quite well across a wide number of conditions, and it is relatively quick to train and to run. It doesn't deal well with varying patterns of light—such as morning, noon, and evening sunshine—or with someone turning lights on or off indoors. This type of global variability can be taken into account by using several different codebook models, one for each condition, and then allowing the condition to control which model is active.

Connected Components for Foreground Cleanup

Before comparing the averaging method to the codebook method, we should pause to discuss ways to clean up the raw segmented image using connected-components analysis. This form of analysis takes in a noisy input mask image; it then uses the morphological operation *open* to shrink areas of small noise to 0 followed by the morphological operation *close* to rebuild the area of surviving components that was lost in opening. Thereafter, we can find the "large enough" contours of the surviving segments and can optionally proceed to take statistics of all such segments. We can then retrieve either the largest contour or all contours of size above some threshold. In the routine that follows, we implement most of the functions that you could want in connected components:

- Whether to approximate the surviving component contours by polygons or by convex hulls

- Setting how large a component contour must be in order not to be deleted

- Setting the maximum number of component contours to return

- Optionally returning the bounding boxes of the surviving component contours

- Optionally returning the centers of the surviving component contours

The connected components header that implements these operations is as follows.

```
/////////////////////////////////////////////////////////
// void find_connected_components(IplImage *mask, int poly1_hull0,
//                                float perimScale, int *num,
//                                CvRect *bbs, CvPoint *centers)
// This cleans up the foreground segmentation mask derived from calls
// to backgroundDiff
//
// mask         Is a grayscale (8-bit depth) "raw" mask image that
//              will be cleaned up
//
// OPTIONAL PARAMETERS:
// poly1_hull0  If set, approximate connected component by
//                  (DEFAULT) polygon, or else convex hull (0)
// perimScale   Len = image (width+height)/perimScale. If contour
//                  len < this, delete that contour (DEFAULT: 4)
// num          Maximum number of rectangles and/or centers to
//                  return; on return, will contain number filled
//                  (DEFAULT: NULL)
// bbs          Pointer to bounding box rectangle vector of
//                  length num. (DEFAULT SETTING: NULL)
// centers      Pointer to contour centers vector of length
//                  num (DEFAULT: NULL)
//
void find_connected_components(
  IplImage* mask,
  int       poly1_hull0 = 1,
  float     perimScale  = 4,
  int*      num         = NULL,
  CvRect*   bbs         = NULL,
  CvPoint*  centers     = NULL
);
```

The function body is listed below. First we declare memory storage for the connected components contour. We then do morphological opening and closing in order to clear out small pixel noise, after which we rebuild the eroded areas that survive the erosion of the opening operation. The routine takes two additional parameters, which here are hardcoded via #define. The defined values work well, and you are unlikely to want to change them. These additional parameters control how simple the boundary of a foreground region should be (higher numbers are more simple) and how many iterations the morphological operators should perform; the higher the number of iterations, the more erosion takes place in opening before dilation in closing.* More erosion eliminates larger regions of blotchy noise at the cost of eroding the boundaries of larger regions. Again, the parameters used in this sample code work well, but there's no harm in experimenting with them if you like.

```
// For connected components:
// Approx.threshold - the bigger it is, the simpler is the boundary
//
```

---

* Observe that the value CVCLOSE_ITR is actually dependent on the resolution. For images of extremely high resolution, leaving this value set to 1 is not likely to yield satisfactory results.

```
#define CVCONTOUR_APPROX_LEVEL  2

// How many iterations of erosion and/or dilation there should be
//
#define CVCLOSE_ITR  1
```

We now discuss the connected-component algorithm itself. The first part of the routine performs the morphological open and closing operations:

```
void find_connected_components(
  IplImage *mask,
  int poly1_hull0,
  float perimScale,
  int *num,
  CvRect *bbs,
  CvPoint *centers
) {

  static CvMemStorage*   mem_storage = NULL;
  static CvSeq*          contours    = NULL;

  //CLEAN UP RAW MASK
  //
  cvMorphologyEx( mask, mask, 0, 0, CV_MOP_OPEN,  CVCLOSE_ITR );
  cvMorphologyEx( mask, mask, 0, 0, CV_MOP_CLOSE, CVCLOSE_ITR );
```

Now that the noise has been removed from the mask, we find all contours:

```
  //FIND CONTOURS AROUND ONLY BIGGER REGIONS
  //
  if( mem_storage==NULL ) {
    mem_storage = cvCreateMemStorage(0);
  } else {
    cvClearMemStorage(mem_storage);
  }

  CvContourScanner scanner = cvStartFindContours(
    mask,
    mem_storage,
    sizeof(CvContour),
    CV_RETR_EXTERNAL,
    CV_CHAIN_APPROX_SIMPLE
  );
```

Next, we toss out contours that are too small and approximate the rest with polygons or convex hulls (whose complexity has already been set by CVCONTOUR_APPROX_LEVEL):

```
  CvSeq* c;
  int numCont = 0;
  while( (c = cvFindNextContour( scanner )) != NULL ) {

    double len = cvContourPerimeter( c );

    // calculate perimeter len threshold:
    //
    double q = (mask->height + mask->width)/perimScale;

    //Get rid of blob if its perimeter is too small:
```

```
      //
      if( len < q ) {
         cvSubstituteContour( scanner, NULL );
      } else {

         // Smooth its edges if its large enough
         //
         CvSeq* c_new;
         if( poly1_hull0 ) {

            // Polygonal approximation
            //
            c_new = cvApproxPoly(
               c,
               sizeof(CvContour),
               mem_storage,
               CV_POLY_APPROX_DP,
               CVCONTOUR_APPROX_LEVEL,
               0
            );

         } else {

            // Convex Hull of the segmentation
            //
            c_new = cvConvexHull2(
               c,
               mem_storage,
               CV_CLOCKWISE,
               1
            );
         }
         cvSubstituteContour( scanner, c_new );
         numCont++;
      }
   }
   contours = cvEndFindContours( &scanner );
```

In the preceding code, `CV_POLY_APPROX_DP` causes the Douglas-Peucker approximation algorithm to be used, and `CV_CLOCKWISE` is the default direction of the convex hull contour.
All this processing yields a list of contours. Before drawing the contours back into the
mask, we define some simple colors to draw:

```
   // Just some convenience variables
   const CvScalar CVX_WHITE = CV_RGB(0xff,0xff,0xff)
   const CvScalar CVX_BLACK = CV_RGB(0x00,0x00,0x00)
```

We use these definitions in the following code, where we first zero out the mask and then
draw the clean contours back into the mask. We also check whether the user wanted to
collect statistics on the contours (bounding boxes and centers):

```
   // PAINT THE FOUND REGIONS BACK INTO THE IMAGE
   //
   cvZero( mask );
   IplImage *maskTemp;
```

```
// CALC CENTER OF MASS AND/OR BOUNDING RECTANGLES
//
if(num != NULL) {

  //User wants to collect statistics
  //
  int N = *num, numFilled = 0, i=0;
  CvMoments moments;
  double M00, M01, M10;
  maskTemp = cvCloneImage(mask);
  for(i=0, c=contours; c != NULL; c = c->h_next,i++ ) {

    if(i < N) {
      // Only process up to *num of them
      //
      cvDrawContours(
        maskTemp,
        c,
        CVX_WHITE,
        CVX_WHITE,
        -1,
        CV_FILLED,
        8
      );

      // Find the center of each contour
      //
      if(centers != NULL) {

        cvMoments(maskTemp,&moments,1);
        M00 = cvGetSpatialMoment(&moments,0,0);
        M10 = cvGetSpatialMoment(&moments,1,0);
        M01 = cvGetSpatialMoment(&moments,0,1);
        centers[i].x = (int)(M10/M00);
        centers[i].y = (int)(M01/M00);
      }

      //Bounding rectangles around blobs
      //
      if(bbs != NULL) {
        bbs[i] = cvBoundingRect(c);
      }
      cvZero(maskTemp);
      numFilled++;
    }
    // Draw filled contours into mask
    //
    cvDrawContours(
      mask,
      c,
      CVX_WHITE,
      CVX_WHITE,
      -1,
      CV_FILLED,
```

```
            8
        );
    }                                       //end looping over contours
    *num = numFilled;
    cvReleaseImage( &maskTemp );
}
```

If the user doesn't need the bounding boxes and centers of the resulting regions in the mask, we just draw back into the mask those cleaned-up contours representing large enough connected components of the background.

```
// ELSE JUST DRAW PROCESSED CONTOURS INTO THE MASK
//
else {
    // The user doesn't want statistics, just draw the contours
    //
    for( c=contours; c != NULL; c = c->h_next ) {
        cvDrawContours(
        mask,
        c,
        CVX_WHITE,
        CVX_BLACK,
        -1,
        CV_FILLED,
        8
        );
    }
}
```

That concludes a useful routine for creating clean masks out of noisy raw masks. Now let's look at a short comparison of the background subtraction methods.

A quick test

We start with an example to see how this really works in an actual video. Let's stick with our video of the tree outside of the window. Recall (Figure 9-1) that at some point a hand passes through the scene. One might expect that we could find this hand relatively easily with a technique such as frame differencing (discussed previously in its own section). The basic idea of frame differencing was to subtract the current frame from a "lagged" frame and then threshold the difference.

Sequential frames in a video tend to be quite similar. Hence one might expect that, if we take a simple difference of the original frame and the lagged frame, we'll not see too much unless there is some foreground object moving through the scene.* But what does "not see too much" mean in this context? Really, it means "just noise." Of course, in practice the problem is sorting out that noise from the signal when a foreground object does come along.

---

* In the context of frame differencing, an object is identified as "foreground" mainly by its velocity. This is reasonable in scenes that are generally static or in which foreground objects are expected to be much closer to the camera than background objects (and thus appear to move faster by virtue of the projective geometry of cameras).

To understand this noise a little better, we will first look at a pair of frames from the video in which there is no foreground object—just the background and the resulting noise. Figure 9-5 shows a typical frame from the video (upper left) and the previous frame (upper right). The figure also shows the results of frame differencing with a threshold value of 15 (lower left). You can see substantial noise from the moving leaves of the tree. Nevertheless, the method of connected components is able to clean up this scattered noise quite well* (lower right). This is not surprising, because there is no reason to expect much spatial correlation in this noise and so its signal is characterized by a large number of very small regions.

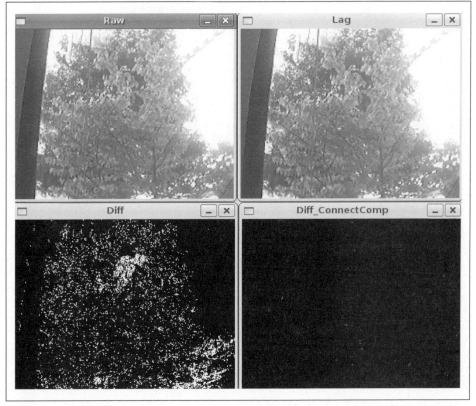

Figure 9-5. Frame differencing: a tree is waving in the background in the current (upper left) and previous (upper right) frame images; the difference image (lower left) is completely cleaned up (lower right) by the connected-components method

Now consider the situation in which a foreground object (our ubiquitous hand) passes through the view of the imager. Figure 9-6 shows two frames that are similar to those in Figure 9-5 except that now the hand is moving across from left to right. As before, the current frame (upper left) and the previous frame (upper right) are shown along

---

* The size threshold for the connected components has been tuned to give zero response in these empty frames. The real question then is whether or not the foreground object of interest (the hand) survives pruning at this size threshold. We will see (Figure 9-6) that it does so nicely.

with the response to frame differencing (lower left) and the fairly good results of the connected-component cleanup (lower right).

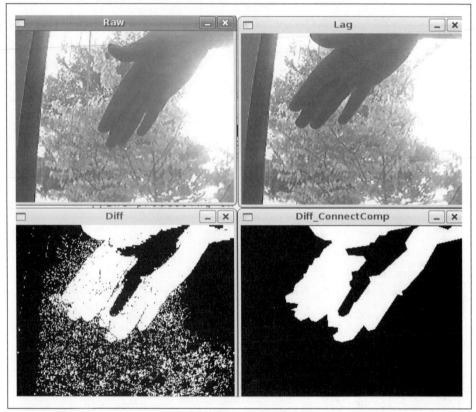

Figure 9-6. Frame difference method of detecting a hand, which is moving left to right as the foreground object (upper two panels); the difference image (lower left) shows the "hole" (where the hand used to be) toward the left and its leading edge toward the right, and the connected-component image (lower right) shows the cleaned-up difference

We can also clearly see one of the deficiencies of frame differencing: it cannot distinguish between the region from where the object moved (the "hole") and where the object is now. Furthermore, in the overlap region there is often a gap because "flesh minus flesh" is 0 (or at least below threshold).

Thus we see that using connected components for cleanup is a powerful technique for rejecting noise in background subtraction. As a bonus, we were also able to glimpse some of the strengths and weaknesses of frame differencing.

Comparing Background Methods

We have discussed two background modeling techniques in this chapter: the average distance method and the codebook method. You might be wondering which method is better, or, at least, when you can get away with using the easy one. In these situations, it's always best to just do a straight bake off* between the available methods.

We will continue with the same tree video that we've been discussing all chapter. In addition to the moving tree, this film has a lot of glare coming off a building to the right and off portions of the inside wall on the left. It is a fairly challenging background to model.

In Figure 9-7 we compare the average difference method at top against the codebook method at bottom; on the left are the raw foreground images and on the right are the cleaned-up connected components. You can see that the average difference method leaves behind a sloppier mask and breaks the hand into two components. This is not so surprising; in Figure 9-2, we saw that using the average difference from the mean as a background model often included pixel values associated with the hand value (shown as a dotted line in that figure). Compare this with Figure 9-4, where codebooks can more accurately model the fluctuations of the leaves and branches and so more precisely identify foreground hand pixels (dotted line) from background pixels. Figure 9-7 confirms not only that the background model yields less noise but also that connected components can generate a fairly accurate object outline.

Watershed Algorithm

In many practical contexts, we would like to segment an image but do not have the benefit of a separate background image. One technique that is often effective in this context is the *watershed algorithm* [Meyer92]. This algorithm converts lines in an image into "mountains" and uniform regions into "valleys" that can be used to help segment objects. The watershed algorithm first takes the gradient of the intensity image; this has the effect of forming valleys or *basins* (the low points) where there is no texture and of forming mountains or *ranges* (high ridges corresponding to edges) where there are dominant lines in the image. It then successively floods basins starting from user-specified (or algorithm-specified) points until these regions meet. Regions that merge across the marks so generated are segmented as belonging together as the image "fills up". In this way, the basins connected to the marker point become "owned" by that marker. We then segment the image into the corresponding marked regions.

More specifically, the watershed algorithm allows a user (or another algorithm!) to mark parts of an object or background that are known to be part of the object or background. The user or algorithm can draw a simple line that effectively tells the watershed algorithm to "group points like these together". The watershed algorithm then segments the image by allowing marked regions to "own" the edge-defined valleys in the gradient image that are connected with the segments. Figure 9-8 clarifies this process.

The function specification of the watershed segmentation algorithm is:

```
void cvWatershed(
  const CvArr* image,
```

---

* For the uninitiated, "bake off" is actually a bona fide term used to describe any challenge or comparison of multiple algorithms on a predetermined data set.

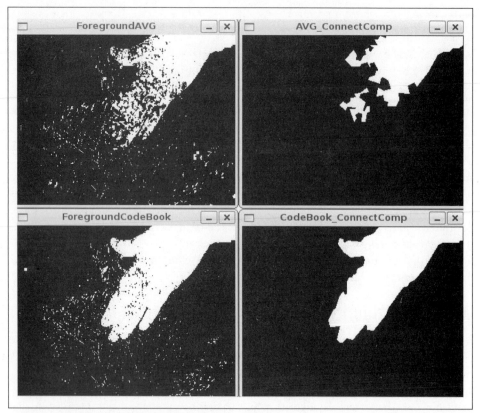

Figure 9-7. With the averaging method (top row), the connected-components cleanup knocks out the fingers (upper right); the codebook method (bottom row) does much better at segmentation and creates a clean connected-component mask (lower right)

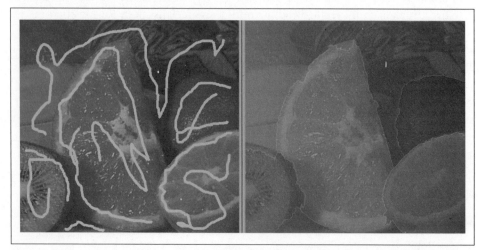

Figure 9-8. Watershed algorithm: after a user has marked objects that belong together (left panel), the algorithm then merges the marked area into segments (right panel)

```
CvArr*      markers
);
```

Here, image is an 8-bit color (three-channel) image and markers is a single-channel integer (IPL_DEPTH_32S) image of the same (*x, y*) dimensions; the value of markers is 0 *except*
where the user (or an algorithm) has indicated by using positive numbers that some
regions belong together. For example, in the left panel of Figure 9-8, the orange might
have been marked with a "1", the lemon with a "2", the lime with a "3", the upper background with "4" and so on. This produces the segmentation you see in the same figure
on the right.

Image Repair by Inpainting

Images are often corrupted by noise. There may be dust or water spots on the lens,
scratches on the older images, or parts of an image that were vandalized. *Inpainting*
[Telea04] is a method for removing such damage by taking the color and texture at the
border of the damaged area and propagating and mixing it inside the damaged area. See
Figure 9-9 for an application that involves the removal of writing from an image.

*Figure 9-9. Inpainting: an image damaged by overwritten text (left panel) is restored by inpainting
(right panel)*

Inpainting works provided the damaged area is not too "thick" and enough of the original texture and color remains around the boundaries of the damage. Figure 9-10 shows
what happens when the damaged area is too large.

The prototype for cvInpaint() is

```
void cvInpaint(
  const CvArr*  src,
  const CvArr*  mask,
  CvArr*        dst,
  double        inpaintRadius,
  int           flags
);
```

Figure 9-10. Inpainting cannot magically restore textures that are completely removed: the navel of the orange has been completely blotted out (left panel); inpainting fills it back in with mostly orange-like texture (right panel)

Here src is an 8-bit single-channel grayscale image or a three-channel color image to be repaired, and mask is an 8-bit single-channel image of the same size as src in which the damaged areas (e.g., the writing seen in the left panel of Figure 9-9) have been marked by nonzero pixels; all other pixels are set to 0 in mask. The output image will be written to dst, which must be the same size and number of channels as src. The inpaintRadius is the area around each inpainted pixel that will be factored into the resulting output color of that pixel. As in Figure 9-10, interior pixels within a thick enough inpainted region may take their color entirely from other inpainted pixels closer to the boundaries. Almost always, one uses a small radius such as 3 because too large a radius will result in a noticeable blur. Finally, the flags parameter allows you to experiment with two different methods of inpainting: CV_INPAINT_NS (Navier-Stokes method), and CV_INPAINT_TELEA (A. Telea's method).

Mean-Shift Segmentation

In Chapter 5 we introduced the function cvPyrSegmentation(). Pyramid segmentation uses a color merge (over a scale that depends on the similarity of the colors to one another) in order to segment images. This approach is based on minimizing the *total energy* in the image; here energy is defined by a *link strength*, which is further defined by *color similarity*. In this section we introduce cvPyrMeanShiftFiltering(), a similar algorithm that is based on mean-shift clustering over color [Comaniciu99]. We'll see the details of the mean-shift algorithm cvMeanShift() in Chapter 10, when we discuss tracking and motion. For now, what we need to know is that mean shift finds the peak of a color-spatial (or other feature) distribution over time. Here, mean-shift segmentation finds the peaks of color distributions over space. The common theme is that both the motion tracking and the color segmentation algorithms rely on the ability of mean shift to find the modes (peaks) of a distribution.

Given a set of multidimensional data points whose dimensions are (*x, y,* blue, green, red), mean shift can find the highest density "clumps" of data in this space by scanning a *window* over the space. Notice, however, that the spatial variables (*x, y*) can have very different ranges from the color magnitude ranges (blue, green, red). Therefore, mean shift needs to allow for different window radii in different dimensions. In this case we should have one radius for the spatial variables (spatialRadius) and one radius for the color magnitudes (colorRadius). As mean-shift windows move, all the points traversed by the windows that converge at a peak in the data become connected or "owned" by that peak. This ownership, radiating out from the densest peaks, forms the segmentation of the image. The segmentation is actually done over a scale pyramid (cvPyrUp(), cvPyrDown()), as described in Chapter 5, so that color clusters at a high level in the pyramid (shrunken image) have their boundaries refined at lower pyramid levels in the pyramid. The function call for cvPyrMeanShiftFiltering() looks like this:

```
void cvPyrMeanShiftFiltering(
    const CvArr*    src,
    CvArr*          dst,
    double          spatialRadius,
    double          colorRadius,
    int             max_level    = 1,
    CvTermCriteria termcrit     = cvTermCriteria(
        CV_TERMCRIT_ITER | CV_TERMCRIT_EPS,
        5,
        1
    )
);
```

In cvPyrMeanShiftFiltering() we have an input image src and an output image dst. Both must be 8-bit, three-channel color images of the same width and height. The spatialRadius and colorRadius define how the mean-shift algorithm averages color and space together to form a segmentation. For a 640-by-480 color image, it works well to set spatialRadius equal to 2 and colorRadius equal to 40. The next parameter of this algorithm is max_level, which describes how many levels of scale pyramid you want used for segmentation. A max_level of 2 or 3 works well for a 640-by-480 color image.

The final parameter is CvTermCriteria, which we saw in Chapter 8. CvTermCriteria is used for all iterative algorithms in OpenCV. The mean-shift segmentation function comes with good defaults if you just want to leave this parameter blank. Otherwise, cvTermCriteria has the following constructor:

```
cvTermCriteria(
    int    type; // CV_TERMCRIT_ITER, CV_TERMCRIT_EPS,
    int    max_iter,
    double epsilon
);
```

Typically we use the cvTermCriteria() function to generate the CvTermCriteria structure that we need. The first argument is either CV_TERMCRIT_ITER or CV_TERMCRIT_EPS, which tells the algorithm that we want to terminate either after some fixed number of iterations or when the convergence metric reaches some small value (respectively). The next two arguments set the values at which one, the other, or both of these criteria should terminate the algorithm. The reason we have both options is because we can set the type to CV_TERMCRIT_ITER | CV_TERMCRIT_EPS to stop when either limit is reached. The parameter max_iter limits the number of iterations if CV_TERMCRIT_ITER is set, whereas epsilon sets the error limit if CV_TERMCRIT_EPS is set. Of course the exact meaning of epsilon depends on the algorithm.

Figure 9-11 shows an example of mean-shift segmentation using the following values:

```
cvPyrMeanShiftFiltering( src, dst, 20, 40, 2);
```

Figure 9-11. Mean-shift segmentation over scale using cvPyrMeanShiftFiltering() with parameters max_level=2, spatialRadius=20, and colorRadius=40; similar areas now have similar values and so can be treated as super pixels, which can speed up subsequent processing significantly

Delaunay Triangulation, Voronoi Tesselation

Delaunay triangulation is a technique invented in 1934 [Delaunay34] for connecting points in a space into triangular groups such that the minimum angle of all the angles in the triangulation is a maximum. This means that Delaunay triangulation tries to avoid long skinny triangles when triangulating points. See Figure 9-12 to get the gist of triangulation, which is done in such a way that any circle that is fit to the points at the vertices of any given triangle contains no other vertices. This is called the *circum-circle property* (panel c in the figure).

For computational efficiency, the Delaunay algorithm invents a far-away outer bounding triangle from which the algorithm starts. Figure 9-12(b) represents the fictitious outer triangle by faint lines going out to its vertex. Figure 9-12(c) shows some examples of the circum-circle property, including one of the circles linking two outer points of the real data to one of the vertices of the fictitious external triangle.

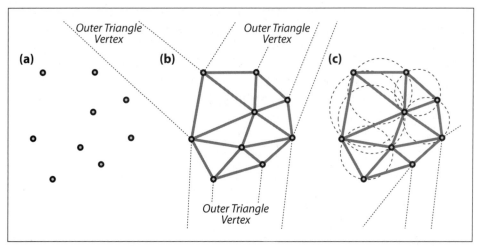

Figure 9-12. Delaunay triangulation: (a) set of points; (b) Delaunay triangulation of the point set with trailers to the outer bounding triangle; (c) example circles showing the circum-circle property

There are now many algorithms to compute Delaunay triangulation; some are very efficient but with difficult internal details. The gist of one of the more simple algorithms is as follows:

1. Add the external triangle and start at one of its vertices (this yields a definitive outer starting point).

2. Add an internal point; then search over all the triangles' circum-circles containing that point and remove those triangulations.

3. Re-triangulate the graph, including the new point in the circum-circles of the just removed triangulations.

4. Return to step 2 until there are no more points to add.

The order of complexity of this algorithm is $O(n^2)$ in the number of data points. The best algorithms are (on average) as low as $O(n \log \log n)$.

Great—but what is it good for? For one thing, remember that this algorithm started with a fictitious outer triangle and so all the real outside points are actually connected to two of that triangle's vertices. Now recall the circum-circle property: circles that are fit through any two of the real outside points and to an external fictitious vertex contain no other inside points. This means that a computer may directly look up exactly which real points form the outside of a set of points by looking at which points are connected to the three outer fictitious vertices. In other words, we can find the convex hull of a set of points almost instantly after a Delaunay triangulation has been done.

We can also find who "owns" the space between points, that is, which coordinates are nearest neighbors to each of the Delaunay vertex points. Thus, using Delaunay triangulation of the original points, you can immediately find the nearest neighbor to a new point. Such a partition is called a *Voronoi tessellation* (see Figure 9-13). This tessellation is the dual image of the Delaunay triangulation, because the Delaunay lines define the distance between existing points and so the Voronoi lines "know" where they must intersect the Delaunay lines in order to keep equal distance between points. These two methods, calculating the convex hull and nearest neighbor, are important basic operations for clustering and classifying points and point sets.

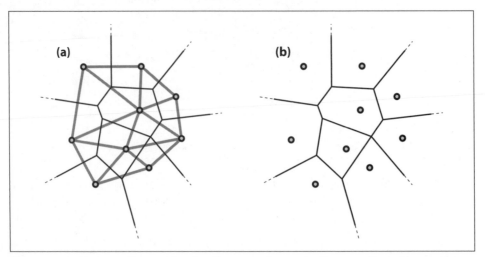

Figure 9-13. Voronoi tessellation, whereby all points within a given Voronoi cell are closer to their Delaunay point than to any other Delaunay point: (a) the Delaunay triangulation in bold with the corresponding Voronoi tessellation in fine lines; (b) the Voronoi cells around each Delaunay point

If you're familiar with 3D computer graphics, you may recognize that Delaunay triangulation is often the basis for representing 3D shapes. If we render an object in three dimensions, we can create a 2D view of that object by its image projection and then use the 2D Delaunay triangulation to analyze and identify this object and/or compare it with a real object. Delaunay triangulation is thus a bridge between computer vision and computer graphics. However, one deficiency of OpenCV (soon to be rectified, we hope; see Chapter 14) is that OpenCV performs Delaunay triangulation only in two dimensions. If we could triangulate point clouds in three dimensions—say, from stereo vision (see Chapter 11)—then we could move seamlessly between 3D computer graphics and computer vision. Nevertheless, 2D Delaunay triangulation is often used in computer vision to register the spatial arrangement of features on an object or a scene for motion tracking, object recognition, or matching views between two different cameras (as in deriving depth from stereo images). Figure 9-14 shows a tracking and recognition application of Delaunay triangulation [Gokturk01; Gokturk02] wherein key facial feature points are spatially arranged according to their triangulation.

Now that we've established the potential usefulness of Delaunay triangulation once given a set of points, how do we derive the triangulation? OpenCV ships with example code for this in the *.../opencv/samples/c/delaunay.c* file. OpenCV refers to Delaunay triangulation as a Delaunay *subdivision*, whose critical and reusable pieces we discuss next.

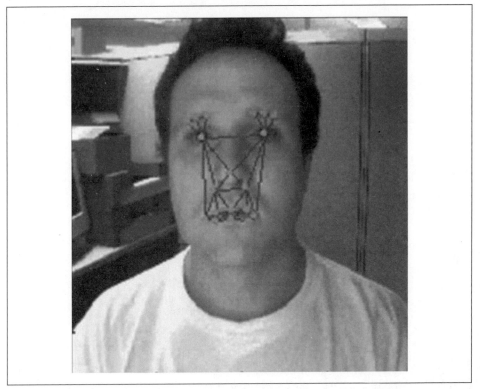

Figure 9-14. Delaunay points can be used in tracking objects; here, a face is tracked using points that are significant in expressions so that emotions may be detected

Creating a Delaunay or Voronoi Subdivision

First we'll need some place to store the Delaunay subdivision in memory. We'll also need an outer bounding box (remember, to speed computations, the algorithm works with a fictitious outer triangle positioned outside a rectangular bounding box). To set this up, suppose the points must be inside a 600-by-600 image:

```
// STORAGE AND STRUCTURE FOR DELAUNAY SUBDIVISION
//
CvRect       rect = { 0, 0, 600, 600 }; //Our outer bounding box
CvMemStorage* storage;                   //Storage for the Delaunay subdivsion
storage = cvCreateMemStorage(0);         //Initialize the storage
CvSubdiv2D*  subdiv;                      //The subdivision itself
subdiv = init_delaunay( storage, rect);  //See this function below
```

The code calls init_delaunay(), which is not an OpenCV function but rather a convenient packaging of a few OpenCV routines:

```
//INITIALIZATION CONVENIENCE FUNCTION FOR DELAUNAY SUBDIVISION
//
CvSubdiv2D* init_delaunay(
  CvMemStorage* storage,
  CvRect rect
```

```
    ) {
      CvSubdiv2D* subdiv;
      subdiv = cvCreateSubdiv2D(
        CV_SEQ_KIND_SUBDIV2D,
        sizeof(*subdiv),
        sizeof(CvSubdiv2DPoint),
        sizeof(CvQuadEdge2D),
        storage
      );
      cvInitSubdivDelaunay2D( subdiv, rect ); //rect sets the bounds
      return subdiv;
    }
```

Next we'll need to know how to insert points. These points must be of type float, 32f:

```
    CvPoint2D32f fp;       //This is our point holder

    for( i = 0; i < as_many_points_as_you_want; i++ ) {

      // However you want to set points
      //
      fp = your_32f_point_list[i];

      cvSubdivDelaunay2DInsert( subdiv, fp );
    }
```

You can convert integer points to 32f points using the convenience macro cvPoint2D32f(double x, double y) or cvPointTo32f(CvPoint point) located in *cxtypes.h*. Now that we can enter points to obtain a Delaunay triangulation, we set and clear the associated Voronoi tessellation with the following two commands:

```
    cvCalcSubdivVoronoi2D( subdiv );   // Fill out Voronoi data in subdiv
    cvClearSubdivVoronoi2D( subdiv );  // Clear the Voronoi from subdiv
```

In both functions, subdiv is of type CvSubdiv2D*. We can now create Delaunay subdivisions of two-dimensional point sets and then add and clear Voronoi tessellations to them. But how do we get at the good stuff inside these structures? We can do this by stepping from edge to point or from edge to edge in subdiv; see Figure 9-15 for the basic maneuvers starting from a given edge and its point of origin. We next find the first edges or points in the subdivision in one of two different ways: (1) by using an external point to locate an edge or a vertex; or (2) by stepping through a sequence of points or edges. We'll first describe how to step around edges and points in the graph and then how to step through the graph.

Navigating Delaunay Subdivisions

Figure 9-15 combines two data structures that we'll use to move around on a subdivision graph. The structure cvQuadEdge2D contains a set of two Delaunay and two Voronoi points and their associated edges (assuming the Voronoi points and edges have been calculated with a prior call to cvCalcSubdivVoronoi2D()); see Figure 9-16. The structure CvSubdiv2DPoint contains the Delaunay edge with its associated vertex point, as shown in Figure 9-17. The quad-edge structure is defined in the code following the figure.

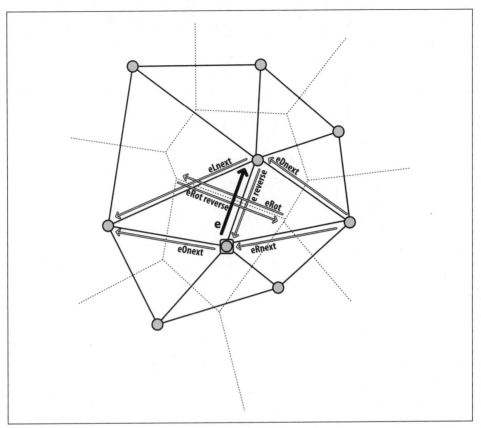

Figure 9-15. Edges relative to a given edge, labeled "e", and its vertex point (marked by a square)

```
// Edges themselves are encoded in long integers. The lower two bits
// are its index (0..3) and upper bits are the quad-edge pointer.
//
typedef long CvSubdiv2DEdge;

// quad-edge structure fields:
//
#define CV_QUADEDGE2D_FIELDS()        /
    int flags;                        /
    struct CvSubdiv2DPoint* pt[4];    /
    CvSubdiv2DEdge  next[4];

typedef struct CvQuadEdge2D {
    CV_QUADEDGE2D_FIELDS()
} CvQuadEdge2D;
```

The Delaunay subdivision point and the associated edge structure is given by:

```
#define CV_SUBDIV2D_POINT_FIELDS() /
    int            flags;          /
    CvSubdiv2DEdge first;          //*The edge "e" in the figures.*/
    CvPoint2D32f   pt;
```

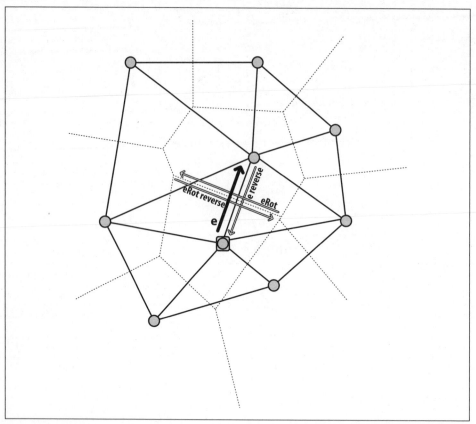

Figure 9-16. Quad edges that may be accessed by cvSubdiv2DRotateEdge() include the Delaunay edge and its reverse (along with their associated vertex points) as well as the related Voronoi edges and points

```
#define CV_SUBDIV2D_VIRTUAL_POINT_FLAG (1 << 30)

typedef struct CvSubdiv2DPoint
{
    CV_SUBDIV2D_POINT_FIELDS()
}
CvSubdiv2DPoint;
```

With these structures in mind, we can now examine the different ways of moving around.

Walking on edges

As indicated by Figure 9-16, we can step around quad edges by using

```
CvSubdiv2DEdge cvSubdiv2DRotateEdge(
  CvSubdiv2DEdge edge,
  int            type
);
```

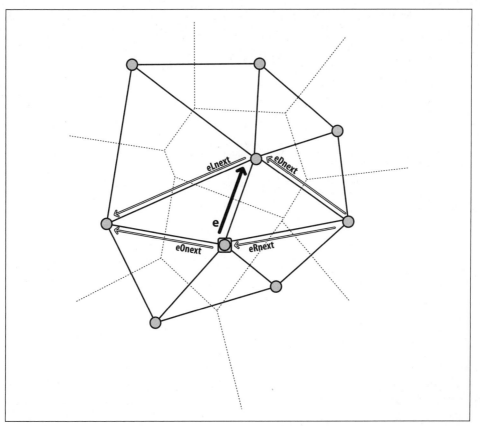

Figure 9-17. A CvSubdiv2DPoint vertex and its associated edge e along with other associated edges that may be accessed via cvSubdiv2DGetEdge()

Given an edge, we can get to the next edge by using the type parameter, which takes one of the following arguments:

- 0, the input edge (e in the figure if e is the input edge)

- 1, the rotated edge (eRot)

- 2, the reversed edge (reversed e)

- 3, the reversed rotated edge (reversed eRot)

Referencing Figure 9-17, we can also get around the Delaunay graph using

```
CvSubdiv2DEdge cvSubdiv2DGetEdge(
   CvSubdiv2DEdge edge,
   CvNextEdgeType type
);
#define cvSubdiv2DNextEdge( edge )        /
   cvSubdiv2DGetEdge(                     /
      edge,                               /
      CV_NEXT_AROUND_ORG                  /
   )
```

Here type specifies one of the following moves:

CV_NEXT_AROUND_ORG

> Next around the edge origin (eOnext in Figure 9-17 if e is the input edge)

CV_NEXT_AROUND_DST

> Next around the edge vertex (eDnext)

CV_PREV_AROUND_ORG

> Previous around the edge origin (reversed eRnext)

CV_PREV_AROUND_DST

> Previous around the edge destination (reversed eLnext)

CV_NEXT_AROUND_LEFT

> Next around the left facet (eLnext)

CV_NEXT_AROUND_RIGHT

> Next around the right facet (eRnext)

CV_PREV_AROUND_LEFT

> Previous around the left facet (reversed eOnext)

CV_PREV_AROUND_RIGHT

> Previous around the right facet (reversed eDnext)

Note that, given an edge associated with a vertex, we can use the convenience macro cvSubdiv2DNextEdge(edge) to find all other edges from that vertex. This is helpful for finding things like the convex hull starting from the vertices of the (fictitious) outer bounding triangle.

The other important movement types are CV_NEXT_AROUND_LEFT and CV_NEXT_AROUND_RIGHT. We can use these to step around a Delaunay triangle if we're on a Delaunay edge or to step around a Voronoi cell if we're on a Voronoi edge.

Points from edges

We'll also need to know how to retrieve the actual points from Delaunay or Voronoi vertices. Each Delaunay or Voronoi edge has two points associated with it: org, its origin point, and dst, its destination point. You may easily obtain these points by using

```
CvSubdiv2DPoint* cvSubdiv2DEdgeOrg( CvSubdiv2DEdge edge );
CvSubdiv2DPoint* cvSubdiv2DEdgeDst( CvSubdiv2DEdge edge );
```

Here are methods to convert CvSubdiv2DPoint to more familiar forms:

```
CvSubdiv2DPoint ptSub;                      //Subdivision vertex point
CvPoint2D32f    pt32f = ptSub->pt;          // to 32f point
CvPoint         pt    = cvPointFrom32f(pt32f); // to an integer point
```

We now know what the subdivision structures look like and how to walk around its points and edges. Let's return to the two methods for getting the first edges or points from the Delaunay/Voronoi subdivision.

Method 1: Use an external point to locate an edge or vertex

The first method is to start with an arbitrary point and then locate that point in the subdivision. This need not be a point that has already been triangulated; it can be any point. The function cvSubdiv2DLocate() fills in one edge and vertex (if desired) of the triangle or Voronoi facet into which that point fell.

```
CvSubdiv2DPointLocation cvSubdiv2DLocate(
    CvSubdiv2D*       subdiv,
    CvPoint2D32f      pt,
    CvSubdiv2DEdge*   edge,
    CvSubdiv2DPoint** vertex = NULL
);
```

Note that these are not necessarily the closest edge or vertex; they just have to be in the triangle or facet. This function's return value tells us where the point landed, as follows.

CV_PTLOC_INSIDE

> The point falls into some facet; *edge will contain one of edges of the facet.

CV_PTLOC_ON_EDGE

> The point falls onto the edge; *edge will contain this edge.

CV_PTLOC_VERTEX

> The point coincides with one of subdivision vertices; *vertex will contain a pointer to the vertex.

CV_PTLOC_OUTSIDE_RECT

> The point is outside the subdivision reference rectangle; the function returns and no pointers are filled.

CV_PTLOC_ERROR

> One of input arguments is invalid.

Method 2: Step through a sequence of points or edges

Conveniently for us, when we create a Delaunay subdivision of a set of points, the first three points and edges form the vertices and sides of the fictitious outer bounding triangle. From there, we may directly access the outer points and edges that form the convex hull of the actual data points. Once we have formed a Delaunay subdivision (call it subdiv), we'll also need to call cvCalcSubdivVoronoi2D(subdiv) in order to calculate the associated Voronoi tessellation. We can then access the three vertices of the outer bounding triangle using

```
CvSubdiv2DPoint* outer_vtx[3];
for( i = 0; i < 3; i++ ) {
  outer_vtx[i] =
    (CvSubdiv2DPoint*)cvGetSeqElem( (CvSeq*)subdiv, I );
}
```

We can similarly obtain the three sides of the outer bounding triangle:

```
CvQuadEdge2D* outer_qedges[3];
for( i = 0; i < 3; i++ ) {
  outer_qedges[i] =
    (CvQuadEdge2D*)cvGetSeqElem( (CvSeq*)(my_subdiv->edges), I );
}
```

Now that we know how to get on the graph and move around, we'll want to know when we're on the outer edge or boundary of the points.

Identifying the bounding triangle or edges on the convex hull and walking the hull

Recall that we used a bounding rectangle rect to initialize the Delaunay triangulation with the call cvInitSubdivDelaunay2D(subdiv, rect). In this case, the following statements hold.

1. If you are on an edge where both the origin and destination points are out of the rect bounds, then that edge is on the fictitious bounding triangle of the subdivision.

2. If you are on an edge with one point inside and one point outside the rect bounds, then the point in bounds is on the convex hull of the set; each point on the convex hull is connected to two vertices of the fictitious outer bounding triangle, and these two edges occur one after another.

From the second condition, you can use the cvSubdiv2DNextEdge() macro to step onto the first edge whose dst point is within bounds. That first edge with both ends in bounds is on the convex hull of the point set, so remember that point or edge. Once on the convex hull, you can then move around the convex hull as follows.

1. Until you have circumnavigated the convex hull, go to the next edge on the hull via cvSubdiv2DRotateEdge(CvSubdiv2DEdge edge, 0).

2. From there, another two calls to the cvSubdiv2DNextEdge() macro will get you on the next edge of the convex hull. Return to step 1.

We now know how to initialize Delaunay and Voronoi subdivisions, how to find the initial edges, and also how to step through the edges and points of the graph. In the next section we present some practical applications.

Usage Examples

We can use cvSubdiv2DLocate() to step around the edges of a Delaunay triangle:

```
void locate_point(
  CvSubdiv2D*   subdiv,
  CvPoint2D32f  fp,
  IplImage*     img,
  CvScalar      active_color
) {
  CvSubdiv2DEdge e;
  CvSubdiv2DEdge e0 = 0;
  CvSubdiv2DPoint* p = 0;
  cvSubdiv2DLocate( subdiv, fp, &e0, &p );
```

```
    if( e0 ) {
      e = e0;
      do // Always 3 edges -- this is a triangulation, after all.
      {
        // [Insert your code here]
        //
        // Do something with e ...
          e = cvSubdiv2DGetEdge(e,CV_NEXT_AROUND_LEFT);
      }
      while( e != e0 );
    }
  }
```

We can also find the closest point to an input point by using

```
CvSubdiv2DPoint* cvFindNearestPoint2D(
  CvSubdiv2D*  subdiv,
  CvPoint2D32f pt
);
```

Unlike cvSubdiv2DLocate(), cvFindNearestPoint2D() will return the nearest vertex point in the Delaunay subdivision. This point is not necessarily on the facet or triangle that the point lands on.

Similarly, we could step around a Voronoi facet (here we draw it) using

```
void draw_subdiv_facet(
  IplImage *img,
  CvSubdiv2DEdge edge
) {

  CvSubdiv2DEdge t = edge;
  int i, count = 0;
  CvPoint* buf = 0;

  // Count number of edges in facet
  do{
      count++;
      t = cvSubdiv2DGetEdge( t, CV_NEXT_AROUND_LEFT );
  } while (t != edge );

  // Gather points
  //
  buf = (CvPoint*)malloc( count * sizeof(buf[0]))
  t = edge;
  for( i = 0; i < count; i++ ) {
      CvSubdiv2DPoint* pt = cvSubdiv2DEdgeOrg( t );
      if( !pt ) break;
      buf[i] = cvPoint( cvRound(pt->pt.x), cvRound(pt->pt.y));
      t = cvSubdiv2DGetEdge( t, CV_NEXT_AROUND_LEFT );
  }

  // Around we go
  //
  if( i == count ){
      CvSubdiv2DPoint* pt = cvSubdiv2DEdgeDst(
```

```
                                    cvSubdiv2DRotateEdge( edge, 1 ));
            cvFillConvexPoly( img, buf, count,
              CV_RGB(rand()&255,rand()&255,rand()&255), CV_AA, 0 );
            cvPolyLine( img, &buf, &count, 1, 1, CV_RGB(0,0,0),
                    1, CV_AA, 0);
            draw_subdiv_point( img, pt->pt, CV_RGB(0,0,0));
        }
      free( buf );
    }
```

Finally, another way to access the subdivision structure is by using a CvSeqReader to step though a sequence of edges. Here's how to step through all Delaunay or Voronoi edges:

```
void visit_edges( CvSubdiv2D* subdiv){

    CvSeqReader  reader;                    //Sequence reader
    int i, total = subdiv->edges->total;      //edge count
    int elem_size = subdiv->edges->elem_size; //edge size

    cvStartReadSeq( (CvSeq*)(subdiv->edges), &reader, 0 );

    cvCalcSubdivVoronoi2D( subdiv ); //Make sure Voronoi exists

    for( i = 0; i < total; i++ ) {

      CvQuadEdge2D* edge = (CvQuadEdge2D*)(reader.ptr);

      if( CV_IS_SET_ELEM( edge )) {

        // Do something with Voronoi and Delaunay edges ...
        //
        CvSubdiv2DEdge voronoi_edge = (CvSubdiv2DEdge)edge + 1;
        CvSubdiv2DEdge delaunay_edge = (CvSubdiv2DEdge)edge;

        // ...OR WE COULD FOCUS EXCLUSIVELY ON VORONOI...

        // left
        //
        voronoi_edge = cvSubdiv2DRotateEdge( edge, 1 );

        // right
        //
        voronoi_edge = cvSubdiv2DRotateEdge( edge, 3 );
      }
      CV_NEXT_SEQ_ELEM( elem_size, reader );
    }
  }
```

Finally, we end with an inline convenience macro: once we find the vertices of a Delaunay triangle, we can find its area by using

```
double cvTriangleArea(
  CvPoint2D32f a,
  CvPoint2D32f b,
  CvPoint2D32f c
)
```

Exercises

1. Using cvRunningAvg(), re-implement the averaging method of background subtraction. In order to do so, learn the running average of the pixel values in the scene to find the mean and the running average of the absolute difference (cvAbsDiff()) as a proxy for the standard deviation of the image.

2. Shadows are often a problem in background subtraction because they can show up as a foreground object. Use the averaging or codebook method of background subtraction to learn the background. Have a person then walk in the foreground. Shadows will "emanate" from the bottom of the foreground object.

   a. Outdoors, shadows are darker and bluer than their surround; use this fact to eliminate them.

   b. Indoors, shadows are darker than their surround; use this fact to eliminate them.

3. The simple background models presented in this chapter are often quite sensitive to their threshold parameters. In Chapter 10 we'll see how to track motion, and this can be used as a "reality" check on the background model and its thresholds. You can also use it when a known person is doing a "calibration walk" in front of the camera: find the moving object and adjust the parameters until the foreground object corresponds to the motion boundaries. We can also use distinct patterns on a calibration object itself (or on the background) for a reality check and tuning guide when we know that a portion of the background has been occluded.

   a. Modify the code to include an autocalibration mode. Learn a background model and then put a brightly colored object in the scene. Use color to find the colored object and then use that object to automatically set the thresholds in the background routine so that it segments the object. Note that you can leave this object in the scene for continuous tuning.

   b. Use your revised code to address the shadow-removal problem of exercise 2.

4. Use background segmentation to segment a person with arms held out. Investigate the effects of the different parameters and defaults in the find_connected_components() routine. Show your results for different settings of:

   a. poly1_hull0

   b. perimScale

   c. CVCONTOUR_APPROX_LEVEL

   d. CVCLOSE_ITR

5. In the 2005 DARPA Grand Challenge robot race, the authors on the Stanford team used a kind of color clustering algorithm to separate road from nonroad. The colors were sampled from a laser-defined trapezoid of road patch in front of the car. Other colors in the scene that were close in color to this patch—and whose connected

component connected to the original trapezoid—were labeled as road. See Figure 9-18, where the watershed algorithm was used to segment the road after using a trapezoid mark inside the road and an inverted "U" mark outside the road. Suppose we could automatically generate these marks. What could go wrong with this method of segmenting the road?

> Hint: Look carefully at Figure 9-8 and then consider that we are trying to extend the road trapezoid by using things that look like what's in the trapezoid.

Figure 9-18. Using the watershed algorithm to identify a road: markers are put in the original image (left), and the algorithm yields the segmented road (right)

6. Inpainting works pretty well for the repair of writing over textured regions. What would happen if the writing obscured a real object edge in a picture? Try it.

7. Although it might be a little slow, try running background segmentation when the video input is first pre-segmented by using `cvPyrMeanShiftFiltering()`. That is, the input stream is first mean-shift segmented and then passed for background learning—and later testing for foreground—by the codebook background segmentation routine.

   a. Show the results compared to not running the mean-shift segmentation.

   b. Try systematically varying the `max_level`, `spatialRadius`, and `colorRadius` of the mean-shift segmentation. Compare those results.

8. How well does inpainting work at fixing up writing drawn over a mean-shift segmented image? Try it for various settings and show the results.

9. Modify the *.../opencv/samples/delaunay.c* code to allow mouse-click point entry (instead of via the existing method where points are selected at a random). Experiment with triangulations on the results.

10. Modify the *delaunay.c* code again so that you can use a keyboard to draw the convex hull of the point set.

11. Do three points in a line have a Delaunay triangulation?

12. Is the triangulation shown in Figure 9-19(a) a Delaunay triangulation? If so, explain your answer. If not, how would you alter the figure so that it is a Delaunay triangulation?

13. Perform a Delaunay triangulation by hand on the points in Figure 9-19(b). For this exercise, you need not add an outer fictitious bounding triangle.

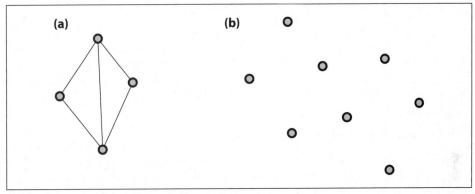

Figure 9-19. Exercise 12 and Exercise 13

CHAPTER 10

Tracking and Motion

The Basics of Tracking

When we are dealing with a video source, as opposed to individual still images, we often have a particular object or objects that we would like to follow through the visual field. In the previous chapter, we saw how to isolate a particular shape, such as a person or an automobile, on a frame-by-frame basis. Now what we'd like to do is understand the motion of this object, a task that has two main components: identification and modeling.

Identification amounts to finding the object of interest from one frame in a subsequent frame of the video stream. Techniques such as moments or color histograms from previous chapters will help us identify the object we seek. Tracking things that we have not yet identified is a related problem. Tracking unidentified objects is important when we wish to determine what is interesting based on its motion—or when an object's motion is precisely what makes it interesting. Techniques for tracking unidentified objects typically involve tracking visually significant key points (more soon on what constitutes "significance"), rather than extended objects. OpenCV provides two methods for achieving this: the Lucas-Kanade* [Lucas81] and Horn-Schunck [Horn81] techniques, which represent what are often referred to as *sparse* or *dense* optical flow respectively.

The second component, modeling, helps us address the fact that these techniques are really just providing us with noisy measurement of the object's actual position. Many powerful mathematical techniques have been developed for estimating the trajectory of an object measured in such a noisy manner. These methods are applicable to two- or three-dimensional models of objects and their locations.

Corner Finding

There are many kinds of local features that one can track. It is worth taking a moment to consider what exactly constitutes such a feature. Obviously, if we pick a point on a large blank wall then it won't be easy to find that same point in the next frame of a video.

---

* Oddly enough, the definitive description of Lucas-Kanade optical flow in a pyramid framework implemented in OpenCV is an unpublished paper by Bouguet [Bouguet04].

If all points on the wall are identical or even very similar, then we won't have much luck tracking that point in subsequent frames. On the other hand, if we choose a point that is unique then we have a pretty good chance of finding that point again. In practice, the point or feature we select should be unique, or nearly unique, and should be parameterizable in such a way that it can be compared to other points in another image. See Figure 10-1.

Figure 10-1. The points in circles are good points to track, whereas those in boxes—even sharply defined edges—are poor choices

Returning to our intuition from the large blank wall, we might be tempted to look for points that have some significant change in them—for example, a strong derivative. It turns out that this is not enough, but it's a start. A point to which a strong derivative is associated may be on an edge of some kind, but it could look like all of the other points along the same edge (see the aperture problem diagrammed in Figure 10-8 and discussed in the section titled "Lucas-Kanade Technique").

However, if strong derivatives are observed in two orthogonal directions then we can hope that this point is more likely to be unique. For this reason, many trackable features are called *corners*. Intuitively, corners—not edges—are the points that contain enough information to be picked out from one frame to the next.

The most commonly used definition of a corner was provided by Harris [Harris88]. This definition relies on the matrix of the second-order derivatives ($\partial^2 x$, $\partial^2 y$, $\partial x\, \partial y$) of the image intensities. We can think of the second-order derivatives of images, taken at all points in the image, as forming new "second-derivative images" or, when combined together, a new *Hessian* image. This terminology comes from the Hessian matrix around a point, which is defined in two dimensions by:

$$H(p) = \begin{bmatrix} \dfrac{\partial^2 I}{\partial x^2} & \dfrac{\partial^2 I}{\partial x\, \partial y} \\[2ex] \dfrac{\partial^2 I}{\partial y\, \partial x} & \dfrac{\partial^2 I}{\partial y^2} \end{bmatrix}_p$$

For the Harris corner, we consider the *autocorrelation matrix* of the second derivative images over a small window around each point. Such a matrix is defined as follows:

$$M(x,y) = \begin{bmatrix} \displaystyle\sum_{-K \le i,j \le K} w_{i,j} I_x^2(x+i,y+j) & \displaystyle\sum_{-K \le i,j \le K} w_{i,j} I_x(x+i,y+j) I_y(x+i,y+j) \\ \displaystyle\sum_{-K \le i,j \le K} w_{i,j} I_x(x+i,y+j) I_y(x+i,y+j) & \displaystyle\sum_{-K \le i,j \le K} w_{i,j} I_y^2(x+i,y+j) \end{bmatrix}$$

(Here $w_{i,j}$ is a weighting term that can be uniform but is often used to create a circular window or Gaussian weighting.) Corners, by Harris's definition, are places in the image where the autocorrelation matrix of the second derivatives has two large eigenvalues. In essence this means that there is texture (or edges) going in at least two separate directions centered around such a point, just as real corners have at least two edges meeting in a point. Second derivatives are useful because they do not respond to uniform gradients.* This definition has the further advantage that, when we consider only the eigenvalues of the autocorrelation matrix, we are considering quantities that are invariant also to rotation, which is important because objects that we are tracking might rotate as well as move. Observe also that these two eigenvalues do more than determine if a point is a good feature to track; they also provide an identifying signature for the point.

Harris's original definition involved taking the determinant of $H(p)$, subtracting the trace of $H(p)$ (with some weighting coefficient), and then comparing this difference to a predetermined threshold. It was later found by Shi and Tomasi [Shi94] that good corners resulted as long as the smaller of the two eigenvalues was greater than a minimum threshold. Shi and Tomasi's method was not only sufficient but in many cases gave more satisfactory results than Harris's method.

The cvGoodFeaturesToTrack() routine implements the Shi and Tomasi definition. This function conveniently computes the second derivatives (using the Sobel operators) that are needed and from those computes the needed eigenvalues. It then returns a list of the points that meet our definition of being good for tracking.

```
void  cvGoodFeaturesToTrack(
    const CvArr*     image,
    CvArr*           eigImage,
    CvArr*           tempImage,
    CvPoint2D32f*    corners,
    int*             corner_count,
    double           quality_level,
    double           min_distance,
    const CvArr*     mask          = NULL,
    int              block_size    = 3,
    int              use_harris    = 0,
    double           k             = 0.4
);
```

---

* A gradient is derived from first derivatives. If first derivatives are uniform (constant), then second derivatives are 0.

In this case, the input image should be an 8-bit or 32-bit (i.e., IPL_DEPTH_8U or IPL_DEPTH_32F) single-channel image. The next two arguments are single-channel 32-bit images of the same size. Both tempImage and eigImage are used as scratch by the algorithm, but the resulting contents of eigImage are meaningful. In particular, each entry there contains the minimal eigenvalue for the corresponding point in the input image. Here corners is an array of 32-bit points (CvPoint2D32f) that contain the result points after the algorithm has run; you must allocate this array before calling cvGoodFeatures ToTrack(). Naturally, since you allocated that array, you only allocated a finite amount of memory. The corner_count indicates the maximum number of points for which there is space to return. After the routine exits, corner_count is overwritten by the number of points that were actually found. The parameter quality_level indicates the minimal acceptable lower eigenvalue for a point to be included as a corner. The actual minimal eigenvalue used for the cutoff is the product of the quality_level and the largest lower eigenvalue observed in the image. Hence, the quality_level should not exceed 1 (a typical value might be 0.10 or 0.01). Once these candidates are selected, a further culling is applied so that multiple points within a small region need not be included in the response. In particular, the min_distance guarantees that no two returned points are within the indicated number of pixels.

The optional mask is the usual image, interpreted as Boolean values, indicating which points should and which points should not be considered as possible corners. If set to NULL, no mask is used. The block_size is the region around a given pixel that is considered when computing the autocorrelation matrix of derivatives. It turns out that it is better to sum these derivatives over a small window than to compute their value at only a single point (i.e., at a block_size of 1). If use_harris is nonzero, then the Harris corner definition is used rather than the Shi-Tomasi definition. If you set use_harris to a nonzero value, then the value k is the weighting coefficient used to set the relative weight given to the trace of the autocorrelation matrix Hessian compared to the determinant of the same matrix.

Once you have called cvGoodFeaturesToTrack(), the result is an array of pixel locations that you hope to find in another similar image. For our current context, we are interested in looking for these features in subsequent frames of video, but there are many other applications as well. A similar technique can be used when attempting to relate multiple images taken from slightly different viewpoints. We will re-encounter this issue when we discuss stereo vision in later chapters.

Subpixel Corners

If you are processing images for the purpose of extracting geometric measurements, as opposed to extracting features for recognition, then you will normally need more resolution than the simple pixel values supplied by cvGoodFeaturesToTrack(). Another way of saying this is that such pixels come with integer coordinates whereas we sometimes require real-valued coordinates—for example, pixel (8.25, 117.16).

One might imagine needing to look for a sharp peak in image values, only to be frustrated by the fact that the peak's location will almost never be in the exact center of a camera pixel element. To overcome this, you might fit a curve (say, a parabola) to the image values and then use a little math to find where the peak occurred between the pixels. Subpixel detection techniques are all about tricks like this (for a review and newer techniques, see Lucchese [Lucchese02] and Chen [Chen05]). Common uses of image measurements are tracking for three-dimensional reconstruction, calibrating a camera, warping partially overlapping views of a scene to stitch them together in the most natural way, and finding an external signal such as precise location of a building in a satellite image.

Subpixel corner locations are a common measurement used in camera calibration or when tracking to reconstruct the camera's path or the three-dimensional structure of a tracked object. Now that we know how to find corner locations on the integer grid of pixels, here's the trick for refining those locations to subpixel accuracy: We use the mathematical fact that the dot product between a vector and an orthogonal vector is 0; this situation occurs at corner locations, as shown in Figure 10-2.

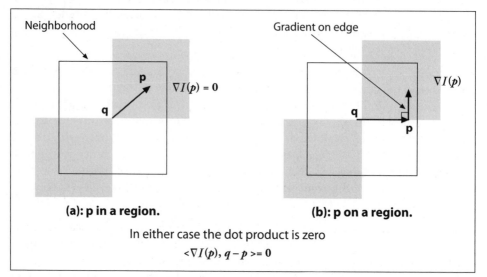

Figure 10-2. Finding corners to subpixel accuracy: (a) the image area around the point p is uniform and so its gradient is 0; (b) the gradient at the edge is orthogonal to the vector q-p along the edge; in either case, the dot product between the gradient at p and the vector q-p is 0 (see text)

In the figure, we assume a starting corner location q that is near the actual subpixel corner location. We examine vectors starting at point q and ending at p. When p is in a nearby uniform or "flat" region, the gradient there is 0. On the other hand, if the vector $q-p$ aligns with an edge then the gradient at p on that edge is orthogonal to the vector $q-p$. In either case, the dot product between the gradient at p and the vector $q-p$ is 0. We can assemble many such pairs of the gradient at a nearby point p and the associated vector $q-p$, set their dot product to 0, and solve this assemblage as a system of equations; the solution will yield a more accurate subpixel location for q, the exact location of the corner.

The function that does subpixel corner finding is cvFindCornerSubPix():

```
void cvFindCornerSubPix(
    const CvArr*    image,
    CvPoint2D32f*   corners,
    int             count,
    CvSize          win,
    CvSize          zero_zone,
    CvTermCriteria  criteria
);
```

The input image is a single-channel, 8-bit, grayscale image. The corners structure contains integer pixel locations, such as those obtained from routines like cvGoodFeatures ToTrack(), which are taken as the initial guesses for the corner locations; count holds how many points there are to compute.

The actual computation of the subpixel location uses a system of dot-product expressions that all equal 0 (see Figure 10-2), where each equation arises from considering a single pixel in the region around p. The parameter win specifies the size of window from which these equations will be generated. This window is centered on the original integer corner location and extends outward in each direction by the number of pixels specified in win (e.g., if win.width = 4 then the search area is actually $4 + 1 + 4 = 9$ pixels wide). These equations form a linear system that can be solved by the inversion of a single autocorrelation matrix (not related to the autocorrelation matrix encountered in our previous discussion of Harris corners). In practice, this matrix is not always invertible owing to small eigenvalues arising from the pixels very close to p. To protect against this, it is common to simply reject from consideration those pixels in the immediate neighborhood of p. The parameter zero_zone defines a window (analogously to win, but always with a smaller extent) that will *not* be considered in the system of constraining equations and thus the autocorrelation matrix. If no such zero zone is desired then this parameter should be set to cvSize(-1,-1).

Once a new location is found for q, the algorithm will iterate using that value as a starting point and will continue until the user-specified termination criterion is reached. Recall that this criterion can be of type CV_TERMCRIT_ITER or of type CV_TERMCRIT_EPS (or both) and is usually constructed with the cvTermCriteria() function. Using CV_TERMCRIT_EPS will effectively indicate the accuracy you require of the subpixel values. Thus, if you specify 0.10 then you are asking for subpixel accuracy down to one tenth of a pixel.

Invariant Features

Since the time of Harris's original paper and the subsequent work by Shi and Tomasi, a great many other types of corners and related local features have been proposed. One widely used type is the SIFT ("scale-invariant feature transform") feature [Lowe04]. Such features are, as their name suggests, scale-invariant. Because SIFT detects the dominant gradient orientation at its location and records its local gradient histogram results with respect to this orientation, SIFT is also rotationally invariant. As a result, SIFT features are relatively well behaved under small affine transformations. Although the SIFT algorithm is not yet implemented as part of the OpenCV library (but see Chapter 14), it is possible to create such an implementation using OpenCV primitives. We will not spend more time on this topic, but it is worth keeping in mind that, given the OpenCV functions we've already discussed, it is possible (albeit less convenient) to create most of the features reported in the computer vision literature (see Chapter 14 for a feature tool kit in development).

Optical Flow

As already mentioned, you may often want to assess motion between two frames (or a sequence of frames) without any other prior knowledge about the content of those frames. Typically, the motion itself is what indicates that something interesting is going on. Optical flow is illustrated in Figure 10-3.

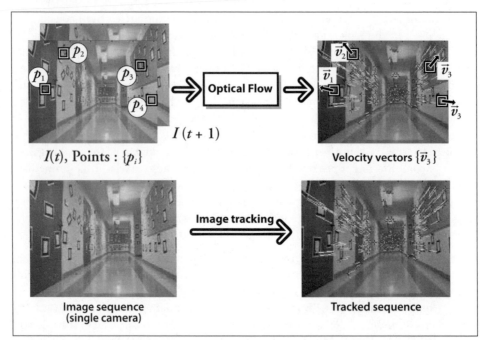

Figure 10-3. Optical flow: target features (upper left) are tracked over time and their movement is converted into velocity vectors (upper right); lower panels show a single image of the hallway (left) and flow vectors (right) as the camera moves down the hall (original images courtesy of Jean-Yves Bouguet)

We can associate some kind of velocity with each pixel in the frame or, equivalently, some displacement that represents the distance a pixel has moved between the previous frame and the current frame. Such a construction is usually referred to as a *dense optical flow*, which associates a velocity with every pixel in an image. The *Horn-Schunck method* [Horn81] attempts to compute just such a velocity field. One seemingly straightforward method—simply attempting to match windows around each pixel from one frame to the next—is also implemented in OpenCV; this is known as *block matching*. Both of these routines will be discussed in the "Dense Tracking Techniques" section.

In practice, calculating dense optical flow is not easy. Consider the motion of a white sheet of paper. Many of the white pixels in the previous frame will simply remain white in the next. Only the edges may change, and even then only those perpendicular to the direction of motion. The result is that dense methods must have some method of interpolating between points that are more easily tracked so as to solve for those points that are more ambiguous. These difficulties manifest themselves most clearly in the high computational costs of dense optical flow.

This leads us to the alternative option, *sparse optical flow*. Algorithms of this nature rely on some means of specifying beforehand the subset of points that are to be tracked. If these points have certain desirable properties, such as the "corners" discussed earlier, then the tracking will be relatively robust and reliable. We know that OpenCV can help us by providing routines for identifying the best features to track. For many practical applications, the computational cost of sparse tracking is so much less than dense tracking that the latter is relegated to only academic interest.*

The next few sections present some different methods of tracking. We begin by considering the most popular sparse tracking technique, *Lucas-Kanade* (LK) optical flow; this method also has an implementation that works with image pyramids, allowing us to track faster motions. We'll then move on to two dense techniques, the Horn-Schunck method and the block matching method.

Lucas-Kanade Method

The Lucas-Kanade (LK) algorithm [Lucas81], as originally proposed in 1981, was an attempt to produce dense results. Yet because the method is easily applied to a subset of the points in the input image, it has become an important sparse technique. The LK algorithm can be applied in a sparse context because it relies only on local information that is derived from some small window surrounding each of the points of interest. This is in contrast to the intrinsically global nature of the Horn and Schunck algorithm (more on this shortly). The disadvantage of using small local windows in Lucas-Kanade is that large motions can move points outside of the local window and thus become impossible for the algorithm to find. This problem led to development of the "pyramidal" LK algorithm, which tracks starting from highest level of an image pyramid (lowest detail) and working down to lower levels (finer detail). Tracking over image pyramids allows large motions to be caught by local windows.

Because this is an important and effective technique, we shall go into some mathematical detail; readers who prefer to forgo such details can skip to the function description and code. However, it is recommended that you at least scan the intervening text and

---

* Black and Anadan have created dense optical flow techniques [Black93; Black96] that are often used in movie production, where, for the sake of visual quality, the movie studio is willing to spend the time necessary to obtain detailed flow information. These techniques are slated for inclusion in later versions of OpenCV (see Chapter 14).

figures, which describe the assumptions behind Lucas-Kanade optical flow, so that you'll have some intuition about what to do if tracking isn't working well.

How Lucas-Kanade works

The basic idea of the LK algorithm rests on three assumptions.

1. *Brightness constancy.* A pixel from the image of an object in the scene does not change in appearance as it (possibly) moves from frame to frame. For grayscale images (LK can also be done in color), this means we assume that the brightness of a pixel does not change as it is tracked from frame to frame.

2. *Temporal persistence or "small movements".* The image motion of a surface patch changes slowly in time. In practice, this means the temporal increments are fast enough relative to the scale of motion in the image that the object does not move much from frame to frame.

3. *Spatial coherence.* Neighboring points in a scene belong to the same surface, have similar motion, and project to nearby points on the image plane.

We now look at how these assumptions, which are illustrated in Figure 10-4, lead us to an effective tracking algorithm. The first requirement, brightness constancy, is just the requirement that pixels in one tracked patch look the same over time:

$$f(x, t) \equiv I(x(t), t) = I(x(t+dt), t+dt)$$

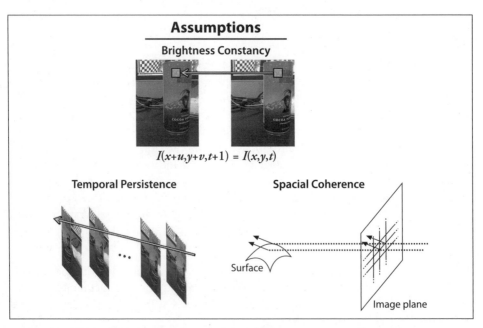

Figure 10-4. Assumptions behind Lucas-Kanade optical flow: for a patch being tracked on an object in a scene, the patch's brightness doesn't change (top); motion is slow relative to the frame rate (lower left); and neighboring points stay neighbors (lower right) (component images courtesy of Michael Black [Black82])

That's simple enough, and it means that our tracked pixel intensity exhibits no change over time:

$$\frac{\partial f(x)}{\partial t} = 0$$

The second assumption, temporal persistence, essentially means that motions are small from frame to frame. In other words, we can view this change as approximating a derivative of the intensity with respect to time (i.e., we assert that the change between one frame and the next in a sequence is *differentially small*). To understand the implications of this assumption, first consider the case of a single spatial dimension.

In this case we can start with our brightness consistency equation, substitute the definition of the brightness $f(x, t)$ while taking into account the implicit dependence of x on t, $I(x(t), t)$, and then apply the chain rule for partial differentiation. This yields:

$$\underbrace{\frac{\partial I}{\partial x}\bigg|_{t}}_{I_x} \underbrace{\left(\frac{\partial x}{\partial t}\right)}_{v} + \underbrace{\frac{\partial I}{\partial t}\bigg|_{x(t)}}_{I_t} = 0$$

where I_x is the spatial derivative across the first image, I_t is the derivative between images over time, and $\mathbf{v}$ is the velocity we are looking for. We thus arrive at the simple equation for optical flow velocity in the simple one-dimensional case:

$$\mathbf{v} = -\frac{I_t}{I_x}$$

Let's now try to develop some intuition for the one-dimensional tracking problem. Consider Figure 10-5, which shows an "edge"—consisting of a high value on the left and a low value on the right—that is moving to the right along the x-axis. Our goal is to identify the velocity $\mathbf{v}$ at which the edge is moving, as plotted in the upper part of Figure 10-5. In the lower part of the figure we can see that our measurement of this velocity is just "rise over run," where the rise is over time and the run is the slope (spatial derivative). The negative sign corrects for the slope of x.

Figure 10-5 reveals another aspect to our optical flow formulation: our assumptions are probably not quite true. That is, image brightness is not really stable; and our time steps (which are set by the camera) are often not as fast relative to the motion as we'd like. Thus, our solution for the velocity is not exact. However, if we are "close enough" then we can iterate to a solution. Iteration is shown in Figure 10-6, where we use our first (inaccurate) estimate of velocity as the starting point for our next iteration and then repeat. Note that we can keep the same spatial derivative in x as computed on the first frame because of the brightness constancy assumption—pixels moving in x do not change. This reuse of the spatial derivative already calculated yields significant computational savings. The time derivative must still be recomputed each iteration and each frame, but

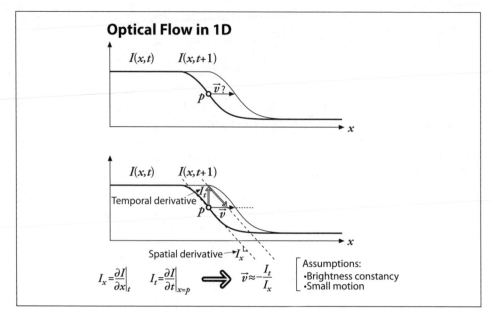

Figure 10-5. Lucas-Kanade optical flow in one dimension: we can estimate the velocity of the moving edge (upper panel) by measuring the ratio of the derivative of the intensity over time divided by the derivative of the intensity over space

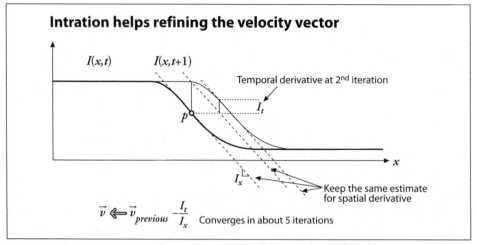

Figure 10-6. Iterating to refine the optical flow solution (Newton's method): using the same two images and the same spatial derivative (slope) we solve again for the time derivative; convergence to a stable solution usually occurs within a few iterations

if we are close enough to start with then these iterations will converge to near exactitude within about five iterations. This is known as *Newton's method*. If our first estimate was not close enough, then Newton's method will actually diverge.

Now that we've seen the one-dimensional solution, let's generalize it to images in two dimensions. At first glance, this seems simple: just add in the y coordinate. Slightly changing notation, we'll call the y component of velocity v and the x component of velocity u; then we have:

$$I_x u + I_y v + I_t = 0$$

Unfortunately, for this single equation there are two unknowns for any given pixel. This means that measurements at the single-pixel level are underconstrained and cannot be used to obtain a unique solution for the two-dimensional motion at that point. Instead, we can only solve for the motion component that is perpendicular or "normal" to the line described by our flow equation. Figure 10-7 presents the mathematical and geometric details.

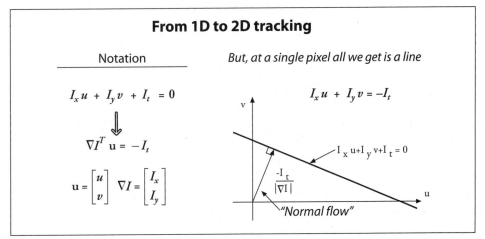

Figure 10-7. Two-dimensional optical flow at a single pixel: optical flow at one pixel is underdetermined and so can yield at most motion, which is perpendicular ("normal") to the line described by the flow equation (figure courtesy of Michael Black)

Normal optical flow results from the *aperture problem*, which arises when you have a small aperture or window in which to measure motion. When motion is detected with a small aperture, you often see only an edge, not a corner. But an edge alone is insufficient to determine exactly how (i.e., in what direction) the entire object is moving; see Figure 10-8.

So then how do we get around this problem that, at one pixel, we cannot resolve the full motion? We turn to the last optical flow assumption for help. If a local patch of pixels moves coherently, then we can easily solve for the motion of the central pixel by using the surrounding pixels to set up a system of equations. For example, if we use a 5-by-5* window of brightness values (you can simply triple this for color-based optical flow) around the current pixel to compute its motion, we can then set up 25 equations as follows.

---

* Of course, the window could be 3-by-3, 7-by-7, or anything you choose. If the window is too large then you will end up violating the coherent motion assumption and will not be able to track well. If the window is too small, you will encounter the aperture problem again.

$$\underbrace{\begin{bmatrix} I_x(p_1) & I_y(p_1) \\ I_x(p_2) & I_y(p_2) \\ \vdots & \vdots \\ I_x(p_{25}) & I_y(p_{25}) \end{bmatrix}}_{\substack{A \\ 25\times2}} \underbrace{\begin{bmatrix} u \\ v \end{bmatrix}}_{\substack{d \\ 2\times1}} = -\underbrace{\begin{bmatrix} I_t(p_1) \\ I_t(p_2) \\ \vdots \\ I_t(p_{25}) \end{bmatrix}}_{\substack{b \\ 2\times1}}$$

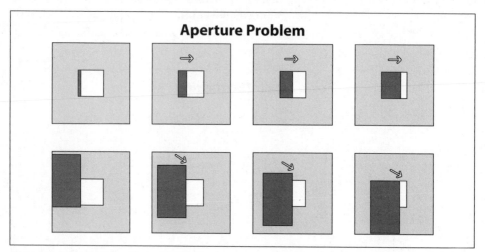

Aperture Problem

Figure 10-8. Aperture problem: through the aperture window (upper row) we see an edge moving to the right but cannot detect the downward part of the motion (lower row)

We now have an overconstrained system for which we can solve provided it contains more than just an edge in that 5-by-5 window. To solve for this system, we set up a least-squares minimization of the equation, whereby $\min\|Ad - b\|^2$ is solved in standard form as:

$$\underbrace{(A^\mathrm{T}A)}_{2\times2}\underbrace{d}_{2\times1} = \underbrace{A^\mathrm{T}b}_{2\times2}$$

From this relation we obtain our u and v motion components. Writing this out in more detail yields:

$$\underbrace{\begin{bmatrix} \sum I_x I_x & \sum I_x I_y \\ \sum I_x I_y & \sum I_y I_y \end{bmatrix}}_{A^\mathrm{T}A} \begin{bmatrix} u \\ v \end{bmatrix} = -\underbrace{\begin{bmatrix} \sum I_x I_t \\ \sum I_y I_t \end{bmatrix}}_{A^\mathrm{T}b}$$

The solution to this equation is then:

$$\begin{bmatrix} u \\ v \end{bmatrix} = (A^\mathrm{T}A)^{-1} A^\mathrm{T}b$$

When can this be solved?—when (A^TA) is invertible. And (A^TA) is invertible when it has full rank (2), which occurs when it has two large eigenvectors. This will happen in image regions that include texture running in at least two directions. In this case, (A^TA) will have the best properties then when the tracking window is centered over a corner region in an image. This ties us back to our earlier discussion of the Harris corner detector. In fact, those corners were "good features to track" (see our previous remarks concerning cvGoodFeaturesToTrack()) for precisely the reason that (A^TA) had two large eigenvectors there! We'll see shortly how all this computation is done for us by the cvCalcOpticalFlowLK() function.

The reader who understands the implications of our assuming small and coherent motions will now be bothered by the fact that, for most video cameras running at 30 Hz, large and noncoherent motions are commonplace. In fact, Lucas-Kanade optical flow by itself does not work very well for exactly this reason: we want a large window to catch large motions, but a large window too often breaks the coherent motion assumption! To circumvent this problem, we can track first over larger spatial scales using an image pyramid and then refine the initial motion velocity assumptions by working our way down the levels of the image pyramid until we arrive at the raw image pixels.

Hence, the recommended technique is first to solve for optical flow at the top layer and then to use the resulting motion estimates as the starting point for the next layer down. We continue going down the pyramid in this manner until we reach the lowest level. Thus we minimize the violations of our motion assumptions and so can track faster and longer motions. This more elaborate function is known as *pyramid Lucas-Kanade optical flow* and is illustrated in Figure 10-9. The OpenCV function that implements Pyramid Lucas-Kanade optical flow is cvCalcOpticalFlowPyrLK(), which we examine next.

Lucas-Kanade code

The routine that implements the nonpyramidal Lucas-Kanade dense optical flow algorithm is:

```
void cvCalcOpticalFlowLK(
    const CvArr*  imgA,
    const CvArr*  imgB,
    CvSize        winSize,
    CvArr*        velx,
    CvArr*        vely
);
```

The result arrays for this OpenCV routine are populated only by those pixels for which it is able to compute the minimum error. For the pixels for which this error (and thus the displacement) cannot be reliably computed, the associated velocity will be set to 0. In most cases, you will not want to use this routine. The following pyramid-based method is better for most situations most of the time.

Pyramid Lucas-Kanade code

We come now to OpenCV's algorithm that computes Lucas-Kanade optical flow in a pyramid, cvCalcOpticalFlowPyrLK(). As we will see, this optical flow function makes use

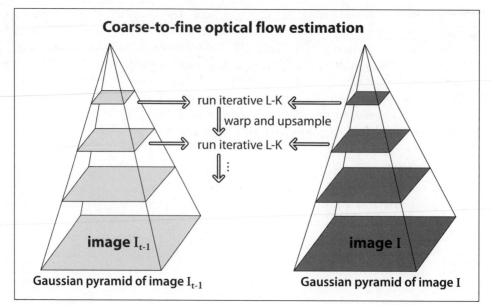

Figure 10-9. Pyramid Lucas-Kanade optical flow: running optical flow at the top of the pyramid first mitigates the problems caused by violating our assumptions of small and coherent motion; the motion estimate from the preceding level is taken as the starting point for estimating motion at the next layer down

of "good features to track" and also returns indications of how well the tracking of each point is proceeding.

```
void cvCalcOpticalFlowPyrLK(
    const CvArr*      imgA,
    const CvArr*      imgB,
    CvArr*            pyrA,
    CvArr*            pyrB,
    CvPoint2D32f*     featuresA,
    CvPoint2D32f*     featuresB,
    int               count,
    CvSize            winSize,
    int               level,
    char*             status,
    float*            track_error,
    CvTermCriteria    criteria,
    int               flags
);
```

This function has a lot of inputs, so let's take a moment to figure out what they all do. Once we have a handle on this routine, we can move on to the problem of which points to track and how to compute them.

The first two arguments of cvCalcOpticalFlowPyrLK() are the initial and final images; both should be single-channel, 8-bit images. The next two arguments are buffers allocated to store the pyramid images. The size of these buffers should be at least (img.width

+ 8)*img.height/3 bytes,* with one such buffer for each of the two input images (pyrA and pyrB). (If these two pointers are set to NULL then the routine will allocate, use, and free the appropriate memory when called, but this is not so good for performance.) The array featuresA contains the points for which the motion is to be found, and featuresB is a similar array into which the computed new locations of the points from featuresA are to be placed; count is the number of points in the featuresA list. The window used for computing the local coherent motion is given by winSize. Because we are constructing an image pyramid, the argument level is used to set the depth of the stack of images. If level is set to 0 then the pyramids are not used. The array status is of length count; on completion of the routine, each entry in status will be either 1 (if the corresponding point was found in the second image) or 0 (if it was not). The track_error parameter is optional and can be turned off by setting it to NULL. If track_error is active then it is an array of numbers, one for each tracked point, equal to the difference between the patch around a tracked point in the first image and the patch around the location to which that point was tracked in the second image. You can use track_error to prune away points whose local appearance patch changes too much as the points move.

The next thing we need is the termination criteria. This is a structure used by many OpenCV algorithms that iterate to a solution:

```
cvTermCriteria(
    int    type,       // CV_TERMCRIT_ITER, CV_TERMCRIT_EPS, or both
    int    max_iter,
    double epsilon
);
```

Typically we use the cvTermCriteria() function to generate the structure we need. The first argument of this function is either CV_TERMCRIT_ITER or CV_TERMCRIT_EPS, which tells the algorithm that we want to terminate either after some number of iterations or when the convergence metric reaches some small value (respectively). The next two arguments set the values at which one, the other, or both of these criteria should terminate the algorithm. The reason we have both options is so we can set the type to CV_TERMCRIT_ITER | CV_TERMCRIT_EPS and thus stop when either limit is reached (this is what is done in most real code).

Finally, flags allows for some fine control of the routine's internal bookkeeping; it may be set to any or all (using bitwise OR) of the following.

CV_LKFLOW_PYR_A_READY

The image pyramid for the first frame is calculated before the call and stored in pyrA.

CV_LKFLOW_PYR_B_READY

The image pyramid for the second frame is calculated before the call and stored in pyrB.

---

* If you are wondering why the funny size, it's because these scratch spaces need to accommodate not just the image itself but the entire pyramid.

CV_LKFLOW_INITIAL_GUESSES
   The array B already contains an initial guess for the feature's coordinates when the
   routine is called.

These flags are particularly useful when handling sequential video. The image pyramids
are somewhat costly to compute, so recomputing them should be avoided whenever
possible. The final frame for the frame pair you just computed will be the initial frame
for the pair that you will compute next. If you allocated those buffers yourself (instead
of asking the routine to do it for you), then the pyramids for each image will be sitting
in those buffers when the routine returns. If you tell the routine that this information is
already computed then it will not be recomputed. Similarly, if you computed the motion
of points from the previous frame then you are in a good position to make good initial
guesses for where they will be in the next frame.

So the basic plan is simple: you supply the images, list the points you want to track in
featuresA, and call the routine. When the routine returns, you check the status array
to see which points were successfully tracked and then check featuresB to find the new
locations of those points.

This leads us back to that issue we put aside earlier: how to decide which features are
good ones to track. Earlier we encountered the OpenCV routine cvGoodFeatures
ToTrack(), which uses the method originally proposed by Shi and Tomasi to solve this
problem in a reliable way. In most cases, good results are obtained by using the combination of cvGoodFeaturesToTrack() and cvCalcOpticalFlowPyrLK(). Of course, you can
also use your own criteria to determine which points to track.

Let's now look at a simple example (Example 10-1) that uses both cvGoodFeaturesToTrack()
and cvCalcOpticalFlowPyrLK(); see also Figure 10-10.

Example 10-1. Pyramid Lucas-Kanade optical flow code

```
// Pyramid L-K optical flow example
//
#include <cv.h>
#include <cxcore.h>
#include <highgui.h>

const int MAX_CORNERS = 500;

int main(int argc, char** argv) {

  // Initialize, load two images from the file system, and
  // allocate the images and other structures we will need for
  // results.
  //
  IplImage* imgA = cvLoadImage("image0.jpg",CV_LOAD_IMAGE_GRAYSCALE);
  IplImage* imgB = cvLoadImage("image1.jpg",CV_LOAD_IMAGE_GRAYSCALE);

  CvSize    img_sz  = cvGetSize( imgA );
  int       win_size = 10;

  IplImage* imgC = cvLoadImage(
```

Example 10-1. Pyramid Lucas-Kanade optical flow code (continued)

```
    "../Data/OpticalFlow1.jpg",
    CV_LOAD_IMAGE_UNCHANGED
);

// The first thing we need to do is get the features
// we want to track.
//
IplImage* eig_image = cvCreateImage( img_sz, IPL_DEPTH_32F, 1 );
IplImage* tmp_image = cvCreateImage( img_sz, IPL_DEPTH_32F, 1 );

int           corner_count = MAX_CORNERS;
CvPoint2D32f* cornersA     = new CvPoint2D32f[ MAX_CORNERS ];

cvGoodFeaturesToTrack(
  imgA,
  eig_image,
  tmp_image,
  cornersA,
  &corner_count,
  0.01,
  5.0,
  0,
  3,
  0,
  0.04
);

cvFindCornerSubPix(
  imgA,
  cornersA,
  corner_count,
  cvSize(win_size,win_size),
  cvSize(-1,-1),
  cvTermCriteria(CV_TERMCRIT_ITER|CV_TERMCRIT_EPS,20,0.03)
);


// Call the Lucas Kanade algorithm
//
char  features_found[ MAX_CORNERS ];
float feature_errors[ MAX_CORNERS ];

CvSize pyr_sz = cvSize( imgA->width+8, imgB->height/3 );

IplImage* pyrA = cvCreateImage( pyr_sz, IPL_DEPTH_32F, 1 );
IplImage* pyrB = cvCreateImage( pyr_sz, IPL_DEPTH_32F, 1 );

CvPoint2D32f* cornersB     = new CvPoint2D32f[ MAX_CORNERS ];

cvCalcOpticalFlowPyrLK(
  imgA,
  imgB,
```

Example 10-1. Pyramid Lucas-Kanade optical flow code (continued)

```
    pyrA,
    pyrB,
    cornersA,
    cornersB,
    corner_count,
    cvSize( win_size,win_size ),
    5,
    features_found,
    feature_errors,
    cvTermCriteria( CV_TERMCRIT_ITER | CV_TERMCRIT_EPS, 20, .3 ),
    0
);

// Now make some image of what we are looking at:
//
for( int i=0; i<corner_count; i++ ) {
  if( features_found[i]==0|| feature_errors[i]>550 ) {
    printf("Error is %f/n",feature_errors[i]);
    continue;
  }
  printf("Got it/n");
  CvPoint p0 = cvPoint(
    cvRound( cornersA[i].x ),
    cvRound( cornersA[i].y )
  );
  CvPoint p1 = cvPoint(
    cvRound( cornersB[i].x ),
    cvRound( cornersB[i].y )
  );
  cvLine( imgC, p0, p1, CV_RGB(255,0,0),2 );
}

cvNamedWindow("ImageA",0);
cvNamedWindow("ImageB",0);
cvNamedWindow("LKpyr_OpticalFlow",0);

cvShowImage("ImageA",imgA);
cvShowImage("ImageB",imgB);
cvShowImage("LKpyr_OpticalFlow",imgC);

cvWaitKey(0);

return 0;
}
```

Dense Tracking Techniques

OpenCV contains two other optical flow techniques that are now seldom used. These routines are typically much slower than Lucas-Kanade; moreover, they (could, but) do not support matching within an image scale pyramid and so cannot track large motions. We will discuss them briefly in this section.

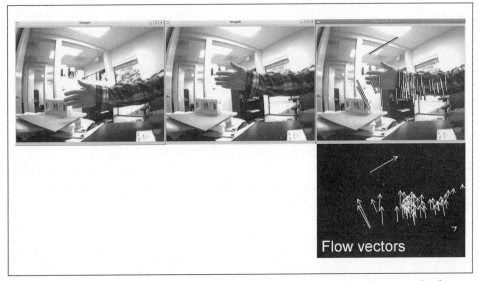

Figure 10-10. Sparse optical flow from pyramid Lucas-Kanade: the center image is one video frame after the left image; the right image illustrates the computed motion of the "good features to track" (lower right shows flow vectors against a dark background for increased visibility)

Horn-Schunck method

The method of Horn and Schunck was developed in 1981 [Horn81]. This technique was one of the first to make use of the brightness constancy assumption and to derive the basic brightness constancy equations. The solution of these equations devised by Horn and Schunck was by hypothesizing a smoothness constraint on the velocities v_x and v_y. This constraint was derived by minimizing the regularized Laplacian of the optical flow velocity components:

$$\frac{\partial}{\partial x}\frac{\partial v_x}{\partial x} - \frac{1}{\alpha} I_x (I_x v_x + I_y v_y + I_t) = 0$$

$$\frac{\partial}{\partial y}\frac{\partial v_y}{\partial y} - \frac{1}{\alpha} I_y (I_x v_x + I_y v_y + I_t) = 0$$

Here α is a constant weighting coefficient known as the *regularization constant*. Larger values of α lead to smoother (i.e., more locally consistent) vectors of motion flow. This is a relatively simple constraint for enforcing smoothness, and its effect is to penalize regions in which the flow is changing in magnitude. As with Lucas-Kanade, the Horn-Schunck technique relies on iterations to solve the differential equations. The function that computes this is:

```
void cvCalcOpticalFlowHS(
    const CvArr*     imgA,
    const CvArr*     imgB,
    int              usePrevious,
    CvArr*           velx,
```

```
CvArr*          vely,
double          lambda,
CvTermCriteria  criteria
);
```

Here imgA and imgB must be 8-bit, single-channel images. The *x* and *y* velocity results will be stored in velx and vely, which must be 32-bit, floating-point, single-channel images. The usePrevious parameter tells the algorithm to use the velx and vely velocities computed from a previous frame as the initial starting point for computing the new velocities. The parameter lambda is a weight related to the *Lagrange multiplier*. You are probably asking yourself: "What Lagrange multiplier?"* The Lagrange multiplier arises when we attempt to minimize (simultaneously) both the motion-brightness equation and the smoothness equations; it represents the relative weight given to the errors in each as we minimize.

Block matching method

You might be thinking: "What's the big deal with optical flow? Just match where pixels in one frame went to in the next frame." This is exactly what others have done. The term "block matching" is a catchall for a whole class of similar algorithms in which the image is divided into small regions called *blocks* [Huang95; Beauchemin95]. Blocks are typically square and contain some number of pixels. These blocks may overlap and, in practice, often do. Block-matching algorithms attempt to divide both the previous and current images into such blocks and then compute the motion of these blocks. Algorithms of this kind play an important role in many video compression algorithms as well as in optical flow for computer vision.

Because block-matching algorithms operate on aggregates of pixels, not on individual pixels, the returned "velocity images" are typically of lower resolution than the input images. This is not always the case; it depends on the severity of the overlap between the blocks. The size of the result images is given by the following formula:

$$W_{result} = \left\lfloor \frac{W_{prev} - W_{block} + W_{shiftsize}}{W_{shiftsize}} \right\rfloor_{floor}$$

$$H_{result} = \left\lfloor \frac{H_{prev} - H_{block} + H_{shiftsize}}{H_{shiftsize}} \right\rfloor_{floor}$$

The implementation in OpenCV uses a spiral search that works out from the location of the original block (in the previous frame) and compares the candidate new blocks with the original. This comparison is a sum of absolute differences of the pixels (i.e., an L1 distance). If a good enough match is found, the search is terminated. Here's the function prototype:

---

* You might even be asking yourself: "What *is* a Lagrange multiplier?". In that case, it may be best to ignore this part of the paragraph and just set lambda equal to 1.

```
void cvCalcOpticalFlowBM(
    const CvArr*    prev,
    const CvArr*    curr,
    CvSize          block_size,
    CvSize          shift_size,
    CvSize          max_range,
    int             use_previous,
    CvArr*          velx,
    CvArr*          vely
);
```

The arguments are straightforward. The prev and curr parameters are the previous and current images; both should be 8-bit, single-channel images. The block_size is the size of the block to be used, and shift_size is the step size between blocks (this parameter controls whether—and, if so, by how much—the blocks will overlap). The max_range parameter is the size of the region around a given block that will be searched for a corresponding block in the subsequent frame. If set, use_previous indicates that the values in velx and vely should be taken as starting points for the block searches.* Finally, velx and vely are themselves 32-bit single-channel images that will store the computed motions of the blocks. As mentioned previously, motion is computed at a block-by-block level and so the coordinates of the result images are for the blocks (i.e., aggregates of pixels), not for the individual pixels of the original image.

Mean-Shift and Camshift Tracking

In this section we will look at two techniques, *mean-shift* and *camshift* (where "camshift" stands for "continuously adaptive mean-shift"). The former is a general technique for data analysis (discussed in Chapter 9 in the context of segmentation) in many applications, of which computer vision is only one. After introducing the general theory of mean-shift, we'll describe how OpenCV allows you to apply it to tracking in images. The latter technique, camshift, builds on mean-shift to allow for the tracking of objects whose size may change during a video sequence.

Mean-Shift

The mean-shift algorithm[†] is a robust method of finding local extrema in the density distribution of a data set. This is an easy process for continuous distributions; in that context, it is essentially just *hill climbing* applied to a density histogram of the data.[‡] For discrete data sets, however, this is a somewhat less trivial problem.

---

* If use_previous==0, then the search for a block will be conducted over a region of max_range distance from the location of the original block. If use_previous!=0, then the center of that search is first displaced by $\Delta x = \text{vel}_x(x, y)$ and $\Delta y = \text{vel}_y(x, y)$.

† Because mean-shift is a fairly deep topic, our discussion here is aimed mainly at developing intuition for the user. For the original formal derivation, see Fukunaga [Fukunaga90] and Comaniciu and Meer [Comaniciu99].

‡ The word "essentially" is used because there is also a scale-dependent aspect of mean-shift. To be exact: mean-shift is equivalent in a continuous distribution to first convolving with the mean-shift kernel and then applying a hill-climbing algorithm.

The descriptor "robust" is used here in its formal statistical sense; that is, mean-shift ignores outliers in the data. This means that it ignores data points that are far away from peaks in the data. It does so by processing only those points within a local window of the data and then moving that window.

The mean-shift algorithm runs as follows.

1. Choose a search window:

   - its initial location;
   - its type (uniform, polynomial, exponential, or Gaussian);
   - its shape (symmetric or skewed, possibly rotated, rounded or rectangular);
   - its size (extent at which it rolls off or is cut off).

2. Compute the window's (possibly weighted) center of mass.

3. Center the window at the center of mass.

4. Return to step 2 until the window stops moving (it always will).*

To give a little more formal sense of what the mean-shift algorithm is: it is related to the discipline of *kernel density estimation*, where by "kernel" we refer to a function that has mostly local focus (e.g., a Gaussian distribution). With enough appropriately weighted and sized kernels located at enough points, one can express a distribution of data entirely in terms of those kernels. Mean-shift diverges from kernel density estimation in that it seeks only to estimate the gradient (direction of change) of the data distribution. When this change is 0, we are at a stable (though perhaps local) peak of the distribution. There might be other peaks nearby or at other scales.

Figure 10-11 shows the equations involved in the mean-shift algorithm. These equations can be simplified by considering a *rectangular* kernel,† which reduces the mean-shift vector equation to calculating the center of mass of the image pixel distribution:

$$x_c = \frac{M_{10}}{M_{00}}, \quad y_c = \frac{M_{01}}{M_{00}}$$

Here the zeroth moment is calculated as:

$$M_{00} = \sum_x \sum_y I(x, y)$$

and the first moments are:

---

* Iterations are typically restricted to some maximum number or to some epsilon change in center shift between iterations; however, they are guaranteed to converge eventually.

† A *rectangular kernel* is a kernel with no falloff with distance from the center, until a single sharp transition to zero value. This is in contrast to the exponential falloff of a Gaussian kernel and the falloff with the square of distance from the center in the commonly used Epanechnikov kernel.

$$M_{10} = \sum_x \sum_y xI(x,y) \quad \text{and} \quad M_{01} = \sum_x \sum_y yI(x,y)$$

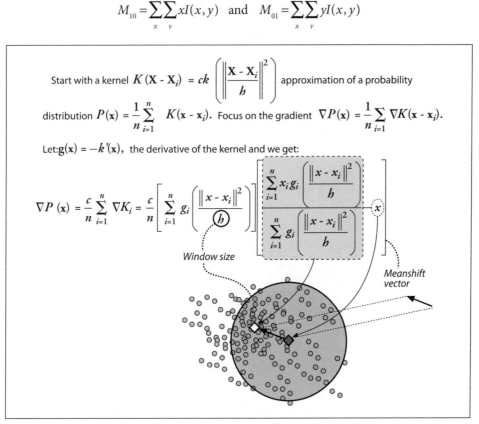

Figure 10-11. Mean-shift equations and their meaning

The mean-shift vector in this case tells us to recenter the mean-shift window over the calculated center of mass within that window. This movement will, of course, change what is "under" the window and so we iterate this recentering process. Such recentering will always converge to a mean-shift vector of 0 (i.e., where no more centering movement is possible). The location of convergence is at a local maximum (peak) of the distribution under the window. Different window sizes will find different peaks because "peak" is fundamentally a scale-sensitive construct.

In Figure 10-12 we see an example of a two-dimensional distribution of data and an initial (in this case, rectangular) window. The arrows indicate the process of convergence on a local mode (peak) in the distribution. Observe that, as promised, this peak finder is statistically robust in the sense that points outside the mean-shift window do not affect convergence—the algorithm is not "distracted" by far-away points.

In 1998, it was realized that this mode-finding algorithm could be used to track moving objects in video [Bradski98a; Bradski98b], and the algorithm has since been greatly extended [Comaniciu03]. The OpenCV function that performs mean-shift is implemented in the context of image analysis. This means in particular that, rather than taking some

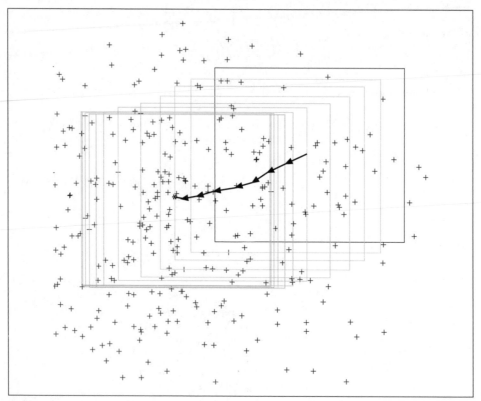

Figure 10-12. Mean-shift algorithm in action: an initial window is placed over a two-dimensional array of data points and is successively recentered over the mode (or local peak) of its data distribution until convergence

arbitrary set of data points (possibly in some arbitrary number of dimensions), the OpenCV implementation of mean-shift expects as input an image representing the density distribution being analyzed. You could think of this image as a two-dimensional histogram measuring the density of points in some two-dimensional space. It turns out that, for vision, this is precisely what you want to do most of the time: it's how you can track the motion of a cluster of interesting features.

```
int cvMeanShift(
    const CvArr*       prob_image,
    CvRect             window,
    CvTermCriteria     criteria,
    CvConnectedComp*   comp
);
```

In cvMeanShift(), the prob_image, which represents the density of probable locations, may be only one channel but of either type (byte or float). The window is set at the initial desired location and size of the kernel window. The termination criteria has been described elsewhere and consists mainly of a maximum limit on number of mean-shift movement iterations and a minimal movement for which we consider the window

---

locations to have converged.* The connected component comp contains the converged search window location in comp->rect, and the sum of all pixels under the window is kept in the comp->area field.

The function cvMeanShift() is one expression of the mean-shift algorithm for rectangular windows, but it may also be used for tracking. In this case, you first choose the feature distribution to represent an object (e.g., color + texture), then start the mean-shift window over the feature distribution generated by the object, and finally compute the chosen feature distribution over the next video frame. Starting from the current window location, the mean-shift algorithm will find the new peak or mode of the feature distribution, which (presumably) is centered over the object that produced the color and texture in the first place. In this way, the mean-shift window tracks the movement of the object frame by frame.

Camshift

A related algorithm is the Camshift tracker. It differs from the meanshift in that the search window adjusts itself in size. If you have well-segmented distributions (say face features that stay compact), then this algorithm will automatically adjust itself for the size of face as the person moves closer to and further from the camera. The form of the Camshift algorithm is:

```
int cvCamShift(
    const CvArr*      prob_image,
    CvRect            window,
    CvTermCriteria    criteria,
    CvConnectedComp*  comp,
    CvBox2D*          box        = NULL
);
```

The first four parameters are the same as for the cvMeanShift() algorithm. The box parameter, if present, will contain the newly resized box, which also includes the orientation of the object as computed via second-order moments. For tracking applications, we would use the resulting resized box found on the previous frame as the window in the next frame.

> Many people think of mean-shift and camshift as tracking using color features, but this is not entirely correct. Both of these algorithms track the distribution of any kind of feature that is expressed in the prob_image; hence they make for very lightweight, robust, and efficient trackers.

Motion Templates

Motion templates were invented in the MIT Media Lab by Bobick and Davis [Bobick96; Davis97] and were further developed jointly with one of the authors [Davis99; Bradski00]. This more recent work forms the basis for the implementation in OpenCV.

---

* Again, mean-shift will always converge, but convergence may be very slow near the local peak of a distribution if that distribution is fairly "flat" there.

Motion templates are an effective way to track general movement and are especially applicable to gesture recognition. Using motion templates requires a silhouette (or part of a silhouette) of an object. Object silhouettes can be obtained in a number of ways.

1. The simplest method of obtaining object silhouettes is to use a reasonably stationary camera and then employ frame-to-frame differencing (as discussed in Chapter 9). This will give you the moving edges of objects, which is enough to make motion templates work.

2. You can use chroma keying. For example, if you have a known background color such as bright green, you can simply take as foreground anything that is not bright green.

3. Another way (also discussed in Chapter 9) is to learn a background model from which you can isolate new foreground objects/people as silhouettes.

4. You can use active silhouetting techniques—for example, creating a wall of near-infrared light and having a near-infrared-sensitive camera look at the wall. Any intervening object will show up as a silhouette.

5. You can use thermal imagers; then any hot object (such as a face) can be taken as foreground.

6. Finally, you can generate silhouettes by using the segmentation techniques (e.g., pyramid segmentation or mean-shift segmentation) described in Chapter 9.

For now, assume that we have a good, segmented object silhouette as represented by the white rectangle of Figure 10-13(A). Here we use white to indicate that all the pixels are set to the floating-point value of the most recent system time stamp. As the rectangle moves, new silhouettes are captured and overlaid with the (new) current time stamp; the new silhouette is the white rectangle of Figure 10-13(B) and Figure 10-13(C). Older motions are shown in Figure 10-13 as successively darker rectangles. These sequentially fading silhouettes record the history of previous movement and thus are referred to as the "motion history image".

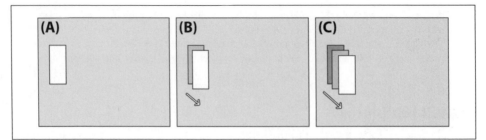

Figure 10-13. Motion template diagram: (A) a segmented object at the current time stamp (white); (B) at the next time step, the object moves and is marked with the (new) current time stamp, leaving the older segmentation boundary behind; (C) at the next time step, the object moves further, leaving older segmentations as successively darker rectangles whose sequence of encoded motion yields the motion history image

Silhouettes whose time stamp is more than a specified duration older than the current system time stamp are set to 0, as shown in Figure 10-14. The OpenCV function that accomplishes this motion template construction is cvUpdateMotionHistory():

```
void cvUpdateMotionHistory(
    const CvArr* silhouette,
    CvArr*       mhi,
    double       timestamp,
    double       duration
);
```

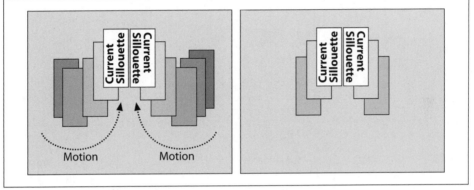

Figure 10-14. Motion template silhouettes for two moving objects (left); silhouettes older than a specified duration are set to 0 (right)

In cvUpdateMotionHistory(), all image arrays consist of single-channel images. The silhouette image is a byte image in which nonzero pixels represent the most recent segmentation silhouette of the foreground object. The mhi image is a floating-point image that represents the motion template (aka motion history image). Here timestamp is the current system time (typically a millisecond count) and duration, as just described, sets how long motion history pixels are allowed to remain in the mhi. In other words, any mhi pixels that are older (less) than timestamp minus duration are set to 0.

Once the motion template has a collection of object silhouettes overlaid in time, we can derive an indication of overall motion by taking the gradient of the mhi image. When we take these gradients (e.g., by using the Scharr or Sobel gradient functions discussed in Chapter 6), some gradients will be large and invalid. Gradients are invalid when older or inactive parts of the mhi image are set to 0, which produces artificially large gradients around the outer edges of the silhouettes; see Figure 10-15(A). Because we know the time-step duration with which we've been introducing new silhouettes into the mhi via cvUpdateMotionHistory(), we know how large our gradients (which are just dx and dy step derivatives) should be. We can therefore use the gradient magnitude to eliminate gradients that are too large, as in Figure 10-15(B). Finally, we can collect a measure of global motion; see Figure 10-15(C). The function that effects parts (A) and (B) of the figure is cvCalcMotionGradient():

```
void cvCalcMotionGradient(
    const CvArr* mhi,
    CvArr* mask,
    CvArr* orientation,
    double delta1,
    double delta2,
    int aperture_size=3
);
```

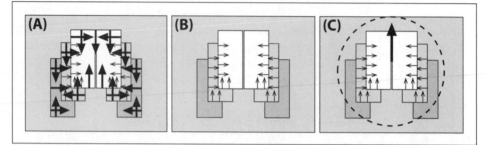

Figure 10-15. Motion gradients of the mhi image: (A) gradient magnitudes and directions; (B) large gradients are eliminated; (C) overall direction of motion is found

In cvCalcMotionGradient(), all image arrays are single-channel. The function input mhi is a floating-point motion history image, and the input variables delta1 and delta2 are (respectively) the minimal and maximal gradient magnitudes allowed. Here, the expected gradient magnitude will be just the average number of time-stamp ticks between each silhouette in successive calls to cvUpdateMotionHistory(); setting delta1 halfway below and delta2 halfway above this average value should work well. The variable aperture_size sets the size in width and height of the gradient operator. These values can be set to -1 (the 3-by-3 CV_SCHARR gradient filter), 3 (the default 3-by-3 Sobel filter), 5 (for the 5-by-5 Sobel filter), or 7 (for the 7-by-7 filter). The function outputs are mask, a single-channel 8-bit image in which nonzero entries indicate where valid gradients were found, and orientation, a floating-point image that gives the gradient direction's angle at each point.

The function cvCalcGlobalOrientation() finds the overall direction of motion as the vector sum of the valid gradient directions.

```
double cvCalcGlobalOrientation(
    const CvArr* orientation,
    const CvArr* mask,
    const CvArr* mhi,
    double      timestamp,
    double      duration
);
```

When using cvCalcGlobalOrientation(), we pass in the orientation and mask image computed in cvCalcMotionGradient() along with the timestamp, duration, and resulting mhi from cvUpdateMotionHistory(); what's returned is the vector-sum global orientation, as in Figure 10-15(C). The `timestamp` together with `duration` tells the routine how much motion to consider from the `mhi` and motion `orientation` images. One could compute the global motion from the center of mass of each of the `mhi` silhouettes, but summing up the precomputed motion vectors is much faster.

We can also isolate regions of the motion template `mhi` image and determine the local motion within that region, as shown in Figure 10-16. In the figure, the `mhi` image is scanned for current silhouette regions. When a region marked with the most current time stamp is found, the region's perimeter is searched for sufficiently recent motion (recent silhouettes) just outside its perimeter. When such motion is found, a downward-stepping flood fill is performed to isolate the local region of motion that "spilled off" the current location of the object of interest. Once found, we can calculate local motion gradient direction in the spill-off region, then remove that region, and repeat the process until all regions are found (as diagrammed in Figure 10-16).

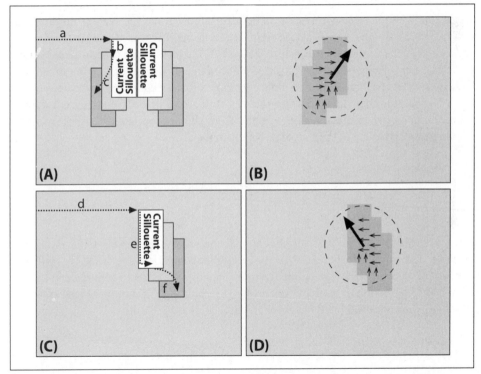

Figure 10-16. Segmenting local regions of motion in the mhi image: (A) scan the mhi image for current silhouettes (a) and, when found, go around the perimeter looking for other recent silhouettes (b); when a recent silhouette is found, perform downward-stepping flood fills (c) to isolate local motion; (B) use the gradients found within the isolated local motion region to compute local motion; (C) remove the previously found region and search for the next current silhouette region (d), scan along it (e), and perform downward-stepping flood fill on it (f); (D) compute motion within the newly isolated region and continue the process (A)-(C) until no current silhouette remains

The function that isolates and computes local motion is cvSegmentMotion():

```
CvSeq* cvSegmentMotion(
    const CvArr*  mhi,
    CvArr*        seg_mask,
    CvMemStorage* storage,
    double        timestamp,
    double        seg_thresh
);
```

In cvSegmentMotion(), the mhi is the single-channel floating-point input. We also pass in storage, a CvMemoryStorage structure allocated via cvCreateMemStorage(). Another input is timestamp, the value of the most current silhouettes in the mhi from which you want to segment local motions. Finally, you must pass in seg_thresh, which is the maximum downward step (from current time to previous motion) that you'll accept as attached motion. This parameter is provided because there might be overlapping silhouettes from recent and much older motion that you don't want to connect together.

It's generally best to set seg_thresh to something like 1.5 times the average difference in silhouette time stamps. This function returns a CvSeq of CvConnectedComp structures, one for each separate motion found, which delineates the local motion regions; it also returns seg_mask, a single-channel, floating-point image in which each region of isolated motion is marked a distinct nonzero number (a zero pixel in seg_mask indicates no motion). To compute these local motions one at a time we call cvCalcGlobalOrientation(), using the appropriate mask region selected from the appropriate CvConnectedComp or from a particular value in the seg_mask; for example,

```
cvCmpS(
  seg_mask,
//  [value_wanted_in_seg_mask],
//  [your_destination_mask],
  CV_CMP_EQ
)
```

Given the discussion so far, you should now be able to understand the *motempl.c* example that ships with OpenCV in the *.../opencv/samples/c/* directory. We will now extract and explain some key points from the update_mhi() function in *motempl.c*. The update_mhi() function extracts templates by thresholding frame differences and then passing the resulting silhouette to cvUpdateMotionHistory():

```
...
cvAbsDiff( buf[idx1], buf[idx2], silh );
cvThreshold( silh, silh, diff_threshold, 1, CV_THRESH_BINARY );
cvUpdateMotionHistory( silh, mhi, timestamp, MHI_DURATION );
...
```

The gradients of the resulting mhi image are then taken, and a mask of valid gradients is produced using cvCalcMotionGradient(). Then CvMemStorage is allocated (or, if it already exists, it is cleared), and the resulting local motions are segmented into CvConnectedComp structures in the CvSeq containing structure seq:

```
...
cvCalcMotionGradient(
```

```
    mhi,
    mask,
    orient,
    MAX_TIME_DELTA,
    MIN_TIME_DELTA,
    3
);

if( !storage )
    storage = cvCreateMemStorage(0);
else
    cvClearMemStorage(storage);

seq = cvSegmentMotion(
    mhi,
    segmask,
    storage,
    timestamp,
    MAX_TIME_DELTA
);
```

A "for" loop then iterates through the seq->total CvConnectedComp structures extracting bounding rectangles for each motion. The iteration starts at -1, which has been designated as a special case for finding the global motion of the whole image. For the local motion segments, small segmentation areas are first rejected and then the orientation is calculated using cvCalcGlobalOrientation(). Instead of using exact masks, this routine restricts motion calculations to regions of interest (ROIs) that bound the local motions; it then calculates where valid motion within the local ROIs was actually found. Any such motion area that is too small is rejected. Finally, the routine draws the motion. Examples of the output for a person flapping their arms is shown in Figure 10-17, where the output is drawn above the raw image for four sequential frames going across in two rows. (For the full code, see .../opencv/samples/c/motempl.c.) In the same sequence, "Y" postures were recognized by the shape descriptors (Hu moments) discussed in Chapter 8, although the shape recognition is not included in the *samples* code.

```
for( i = -1; i < seq->total; i++ ) {
    if( i < 0 ) { // case of the whole image
//        ...[does the whole image]...
    else { // i-th motion component
        comp_rect = ((CvConnectedComp*)cvGetSeqElem( seq, i ))->rect;
//            [reject very small components]...
    }
    ...[set component ROI regions]...
    angle = cvCalcGlobalOrientation( orient, mask, mhi,
                                     timestamp, MHI_DURATION);
    ...[find regions of valid motion]...
    ...[reset ROI regions]...
    ...[skip small valid motion regions]...
    ...[draw the motions]...
    }
```

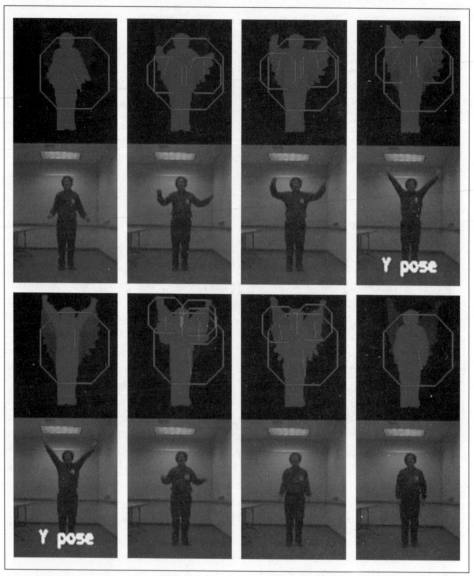

Figure 10-17. Results of motion template routine: going across and top to bottom, a person moving and the resulting global motions indicated in large octagons and local motions indicated in small octagons; also, the "Y" pose can be recognized via shape descriptors (Hu moments)

Estimators

Suppose we are tracking a person who is walking across the view of a video camera. At each frame we make a determination of the location of this person. This could be done any number of ways, as we have seen, but in each case we find ourselves with an estimate of the position of the person at each frame. This estimation is not likely to be extremely accurate. The reasons for this are many. They may include inaccuracies in the sensor, approximations in earlier processing stages, issues arising from occlusion or shadows, or the apparent changing of shape when a person is walking due to their legs and arms swinging as they move. Whatever the source, we expect that these measurements will vary, perhaps somewhat randomly, about the "actual" values that might be received from an idealized sensor. We can think of all these inaccuracies, taken together, as simply adding noise to our tracking process.

We'd like to have the capability of estimating the motion of this person in a way that makes maximal use of the measurements we've made. Thus, the cumulative effect of our many measurements could allow us to detect the part of the person's observed trajectory that does not arise from noise. The key additional ingredient is a *model* for the person's motion. For example, we might model the person's motion with the following statement: "A person enters the frame at one side and walks across the frame at constant velocity." Given this model, we can ask not only where the person is but also what parameters of the model are supported by our observations.

This task is divided into two phases (see Figure 10-18). In the first phase, typically called the *prediction phase*, we use information learned in the past to further refine our model for what the next location of the person (or object) will be. In the second phase, the *correction phase*, we make a measurement and then reconcile that measurement with the predictions based on our previous measurements (i.e., our model).

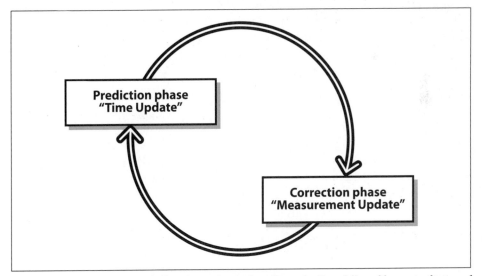

Figure 10-18. Two-phase estimator cycle: prediction based on prior data followed by reconciliation of the newest measurement

The machinery for accomplishing the two-phase estimation task falls generally under the heading of *estimators*, with the *Kalman filter* [Kalman60] being the most widely used technique. In addition to the Kalman filter, another important method is the *condensation algorithm*, which is a computer-vision implementation of a broader class of methods known as *particle filters*. The primary difference between the Kalman filter and the condensation algorithm is how the state probability density is described. We will explore the meaning of this distinction in the following sections.

The Kalman Filter

First introduced in 1960, the Kalman filter has risen to great prominence in a wide variety of signal processing contexts. The basic idea behind the Kalman filter is that, under a strong but reasonable* set of assumptions, it will be possible—given a history of measurements of a system—to build a model for the state of the system that maximizes the a posteriori† probability of those previous measurements. For a good introduction, see Welsh and Bishop [Welsh95]. In addition, we can maximize the a posteriori probability without keeping a long history of the previous measurements themselves. Instead, we iteratively update our model of a system's state and keep only that model for the next iteration. This greatly simplifies the computational implications of this method.

Before we go into the details of what this all means in practice, let's take a moment to look at the assumptions we mentioned. There are three important assumptions required in the theoretical construction of the Kalman filter: (1) the system being modeled is linear, (2) the noise that measurements are subject to is "white", and (3) this noise is also Gaussian in nature. The first assumption means (in effect) that the state of the system at time k can be modeled as some matrix multiplied by the state at time $k-1$. The additional assumptions that the noise is both white and Gaussian means that the noise is not correlated in time and that its amplitude can be accurately modeled using only an average and a covariance (i.e., the noise is completely described by its first and second moments). Although these assumptions may seem restrictive, they actually apply to a surprisingly general set of circumstances.‡

What does it mean to "maximize the a posteriori probability of those previous measurements"? It means that the new model we construct after making a measurement—taking into account both our previous model with its uncertainty and the new measurement with its uncertainty—is the model that has the highest probability of being correct. For our purposes, this means that the Kalman filter is, given the three assumptions, the best way to combine data from different sources or from the same source at different times. We start with what we know, we obtain new information, and then we decide to change

---

* Here by "reasonable" we mean something like "sufficiently unrestrictive that the method is useful for a reasonable variety of actual problems arising in the real world". "Reasonable" just seemed like less of a mouthful.

† The modifier "a posteriori" is academic jargon for "with hindsight". Thus, when we say that such and such a distribution "maximizes the a posteriori probability", what we mean is that that distribution, which is essentially a possible explanation of "what really happened", is actually the most likely one given the data we have observed . . . you know, looking back on it all in retrospect.

‡ OK, one more footnote. We actually slipped in another assumption here, which is that the initial distribution also must be Gaussian in nature. Often in practice the initial state is known exactly, or at least we treat it like it is, and so this satisfies our requirement. If the initial state were (for example) a 50-50 chance of being either in the bedroom or the bathroom, then we'd be out of luck and would need something more sophisticated than a single Kalman filter.

---

what we know based on how certain we are about the old and new information using a weighted combination of the old and the new.

Let's work all this out with a little math for the case of one-dimensional motion. You can skip the next section if you want, but linear systems and Gaussians are so friendly that Dr. Kalman might be upset if you didn't at least give it a try.

Some Kalman math

So what's the gist of the Kalman filter?—*information fusion*. Suppose you want to know where some point is on a line (our one-dimensional scenario).* As a result of noise, you have two unreliable (in a Gaussian sense) reports about where the object is: locations x_1 and x_2. Because there is Gaussian uncertainty in these measurements, they have means of $\bar{x}_1$ and $\bar{x}_2$ together with standard deviations σ_1 and σ_2. The standard deviations are, in fact, expressions of our uncertainty regarding how good our measurements are. The probability distribution as a function of location is the *Gaussian distribution*:

$$p_i(x) = \frac{1}{\sigma_i \sqrt{2\pi}} \exp\left(-\frac{(x-\bar{x}_i)^2}{2\sigma_i^2}\right) \quad (i=1, 2)$$

given two such measurements, each with a Gaussian probability distribution, we would expect that the probability density for some value of x given both measurements would be proportional to $p(x) = p_1(x)\, p_2(x)$. It turns out that this product is another Gaussian distribution, and we can compute the mean and standard deviation of this new distribution as follows. Given that

$$p_{12}(x) \propto \exp\left(-\frac{(x-\bar{x}_1)^2}{2\sigma_1^2}\right) \exp\left(-\frac{(x-\bar{x}_2)^2}{2\sigma_2^2}\right) = \exp\left(-\frac{(x-\bar{x}_1)^2}{2\sigma_1^2} - \frac{(x-\bar{x}_2)^2}{2\sigma_2^2}\right)$$

Given also that a Gaussian distribution is maximal at the average value, we can find that average value simply by computing the derivative of $p(x)$ with respect to x. Where a function is maximal its derivative is 0, so

$$\frac{dp_{12}}{dx}\bigg|_{\bar{x}_{12}} = -\left[\frac{\bar{x}_{12}-\bar{x}_1}{\sigma_1^2} + \frac{\bar{x}_{12}-\bar{x}_2}{\sigma_2^2}\right] \cdot p_{12}(\bar{x}_{12}) = 0$$

Since the probability distribution function $p(x)$ is never 0, it follows that the term in brackets must be 0. Solving that equation for x gives us this very important relation:

$$\bar{x}_{12} = \left(\frac{\sigma_2^2}{\sigma_1^2 + \sigma_2^2}\right) x_1 + \left(\frac{\sigma_1^2}{\sigma_1^2 + \sigma_2^2}\right) x_2$$

---

* For a more detailed explanation that follows a similar trajectory, the reader is referred to J. D. Schutter, J. De Geeter, T. Lefebvre, and H. Bruyninckx, "Kalman Filters: A Tutorial" (*http://citeseer.ist.psu.edu/443226.html*).

Thus, the new mean value $\bar{x}_{12}$ is just a weighted combination of the two measured means, where the weighting is determined by the relative uncertainties of the two measurements. Observe, for example, that if the uncertainty σ_2 of the second measurement is particularly large, then the new mean will be essentially the same as the mean x_1 for the more certain previous measurement.

With the new mean $\bar{x}_{12}$ in hand, we can substitute this value into our expression for $p_{12}(x)$ and, after substantial rearranging,* identify the uncertainty σ_{12}^2 as:

$$\sigma_{12}^2 = \frac{\sigma_1^2 \sigma_2^2}{\sigma_1^2 + \sigma_2^2}.$$

At this point, you are probably wondering what this tells us. Actually, it tells us a lot. It says that when we make a new measurement with a new mean and uncertainty, we can combine that measurement with the mean and uncertainty we already have to obtain a new state that is characterized by a still newer mean and uncertainty. (We also now have numerical expressions for these things, which will come in handy momentarily.)

This property that two Gaussian measurements, when combined, are equivalent to a single Gaussian measurement (with a computable mean and uncertainty) will be the most important feature for us. It means that when we have M measurements, we can combine the first two, then the third with the combination of the first two, then the fourth with the combination of the first three, and so on. This is what happens with tracking in computer vision; we obtain one measure followed by another followed by another.

Thinking of our measurements (x_i, σ_i) as time steps, we can compute the current state of our estimation $(\hat{x}_i, \hat{\sigma}_i)$ as follows. At time step 1, we have only our first measure $x_1 = x_1$ and its uncertainty $\hat{\sigma}_1^2 = \sigma_1^2$. Substituting this in our optimal estimation equations yields an iteration equation:

$$\hat{x}_2 = \frac{\sigma_2^2}{\hat{\sigma}_1^2 + \sigma_2^2} x_1 + \frac{\sigma_1^2}{\hat{\sigma}_1^2 + \sigma_2^2} x_2$$

Rearranging this equation gives us the following useful form:

$$\hat{x}_2 = \hat{x}_1 + \frac{\hat{\sigma}_1^2}{\hat{\sigma}_1^2 + \sigma_2^2} (x_2 - \hat{x}_1)$$

Before we worry about just what this is useful for, we should also compute the analogous equation for $\hat{\sigma}_2^2$. First, after substituting $\hat{\sigma}_1^2 = \sigma_1^2$ we have:

---

* The rearranging is a bit messy. If you want to verify all this, it is much easier to (1) start with the equation for the Gaussian distribution $p_{12}(x)$ in terms of $\bar{x}_{12}$ and σ_{12}, (2) substitute in the equations that relate $\bar{x}_{12}$ to $\bar{x}_1$ and $\bar{x}_2$ and those that relate σ_{12} to σ_1 and σ_2, and (3) verify that the result can be separated into the product of the Gaussians with which we started.

$$\hat{\sigma}_2^2 = \frac{\sigma_2^2 \hat{\sigma}_1^2}{\hat{\sigma}_1^2 + \sigma_2^2}$$

A rearrangement similar to what we did for $\hat{x}_2$ yields an iterative equation for estimating variance given a new measurement:

$$\hat{\sigma}_2^2 = \left(1 - \frac{\hat{\sigma}_1^2}{\hat{\sigma}_1^2 + \sigma_2^2}\right)\hat{\sigma}_1^2$$

In their current form, these equations allow us to separate clearly the "old" information (what we knew before a new measurement was made) from the "new" information (what our latest measurement told us). The new information $(x_2 - \hat{x}_1)$, seen at time step 2, is called the *innovation*. We can also see that our optimal iterative update factor is now:

$$K = \frac{\hat{\sigma}_1^2}{\hat{\sigma}_1^2 + \sigma_2^2}$$

This factor is known as the *update gain*. Using this definition for K, we obtain the following convenient recursion form:

$$\hat{x}_2 = \hat{x}_1 + K(x_2 - \hat{x}_1)$$

$$\hat{\sigma}_2^2 = (1 - K)\hat{\sigma}_1^2$$

In the Kalman filter literature, if the discussion is about a general series of measurements then our second time step "2" is usually denoted k and the first time step is thus $k - 1$.

Systems with dynamics

In our simple one-dimensional example, we considered the case of an object being located at some point x, and a series of successive measurements of that point. In that case we did not specifically consider the case in which the object might actually be moving in between measurements. In this new case we will have what is called the *prediction phase*. During the prediction phase, we use what we know to figure out where we expect the system to be before we attempt to integrate a new measurement.

In practice, the prediction phase is done immediately after a new measurement is made, but before the new measurement is incorporated into our estimation of the state of the system. An example of this might be when we measure the position of a car at time t, then again at time $t + dt$. If the car has some velocity v, then we do not just incorporate the second measurement directly. We first *fast-forward* our model based on what we knew at time t so that we have a model not only of the system at time t but also of the system at time $t + dt$, the instant before the new information is incorporated. In this way, the new information, acquired at time $t + dt$, is fused not with the old model of the system, but with the old model of the system projected forward to time $t + dt$. This is the meaning of the cycle depicted in Figure 10-18. In the context of Kalman filters, there are three kinds of motion that we would like to consider.

The first is *dynamical motion*. This is motion that we expect as a direct result of the state of the system when last we measured it. If we measured the system to be at position x with some velocity v at time t, then at time $t + dt$ we would expect the system to be located at position $x + v * dt$, possibly still with velocity.

The second form of motion is called *control motion*. Control motion is motion that we expect because of some external influence applied to the system of which, for whatever reason, we happen to be aware. As the name implies, the most common example of control motion is when we are estimating the state of a system that we ourselves have some control over, and we know what we did to bring about the motion. This is particularly the case for robotic systems where the control is the system telling the robot to (for example) accelerate or go forward. Clearly, in this case, if the robot was at x and moving with velocity v at time t, then at time $t + dt$ we expect it to have moved not only to $x + v * dt$ (as it would have done without the control), but also a little farther, since we did tell it to accelerate.

The final important class of motion is *random motion*. Even in our simple one-dimensional example, if whatever we were looking at had a possibility of moving on its own for whatever reason, we would want to include random motion in our prediction step. The effect of such random motion will be to simply increase the variance of our state estimate with the passage of time. Random motion includes any motions that are not known or under our control. As with everything else in the Kalman filter framework, however, there is an assumption that this random motion is either Gaussian (i.e., a kind of random walk) or that it can at least be modeled effectively as Gaussian.

Thus, to include dynamics in our simulation model, we would first do an "update" step before including a new measurement. This update step would include first applying any knowledge we have about the motion of the object according to its prior state, applying any additional information resulting from actions that we ourselves have taken or that we know to have been taken on the system from another outside agent, and, finally, incorporating our notion of random events that might have changed the state of the system since we last measured it. Once those factors have been applied, we can then incorporate our next new measurement.

In practice, the dynamical motion is particularly important when the "state" of the system is more complex than our simulation model. Often when an object is moving, there are multiple components to the "state" such as the position as well as the velocity. In this case, of course, the state evolves according to the velocity that we believe it to have. Handling systems with multiple components to the state is the topic of the next section. We will develop a little more sophisticated notation as well to handle these new aspects of the situation.

Kalman equations

We can now generalize these motion equations in our toy model. Our more general discussion will allow us to factor in any model that is a linear function F of the object's state. Such a model might consider combinations of the first and second derivatives of the previous motion, for example. We'll also see how to allow for a control input u_k to our model. Finally, we will allow for a more realistic observation model z in which we might measure only some of the model's state variables and in which the measurements may be only indirectly related to the state variables.*

To get started, let's look at how K, the gain in the previous section, affects the estimates. If the uncertainty of the new measurement is very large, then the new measurement essentially contributes nothing and our equations reduce to the combined result being the same as what we already knew at time $k - 1$. Conversely, if we start out with a large variance in the original measurement and then make a new, more accurate measurement, then we will "believe" mostly the new measurement. When both measurements are of equal certainty (variance), the new expected value is exactly between them. All of these remarks are in line with our reasonable expectations.

Figure 10-19 shows how our uncertainty evolves over time as we gather new observations.

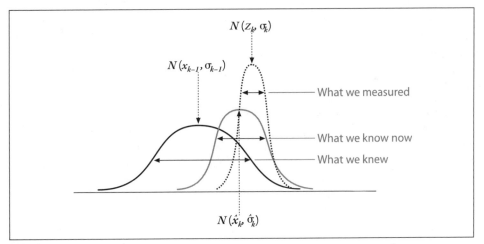

Figure 10-19. Combining our prior knowledge $N(x_{k-1}, \sigma_{k-1})$ with our measurement observation $N(z_k, \sigma_k)$; the result is our new estimate $N(\hat{x}_k, \hat{\sigma}_k)$

This idea of an update that is sensitive to uncertainty can be generalized to many state variables. The simplest example of this might be in the context of video tracking, where objects can move in two or three dimensions. In general, the state might contain

---

* Observe the change in notation from x_k to z_k. The latter is standard in the literature and is intended to clarify that z_k is a *general* measurement, possibly of multiple parameters of the model, and not just (and sometimes not even) the position x_k.

additional elements, such as the velocity of an object being tracked. In any of these general cases, we will need a bit more notation to keep track of what we are talking about. We will generalize the description of the state at time step k to be the following function of the state at time step $k - 1$:

$$x_k = Fx_{k-1} + Bu_k + w_k$$

Here x_k is now an n-dimensional vector of state components and F is an n-by-n matrix, sometimes called the *transfer matrix*, that multiplies x_{k-1}. The vector u_k is new. It's there to allow external controls on the system, and it consists of a c-dimensional vector referred to as the *control inputs*; B is an n-by-c matrix that relates these control inputs to the state change.* The variable w_k is a random variable (usually called the *process noise*) associated with random events or forces that directly affect the actual state of the system. We assume that the components of w_k have Gaussian distribution $N(0, Q_k)$ for some n-by-n covariance matrix Q_k (Q is allowed to vary with time, but often it does not).

In general, we make measurements z_k that may or may not be direct measurements of the state variable x_k. (For example, if you want to know how fast a car is moving then you could either measure its speed with a radar gun or measure the sound coming from its tailpipe; in the former case, z_k will be x_k with some added measurement noise, but in the latter case, the relationship is not direct in this way.) We can summarize this situation by saying that we measure the m-dimensional vector of measurements z_k given by:

$$z_k = H_k x_k + v_k$$

Here H_k is an m-by-n matrix and v_k is the measurement error, which is also assumed to have Gaussian distributions $N(0, R_k)$ for some m-by-m covariance matrix R_k.†

Before we get totally lost, let's consider a particular realistic situation of taking measurements on a car driving in a parking lot. We might imagine that the state of the car could be summarized by two position variables, x and y, and two velocities, v_k and v_y. These four variables would be the elements of the state vector x_k. This suggests that the correct form for F is:

$$x_k = \begin{bmatrix} x \\ y \\ v_x \\ v_y \end{bmatrix}_k, \quad F = \begin{bmatrix} 1 & 0 & dt & 0 \\ 0 & 1 & 0 & dt \\ 0 & 0 & 1 & 0 \\ 0 & 0 & 0 & 1 \end{bmatrix}$$

---

* The astute reader, or one who already knows something about Kalman filters, will notice another important assumption we slipped in—namely, that there is a linear relationship (via matrix multiplication) between the controls u_k and the change in state. In practical applications, this is often the first assumption to break down.

† The k in these terms allows them to vary with time but does not require this. In actual practice, it's common for H and R not to vary with time.

However, when using a camera to make measurements of the car's state, we probably measure only the position variables:

$$z_k = \begin{bmatrix} z_x \\ z_y \end{bmatrix}_k$$

This implies that the structure of H is something like:

$$H = \begin{bmatrix} 1 & 0 \\ 0 & 1 \\ 0 & 0 \\ 0 & 0 \end{bmatrix}$$

In this case, we might not really believe that the velocity of the car is constant and so would assign a value of Q_k to reflect this. We would choose R_k based on our estimate of how accurately we have measured the car's position using (for example) our image analysis techniques on a video stream.

All that remains now is to plug these expressions into the generalized forms of the update equations. The basic idea is the same, however. First we compute the a priori estimate x_k^- of the state. It is relatively common (though not universal) in the literature to use the superscript minus sign to mean "at the time immediately prior to the new measurement"; we'll adopt that convention here as well. This a priori estimate is given by:

$$x_k^- = Fx_{k-1} + Bu_{k-1} + w_k$$

Using P_k^- to denote the error covariance, the a priori estimate for this covariance at time k is obtained from the value at time $k - 1$ by:

$$P_k^- = FP_{k-1}F^T + Q_{k-1}$$

This equation forms the basis of the predictive part of the estimator, and it tells us "what we expect" based on what we've already seen. From here we'll state (without derivation) what is often called the *Kalman gain* or the *blending factor*, which tells us how to weight new information against what we think we already know:

$$K_k = P_k^- H_k^T (H_k P_k^- H_k^T + R_k)^{-1}$$

Though this equation looks intimidating, it's really not so bad. We can understand it more easily by considering various simple cases. For our one-dimensional example in which we measured one position variable directly, H_k is just a 1-by-1 matrix containing only a 1! Thus, if our measurement error is σ_{k+1}^2, then R_k is also a 1-by-1 matrix containing that value. Similarly, P_k is just the variance σ_k^2. So that big equation boils down to just this:

$$K = \frac{\sigma_k^2}{\sigma_k^2 + \sigma_{k+1}^2}$$

Note that this is exactly what we thought it would be. The gain, which we first saw in the previous section, allows us to optimally compute the updated values for x_k and P_k when a new measurement is available:

$$x_k = x_k^- + K_k(z_k^- - H_k x_k^-)$$

$$P_k = (I - K_k H_k)P_k^-$$

Once again, these equations look intimidating at first; but in the context of our simple one-dimensional discussion, it's really not as bad as it looks. The optimal weights and gains are obtained by the same methodology as for the one-dimensional case, except this time we minimize the uncertainty of our position state x by setting to 0 the partial derivatives with respect to x before solving. We can show the relationship with the simpler one-dimensional case by first setting $F = I$ (where I is the identity matrix), $B = 0$, and $Q = 0$. The similarity to our one-dimensional filter derivation is then revealed by making the following substitutions in our more general equations: $x_k \leftarrow \hat{x}_2$, $x_k^- \leftarrow \hat{x}_1$, $K_k \leftarrow K$, $z_k \leftarrow x_2$, $H_k \leftarrow 1$, $P_k \leftarrow \hat{\sigma}_2^2$, $I \leftarrow 1$, $P_k^- \leftarrow \hat{\sigma}_1^2$, and $R_k \leftarrow \sigma_2^2$.

OpenCV and the Kalman filter

With all of this at our disposal, you might feel that we don't need OpenCV to do anything for us or that we desperately need OpenCV to do all of this for us. Fortunately, OpenCV is amenable to either interpretation. It provides four functions that are directly related to working with Kalman filters.

```
cvCreateKalman(
    int        nDynamParams,
    int        nMeasureParams,
    int        nControlParams
);
cvReleaseKalman(
    CvKalman** kalman
);
```

The first of these generates and returns to us a pointer to a CvKalman data structure, and the second deletes that structure.

```
typedef struct CvKalman {
    int MP;                        // measurement vector dimensions
    int DP;                        // state vector dimensions
    int CP;                        // control vector dimensions
    CvMat* state_pre;              // predicted state:
                                   //   x_k = F x_k-1 + B u_k
    CvMat* state_post;             // corrected state:
                                   //   x_k = x_k' + K_k (z_k'- H x_k')
    CvMat* transition_matrix;      // state transition matrix
                                   //   F
    CvMat* control_matrix;         // control matrix
                                   //   B
                                   //   (not used if there is no control)
    CvMat* measurement_matrix;     // measurement matrix
                                   //   H
```

```
    CvMat* process_noise_cov;        // process noise covariance
                                     //   Q
    CvMat* measurement_noise_cov;    // measurement noise covariance
                                     //   R
    CvMat* error_cov_pre;            // prior error covariance:
                                     //   (P_k'=F P_k-1 Ft) + Q
    CvMat* gain;                     // Kalman gain matrix:
                                     //   K_k = P_k' H^T (H P_k' H^T + R)^-1
    CvMat* error_cov_post;           // posteriori error covariance
                                     //   P_k = (I - K_k H) P_k'
    CvMat* temp1;                    // temporary matrices
    CvMat* temp2;
    CvMat* temp3;
    CvMat* temp4;
    CvMat* temp5;
  } CvKalman;
```

The next two functions implement the Kalman filter itself. Once the data is in the structure, we can compute the prediction for the next time step by calling cvKalmanPredict() and then integrate our new measurements by calling cvKalmanCorrect(). After running each of these routines, we can read the state of the system being tracked. The result of cvKalmanCorrect() is in state_post, and the result of cvKalmanPredict() is in state_pre.

```
    cvKalmanPredict(
        CvKalman*  kalman,
        const      CvMat* control = NULL
    );
    cvKalmanCorrect(
        CvKalman*  kalman,
        CvMat*     measured
    );
```

Kalman filter example code

Clearly it is time for a good example. Let's take a relatively simple one and implement it explicitly. Imagine that we have a point moving around in a circle, like a car on a race track. The car moves with mostly constant velocity around the track, but there is some variation (i.e., process noise). We measure the location of the car using a method such as tracking it via our vision algorithms. This generates some (unrelated and probably different) noise as well (i.e., measurement noise).

So our model is quite simple: the car has a position and an angular velocity at any moment in time. Together these factors form a two-dimensional state vector x_k. However, our measurements are only of the car's position and so form a one-dimensional "vector" z_k.

We'll write a program (Example 10-2) whose output will show the car circling around (in red) as well as the measurements we make (in yellow) and the location predicted by the Kalman filter (in white).

We begin with the usual calls to include the library header files. We also define a macro that will prove useful when we want to transform the car's location from angular to Cartesian coordinates so we can draw on the screen.

Example 10-2. Kalman filter sample code

```
//  Use Kalman Filter to model particle in circular trajectory.
//
#include "cv.h"
#include "highgui.h"
#include "cvx_defs.h"

#define phi2xy(mat)                                                          /
  cvPoint( cvRound(img->width/2 + img->width/3*cos(mat->data.fl[0])),       /
    cvRound( img->height/2 - img->width/3*sin(mat->data.fl[0])) )

int main(int argc, char** argv) {

  // Initialize, create Kalman Filter object, window, random number
  // generator etc.
  //
  cvNamedWindow( "Kalman", 1 );
. . . continued below
```

Next, we will create a random-number generator, an image to draw to, and the Kalman filter structure. Notice that we need to tell the Kalman filter how many dimensions the state variables are (2) and how many dimensions the measurement variables are (1).

```
 . . . continued from above
    CvRandState rng;
    cvRandInit( &rng, 0, 1, -1, CV_RAND_UNI );

    IplImage* img = cvCreateImage( cvSize(500,500), 8, 3 );
    CvKalman* kalman = cvCreateKalman( 2, 1, 0 );
  . . . continued below
```

Once we have these building blocks in place, we create a matrix (really a vector, but in OpenCV we call everything a matrix) for the state x_k, the process noise w_k, the measurements z_k, and the all-important transition matrix *F*. The state needs to be initialized to something, so we fill it with some reasonable random numbers that are narrowly distributed around zero.

The transition matrix is crucial because it relates the state of the system at time *k* to the state at time *k* + 1. In this case, the transition matrix will be 2-by-2 (since the state vector is two-dimensional). It is, in fact, the transition matrix that gives meaning to the components of the state vector. We view x_k as representing the angular position of the car (φ) and the car's angular velocity (ω). In this case, the transition matrix has the components [[1, *dt*], [0, 1]]. Hence, after multiplying by *F*, the state (φ, ω) becomes ($\varphi + \omega\ dt$, ω)—that is, the angular velocity is unchanged but the angular position increases by an amount equal to the angular velocity multiplied by the time step. In our example we choose dt=1.0 for convenience, but in practice we'd need to use something like the time between sequential video frames.

```
 . . . continued from above
    // state is (phi, delta_phi) - angle and angular velocity
    // Initialize with random guess.
```

```
        //
        CvMat* x_k = cvCreateMat( 2, 1, CV_32FC1 );
        cvRandSetRange( &rng, 0, 0.1, 0 );
        rng.disttype = CV_RAND_NORMAL;
        cvRand( &rng, x_k );

        // process noise
        //
        CvMat* w_k = cvCreateMat( 2, 1, CV_32FC1 );

        // measurements, only one parameter for angle
        //
        CvMat* z_k = cvCreateMat( 1, 1, CV_32FC1 );
        cvZero( z_k );

        // Transition matrix 'F' describes relationship between
        // model parameters at step k and at step k+1 (this is
        // the "dynamics" in our model)
        //
        const float F[] = { 1, 1, 0, 1 };
        memcpy( kalman->transition_matrix->data.fl, F, sizeof(F));
. . . continued below
```

The Kalman filter has other internal parameters that must be initialized. In particular, the 1-by-2 measurement matrix H is initialized to [1, 0] by a somewhat unintuitive use of the identity function. The covariance of process noise and of measurement noise are set to reasonable but interesting values (you can play with these yourself), and we initialize the posterior error covariance to the identity as well (this is required to guarantee the meaningfulness of the first iteration; it will subsequently be overwritten).

Similarly, we initialize the posterior state (of the hypothetical step previous to the first one!) to a random value since we have no information at this time.

```
. . . continued from above
        // Initialize other Kalman filter parameters.
        //
        cvSetIdentity( kalman->measurement_matrix,      cvRealScalar(1) );
        cvSetIdentity( kalman->process_noise_cov,       cvRealScalar(1e-5) );
        cvSetIdentity( kalman->measurement_noise_cov,   cvRealScalar(1e-1) );
        cvSetIdentity( kalman->error_cov_post,          cvRealScalar(1));

        // choose random initial state
        //
        cvRand( &rng, kalman->state_post );

        while( 1 ) {
. . . continued below
```

Finally we are ready to start up on the actual dynamics. First we ask the Kalman filter to predict what it thinks this step will yield (i.e., before giving it any new information); we call this y_k. Then we proceed to generate the new value of z_k (the measurement) for this iteration. By definition, this value is the "real" value x_k multiplied by the measurement matrix H with the random measurement noise added. We must remark here that, in anything but a toy application such as this, you would not generate z_k from x_k; instead, a generating function would arise from the state of the world or your sensors. In this simulated case, we generate the measurements from an underlying "real" data model by adding random noise ourselves; this way, we can see the effect of the Kalman filter.

```
. . . continued from above
  // predict point position
  const CvMat* y_k = cvKalmanPredict( kalman, 0 );

  // generate measurement (z_k)
  //
  cvRandSetRange(
    &rng,
    0,
    sqrt(kalman->measurement_noise_cov->data.fl[0]),
    0
  );
  cvRand( &rng, z_k );
  cvMatMulAdd( kalman->measurement_matrix, x_k, z_k, z_k );
. . . continued below
```

Draw the three points corresponding to the observation we synthesized previously, the location predicted by the Kalman filter, and the underlying state (which we happen to know in this simulated case).

```
. . . continued from above
  // plot points (eg convert to planar coordinates and draw)
  //
  cvZero( img );
  cvCircle( img, phi2xy(z_k), 4, CVX_YELLOW );   // observed state
  cvCircle( img, phi2xy(y_k), 4, CVX_WHITE, 2 ); // "predicted" state
  cvCircle( img, phi2xy(x_k), 4, CVX_RED );      // real state
  cvShowImage( "Kalman", img );
. . . continued below
```

At this point we are ready to begin working toward the next iteration. The first thing to do is again call the Kalman filter and inform it of our newest measurement. Next we will generate the process noise. We then use the transition matrix *F* to time-step x_k forward one iteration and then add the process noise we generated; now we are ready for another trip around.

```
. . . continued from above
  // adjust Kalman filter state
  //
  cvKalmanCorrect( kalman, z_k );

  // Apply the transition matrix 'F' (e.g., step time forward)
  // and also apply the "process" noise w_k.
  //
  cvRandSetRange(
    &rng,
    0,
    sqrt(kalman->process_noise_cov->data.fl[0]),
    0
```

```
        );
        cvRand( &rng, w_k );
        cvMatMulAdd( kalman->transition_matrix, x_k, w_k, x_k );

        // exit if user hits 'Esc'
        if( cvWaitKey( 100 ) == 27 ) break;
    }

    return 0;
}
```

As you can see, the Kalman filter part was not that complicated; half of the required code was just generating some information to push into it. In any case, we should summarize everything we've done, just to be sure it all makes sense.

We started out by creating matrices to represent the state of the system and the measurements we would make. We defined both the transition and measurement matrices and then initialized the noise covariances and other parameters of the filter.

After initializing the state vector to a random value, we called the Kalman filter and asked it to make its first prediction. Once we read out that prediction (which was not very meaningful this first time through), we drew to the screen what was predicted. We also synthesized a new observation and drew that on the screen for comparison with the filter's prediction. Next we passed the filter new information in the form of that new measurement, which it integrated into its internal model. Finally, we synthesized a new "real" state for the model so that we could iterate through the loop again.

Running the code, the little red ball orbits around and around. The little yellow ball appears and disappears about the red ball, representing the noise that the Kalman filter is trying to "see through". The white ball rapidly converges down to moving in a small space around the red ball, showing that the Kalman filter has given a reasonable estimate of the motion of the particle (the car) within the framework of our model.

One topic that we did not address in our example is the use of control inputs. For example, if this were a radio-controlled car and we had some knowledge of what the person with the controller was doing, then we could include that information into our model. In that case it might be that the velocity is being set by the controller. We'd then need to supply the matrix B (kalman->control_matrix) and also to provide a second argument for cvKalmanPredict() to accommodate the control vector u.

A Brief Note on the Extended Kalman Filter

You might have noticed that requiring the dynamics of the system to be linear in the underlying parameters is quite restrictive. It turns out that the Kalman filter is still useful to us when the dynamics are nonlinear, and the OpenCV Kalman Filter routines remain useful as well.

Recall that "linear" meant (in effect) that the various steps in the definition of the Kalman filter could be represented with matrices. When might this not be the case? There are actually many possibilities. For example, suppose our control measure is the amount by which our car's gas pedal is depressed: the relationship between the car's velocity and the gas pedal's depression is not a linear one. Another common problem is a force on the car that is more naturally expressed in Cartesian coordinates while the motion of the car (as in our example) is more naturally expressed in polar coordinates. This might arise if our car were instead a boat moving in circles but in a uniform water current and heading some particular direction.

In all such cases, the Kalman filter is not, by itself, sufficient. One way to handle these nonlinearities (or at least attempt to handle them) is to *linearize* the relevant processes (e.g., the update F or the control input response B). Thus, we'd need to compute new values for F and B, at every time step, based on the state x. These values would only approximate the real update and control functions in the vicinity of the particular value of x, but in practice this is often sufficient. This extension to the Kalman filter is known simply enough as the *extended Kalman filter* [Schmidt66].

OpenCV does not provide any specific routines to implement this, but none are actually needed. All we have to do is recompute and reset the values of kalman->update_matrix and kalman->control_matrix before each update. The Kalman filter has since been more elegantly extended to nonlinear systems in a formulation called the *unscented particle filter* [Merwe00]. A very good overview of the entire field of Kalman filtering, including the latest advances, is given in [Thrun05].

The Condensation Algorithm

The Kalman filter models a single hypothesis. Because the underlying model of the probability distribution for that hypothesis is unimodal Gaussian, it is not possible to represent multiple hypotheses simultaneously using the Kalman filter. A somewhat more advanced technique known as the *condensation algorithm* [Isard98], which is based on a broader class of estimators called *particle filters*, will allow us to address this issue.

To understand the purpose of the condensation algorithm, consider the hypothesis that an object is moving with constant speed (as modeled by the Kalman filter). Any data measured will, in essence, be integrated into the model as if it supports this hypothesis. Consider now the case of an object moving behind an occlusion. Here we do not know what the object is doing; it might be continuing at constant speed, it might have stopped and/or reversed direction. The Kalman filter cannot represent these multiple possibilities other than by simply broadening the uncertainty associated with the (Gaussian) distribution of the object's location. The Kalman filter, since it is necessarily Gaussian, cannot represent such multimodal distributions.

As with the Kalman filter, we have two routines for (respectively) creating and destroying the data structure used to represent the condensation filter. The only difference is that in this case the creation routine cvCreateConDensation() has an extra parameter. The value entered for this parameter sets the number of hypotheses (i.e., "particles") that the filter will maintain at any given time. This number should be relatively large (50 or 100; perhaps more for complicated situations) because the collection of these individual hypotheses takes the place of the parameterized Gaussian probability distribution of the Kalman filter. See Figure 10-20.

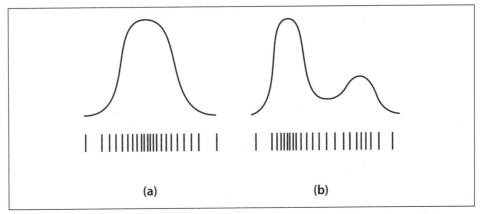

Figure 10-20. Distributions that can (panel a) and cannot (panel b) be represented as a continuous Gaussian distribution parameterizable by a mean and an uncertainty; both distributions can alternatively be represented by a set of particles whose density approximates the represented distribution

```
CvConDensation* cvCreateConDensation(
    int dynam_params,
    int measure_params,
    int sample_count
);

void cvReleaseConDensation(
    CvConDensation** condens
);
```

This data structure has the following internal elements:

```
)typedef struct CvConDensation
{
    int     MP;             // Dimension of measurement vector
    int     DP;             // Dimension of state vector
    float*  DynamMatr;      // Matrix of the linear Dynamics system
    float*  State;          // Vector of State
    int     SamplesNum;     // Number of Samples
    float** flSamples;      // array of the Sample Vectors
    float** flNewSamples;   // temporary array of the Sample Vectors
    float*  flConfidence;   // Confidence for each Sample
    float*  flCumulative;   // Cumulative confidence
    float*  Temp;           // Temporary vector
    float*  RandomSample;   // RandomVector to update sample set
    CvRandState* RandS;     // Array of structures to generate random vectors
} CvConDensation;
```

Once we have allocated the condensation filter's data structure, we need to initialize that structure. We do this with the routine cvConDensInitSampleSet(). While creating the CvConDensation structure we indicated how many particles we'd have, and for each particle we also specified some number of dimensions. Initializing all of these particles could be quite a hassle.* Fortunately, `cvConDensInitSampleSet()` does this for us in a convenient way; we need only specify the ranges for each dimension.

```
void cvConDensInitSampleSet(
    CvConDensation* condens,
    CvMat*          lower_bound,
    CvMat*          upper_bound
);
```

This routine requires that we initialize two `CvMat` structures. Both are vectors (meaning that they have only one column), and each has as many entries as the number of dimensions in the system state. These vectors are then used to set the ranges that will be used to initialize the sample vectors in the `CvConDensation` structure.

The following code creates two matrices of size `Dim` and initializes them to -1 and +1, respectively. When `cvConDensInitSampleSet()` is called, the initial sample set will be initialized to random numbers each of which falls within the (in this case, identical) interval from -1 to +1. Thus, if `Dim` were three then we would be initializing the filter with particles uniformly distributed inside of a cube centered at the origin and with sides of length 2.

```
CvMat LB = cvMat(Dim,1,CV_MAT32F,NULL);
CvMat UB = cvMat(Dim,1,CV_MAT32F,NULL);
cvmAlloc(&LB);
cvmAlloc(&UB);
ConDens = cvCreateConDensation(Dim, Dim,SamplesNum);
for( int i = 0; i<Dim; i++) {
    LB.data.fl[i] = -1.0f;
    UB.data.fl[i] =  1.0f;
}
cvConDensInitSampleSet(ConDens,&LB,&UB);
```

Finally, our last routine allows us to update the condensation filter state:

```
void cvConDensUpdateByTime( CvConDensation* condens );
```

There is a little more to using this routine than meets the eye. In particular, we must update the confidences of all of the particles in light of whatever new information has become available since the previous update. Sadly, there is no convenient routine for doing this in OpenCV. The reason is that the relationship between the new confidence for a particle and the new information depends on the context. Here is an example of such an update, which applies a simple[†] update to the confidence of each particle in the filter.

```
// Update the confidences on all of the particles in the filter
// based on a new measurement M[]. Here M has the dimensionality of
// the particles in the filter.
//
void CondProbDens(
    CvConDensation* CD,
    float* M
```

---

* Of course, if you know about particle filters then you know that this is where we could initialize the filter with our prior knowledge (or prior assumptions) about the state of the system. The function that initializes the filter is just to help you generate a uniform distribution of points (i.e., a *flat* prior).

† The attentive reader will notice that this update actually implies a Gaussian probability distribution, but of course you could have a much more complicated update for your particular context.

```
) {
    for( int i=0; i<CD->SamplesNum; i++ ) {
        float p = 1.0f;
        for( int j=0; j<CD->DP; j++ ) {
            p *= (float) exp(
                -0.05*(M[j] - CD->flSamples[i][j])*(M[j]-CD->flSamples[i][j])
            );
        }
        CD->flConfidence[i] = Prob;
    }
}
```

Once you have updated the confidences, you can then call cvCondensUpdateByTime() in order to update the particles. Here "updating" means *resampling*, which is to say that a new set of particles will be generated in accordance with the computed confidences. After updating, all of the confidences will again be exactly 1.0f, but the distribution of particles will now include the previously modified confidences directly into the density of particles in the next iteration.

Exercises

There are sample code routines in the *.../opencv/samples/c/* directory that demonstrate many of the algorithms discussed in this chapter:

- *lkdemo.c* (optical flow)
- *camshiftdemo.c* (mean-shift tracking of colored regions)
- *motempl.c* (motion template)
- *kalman.c* (Kalman filter)

1. The covariance Hessian matrix used in cvGoodFeaturesToTrack() is computed over some square region in the image set by block_size in that function.

    a. Conceptually, what happens when block size increases? Do we get more or fewer "good features"? Why?

    b. Dig into the *lkdemo.c* code, search for cvGoodFeaturesToTrack(), and try playing with the block_size to see the difference.

2. Refer to Figure 10-2 and consider the function that implements subpixel corner finding, cvFindCornerSubPix().

    a. What would happen if, in Figure 10-2, the checkerboard were twisted so that the straight dark-light lines formed curves that met in a point? Would subpixel corner finding still work? Explain.

    b. If you expand the window size around the twisted checkerboard's corner point (after expanding the win and zero_zone parameters), does subpixel corner finding become more accurate or does it rather begin to diverge? Explain your answer.

3. *Optical flow*

   a. Describe an object that would be better tracked by block matching than by Lucas-Kanade optical flow.

   b. Describe an object that would be better tracked by Lucas-Kanade optical flow than by block matching.

4. Compile *lkdemo.c*. Attach a web camera (or use a previously captured sequence of a textured moving object). In running the program, note that "r" autoinitializes tracking, "c" clears tracking, and a mouse click will enter a new point or turn off an old point. Run *lkdemo.c* and initialize the point tracking by typing "r". Observe the effects.

   a. Now go into the code and remove the subpixel point placement function cvFindCornerSubPix(). Does this hurt the results? In what way?

   b. Go into the code again and, in place of cvGoodFeaturesToTrack(), just put down a grid of points in an ROI around the object. Describe what happens to the points and why.

   > Hint: Part of what happens is a consequence of the aperture problem— given a fixed window size and a line, we can't tell how the line is moving.

5. Modify the *lkdemo.c* program to create a program that performs simple image stabilization for moderately moving cameras. Display the stabilized results in the center of a much larger window than the one output by your camera (so that the frame may wander while the first points remain stable).

6. Compile and run *camshiftdemo.c* using a web camera or color video of a moving colored object. Use the mouse to draw a (tight) box around the moving object; the routine will track it.

   a. In *camshiftdemo.c*, replace the cvCamShif() routine with cvMeanShift(). Describe situations where one tracker will work better than another.

   b. Write a function that will put down a grid of points in the initial cvMeanShift() box. Run both trackers at once.

   c. How can these two trackers be used together to make tracking more robust? Explain and/or experiment.

7. Compile and run the motion template code *motempl.c* with a web camera or using a previously stored movie file.

   a. Modify *motempl.c* so that it can do simple gesture recognition.

   b. If the camera was moving, explain how to use your motion stabilization code from exercise 5 to enable motion templates to work also for moderately moving cameras.

8. Describe how you can track circular (nonlinear) motion using a linear state model (not extended) Kalman filter.

Hint: How could you preprocess this to get back to linear dynamics?

9. Use a motion model that posits that the current state depends on the previous state's location and velocity. Combine the *lkdemo.c* (using only a few click points) with the Kalman filter to track Lucas-Kanade points better. Display the uncertainty around each point. Where does this tracking fail?

Hint: Use Lucas-Kanade as the observation model for the Kalman filter, and adjust noise so that it tracks. Keep motions reasonable.

10. A Kalman filter depends on linear dynamics and on Markov independence (i.e., it assumes the current state depends only on the immediate past state, not on all past states). Suppose you want to track an object whose movement is related to its previous location and its previous velocity but that you mistakenly include a dynamics term only for state dependence on the previous location—in other words, forgetting the previous velocity term.

   a. Do the Kalman assumptions still hold? If so, explain why; if not, explain how the assumptions were violated.

   b. How can a Kalman filter be made to still track when you forget some terms of the dynamics?

   Hint: Think of the noise model.

11. Use a web cam or a movie of a person waving two brightly colored objects, one in each hand. Use condensation to track both hands.

CHAPTER 11

Camera Models and Calibration

Vision begins with the detection of light from the world. That light begins as rays emanating from some source (e.g., a light bulb or the sun), which then travels through space until striking some object. When that light strikes the object, much of the light is absorbed, and what is not absorbed we perceive as the color of the light. Reflected light that makes its way to our eye (or our camera) is collected on our retina (or our imager). The geometry of this arrangement—particularly of the ray's travel from the object, through the lens in our eye or camera, and to the retina or imager—is of particular importance to practical computer vision.

A simple but useful model of how this happens is the pinhole camera model.* A *pinhole* is an imaginary wall with a tiny hole in the center that blocks all rays except those passing through the tiny aperture in the center. In this chapter, we will start with a pinhole camera model to get a handle on the basic geometry of projecting rays. Unfortunately, a real pinhole is not a very good way to make images because it does not gather enough light for rapid exposure. This is why our eyes and cameras use lenses to gather more light than what would be available at a single point. The downside, however, is that gathering more light with a lens not only forces us to move beyond the simple geometry of the pinhole model but also introduces distortions from the lens itself.

In this chapter we will learn how, using *camera calibration*, to correct (mathematically) for the main deviations from the simple pinhole model that the use of lenses imposes on us. Camera calibration is important also for relating camera measurements with measurements in the real, three-dimensional world. This is important because scenes are not only three-dimensional; they are also physical spaces with physical units. Hence, the relation between the camera's natural units (pixels) and the units of the

---

* Knowledge of lenses goes back at least to Roman times. The pinhole camera model goes back at least 987 years to al-Hytham [1021] and is the classic way of introducing the geometric aspects of vision. Mathematical and physical advances followed in the 1600s and 1700s with Descartes, Kepler, Galileo, Newton, Hooke, Euler, Fermat, and Snell (see O'Connor [O'Connor02]). Some key modern texts for geometric vision include those by Trucco [Trucco98], Jaehne (also sometimes spelled Jähne) [Jaehne95; Jaehne97], Hartley and Zisserman [Hartley06], Forsyth and Ponce [Forsyth03], Shapiro and Stockman [Shapiro02], and Xu and Zhang [Xu96].

physical world (e.g., meters) is a critical component in any attempt to reconstruct a three-dimensional scene.

The process of camera calibration gives us both a model of the camera's geometry and a *distortion* model of the lens. These two informational models define the *intrinsic parameters* of the camera. In this chapter we use these models to correct for lens distortions; in Chapter 12, we will use them to interpret a physical scene.

We shall begin by looking at camera models and the causes of lens distortion. From there we will explore the *homography transform*, the mathematical instrument that allows us to capture the effects of the camera's basic behavior and of its various distortions and corrections. We will take some time to discuss exactly how the transformation that characterizes a particular camera can be calculated mathematically. Once we have all this in hand, we'll move on to the OpenCV function that does most of this work for us.

Just about all of this chapter is devoted to building enough theory that you will truly understand what is going into (and what is coming out of) the OpenCV function cvCalibrateCamera2() as well as what that function is doing "under the hood". This is important stuff if you want to use the function responsibly. Having said that, if you are already an expert and simply want to know how to use OpenCV to do what you already understand, jump right ahead to the "Calibration Function" section and get to it.

Camera Model

We begin by looking at the simplest model of a camera, the pinhole camera model. In this simple model, light is envisioned as entering from the scene or a distant object, but only a single ray enters from any particular point. In a physical pinhole camera, this point is then "projected" onto an imaging surface. As a result, the image on this *image plane* (also called the *projective plane*) is always in focus, and the size of the image relative to the distant object is given by a single parameter of the camera: its *focal length*. For our idealized pinhole camera, the distance from the pinhole aperture to the screen is precisely the focal length. This is shown in Figure 11-1, where *f* is the focal length of the camera, *Z* is the distance from the camera to the object, *X* is the length of the object, and *x* is the object's image on the imaging plane. In the figure, we can see by similar triangles that $-x/f = X/Z$, or

$$-x = f\frac{X}{Z}$$

We shall now rearrange our pinhole camera model to a form that is equivalent but in which the math comes out easier. In Figure 11-2, we swap the pinhole and the image plane.* The main difference is that the object now appears rightside up. The point in the pinhole is reinterpreted as the *center of projection*. In this way of looking at things, every

---

* Typical of such mathematical abstractions, this new arrangement is not one that can be built physically; the image plane is simply a way of thinking of a "slice" through all of those rays that happen to strike the center of projection. This arrangement is, however, much easier to draw and do math with.

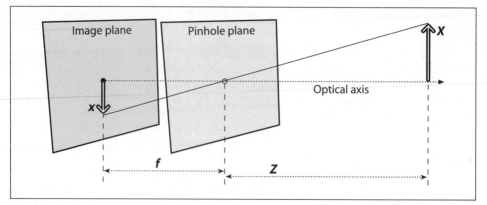

Figure 11-1. Pinhole camera model: a pinhole (the pinhole aperture) lets through only those light rays that intersect a particular point in space; these rays then form an image by "projecting" onto an image plane

ray leaves a point on the distant object and heads for the center of projection. The point at the intersection of the image plane and the optical axis is referred to as the *principal point*. On this new frontal image plane (see Figure 11-2), which is the equivalent of the old projective or image plane, the image of the distant object is exactly the same size as it was on the image plane in Figure 11-1. The image is generated by intersecting these rays with the image plane, which happens to be exactly a distance f from the center of projection. This makes the similar triangles relationship $x/f = X/Z$ more directly evident than before. The negative sign is gone because the object image is no longer upside down.

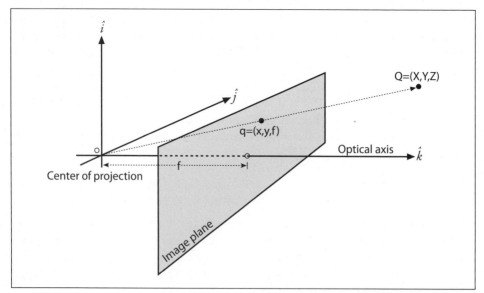

Figure 11-2. A point $Q = (X, Y, Z)$ is projected onto the image plane by the ray passing through the center of projection, and the resulting point on the image is $q = (z, y, f)$; the image plane is really just the projection screen "pushed" in front of the pinhole (the math is equivalent but simpler this way)

You might think that the principle point is equivalent to the center of the imager; yet this would imply that some guy with tweezers and a tube of glue was able to attach the imager in your camera to micron accuracy. In fact, the center of the chip is usually not on the optical axis. We thus introduce two new parameters, c_x and c_y, to model a possible displacement (away from the optic axis) of the center of coordinates on the projection screen. The result is that a relatively simple model in which a point Q in the physical world, whose coordinates are (X, Y, Z), is projected onto the screen at some pixel location given by (x_{screen}, y_{screen}) in accordance with the following equations:*

$$x_{screen} = f_x\left(\frac{X}{Z}\right) + c_x, \qquad y_{screen} = f_y\left(\frac{Y}{Z}\right) + c_y$$

Note that we have introduced two different focal lengths; the reason for this is that the individual pixels on a typical low-cost imager are rectangular rather than square. The focal length f_x (for example) is actually the product of the physical focal length of the lens and the size s_x of the individual imager elements (this should make sense because s_x has units of pixels per millimeter[†] while F has units of millimeters, which means that f_x is in the required units of pixels). Of course, similar statements hold for f_y and s_y. It is important to keep in mind, though, that s_x and s_y cannot be measured directly via any camera calibration process, and neither is the physical focal length F directly measurable. Only the combinations $f_x = Fs_x$ and $f_y = Fs_y$ can be derived without actually dismantling the camera and measuring its components directly.

Basic Projective Geometry

The relation that maps the points Q_i in the physical world with coordinates (X_i, Y_i, Z_i) to the points on the projection screen with coordinates (x_i, y_i) is called a *projective transform*. When working with such transforms, it is convenient to use what are known as *homogeneous coordinates*. The homogeneous coordinates associated with a point in a projective space of dimension n are typically expressed as an $(n + 1)$-dimensional vector (e.g., x, y, z becomes x, y, z, w), with the additional restriction that any two points whose values are proportional are equivalent. In our case, the image plane is the projective space and it has two dimensions, so we will represent points on that plane as three-dimensional vectors $q = (q_1, q_2, q_3)$. Recalling that all points having proportional values in the projective space are equivalent, we can recover the actual pixel coordinates by dividing through by q_3. This allows us to arrange the parameters that define our camera (i.e., $f_x, f_y, c_x,$ and c_y) into a single 3-by-3 matrix, which we will call the *camera intrinsics matrix* (the approach OpenCV takes to camera intrinsics is derived from Heikkila and

---

* Here the subscript "screen" is intended to remind you that the coordinates being computed are in the coordinate system of the screen (i.e., the imager). The difference between (x_{screen}, y_{screen}) in the equation and (x, y) in Figure 11-2 is precisely the point of c_x and c_y. Having said that, we will subsequently drop the "screen" subscript and simply use lowercase letters to describe coordinates on the imager.

† Of course, "millimeter" is just a stand-in for any physical unit you like. It could just as easily be "meter," "micron," or "furlong." The point is that s_x converts physical units to pixel units.

Silven [Heikkila97]). The projection of the points in the physical world into the camera is now summarized by the following simple form:

$$q = MQ, \quad \text{where} \quad q = \begin{bmatrix} x \\ y \\ w \end{bmatrix}, \quad M = \begin{bmatrix} f_x & 0 & c_x \\ 0 & f_y & c_y \\ 0 & 0 & 1 \end{bmatrix}, \quad Q = \begin{bmatrix} X \\ Y \\ Z \end{bmatrix}$$

Multiplying this out, you will find that $w = Z$ and so, since the point q is in homogeneous coordinates, we should divide through by w (or Z) in order to recover our earlier definitions. (The minus sign is gone because we are now looking at the noninverted image on the projective plane in front of the pinhole rather than the inverted image on the projection screen behind the pinhole.)

While we are on the topic of homogeneous coordinates, there is a function in the OpenCV library which would be appropriate to introduce here: cvConvertPointsHomogenious()* is handy for converting to and from homogeneous coordinates; it also does a bunch of other useful things.

```
void cvConvertPointsHomogenious(
    const CvMat* src,
    CvMat*       dst
);
```

Don't let the simple arguments fool you; this routine does a whole lot of useful stuff. The input array src can be M_{scr}-by-N or N-by-M_{scr} (for $M_{scr} = 2$, 3, or 4); it can also be 1-by-N or N-by-1, with the array having $M_{scr} = 2$, 3, or 4 channels (N can be any number; it is essentially the number of points that you have stuffed into the matrix src for conversion). The output array dst can be any of these types as well, with the additional restriction that the dimensionality M_{dst} must be equal to M_{scr}, $M_{scr} - 1$, or $M_{scr} + 1$.

When the input dimension M_{scr} is equal to the output dimension M_{dst}, the data is simply copied (and, if necessary, transposed). If $M_{scr} > M_{dst}$, then the elements in dst are computed by dividing all but the last elements of the corresponding vector from src by the last element of that same vector (i.e., src is assumed to contain homogeneous coordinates). If $M_{scr} < M_{dst}$, then the points are copied but with a 1 being inserted into the final coordinate of every vector in the dst array (i.e., the vectors in src are extended to homogeneous coordinates). In these cases, just as in the trivial case of $M_{scr} = M_{dst}$, any necessary transpositions are also done.

One word of warning about this function is that there can be cases (when $N < 5$) where the input and output dimensionality are ambiguous. In this event, the function will throw an error. If you find yourself in this situation, you can just pad out the matrices with some bogus values. Alternatively, the user may pass multichannel N-by-1 or 1-by-N matrices, where the number of channels is M_{scr} (M_{dst}). The function cvReshape() can be used to convert single-channel matrices to multichannel ones without copying any data.

---

* Yes, "Homogenious" in the function name is misspelled.

With the ideal pinhole, we have a useful model for some of the three-dimensional geometry of vision. Remember, however, that very little light goes through a pinhole; thus, in practice such an arrangement would make for very slow imaging while we wait for enough light to accumulate on whatever imager we are using. For a camera to form images at a faster rate, we must gather a lot of light over a wider area and bend (i.e., focus) that light to converge at the point of projection. To accomplish this, we use a lens. A lens can focus a large amount of light on a point to give us fast imaging, but it comes at the cost of introducing distortions.

Lens Distortions

In theory, it is possible to define a lens that will introduce no distortions. In practice, however, no lens is perfect. This is mainly for reasons of manufacturing; it is much easier to make a "spherical" lens than to make a more mathematically ideal "parabolic" lens. It is also difficult to mechanically align the lens and imager exactly. Here we describe the two main lens distortions and how to model them.* *Radial distortions* arise as a result of the shape of lens, whereas *tangential distortions* arise from the assembly process of the camera as a whole.

We start with radial distortion. The lenses of real cameras often noticeably distort the location of pixels near the edges of the imager. This bulging phenomenon is the source of the "barrel" or "fish-eye" effect (see the room-divider lines at the top of Figure 11-12 for a good example). Figure 11-3 gives some intuition as to why radial distortion occurs. With some lenses, rays farther from the center of the lens are bent more than those closer in. A typical inexpensive lens is, in effect, stronger than it ought to be as you get farther from the center. Barrel distortion is particularly noticeable in cheap web cameras but less apparent in high-end cameras, where a lot of effort is put into fancy lens systems that minimize radial distortion.

For radial distortions, the distortion is 0 at the (optical) center of the imager and increases as we move toward the periphery. In practice, this distortion is small and can be characterized by the first few terms of a Taylor series expansion around $r = 0$.† For cheap web cameras, we generally use the first two such terms; the first of which is conventionally called k_1 and the second k_2. For highly distorted cameras such as fish-eye lenses we can use a third radial distortion term k_3. In general, the radial location of a point on the imager will be rescaled according to the following equations:

---

* The approach to modeling lens distortion taken here derives mostly from Brown [Brown71] and earlier Fryer and Brown [Fryer86].

† If you don't know what a Taylor series is, don't worry too much. The Taylor series is a mathematical technique for expressing a (potentially) complicated function in the form of a polynomial of similar value to the approximated function in at least a small neighborhood of some particular point (the more terms we include in the polynomial series, the more accurate the approximation). In our case we want to expand the distortion function as a polynomial in the neighborhood of $r = 0$. This polynomial takes the general form $f(r) = a_0 + a_1 r + a_2 r^2 + \cdots$, but in our case the fact that $f(r) = 0$ at $r = 0$ implies $a_0 = 0$. Similarly, because the function must be symmetric in r, only the coefficients of even powers of r will be nonzero. For these reasons, the only parameters that are necessary for characterizing these radial distortions are the coefficients of r^2, r^4, and (sometimes) r^6.

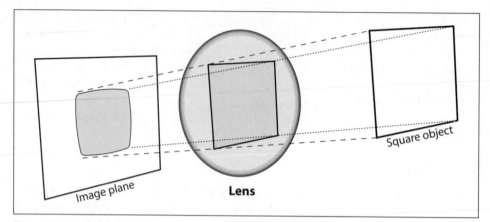

Figure 11-3. Radial distortion: rays farther from the center of a simple lens are bent too much compared to rays that pass closer to the center; thus, the sides of a square appear to bow out on the image plane (this is also known as barrel distortion)

$$x_{corrected} = x(1 + k_1 r^2 + k_2 r^4 + k_3 r^6)$$

$$y_{corrected} = y(1 + k_1 r^2 + k_2 r^4 + k_3 r^6)$$

Here, (x, y) is the original location (on the imager) of the distorted point and $(x_{corrected},$ $y_{corrected})$ is the new location as a result of the correction. Figure 11-4 shows displacements of a rectangular grid that are due to radial distortion. External points on a front-facing rectangular grid are increasingly displaced inward as the radial distance from the optical center increases.

The second-largest common distortion is *tangential distortion*. This distortion is due to manufacturing defects resulting from the lens not being exactly parallel to the imaging plane; see Figure 11-5.

Tangential distortion is minimally characterized by two additional parameters, p_1 and p_2, such that:*

$$x_{corrected} = x + [2p_1 y + p_2(r^2 + 2x^2)]$$

$$y_{corrected} = y + [p_1(r^2 + 2y^2) + 2p_2 x]$$

Thus in total there are five distortion coefficients that we require. Because all five are necessary in most of the OpenCV routines that use them, they are typically bundled into one *distortion vector*; this is just a 5-by-1 matrix containing k_1, k_2, p_1, p_2, and k_3 (in that order). Figure 11-6 shows the effects of tangential distortion on a front-facing external rectangular grid of points. The points are displaced elliptically as a function of location and radius.

---

* The derivation of these equations is beyond the scope of this book, but the interested reader is referred to the "plumb bob" model; see D. C. Brown, "Decentering Distortion of Lenses", *Photometric Engineering* 32(3) (1966), 444–462.

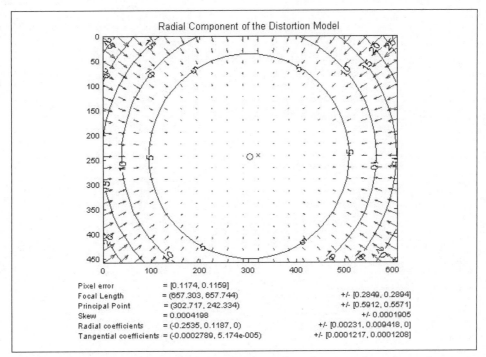

Figure 11-4. Radial distortion plot for a particular camera lens: the arrows show where points on an external rectangular grid are displaced in a radially distorted image (courtesy of Jean-Yves Bouguet)

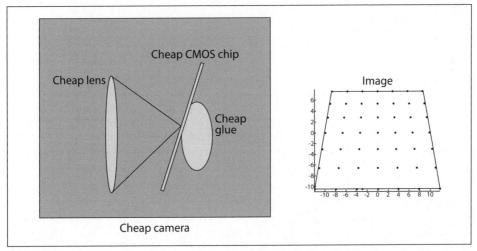

Figure 11-5. Tangential distortion results when the lens is not fully parallel to the image plane; in cheap cameras, this can happen when the imager is glued to the back of the camera (image courtesy of Sebastian Thrun)

There are many other kinds of distortions that occur in imaging systems, but they typically have lesser effects than radial and tangential distortions. Hence neither we nor OpenCV will deal with them further.

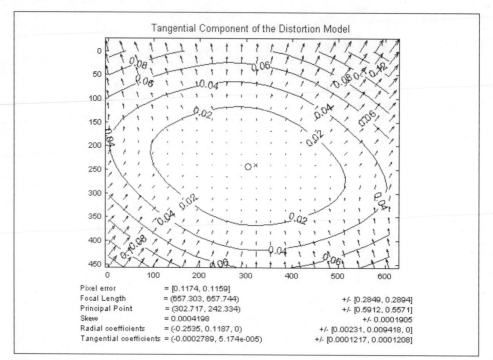

Figure 11-6. *Tangential distortion plot for a particular camera lens: the arrows show where points on an external rectangular grid are displaced in a tangentially distorted image (courtesy of Jean-Yves Bouguet)*

Calibration

Now that we have some idea of how we'd describe the intrinsic and distortion properties of a camera mathematically, the next question that naturally arises is how we can use OpenCV to compute the intrinsics matrix and the distortion vector.*

OpenCV provides several algorithms to help us compute these intrinsic parameters. The actual calibration is done via cvCalibrateCamera2(). In this routine, the method of calibration is to target the camera on a known structure that has many individual and identifiable points. By viewing this structure from a variety of angles, it is possible to then compute the (relative) location and orientation of the camera at the time of each image as well as the intrinsic parameters of the camera (see Figure 11-9 in the "Chessboards" section). In order to provide multiple views, we rotate and translate the object, so let's pause to learn a little more about rotation and translation.

---

* For a great online tutorial of camera calibration, see Jean-Yves Bouguet's calibration website (*http://www.vision.caltech.edu/bouguetj/calib_doc*).

Rotation Matrix and Translation Vector

For each image the camera takes of a particular object, we can describe the *pose* of the object relative to the camera coordinate system in terms of a rotation and a translation; see Figure 11-7.

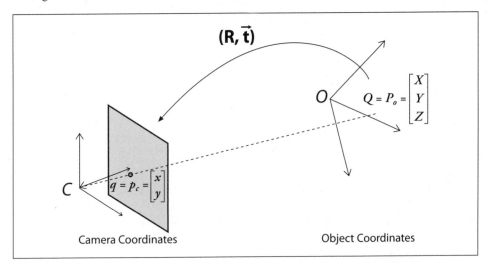

*Figure 11-7. Converting from object to camera coordinate systems: the point P on the object is seen as point p on the image plane; the point p is related to point P by applying a rotation matrix R and a translation vector **t** to P*

In general, a rotation in any number of dimensions can be described in terms of multiplication of a coordinate vector by a square matrix of the appropriate size. Ultimately, a rotation is equivalent to introducing a new description of a point's location in a different coordinate system. Rotating the coordinate system by an angle θ is equivalent to counterrotating our target point around the origin of that coordinate system by the same angle θ. The representation of a two-dimensional rotation as matrix multiplication is shown in Figure 11-8. Rotation in three dimensions can be decomposed into a two-dimensional rotation around each axis in which the pivot axis measurements remain constant. If we rotate around the x-, y-, and z-axes in sequence* with respective rotation angles ψ, φ, and θ, the result is a total rotation matrix R that is given by the product of the three matrices $R_x(\psi)$, $R_y(\varphi)$, and $R_z(\theta)$, where:

$$R_x(\psi) = \begin{bmatrix} 1 & 0 & 0 \\ 0 & \cos\psi & \sin\psi \\ 0 & -\sin\psi & \cos\psi \end{bmatrix}$$

---

* Just to be clear: the rotation we are describing here is first around the z-axis, then around the *new* position of the y-axis, and finally around the *new* position of the x-axis.

$$R_y(\varphi) = \begin{bmatrix} \cos\varphi & 0 & -\sin\varphi \\ 0 & 1 & 0 \\ \sin\varphi & 0 & \cos\varphi \end{bmatrix}$$

$$R_z(\theta) = \begin{bmatrix} \cos\theta & \sin\theta & 0 \\ -\sin\theta & \cos\theta & 0 \\ 0 & 0 & 1 \end{bmatrix}$$

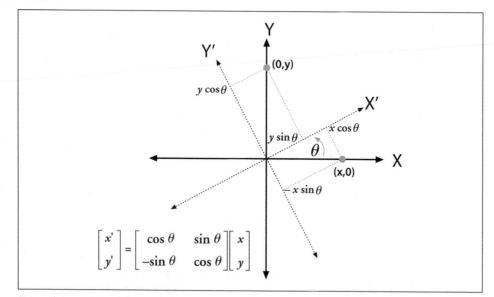

Figure 11-8. *Rotating points by θ (in this case, around the Z-axis) is the same as counterrotating the coordinate axis by θ; by simple trigonometry, we can see how rotation changes the coordinates of a point*

Thus, $R = R_z(\theta)$, $R_y(\varphi)$, $R_x(\psi)$. The rotation matrix R has the property that its inverse is its transpose (we just rotate back); hence we have $R^T R = R R^T = I$, where I is the identity matrix consisting of 1s along the diagonal and 0s everywhere else.

The *translation vector* is how we represent a shift from one coordinate system to another system whose origin is displaced to another location; in other words, the translation vector is just the offset from the origin of the first coordinate system to the origin of the second coordinate system. Thus, to shift from a coordinate system centered on an object to one centered at the camera, the appropriate translation vector is simply $T = \text{origin}_{\text{object}} - \text{origin}_{\text{camera}}$. We then have (with reference to Figure 11-7) that a point in the object (or world) coordinate frame P_o has coordinates P_c in the camera coordinate frame:

$$P_c = R(P_o - T)$$

Combining this equation for P_c above with the camera intrinsic corrections will form the basic system of equations that we will be asking OpenCV to solve. The solution to these equations will be the camera calibration parameters we seek.

We have just seen that a three-dimensional rotation can be specified with three angles and that a three-dimensional translation can be specified with the three parameters (x, y, z); thus we have six parameters so far. The OpenCV intrinsics matrix for a camera has four parameters (f_x, f_y, c_x, and c_y), yielding a grand total of ten parameters that must be solved for each view (but note that the camera intrinsic parameters stay the same between views). Using a planar object, we'll soon see that each view fixes eight parameters. Because the six parameters of rotation and translation change between views, for each view we have constraints on two additional parameters that we use to resolve the camera intrinsic matrix. We'll then need at least two views to solve for all the geometric parameters.

We'll provide more details on the parameters and their constraints later in the chapter, but first we discuss the *calibration object*. The calibration object used in OpenCV is a flat grid of alternating black and white squares that is usually called a "chessboard" (even though it needn't have eight squares, or even an equal number of squares, in each direction).

Chessboards

In principle, any appropriately characterized object could be used as a calibration object, yet the practical choice is a regular pattern such as a chessboard.* Some calibration methods in the literature rely on three-dimensional objects (e.g., a box covered with markers), but flat chessboard patterns are much easier to deal with; it is difficult to make (and to store and distribute) precise 3D calibration objects. OpenCV thus opts for using multiple views of a planar object (a chessboard) rather than one view of a specially constructed 3D object. We use a pattern of alternating black and white squares (see Figure 11-9), which ensures that there is no bias toward one side or the other in measurement. Also, the resulting grid corners lend themselves naturally to the subpixel localization function discussed in Chapter 10.

Given an image of a chessboard (or a person holding a chessboard, or any other scene with a chessboard and a reasonably uncluttered background), you can use the OpenCV function cvFindChessboardCorners() to locate the corners of the chessboard.

```
int cvFindChessboardCorners(
  const void*    image,
  CvSize         pattern_size,
  CvPoint2D32f*  corners,
  int*           corner_count = NULL,
  int            flags        = CV_CALIB_CB_ADAPTIVE_THRESH
);
```

---

* The specific use of this calibration object—and much of the calibration approach itself—comes from Zhang [Zhang99; Zhang00] and Sturm [Sturm99].

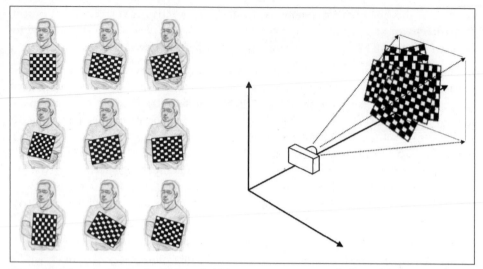

Figure 11-9. Images of a chessboard being held at various orientations (left) provide enough information to completely solve for the locations of those images in global coordinates (relative to the camera) and the camera intrinsics

This function takes as arguments a single image containing a chessboard. This image must be an 8-bit grayscale (single-channel) image. The second argument, pattern_size, indicates how many corners are in each row and column of the board. This count is the number of *interior* corners; thus, for a standard chess game board the correct value would be cvSize(7,7).* The next argument, corners, is a pointer to an array where the corner locations can be recorded. This array must be preallocated and, of course, must be large enough for all of the corners on the board (49 on a standard chess game board). The individual values are the locations of the corners in pixel coordinates. The corner_ count argument is optional; if non-NULL, it is a pointer to an integer where the number of corners found can be recorded. If the function is successful at finding all of the corners,[†] then the return value will be a nonzero number. If the function fails, 0 will be returned. The final flags argument can be used to implement one or more additional filtration steps to help find the corners on the chessboard. Any or all of the arguments may be combined using a Boolean OR.

CV_CALIB_CB_ADAPTIVE_THRESH

> The default behavior of cvFindChessboardCorners() is first to threshold the image based on average brightness, but if this flag is set then an adaptive threshold will be used instead.

---

* In practice, it is often more convenient to use a chessboard grid that is asymmetric and of even and odd dimensions—for example, (5, 6). Using such even-odd asymmetry yields a chessboard that has only one symmetry axis, so the board orientation can always be defined uniquely.

† Actually, the requirement is slightly stricter: not only must all the corners be found, they must also be ordered into rows and columns as expected. Only if the corners can be found and ordered correctly will the return value of the function be nonzero.

CV_CALIB_CB_NORMALIZE_IMAGE

> If set, this flag causes the image to be normalized via cvEqualizeHist() before the thresholding is applied.

CV_CALIB_CB_FILTER_QUADS

> Once the image is thresholded, the algorithm attempts to locate the quadrangles resulting from the perspective view of the black squares on the chessboard. This is an approximation because the lines of each edge of a quadrangle are assumed to be straight, which isn't quite true when there is radial distortion in the image. If this flag is set, then a variety of additional constraints are applied to those quadrangles in order to reject false quadrangles.

Subpixel corners

The corners returned by cvFindChessboardCorners() are only approximate. What this means in practice is that the locations are accurate only to within the limits of our imaging device, which means accurate to within one pixel. A separate function must be used to compute the exact locations of the corners (given the approximate locations and the image as input) to subpixel accuracy. This function is the same cvFindCornerSubPix() function that we used for tracking in Chapter 10. It should not be surprising that this function can be used in this context, since the chessboard interior corners are simply a special case of the more general Harris corners; the chessboard corners just happen to be particularly easy to find and track. Neglecting to call subpixel refinement after you first locate the corners can cause substantial errors in calibration.

Drawing chessboard corners

Particularly when debugging, it is often desirable to draw the found chessboard corners onto an image (usually the image that we used to compute the corners in the first place); this way, we can see whether the projected corners match up with the observed corners. Toward this end, OpenCV provides a convenient routine to handle this common task. The function cvDrawChessboardCorners() draws the corners found by cvFindChessboard-Corners() onto an image that you provide. If not all of the corners were found, the available corners will be represented as small red circles. If the entire pattern was found, then the corners will be painted into different colors (each row will have its own color) and connected by lines representing the identified corner order.

```
void cvDrawChessboardCorners(
    CvArr*          image,
    CvSize          pattern_size,
    CvPoint2D32f*   corners,
    int             count,
    int             pattern_was_found
);
```

The first argument to cvDrawChessboardCorners() is the image to which the drawing will be done. Because the corners will be represented as colored circles, this must be an 8-bit color image; in most cases, this will be a copy of the image you gave to cvFindChessboardCorners() (but you must convert it to a three-channel image yourself).

The next two arguments, pattern_size and corners, are the same as the corresponding arguments for cvFindChessboardCorners(). The argument count is an integer equal to the number of corners. Finally the argument pattern_was_found indicates whether the entire chessboard pattern was successfully found; this can be set to the return value from cvFindChessboardCorners(). Figure 11-10 shows the result of applying cvDrawChessboardCorners() to a chessboard image.

Figure 11-10. Result of cvDrawChessboardCorners(); once you find the corners using cvFindChessboardCorners(), you can project where these corners were found (small circles on corners) and in what order they belong (as indicated by the lines between circles)

We now turn to what a planar object can do for us. Points on a plane undergo perspective transform when viewed through a pinhole or lens. The parameters for this transform are contained in a 3-by-3 *homography* matrix, which we describe next.

Homography

In computer vision, we define *planar homography* as a projective mapping from one plane to another.* Thus, the mapping of points on a two-dimensional planar surface to

---

* The term "homography" has different meanings in different sciences; for example, it has a somewhat more general meaning in mathematics. The homographies of greatest interest in computer vision are a subset of the other, more general, meanings of the term.

the imager of our camera is an example of planar homography. It is possible to express this mapping in terms of matrix multiplication if we use homogeneous coordinates to express both the viewed point Q and the point q on the imager to which Q is mapped. If we define:

$$\tilde{Q} = \begin{bmatrix} X & Y & Z & 1 \end{bmatrix}^{\mathrm{T}}$$
$$\tilde{q} = \begin{bmatrix} x & y & 1 \end{bmatrix}^{\mathrm{T}}$$

then we can express the action of the homography simply as:

$$\tilde{q} = sH\tilde{Q}$$

Here we have introduced the parameter s, which is an arbitrary scale factor (intended to make explicit that the homography is defined only up to that factor). It is conventionally factored out of H, and we'll stick with that convention here.

With a little geometry and some matrix algebra, we can solve for this transformation matrix. The most important observation is that H has two parts: the physical transformation, which essentially locates the object plane we are viewing; and the projection, which introduces the camera intrinsics matrix. See Figure 11-11.

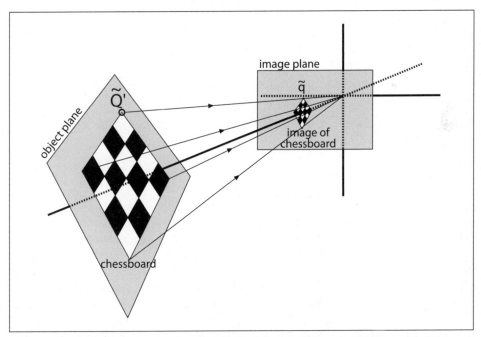

Figure 11-11. View of a planar object as described by homography: a mapping—from the object plane to the image plane—that simultaneously comprehends the relative locations of those two planes as well as the camera projection matrix

The physical transformation part is the sum of the effects of some rotation R and some translation $\mathbf{t}$ that relate the plane we are viewing to the image plane. Because we are working in homogeneous coordinates, we can combine these within a single matrix as follows:*

$$W = \begin{bmatrix} R & \mathbf{t} \end{bmatrix}$$

Then, the action of the camera matrix M (which we already know how to express in projective coordinates) is multiplied by $W\tilde{Q}$; this yields:

$$\tilde{q} = sMW\tilde{Q}, \quad \text{where} \quad M = \begin{bmatrix} f_x & 0 & c_x \\ 0 & f_y & c_y \\ 0 & 0 & 1 \end{bmatrix}$$

It would seem that we are done. However, it turns out that in practice our interest is not the coordinate $\tilde{Q}$, which is defined for all of space, but rather a coordinate $\tilde{Q}'$, which is defined only on the plane we are looking at. This allows for a slight simplification.

Without loss of generality, we can choose to define the object plane so that $Z = 0$. We do this because, if we also break up the rotation matrix into three 3-by-1 columns (i.e., $R = [r_1 \ r_2 \ r_3]$), then one of those columns is not needed. In particular:

$$\begin{bmatrix} x \\ y \\ 1 \end{bmatrix} = sM \begin{bmatrix} r_1 & r_2 & r_3 & \mathbf{t} \end{bmatrix} \begin{bmatrix} X \\ Y \\ 0 \\ 1 \end{bmatrix} = sM \begin{bmatrix} r_1 & r_2 & \mathbf{t} \end{bmatrix} \begin{bmatrix} X \\ Y \\ 1 \end{bmatrix}$$

The homography matrix H that maps a planar object's points onto the imager is then described completely by $H = sM[r_1 \ r_2 \ \mathbf{t}]$, where:

$$\tilde{q} = sH\tilde{Q}'$$

Observe that H is now a 3-by-3 matrix.

OpenCV uses the preceding equations to compute the homography matrix. It uses multiple images of the same object to compute both the individual translations and rotations for each view as well as the intrinsics (which are the same for all views). As we have discussed, rotation is described by three angles and translation is defined by three offsets; hence there are six unknowns for each view. This is OK, because a known planar object (such as our chessboard) gives us eight equations—that is, the mapping of a square into a quadrilateral can be described by four (x, y) points. Each new frame gives us eight equations at the cost of six new extrinsic unknowns, so given enough images we should be able to compute any number of intrinsic unknowns (more on this shortly).

---

* Here $W = [R \ \mathbf{t}]$ is a 3-by-4 matrix whose first three columns comprise the nine entries of R and whose last column consists of the three-component vector $\mathbf{t}$.

The homography matrix H relates the positions of the points on a source image plane to the points on the destination image plane (usually the imager plane) by the following simple equations:

$$p_{dst} = Hp_{src}, \quad p_{src} = H^{-1}p_{dst}$$

$$p_{dst} = \begin{bmatrix} x_{dst} \\ y_{dst} \\ 1 \end{bmatrix}, \quad p_{src} = \begin{bmatrix} x_{src} \\ y_{src} \\ 1 \end{bmatrix}$$

Notice that we can compute H without knowing anything about the camera intrinsics. In fact, computing multiple homographies from multiple views is the method OpenCV uses to solve for the camera intrinsics, as we'll see.

OpenCV provides us with a handy function, cvFindHomography(), which takes a list of correspondences and returns the homography matrix that best describes those correspondences. We need a minimum of four points to solve for H, but we can supply many more if we have them* (as we will with any chessboard bigger than 3-by-3). Using more points is beneficial, because invariably there will be noise and other inconsistencies whose effect we would like to minimize.

```
void cvFindHomography(
    const CvMat* src_points,
    const CvMat* dst_points,
    CvMat*       homography
);
```

The input arrays src_points and dst_points can be either N-by-2 matrices or N-by-3 matrices. In the former case the points are pixel coordinates, and in the latter they are expected to be homogeneous coordinates. The final argument, homography, is just a 3-by-3 matrix to be filled by the function in such a way that the back-projection error is minimized. Because there are only eight free parameters in the homography matrix, we chose a normalization where $H_{33} = 1$. Scaling the homography could be applied to the ninth homography parameter, but usually scaling is instead done by multiplying the entire homography matrix by a scale factor.

Camera Calibration

We finally arrive at camera calibration for camera intrinsics and distortion parameters. In this section we'll learn how to compute these values using cvCalibrateCamera2() and also how to use these models to correct distortions in the images that the calibrated camera would have otherwise produced. First we say a little more about how many views of a chessboard are necessary in order to solve for the intrinsics and distortion. Then we'll offer a high-level overview of how OpenCV actually solves this system before moving on to the code that makes it all easy to do.

---

* Of course, an exact solution is guaranteed only when there are four correspondences. If more are provided, then what's computed is a solution that is optimal in the sense of least-squares error.

How many chess corners for how many parameters?

It will prove instructive to review our unknowns. That is, how many parameters are we attempting to solve for through calibration? In the OpenCV case, we have four *intrinsic* parameters (f_x, f_y, c_x, c_y) and five *distortion* parameters: three radial (k_1, k_2, k_3) and two tangential (p_1, p_2). Intrinsic parameters are directly tied to the 3D geometry (and hence the extrinsic parameters) of where the chessboard is in space; distortion parameters are tied to the 2D geometry of how the pattern of points gets distorted, so we deal with the constraints on these two classes of parameters separately. Three corner points in a known pattern yielding six pieces of information are (in principle) all that is needed to solve for our five distortion parameters (of course, we use much more for robustness). Thus, one view of a chessboard is all that we need to compute our distortion parameters. The same chessboard view could also be used in our intrinsics computation, which we consider next, starting with the extrinsic parameters. For the *extrinsic* parameters we'll need to know where the chessboard is. This will require three rotation parameters (ψ, ϕ, θ) and three translation parameters (T_x, T_y, T_z) for a total of six per view of the chessboard, because in each image the chessboard will move. Together, the four intrinsic and six extrinsic parameters make for ten altogether that we must solve for each view.

Let's say we have N corners and K images of the chessboard (in different positions). How many views and corners must we see so that there will be enough constraints to solve for all these parameters?

- K images of the chessboard provide $2NK$ constraints (we use the multiplier 2 because each point on the image has both an x and a y coordinate).

- Ignoring the distortion parameters for the moment, we have 4 intrinsic parameters and $6K$ extrinsic parameters (since we need to find the 6 parameters of the chessboard location in each of the K views).

- Solving then requires that $2NK \geq 6K + 4$ hold (or, equivalently, $(N - 3) K \geq 2$).

It seems that if $N = 5$ then we need only $K = 1$ image, but watch out! For us, K (the number of images) must be more than 1. The reason for requiring $K > 1$ is that we're using chessboards for calibration to fit a homography matrix for each of the K views. As discussed previously, a homography can yield at most eight parameters from four (x, y) pairs. This is because only four points are needed to express everything that a planar perspective view can do: it can stretch a square in four different directions at once, turning it into any quadrilateral (see the perspective images in Chapter 6). So, no matter how many corners we detect on a plane, we only get four corners' worth of information. Per chessboard view, then, the equation can give us only four corners of information or $(4 - 3) K > 1$, which means $K > 1$. This implies that two views of a 3-by-3 chessboard (counting only internal corners) are the minimum that could solve our calibration problem. Consideration for noise and numerical stability is typically what requires the collection of more images of a larger chessboard. In practice, for high-quality results, you'll need at least ten images of a 7-by-8 or larger chessboard (and that's only if you move the chessboard enough between images to obtain a "rich" set of views).

What's under the hood?

This subsection is for those who want to go deeper; it can be safely skipped if you just want to call the calibration functions. If you are still with us, the question remains: how is all this mathematics used for calibration? Although there are many ways to solve for the camera parameters, OpenCV chose one that works well on planar objects. The algorithm OpenCV uses to solve for the focal lengths and offsets is based on Zhang's method [Zhang00], but OpenCV uses a different method based on Brown [Brown71] to solve for the distortion parameters.

To get started, we pretend that there is no distortion in the camera while solving for the other calibration parameters. For each view of the chessboard, we collect a homography H as described previously. We'll write H out as column vectors, $H = [h_1\ h_2\ h_3]$, where each h is a 3-by-1 vector. Then, in view of the preceding homography discussion, we can set H equal to the camera intrinsics matrix M multiplied by a combination of the first two rotation matrix columns, r_1 and r_2, and the translation vector $\mathbf{t}$; after including the scale factor s, this yields:

$$H = \begin{bmatrix} h_1 & h_2 & h_3 \end{bmatrix} = sM \begin{bmatrix} r_1 & r_2 & \mathbf{t} \end{bmatrix}$$

Reading off these equations, we have:

$$h_1 = sMr_1 \quad \text{or} \quad r_1 = \lambda M^{-1} h_1$$

$$h_2 = sMr_2 \quad \text{or} \quad r_2 = \lambda M^{-1} h_2$$

$$h_3 = sMt \quad \text{or} \quad t = \lambda M^{-1} h_3$$

Here, $\lambda = 1/s$.

The rotation vectors are orthogonal to each other by construction, and since the scale is extracted it follows that r_1 and r_2 are orthonormal. Orthonormal implies two things: the rotation vector's dot product is 0, and the vectors' magnitudes are equal. Starting with the dot product, we have:

$$r_1^\mathrm{T} r_2 = 0$$

For any vectors a and b we have $(ab)^\mathrm{T} = b^\mathrm{T} a^\mathrm{T}$, so we can substitute for r_1 and r_2 to derive our first constraint:

$$h_1^\mathrm{T} M^{-\mathrm{T}} M^{-1} h_2 = 0$$

where $A^{-\mathrm{T}}$ is shorthand for $(A^{-1})^\mathrm{T}$. We also know that the magnitudes of the rotation vectors are equal:

$$\|r_1\| = \|r_2\| \quad \text{or} \quad r_1^\mathrm{T} r_1 = r_2^\mathrm{T} r_2$$

Substituting for r_1 and r_2 yields our second constraint:

$$h_1^\mathrm{T} M^{-\mathrm{T}} M^{-1} h_1 = h_2^\mathrm{T} M^{-\mathrm{T}} M^{-1} h_2$$

To make things easier, we set $B = M^{-T}M^{-1}$. Writing this out, we have:

$$B = M^{-T}M^{-1} = \begin{bmatrix} B_{11} & B_{12} & B_{13} \\ B_{12} & B_{22} & B_{23} \\ B_{13} & B_{23} & B_{33} \end{bmatrix}$$

It so happens that this matrix B has a general closed-form solution:

$$B = \begin{bmatrix} \dfrac{1}{f_x^2} & 0 & \dfrac{-c_x}{f_x^2} \\[2ex] 0 & \dfrac{1}{f_y^2} & \dfrac{-c_y}{f_y^2} \\[2ex] \dfrac{-c_x}{f_x^2} & \dfrac{-c_y}{f_y^2} & \dfrac{c_x^2}{f_x^2} + \dfrac{c_y^2}{f_y^2} + 1 \end{bmatrix}$$

Using the B-matrix, both constraints have the general form $h_i^T B h_j$ in them. Let's multiply this out to see what the components are. Because B is symmetric, it can be written as one six-dimensional vector dot product. Arranging the necessary elements of B into the new vector b, we have:

$$h_i^T B h_j = v_{ij}^T b = \begin{bmatrix} h_{i1}h_{j1} \\ h_{i1}h_{j2} + h_{i2}h_{j1} \\ h_{i2}h_{j2} \\ h_{i3}h_{j1} + h_{i1}h_{j3} \\ h_{i3}h_{j2} + h_{i2}h_{j3} \\ h_{i3}h_{j3} \end{bmatrix}^T \begin{bmatrix} B_{11} \\ B_{12} \\ B_{22} \\ B_{13} \\ B_{23} \\ B_{33} \end{bmatrix}^T$$

Using this definition for v_{ij}^T, our two constraints may now be written as:

$$\begin{bmatrix} v_{12}^T \\ (v_{11} - v_{22})^T \end{bmatrix} b = 0$$

If we collect K images of chessboards together, then we can stack K of these equations together:

$$Vb = 0$$

where V is a $2K$-by-6 matrix. As before, if $K \geq 2$ then this equation can be solved for our $b = [B_{11}, B_{12}, B_{22}, B_{13}, B_{23}, B_{33}]^T$. The camera intrinsics are then pulled directly out of our closed-form solution for the B-matrix:

$$f_x = \sqrt{\lambda / B_{11}}$$

$$f_y = \sqrt{\lambda B_{11} / (B_{11} B_{22} - B_{12}^2)}$$

$$c_x = -B_{13} f_x^2 / \lambda$$

$$c_y = (B_{12} B_{13} - B_{11} B_{23}) / (B_{11} B_{22} - B_{12}^2)$$

where:

$$\lambda = B_{33} - (B_{13}^2 + c_y (B_{12} B_{13} - B_{11} B_{23})) / B_{11}$$

The extrinsics (rotation and translation) are then computed from the equations we read off of the homography condition:

$$r_1 = \lambda M^{-1} h_1$$

$$r_2 = \lambda M^{-1} h_2$$

$$r_3 = r_1 \times r_2$$

$$t = \lambda M^{-1} h_3$$

Here the scaling parameter is determined from the orthonormality condition $\lambda = 1/\|M^{-1} h_1\|$.

Some care is required because, when we solve using real data and put the r-vectors together ($R = [r_1 \ r_2 \ r_3]$), we will not end up with an exact rotation matrix for which $R^T R = R R^T = I$ holds.

To get around this problem, the usual trick is to take the singular value decomposition (SVD) of R. As discussed in Chapter 3, SVD is a method of factoring a matrix into two orthonormal matrices, U and V, and a middle matrix D of scale values on its diagonal. This allows us to turn R into $R = UDV^T$. Because R is itself orthonormal, the matrix D must be the identity matrix I such that $R = UIV^T$. We can thus "coerce" our computed R into being a rotation matrix by taking R's singular value decomposition, setting its D matrix to the identity matrix, and multiplying by the SVD again to yield our new, conforming rotation matrix R'.

Despite all this work, we have not yet dealt with lens distortions. We use the camera intrinsics found previously—together with the distortion parameters set to 0—for our initial guess to start solving a larger system of equations.

The points we "perceive" on the image are really in the wrong place owing to distortion. Let (x_p, y_p) be the point's location if the pinhole camera were perfect and let (x_d, y_d) be its distorted location; then:

$$\begin{bmatrix} x_p \\ y_p \end{bmatrix} = \begin{bmatrix} f_x X^W / Z^W + c_x \\ f_y X^W / Z^W + c_y \end{bmatrix}$$

We use the results of the calibration without distortion via the following substitution:

$$
\begin{bmatrix} x_p \\ y_p \end{bmatrix} = (1 + k_1 r^2 + k_2 r^4 + k_3 r^6) \begin{bmatrix} x_d \\ y_d \end{bmatrix} + \begin{bmatrix} 2 p_1 x_d y_d + p_2 (r^2 + 2 x_d^2) \\ p_1 (r^2 + 2 y_d^2) + 2 p_2 x_d y_d \end{bmatrix}
$$

A large list of these equations are collected and solved to find the distortion parameters, after which the intrinsics and extrinsics are reestimated. That's the heavy lifting that the single function cvCalibrateCamera2()* does for you!

Calibration function

Once we have the corners for several images, we can call cvCalibrateCamera2(). This routine will do the number crunching and give us the information we want. In particular, the results we receive are the *camera intrinsics matrix*, the *distortion coefficients*, the *rotation vectors*, and the *translation vectors*. The first two of these constitute the intrinsic parameters of the camera, and the latter two are the extrinsic measurements that tell us where the objects (i.e., the chessboards) were found and what their orientations were. The distortion coefficients $(k_1, k_2, p_1, p_2,$ and $k_3)^\dagger$ are the coefficients from the radial and tangential distortion equations we encountered earlier; they help us when we want to correct that distortion away. The camera intrinsic matrix is perhaps the most interesting final result, because it is what allows us to transform from 3D coordinates to the image's 2D coordinates. We can also use the camera matrix to do the reverse operation, but in this case we can only compute a line in the three-dimensional world to which a given image point must correspond. We will return to this shortly.

Let's now examine the camera calibration routine itself.

```
void cvCalibrateCamera2(
    CvMat*     object_points,
    CvMat*     image_points,
    int*       point_counts,
    CvSize     image_size,
    CvMat*     intrinsic_matrix,
    CvMat*     distortion_coeffs,
    CvMat*     rotation_vectors    = NULL,
    CvMat*     translation_vectors = NULL,
    int        flags               = 0
);
```

When calling cvCalibrateCamera2(), there are many arguments to keep straight. Yet we've covered (almost) all of them already, so hopefully they'll make sense.

---

* The cvCalibrateCamera2() function is used internally in the stereo calibration functions we will see in Chapter 12. For stereo calibration, we'll be calibrating two cameras at the same time and will be looking to relate them together through a rotation matrix and a translation vector.

† The third radial distortion component k_3 comes last because it was a late addition to OpenCV to allow better correction to highly distorted fish eye type lenses and should only be used in such cases. We will see momentarily that k_3 can be set to 0 by first initializing it to 0 and then setting the flag to CV_CALIB_FIX_K3.

The first argument is the object_points, which is an N-by-3 matrix containing the physical coordinates of each of the K points on each of the M images of the object (i.e., $N = K \times M$). These points are located in the coordinate frame attached to the object.[*] This argument is a little more subtle than it appears in that your manner of describing the points on the object will implicitly define your physical units and the structure of your coordinate system hereafter. In the case of a chessboard, for example, you might define the coordinates such that all of the points on the chessboard had a z-value of 0 while the x- and y-coordinates are measured in centimeters. Had you chosen inches, all computed parameters would then (implicitly) be in inches. Similarly if you had chosen all the x-coordinates (rather than the z-coordinates) to be 0, then the implied location of the chessboards relative to the camera would be largely in the x-direction rather than the z-direction. The squares define one unit, so that if, for example, your squares are 90 mm on each side, your camera world, object and camera coordinate units would be in mm/90. In principle you can use an object other than a chessboard, so it is not really necessary that all of the object points lie on a plane, but this is usually the easiest way to calibrate a camera.[†] In the simplest case, we simply define each square of the chessboard to be of dimension one "unit" so that the coordinates of the corners on the chessboard are just integer corner rows and columns. Defining S_{width} as the number of squares across the width of the chessboard and S_{height} as the number of squares over the height:

$$(0,0),(0,1),(0,2),\dots,(1,0),(2,0),\dots,(1,1),\dots,(S_{width}-1,S_{height}-1)$$

The second argument is the image_points, which is an N-by-2 matrix containing the pixel coordinates of all the points supplied in object_points. If you are performing a calibration using a chessboard, then this argument consists simply of the return values for the M calls to cvFindChessboardCorners() but now rearranged into a slightly different format.

The argument point_counts indicates the number of points in each image; this is supplied as an M-by-1 matrix. The image_size is just the size, in pixels, of the images from which the image points were extracted (e.g., those images of yourself waving a chessboard around).

The next two arguments, intrinsic_matrix and distortion_coeffs, constitute the intrinsic parameters of the camera. These arguments can be both outputs (filling them in is the main reason for calibration) and inputs. When used as inputs, the values in these matrices when the function is called will affect the computed result. Which of these matrices will be used as input will depend on the flags parameter; see the following discussion. As we discussed earlier, the intrinsic matrix completely specifies the behavior

---

[*] Of course, it's normally the same object in every image, so the N points described are actually M repeated listings of the locations of the K points on a single object.

[†] At the time of this writing, automatic initialization of the intrinsic matrix before the optimization algorithm runs has been implemented only for planar calibration objects. This means that if you have a nonplanar object then you must provide a starting guess for the principal point and focal lengths (see CV_CALIB_USE_INTRINSIC_GUESS to follow).

of the camera in our ideal camera model, while the distortion coefficients characterize much of the camera's nonideal behavior. The camera matrix is always 3-by-3 and the distortion coefficients always number five, so the distortion_coeffs argument should be a pointer to a 5-by-1 matrix (they will be recorded in the order k_1, k_2, p_1, p_2, k_3).

Whereas the previous two arguments summarized the camera's intrinsic information, the next two summarize the extrinsic information. That is, they tell us where the calibration objects (e.g., the chessboards) were located relative to the camera in each picture. The locations of the objects are specified by a rotation and a translation.[*] The rotations, rotation_vectors, are defined by M three-component vectors arranged into an M-by-3 matrix (where M is the number of images). Be careful, these are not in the form of the 3-by-3 rotation matrix we discussed previously; rather, each vector represents an axis in three-dimensional space in the camera coordinate system around which the chessboard was rotated and where the length or magnitude of the vector encodes the counterclockwise angle of the rotation. Each of these rotation vectors can be converted to a 3-by-3 rotation matrix by calling cvRodrigues2(), which is described in its own section to follow. The translations, translation_vectors, are similarly arranged into a second M-by-3 matrix, again in the camera coordinate system. As stated before, the units of the camera coordinate system are exactly those assumed for the chessboard. That is, if a chessboard square is 1 inch by 1 inch, the units are inches.

Finding parameters through optimization can be somewhat of an art. Sometimes trying to solve for all parameters at once can produce inaccurate or divergent results if your initial starting position in parameter space is far from the actual solution. Thus, it is often better to "sneak up" on the solution by getting close to a good parameter starting position in stages. For this reason, we often hold some parameters fixed, solve for other parameters, then hold the other parameters fixed and solve for the original and so on. Finally, when we think all of our parameters are close to the actual solution, we use our close parameter setting as the starting point and solve for everything at once. OpenCV allows you this control through the flags setting. The flags argument allows for some finer control of exactly how the calibration will be performed. The following values may be combined together with a Boolean OR operation as needed.

CV_CALIB_USE_INTRINSIC_GUESS
> Normally the intrinsic matrix is computed by cvCalibrateCamera2() with no additional information. In particular, the initial values of the parameters c_x and c_y (the image center) are taken directly from the image_size argument. If this argument is set, then intrinsic_matrix is assumed to contain valid values that will be used as an initial guess to be further optimized by cvCalibrateCamera2().

---

[*] You can envision the chessboard's location as being expressed by (1) "creating" a chessboard at the origin of your camera coordinates, (2) rotating that chessboard by some amount around some axis, and (3) moving that oriented chessboard to a particular place. For those who have experience with systems like OpenGL, this should be a familiar construction.

CV_CALIB_FIX_PRINCIPAL_POINT

This flag can be used with or without CV_CALIB_USE_INTRINSIC_GUESS. If used without, then the principle point is fixed at the center of the image; if used with, then the principle point is fixed at the supplied initial value in the intrinsic_matrix.

CV_CALIB_FIX_ASPECT_RATIO

If this flag is set, then the optimization procedure will only vary f_x and f_y together and will keep their ratio fixed to whatever value is set in the intrinsic_matrix when the calibration routine is called. (If the CV_CALIB_USE_INTRINSIC_GUESS flag is not also set, then the values of f_x and f_y in intrinsic_matrix can be any arbitrary values and only their ratio will be considered relevant.)

CV_CALIB_FIX_FOCAL_LENGTH

This flag causes the optimization routine to just use the f_x and f_y that were passed in in the intrinsic_matrix.

CV_CALIB_FIX_K1, CV_CALIB_FIX_K2 and CV_CALIB_FIX_K3

Fix the radial distortion parameters k_1, k_2, and k_3. The radial parameters may be set in any combination by adding these flags together. In general, the last parameter should be fixed to 0 unless you are using a fish-eye lens.

CV_CALIB_ZERO_TANGENT_DIST:

This flag is important for calibrating high-end cameras which, as a result of precision manufacturing, have very little tangential distortion. Trying to fit parameters that are near 0 can lead to noisy spurious values and to problems of numerical stability. Setting this flag turns off fitting the tangential distortion parameters p_1 and p_2, which are thereby both set to 0.

Computing extrinsics only

In some cases you will already have the intrinsic parameters of the camera and therefore need only to compute the location of the object(s) being viewed. This scenario clearly differs from the usual camera calibration, but it is nonetheless a useful task to be able to perform.

```
void cvFindExtrinsicCameraParams2(
    const CvMat* object_points,
    const CvMat* image_points,
    const CvMat* intrinsic_matrix,
    const CvMat* distortion_coeffs,
    CvMat*       rotation_vector,
    CvMat*       translation_vector
);
```

The arguments to cvFindExtrinsicCameraParams2() are identical to the corresponding arguments for cvCalibrateCamera2() with the exception that the intrinsic matrix and the distortion coefficients are being supplied rather than computed. The rotation output is in the form of a 1-by-3 or 3-by-1 rotation_vector that represents the 3D axis around which the chessboard or points were rotated, and the vector magnitude or length represents the counterclockwise angle of rotation. This rotation vector can be converted into the 3-by-3 rotation matrix we've discussed before via the cvRodrigues2() function. The translation vector is the offset in camera coordinates to where the chessboard origin is located.

Undistortion

As we have alluded to already, there are two things that one often wants to do with a calibrated camera. The first is to correct for distortion effects, and the second is to construct three-dimensional representations of the images it receives. Let's take a moment to look at the first of these before diving into the more complicated second task in Chapter 12.

OpenCV provides us with a ready-to-use undistortion algorithm that takes a raw image and the distortion coefficients from cvCalibrateCamera2() and produces a corrected image (see Figure 11-12). We can access this algorithm either through the function cvUndistort2(), which does everything we need in one shot, or through the pair of routines cvInitUndistortMap() and cvRemap(), which allow us to handle things a little more efficiently for video or other situations where we have many images from the same camera.*

Figure 11-12. Camera image before undistortion (left) and after undistortion (right)

The basic method is to compute a *distortion map*, which is then used to correct the image. The function cvInitUndistortMap() computes the distortion map, and cvRemap() can be used to apply this map to an arbitrary image.† The function cvUndistort2() does one after the other in a single call. However, computing the distortion map is a time-consuming operation, so it's not very smart to keep calling cvUndistort2() if the distortion map is not changing. Finally, if we just have a list of 2D points, we can call the function cvUndistortPoints() to transform them from their original coordinates to their undistorted coordinates.

---

* We should take a moment to clearly make a distinction here between *undistortion*, which mathematically removes lens distortion, and *rectification*, which mathematically aligns the images with respect to each other.

† We first encountered cvRemap() in the context of image transformations (Chapter 6).

```
// Undistort images
void cvInitUndistortMap(
  const CvMat*    intrinsic_matrix,
  const CvMat*    distortion_coeffs,
  cvArr*          mapx,
  cvArr*          mapy
);
void cvUndistort2(
  const CvArr*    src,
  CvArr*          dst,
  const cvMat*    intrinsic_matrix,
  const cvMat*    distortion_coeffs
);
// Undistort a list of 2D points only
void cvUndistortPoints(
  const CvMat* _src,
  CvMat*         dst,
  const CvMat*  intrinsic_matrix,
  const CvMat*  distortion_coeffs,
  const CvMat*  R  = 0,
  const CvMat*  Mr = 0;
);
```

The function cvInitUndistortMap() computes the distortion map, which relates each point in the image to the location where that point is mapped. The first two arguments are the camera intrinsic matrix and the distortion coefficients, both in the expected form as received from cvCalibrateCamera2(). The resulting distortion map is represented by two separate 32-bit, single-channel arrays: the first gives the *x*-value to which a given point is to be mapped and the second gives the *y*-value. You might be wondering why we don't just use a single two-channel array instead. The reason is so that the results from cvUnitUndistortMap() can be passed directly to cvRemap().

The function cvUndistort2() does all this in a single pass. It takes your initial (distorted image) as well as the camera's intrinsic matrix and distortion coefficients, and then outputs an undistorted image of the same size. As mentioned previously, cvUndistortPoints() is used if you just have a list of 2D point coordinates from the original image and you want to compute their associated undistorted point coordinates. It has two extra parameters that relate to its use in stereo rectification, discussed in Chapter 12. These parameters are R, the rotation matrix between the two cameras, and Mr, the camera intrinsic matrix of the rectified camera (only really used when you have two cameras as per Chapter 12). The rectified camera matrix Mr can have dimensions of 3-by-3 or 3-by-4 deriving from the first three or four columns of cvStereoRectify()'s return value for camera matrices P1 or P2 (for the left or right camera; see Chapter 12). These parameters are by default NULL, which the function interprets as identity matrices.

Putting Calibration All Together

OK, now it's time to put all of this together in an example. We'll present a program that performs the following tasks: it looks for chessboards of the dimensions that the user specified, grabs as many full images (i.e., those in which it can find all the chessboard corners) as the user requested, and computes the camera intrinsics and distortion parameters. Finally, the program enters a display mode whereby an undistorted version of the camera image can be viewed; see Example 11-1. When using this algorithm, you'll want to substantially change the chessboard views between successful captures. Otherwise, the matrices of points used to solve for calibration parameters may form an ill-conditioned (rank deficient) matrix and you will end up with either a bad solution or no solution at all.

Example 11-1. Reading a chessboard's width and height, reading and collecting the requested number of views, and calibrating the camera

```
// calib.cpp
// Calling convention:
// calib board_w board_h number_of_views
//
// Hit 'p' to pause/unpause, ESC to quit
//
#include <cv.h>
#include <highgui.h>
#include <stdio.h>
#include <stdlib.h>

int n_boards = 0; //Will be set by input list
const int board_dt = 20; //Wait 20 frames per chessboard view
int board_w;
int board_h;

int main(int argc, char* argv[]) {

    if(argc != 4){
      printf("ERROR: Wrong number of input parameters\n");
      return -1;
    }
    board_w   = atoi(argv[1]);
    board_h   = atoi(argv[2]);
    n_boards = atoi(argv[3]);
    int board_n  = board_w * board_h;
    CvSize board_sz = cvSize( board_w, board_h );
    CvCapture* capture = cvCreateCameraCapture( 0 );
    assert( capture );

    cvNamedWindow( "Calibration" );
    //ALLOCATE STORAGE
    CvMat* image_points      = cvCreateMat(n_boards*board_n,2,CV_32FC1);
    CvMat* object_points     = cvCreateMat(n_boards*board_n,3,CV_32FC1);
    CvMat* point_counts      = cvCreateMat(n_boards,1,CV_32SC1);
    CvMat* intrinsic_matrix  = cvCreateMat(3,3,CV_32FC1);
    CvMat* distortion_coeffs = cvCreateMat(5,1,CV_32FC1);

    CvPoint2D32f* corners = new CvPoint2D32f[ board_n ];
    int corner_count;
    int successes = 0;
    int step, frame = 0;
```

```
IplImage *image = cvQueryFrame( capture );
IplImage *gray_image = cvCreateImage(cvGetSize(image),8,1);//subpixel

// CAPTURE CORNER VIEWS LOOP UNTIL WE'VE GOT n_boards
// SUCCESSFUL CAPTURES (ALL CORNERS ON THE BOARD ARE FOUND)
//
while(successes < n_boards) {
  //Skip every board_dt frames to allow user to move chessboard
  if(frame++ % board_dt == 0) {
    //Find chessboard corners:
    int found = cvFindChessboardCorners(
            image, board_sz, corners, &corner_count,
            CV_CALIB_CB_ADAPTIVE_THRESH | CV_CALIB_CB_FILTER_QUADS
    );

    //Get Subpixel accuracy on those corners
    cvCvtColor(image, gray_image, CV_BGR2GRAY);
    cvFindCornerSubPix(gray_image, corners, corner_count,
              cvSize(11,11),cvSize(-1,-1), cvTermCriteria(
              CV_TERMCRIT_EPS+CV_TERMCRIT_ITER, 30, 0.1 ));

    //Draw it
    cvDrawChessboardCorners(image, board_sz, corners,
              corner_count, found);
    cvShowImage( "Calibration", image );

    // If we got a good board, add it to our data
    if( corner_count == board_n ) {
      step = successes*board_n;
      for( int i=step, j=0; j<board_n; ++i,++j ) {
        CV_MAT_ELEM(*image_points, float,i,0) = corners[j].x;
        CV_MAT_ELEM(*image_points, float,i,1) = corners[j].y;
        CV_MAT_ELEM(*object_points,float,i,0) = j/board_w;
        CV_MAT_ELEM(*object_points,float,i,1) = j%board_w;
        CV_MAT_ELEM(*object_points,float,i,2) = 0.0f;
      }
      CV_MAT_ELEM(*point_counts, int,successes,0) = board_n;
      successes++;
    }
  } //end skip board_dt between chessboard capture

  //Handle pause/unpause and ESC
  int c = cvWaitKey(15);
  if(c == 'p'){
    c = 0;
    while(c != 'p' && c != 27){
        c = cvWaitKey(250);
    }
  }
  if(c == 27)
      return 0;
```

Example 11-1. Reading a chessboard's width and height, reading and collecting the requested number of views, and calibrating the camera (continued)

```
    image = cvQueryFrame( capture ); //Get next image
} //END COLLECTION WHILE LOOP.

//ALLOCATE MATRICES ACCORDING TO HOW MANY CHESSBOARDS FOUND
CvMat* object_points2  = cvCreateMat(successes*board_n,3,CV_32FC1);
CvMat* image_points2   = cvCreateMat(successes*board_n,2,CV_32FC1);
CvMat* point_counts2   = cvCreateMat(successes,1,CV_32SC1);
//TRANSFER THE POINTS INTO THE CORRECT SIZE MATRICES
//Below, we write out the details in the next two loops. We could
//instead have written:
//image_points->rows = object_points->rows  = \
//successes*board_n; point_counts->rows = successes;
//
for(int i = 0; i<successes*board_n; ++i) {
    CV_MAT_ELEM( *image_points2, float, i, 0) =
            CV_MAT_ELEM( *image_points, float, i, 0);
    CV_MAT_ELEM( *image_points2, float,i,1) =
            CV_MAT_ELEM( *image_points, float, i, 1);
    CV_MAT_ELEM(*object_points2, float, i, 0) =
            CV_MAT_ELEM( *object_points, float, i, 0) ;
    CV_MAT_ELEM( *object_points2, float, i, 1) =
            CV_MAT_ELEM( *object_points, float, i, 1) ;
    CV_MAT_ELEM( *object_points2, float, i, 2) =
            CV_MAT_ELEM( *object_points, float, i, 2) ;
}
for(int i=0; i<successes; ++i){ //These are all the same number
  CV_MAT_ELEM( *point_counts2, int, i, 0) =
            CV_MAT_ELEM( *point_counts, int, i, 0);
}
cvReleaseMat(&object_points);
cvReleaseMat(&image_points);
cvReleaseMat(&point_counts);

// At this point we have all of the chessboard corners we need.
// Initialize the intrinsic matrix such that the two focal
// lengths have a ratio of 1.0
//
CV_MAT_ELEM( *intrinsic_matrix, float, 0, 0 ) = 1.0f;
CV_MAT_ELEM( *intrinsic_matrix, float, 1, 1 ) = 1.0f;

//CALIBRATE THE CAMERA!
cvCalibrateCamera2(
    object_points2, image_points2,
    point_counts2,  cvGetSize( image ),
    intrinsic_matrix, distortion_coeffs,
    NULL, NULL,0   //CV_CALIB_FIX_ASPECT_RATIO
);

// SAVE THE INTRINSICS AND DISTORTIONS
cvSave("Intrinsics.xml",intrinsic_matrix);
cvSave("Distortion.xml",distortion_coeffs);
```

Example 11-1. Reading a chessboard's width and height, reading and collecting the requested number of views, and calibrating the camera (continued)

```
// EXAMPLE OF LOADING THESE MATRICES BACK IN:
CvMat *intrinsic = (CvMat*)cvLoad("Intrinsics.xml");
CvMat *distortion = (CvMat*)cvLoad("Distortion.xml");

// Build the undistort map that we will use for all
// subsequent frames.
//
IplImage* mapx = cvCreateImage( cvGetSize(image), IPL_DEPTH_32F, 1 );
IplImage* mapy = cvCreateImage( cvGetSize(image), IPL_DEPTH_32F, 1 );
cvInitUndistortMap(
  intrinsic,
  distortion,
  mapx,
  mapy
);
// Just run the camera to the screen, now showing the raw and
// the undistorted image.
//
cvNamedWindow( "Undistort" );
while(image) {
  IplImage *t = cvCloneImage(image);
  cvShowImage( "Calibration", image ); // Show raw image
  cvRemap( t, image, mapx, mapy );     // Undistort image
  cvReleaseImage(&t);
  cvShowImage("Undistort", image);     // Show corrected image

  //Handle pause/unpause and ESC
  int c = cvWaitKey(15);
  if(c == 'p') {
     c = 0;
     while(c != 'p' && c != 27) {
          c = cvWaitKey(250);
     }
  }
  if(c == 27)
      break;
  image = cvQueryFrame( capture );
}

  return 0;
}
```

Rodrigues Transform

When dealing with three-dimensional spaces, one most often represents rotations in that space by 3-by-3 matrices. This representation is usually the most convenient because multiplication of a vector by this matrix is equivalent to rotating the vector in some way. The downside is that it can be difficult to intuit just what 3-by-3 matrix goes with what rotation. An alternate and somewhat easier-to-visualize* representation for a rotation is in the form of a vector about which the rotation operates together with a single angle. In this case it is standard practice to use only a single vector whose direction encodes the direction of the axis to be rotated around and to use the size of the vector to encode the amount of rotation in a counterclockwise direction. This is easily done because the direction can be equally well represented by a vector of any magnitude; hence we can choose the magnitude of our vector to be equal to the magnitude of the rotation. The relationship between these two representations, the matrix and the vector, is captured by the Rodrigues transform.† Let r be the three-dimensional vector $r = [r_x \; r_y \; r_z]$; this vector implicitly defines θ, the magnitude of the rotation by the length (or magnitude) of r. We can then convert from this axis-magnitude representation to a rotation matrix R as follows:

$$R = \cos(\theta) \cdot I + (1 - \cos(\theta)) \cdot rr^{\mathrm{T}} + \sin(\theta) \cdot \begin{bmatrix} 0 & -r_z & r_y \\ r_z & 0 & -r_x \\ r_y & r_x & 0 \end{bmatrix}$$

We can also go from a rotation matrix back to the axis-magnitude representation by using:

$$\sin(\theta) \cdot \begin{bmatrix} 0 & -r_z & r_y \\ r_z & 0 & -r_x \\ r_y & r_x & 0 \end{bmatrix} = \frac{(R - R^{\mathrm{T}})}{2}$$

Thus we find ourselves in the situation of having one representation (the matrix representation) that is most convenient for computation and another representation (the Rodrigues representation) that is a little easier on the brain. OpenCV provides us with a function for converting from either representation to the other.

```
void  cvRodrigues2(
    const CvMat*  src,
    CvMat*        dst,
    CvMat*        jacobian = NULL
);
```

Suppose we have the vector r and need the corresponding rotation matrix representation R; we set src to be the 3-by-1 vector r and dst to be the 3-by-3 rotation matrix R. Conversely, we can set src to be a 3-by-3 rotation matrix R and dst to be a 3-by-1 vector r. In either case, cvRodrigues2() will do the right thing. The final argument is optional. If jacobian is not NULL, then it should be a pointer to a 3-by-9 or a 9-by-3 matrix that will

---

* This "easier" representation is not just for humans. Rotation in 3D space has only three components. For numerical optimization procedures, it is more efficient to deal with the three components of the Rodrigues representation than with the nine components of a 3-by-3 rotation matrix.

† Rodrigues was a 19th-century French mathematician.

be filled with the partial derivatives of the output array components with respect to the input array components. The jacobian outputs are mainly used for the internal optimization algorithms of cvFindExtrinsicCameraParameters2() and cvCalibrateCamera2(); your use of the jacobian function will mostly be limited to converting the outputs of cvFindExtrinsicCameraParameters2() and cvCalibrateCamera2() from the Rodrigues format of 1-by-3 or 3-by-1 axis-angle vectors to rotation matrices. For this, you can leave jacobian set to NULL.

Exercises

1. Use Figure 11-2 to derive the equations $x = f_x \cdot (X/Z) + c_x$ and $y - f_y \cdot (Y/Z) + c_y$ using similar triangles with a center-position offset.

2. Will errors in estimating the true center location (c_x, c_y) affect the estimation of other parameters such as focus?

    Hint: See the $q = MQ$ equation.

3. Draw an image of a square:

    a. Under radial distortion.

    b. Under tangential distortion.

    c. Under both distortions.

4. Refer to Figure 11-13. For perspective views, explain the following.

    a. Where does the "line at infinity" come from?

    b. Why do parallel lines on the object plane converge to a point on the image plane?

    c. Assume that the object and image planes are perpendicular to one another. On the object plane, starting at a point p_1, move 10 units directly away from the image plane to p_2. What is the corresponding movement distance on the image plane?

5. Figure 11-3 shows the outward-bulging "barrel distortion" effect of radial distortion, which is especially evident in the left panel of Figure 11-12. Could some lenses generate an inward-bending effect? How would this be possible?

6. Using a cheap web camera or cell phone, create examples of radial and tangential distortion in images of concentric squares or chessboards.

7. Calibrate the camera in exercise 6. Display the pictures before and after undistortion.

8. Experiment with numerical stability and noise by collecting many images of chessboards and doing a "good" calibration on all of them. Then see how the calibration parameters change as you reduce the number of chessboard images. Graph your results: camera parameters as a function of number of chessboard images.

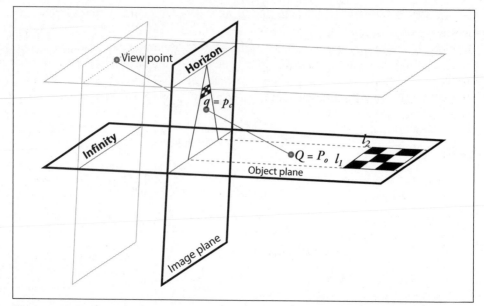

Figure 11-13. Homography diagram showing intersection of the object plane with the image plane and a viewpoint representing the center of projection

9. With reference to exercise 8, how do calibration parameters change when you use (say) 10 images of a 3-by-5, a 4-by-6, and a 5-by-7 chessboard? Graph the results.

10. High-end cameras typically have systems of lens that correct physically for distortions in the image. What might happen if you nevertheless use a multiterm distortion model for such a camera?

    Hint: This condition is known as *overfitting*.

11. *Three-dimensional joystick trick*. Calibrate a camera. Using video, wave a chessboard around and use cvFindExtrinsicCameraParams2() as a 3D joystick. Remember that cvFindExtrinsicCameraParams2() outputs rotation as a 3-by-1 or 1-by-3 vector axis of rotation, where the magnitude of the vector represents the counterclockwise angle of rotation along with a 3D translation vector.

    a. Output the chessboard's axis and angle of the rotation along with where it is (i.e., the translation) in real time as you move the chessboard around. Handle cases where the chessboard is not in view.

    b. Use cvRodrigues2() to translate the output of cvFindExtrinsicCameraParams2() into a 3-by-3 rotation matrix and a translation vector. Use this to animate a simple 3D stick figure of an airplane rendered back into the image in real time as you move the chessboard in view of the video camera.

CHAPTER 12

Projection and 3D Vision

In this chapter we'll move into three-dimensional vision, first with projections and then with multicamera stereo depth perception. To do this, we'll have to carry along some of the concepts from Chapter 11. We'll need the *camera instrinsics* matrix *M*, the *distortion coefficients*, the rotation matrix *R*, the translation vector *T*, and especially the *homography matrix H*.

We'll start by discussing projection into the 3D world using a calibrated camera and reviewing affine and projective transforms (which we first encountered in Chapter 6); then we'll move on to an example of how to get a bird's-eye view of a ground plane.* We'll also discuss POSIT, an algorithm that allows us to find the 3D pose (position and rotation) of a known 3D object in an image.

We will then move into the three-dimensional geometry of multiple images. In general, there is no reliable way to do calibration or to extract 3D information without multiple images. The most obvious case in which we use multiple images to reconstruct a three-dimensional scene is *stereo vision*. In stereo vision, features in two (or more) images taken at the same time from separate cameras are matched with the corresponding features in the other images, and the differences are analyzed to yield depth information. Another case is *structure from motion*. In this case we may have only a single camera, but we have multiple images taken at different times and from different places. In the former case we are primarily interested in *disparity effects* (triangulation) as a means of computing distance. In the latter, we compute something called the *fundamental matrix* (relates two different views together) as the source of our scene understanding. Let's get started with projection.

Projections

Once we have calibrated the camera (see Chapter 11), it is possible to unambiguously project points in the physical world to points in the image. This means that, given a location in the three-dimensional physical coordinate frame attached to the camera, we

---

* This is a recurrent problem in robotics as well as many other vision applications.

can compute where on the imager, in pixel coordinates, an external 3D point should appear. This transformation is accomplished by the OpenCV routine cvProjectPoints2().

```
void cvProjectPoints2(
    const CvMat* object_points,
    const CvMat* rotation_vector,
    const CvMat* translation_vector,
    const CvMat* intrinsic_matrix,
    const CvMat* distortion_coeffs,
    CvMat*       image_points,
    CvMat*       dpdrot        = NULL,
    CvMat*       dpdt          = NULL,
    CvMat*       dpdf          = NULL,
    CvMat*       dpdc          = NULL,
    CvMat*       dpddist       = NULL,
    double       aspectRatio   = 0
);
```

At first glance the number of arguments might be a little intimidating, but in fact this is a simple function to use. The cvProjectPoints2() routine was designed to accommodate the (very common) circumstance where the points you want to project are located on some rigid body. In this case, it is natural to represent the points not as just a list of locations in the camera coordinate system but rather as a list of locations in the object's own body centered coordinate system; then we can add a rotation and a translation to specify the relationship between the object coordinates and the camera's coordinate system. In fact, cvProjectPoints2() is used internally in cvCalibrateCamera2(), and of course this is the way cvCalibrateCamera2() organizes its own internal operation. All of the optional arguments are primarily there for use by cvCalibrateCamera2(), but sophisticated users might find them handy for their own purposes as well.

The first argument, object_points, is the list of points you want projected; it is just an N-by-3 matrix containing the point locations. You can give these in the object's own local coordinate system and then provide the 3-by-1 matrices rotation_vector* and translation_vector to relate the two coordinates. If in your particular context it is easier to work directly in the camera coordinates, then you can just give object_points in that system and set both rotation_vector and translation_vector to contain 0s.†

The intrinsic_matrix and distortion_coeffs are just the camera intrinsic information and the distortion coefficients that come from cvCalibrateCamera2() discussed in Chapter 11. The image_points argument is an N-by-2 matrix into which the results of the computation will be written.

Finally, the long list of optional arguments dpdrot, dpdt, dpdf, dpdc, and dpddist are all Jacobian matrices of partial derivatives. These matrices relate the image points to each of the different input parameters. In particular: dpdrot is an N-by-3 matrix of partial derivatives of image points with respect to components of the rotation vector; dpdt is an

---

* The "rotation vector" is in the usual Rodrigues representation.

† Remember that this rotation vector is an axis-angle representation of the rotation, so being set to all 0s means it has zero magnitude and thus "no rotation".

N-by-3 matrix of partial derivatives of image points with respect to components of the translation vector; dpdf is an *N*-by-2 matrix of partial derivatives of image points with respect to f_x and f_y; dpdc is an *N*-by-2 matrix of partial derivatives of image points with respect to c_x and c_y; and dpddist is an *N*-by-4 matrix of partial derivatives of image points with respect to the distortion coefficients. In most cases, you will just leave these as NULL, in which case they will not be computed. The last parameter, aspectRatio, is also optional; it is used for derivatives only when the aspect ratio is fixed in cvCalibrateCamera2() or cvStereoCalibrate(). If this parameter is not 0 then the derivatives dpdf are adjusted.

Affine and Perspective Transformations

Two transformations that come up often in the OpenCV routines we have discussed—as well as in other applications you might write yourself—are the affine and perspective transformations. We first encountered these in Chapter 6. As implemented in OpenCV, these routines affect either lists of points or entire images, and they map points on one location in the image to a different location, often performing subpixel interpolation along the way. You may recall that an affine transform can produce any parallelogram from a rectangle; the perspective transform is more general and can produce any trapezoid from a rectangle.

The *perspective transformation* is closely related to the *perspective projection*. Recall that the perspective projection maps points in the three-dimensional physical world onto points on the two-dimensional image plane along a set of projection lines that all meet at a single point called *the center of projection*. The perspective transformation, which is a specific kind of *homography*,* relates two different images that are alternative projections of the same three-dimensional object onto two different *projective planes* (and thus, for nondegenerate configurations such as the plane physically intersecting the 3D object, typically to two different centers of projection).

These projective transformation-related functions were discussed in detail in Chapter 6; for convenience, we summarize them here in Table 12-1.

Table 12-1. Affine and perspective transform functions

Function	Use
cvTransform()	Affine transform a list of points
cvWarpAffine()	Affine transform a whole image
cvGetAffineTransform()	Fill in affine transform matrix parameters
cv2DRotationMatrix()	Fill in affine transform matrix parameters
cvGetQuadrangleSubPix()	Low-overhead whole image affine transform
cvPerspectiveTransform()	Perspective transform a list of points
cvWarpPerspective()	Perspective transform a whole image
cvGetPerspectiveTransform()	Fill in perspective transform matrix parameters

---

* Recall from Chapter 11 that this special kind of homography is known as *planar homography*.

Bird's-Eye View Transform Example

A common task in robotic navigation, typically used for planning purposes, is to convert the robot's camera view of the scene into a top-down "bird's-eye" view. In Figure 12-1, a robot's view of a scene is turned into a bird's-eye view so that it can be subsequently overlaid with an alternative representation of the world created from scanning laser range finders. Using what we've learned so far, we'll look in detail about how to use our calibrated camera to compute such a view.

Figure 12-1. Bird's-eye view: A camera on a robot car looks out at a road scene where laser range finders have identified a region of "road" in front of the car and marked it with a box (top); vision algorithms have segmented the flat, roadlike areas (center); the segmented road areas are converted to a bird's-eye view and merged with the bird's-eye view laser map (bottom)

To get a bird's-eye view,* we'll need our camera intrinsics and distortion matrices from the calibration routine. Just for the sake of variety, we'll read the files from disk. We put a chessboard on the floor and use that to obtain a ground plane image for a robot cart; we then remap that image into a bird's-eye view. The algorithm runs as follows.

1. Read the intrinsics and distortion models for the camera.

2. Find a known object on the ground plane (in this case, a chessboard). Get at least four points at subpixel accuracy.

3. Enter the found points into cvGetPerspectiveTransform() (see Chapter 6) to compute the homography matrix *H* for the ground plane view.

4. Use cvWarpPerspective( ) (again, see Chapter 6) with the flags CV_INTER_LINEAR + CV_WARP_INVERSE_MAP + CV_WARP_FILL_OUTLIERS to obtain a frontal parallel (bird's-eye) view of the ground plane.

Example 12-1 shows the full working code for bird's-eye view.

Example 12-1. Bird's-eye view

```
//Call:
//  birds-eye board_w board_h instrinics distortion image_file
// ADJUST VIEW HEIGHT using keys 'u' up, 'd' down. ESC to quit.
//

int main(int argc, char* argv[]) {
  if(argc != 6) return -1;

  // INPUT PARAMETERS:
  //
  int       board_w     = atoi(argv[1]);
  int       board_h     = atoi(argv[2]);
  int       board_n     = board_w * board_h;
  CvSize    board_sz    = cvSize( board_w, board_h );
  CvMat*    intrinsic   = (CvMat*)cvLoad(argv[3]);
  CvMat*    distortion  = (CvMat*)cvLoad(argv[4]);
  IplImage* image       = 0;
  IplImage* gray_image  = 0;
  if( (image = cvLoadImage(argv[5])) == 0 ) {
    printf("Error: Couldn't load %s\n",argv[5]);
    return -1;
  }
  gray_image = cvCreateImage( cvGetSize(image), 8, 1 );
  cvCvtColor(image, gray_image, CV_BGR2GRAY );

  // UNDISTORT OUR IMAGE
  //
  IplImage* mapx = cvCreateImage( cvGetSize(image), IPL_DEPTH_32F, 1 );
  IplImage* mapy = cvCreateImage( cvGetSize(image), IPL_DEPTH_32F, 1 );
```

---

* The bird's-eye view technique also works for transforming perspective views of any plane (e.g., a wall or ceiling) into frontal parallel views.

Example 12-1. Bird's-eye view (continued)

```
//This initializes rectification matrices
//
cvInitUndistortMap(
  intrinsic,
  distortion,
  mapx,
  mapy
);
IplImage *t = cvCloneImage(image);

// Rectify our image
//
cvRemap( t, image, mapx, mapy );

// GET THE CHESSBOARD ON THE PLANE
//
cvNamedWindow("Chessboard");
CvPoint2D32f* corners = new CvPoint2D32f[ board_n ];
int corner_count = 0;
int found = cvFindChessboardCorners(
  image,
  board_sz,
  corners,
  &corner_count,
  CV_CALIB_CB_ADAPTIVE_THRESH | CV_CALIB_CB_FILTER_QUADS
);
if(!found){
  printf("Couldn't aquire chessboard on %s, "
    "only found %d of %d corners\n",
    argv[5],corner_count,board_n
  );
  return -1;
}
//Get Subpixel accuracy on those corners:
cvFindCornerSubPix(
  gray_image,
  corners,
  corner_count,
  cvSize(11,11),
  cvSize(-1,-1),
  cvTermCriteria( CV_TERMCRIT_EPS | CV_TERMCRIT_ITER, 30, 0.1 )
);

//GET THE IMAGE AND OBJECT POINTS:
// We will choose chessboard object points as (r,c):
// (0,0), (board_w-1,0), (0,board_h-1), (board_w-1,board_h-1).
//
CvPoint2D32f objPts[4], imgPts[4];
objPts[0].x = 0;          objPts[0].y = 0;
objPts[1].x = board_w-1; objPts[1].y = 0;
objPts[2].x = 0;          objPts[2].y = board_h-1;
objPts[3].x = board_w-1; objPts[3].y = board_h-1;
imgPts[0]   = corners[0];
```

Example 12-1. Bird's-eye view (continued)

```
imgPts[1]    = corners[board_w-1];
imgPts[2]    = corners[(board_h-1)*board_w];
imgPts[3]    = corners[(board_h-1)*board_w + board_w-1];

// DRAW THE POINTS in order: B,G,R,YELLOW
//
cvCircle( image, cvPointFrom32f(imgPts[0]), 9, CV_RGB(0,0,255),    3);
cvCircle( image, cvPointFrom32f(imgPts[1]), 9, CV_RGB(0,255,0),    3);
cvCircle( image, cvPointFrom32f(imgPts[2]), 9, CV_RGB(255,0,0),    3);
cvCircle( image, cvPointFrom32f(imgPts[3]), 9, CV_RGB(255,255,0),  3);

// DRAW THE FOUND CHESSBOARD
//
cvDrawChessboardCorners(
  image,
  board_sz,
  corners,
  corner_count,
  found
);
cvShowImage( "Chessboard", image );

// FIND THE HOMOGRAPHY
//
CvMat *H = cvCreateMat( 3, 3, CV_32F);
cvGetPerspectiveTransform( objPts, imgPts, H);

// LET THE USER ADJUST THE Z HEIGHT OF THE VIEW
//
float Z = 25;
int key = 0;
IplImage *birds_image = cvCloneImage(image);
cvNamedWindow("Birds_Eye");

// LOOP TO ALLOW USER TO PLAY WITH HEIGHT:
//
// escape key stops
//
while(key != 27) {
  // Set the height
  //
  CV_MAT_ELEM(*H,float,2,2) = Z;

  // COMPUTE THE FRONTAL PARALLEL OR BIRD'S-EYE VIEW:
  // USING HOMOGRAPHY TO REMAP THE VIEW
  //
  cvWarpPerspective(
    image,
    birds_image,
    H,
    CV_INTER_LINEAR | CV_WARP_INVERSE_MAP | CV_WARP_FILL_OUTLIERS
  );
  cvShowImage( "Birds_Eye", birds_image );
```

Example 12-1. Bird's-eye view (continued)

```
    key = cvWaitKey();
    if(key == 'u') Z += 0.5;
    if(key == 'd') Z -= 0.5;
}

    cvSave("H.xml",H); //We can reuse H for the same camera mounting
    return 0;
}
```

Once we have the homography matrix and the height parameter set as we wish, we could then remove the chessboard and drive the cart around, making a bird's-eye view video of the path, but we'll leave that as an exercise for the reader. Figure 12-2 shows the input at left and output at right for the bird's-eye view code.

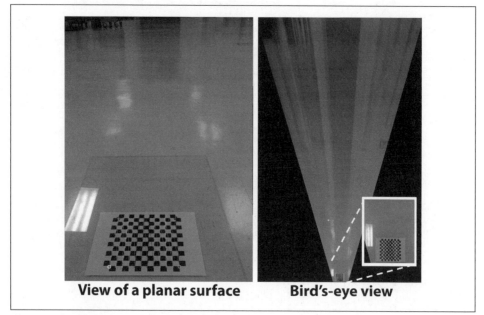

View of a planar surface **Bird's-eye view**

Figure 12-2. Bird's-eye view example

POSIT: 3D Pose Estimation

Before moving on to stereo vision, we should visit a useful algorithm that can estimate the positions of known objects in three dimensions. POSIT (aka "Pose from Orthography and Scaling with Iteration") is an algorithm originally proposed in 1992 for computing the pose (the position T and orientation R described by six parameters [DeMenthon92]) of a 3D object whose exact dimensions are known. To compute this pose, we must find on the image the corresponding locations of at least four non-coplanar points on the surface of that object. The first part of the algorithm, *pose from orthography and scaling*

(POS), assumes that the points on the object are all at effectively the same depth* and that size variations from the original model are due solely to scaling with distance from the camera. In this case there is a closed-form solution for that object's 3D pose based on scaling. The assumption that the object points are all at the same depth effectively means that the object is far enough away from the camera that we can neglect any internal depth differences within the object; this assumption is known as the *weak-perspective* approximation.

Given that we know the camera intrinsics, we can find the perspective scaling of our known object and thus compute its approximate pose. This computation will not be very accurate, but we can then project where our four observed points would go if the true 3D object were at the pose we calculated through POS. We then start all over again with these new point positions as the inputs to the POS algorithm. This process typically converges within four or five iterations to the true object pose—hence the name "POS algorithm *with iteration*". Remember, though, that all of this assumes that the internal depth of the object is in fact small compared to the distance away from the camera. If this assumption is not true, then the algorithm will either not converge or will converge to a "bad pose". The OpenCV implementation of this algorithm will allow us to track more than four (non-coplanar) points on the object to improve pose estimation accuracy.

The POSIT algorithm in OpenCV has three associated functions: one to allocate a data structure for the pose of an individual object, one to de-allocate the same data structure, and one to actually implement the algorithm.

```
CvPOSITObject* cvCreatePOSITObject(
    CvPoint3D32f* points,
    int           point_count
);
void cvReleasePOSITObject(
    CvPOSITObject** posit_object
);
```

The cvCreatePOSITObject() routine just takes points (a set of three-dimensional points) and point_count (an integer indicating the number of points) and returns a pointer to an allocated POSIT object structure. Then cvReleasePOSITObject() takes a pointer to such a structure pointer and de-allocates it (setting the pointer to NULL in the process).

```
void cvPOSIT(
    CvPOSITObject* posit_object,
    CvPoint2D32f*  image_points,
    double         focal_length,
    CvTermCriteria criteria,
    float*         rotation_matrix,
    float*         translation_vector
);
```

---

* The construction finds a reference plane through the object that is parallel to the image plane; this plane through the object then has a single distance Z from the image plane. The 3D points on the object are first projected to this plane through the object and then projected onto the image plane using perspective projection. The result is scaled orthographic projection, and it makes relating object size to depth particularly easy.

Now, on to the POSIT function itself. The argument list to cvPOSIT() is different stylistically than most of the other functions we have seen in that it uses the "old style" arguments common in earlier versions of OpenCV.* Here posit_object is just a pointer to the POSIT object that you are trying to track, and image_points is a list of the locations of the corresponding points in the image plane (notice that these are 32-bit values, thus allowing for subpixel locations). The current implementation of cvPOSIT() assumes square pixels and thus allows only a single value for the focal_length parameter instead of one in the x and one in the y directions. Because cvPOSIT() is an iterative algorithm, it requires a termination criteria: criteria is of the usual form and indicates when the fit is "good enough". The final two parameters, rotation_matrix and translation_vector, are analogous to the same arguments in earlier routines; observe, however, that these are pointers to float and so are just the data part of the matrices you would obtain from calling (for example) cvCalibrateCamera2(). In this case, given a matrix M, you would want to use something like M->data.fl as an argument to cvPOSIT().

When using POSIT, keep in mind that the algorithm does not benefit from additional surface points that are coplanar with other points already on the surface. Any point lying on a plane defined by three other points will not contribute anything useful to the algorithm. In fact, extra coplanar points can cause degeneracies that hurt the algorithm's performance. Extra non-coplanar points will help the algorithm. Figure 12-3 shows the POSIT algorithm in use with a toy plane [Tanguay00]. The plane has marking lines on it, which are used to define four non-coplanar points. These points were fed into cvPOSIT(), and the resulting rotation_matrix and translation_vector are used to control a flight simulator.

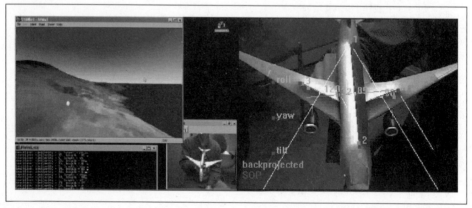

Figure 12-3. POSIT algorithm in use: four non-coplanar points on a toy jet are used to control a flight simulator

---

* You might have noticed that many function names end in "2". More often than not, this is because the function in the current release in the library has been modified from its older incarnation to use the newer style of arguments.

Stereo Imaging

Now we are in a position to address *stereo imaging*.* We all are familiar with the stereo imaging capability that our eyes give us. To what degree can we emulate this capability in computational systems? Computers accomplish this task by finding correspondences between points that are seen by one imager and the same points as seen by the other imager. With such correspondences and a known baseline separation between cameras, we can compute the 3D location of the points. Although the search for corresponding points can be computationally expensive, we can use our knowledge of the geometry of the system to narrow down the search space as much as possible. In practice, stereo imaging involves four steps when using two cameras.

1. Mathematically remove radial and tangential lens distortion; this is called *undistortion* and is detailed in Chapter 11. The outputs of this step are undistorted images.

2. Adjust for the angles and distances between cameras, a process called *rectification*. The outputs of this step are images that are row-aligned[†] and rectified.

3. Find the same features in the left and right[‡] camera views, a process known as *correspondence*. The output of this step is a *disparity* map, where the disparities are the differences in x-coordinates on the image planes of the same feature viewed in the left and right cameras: $x^l - x^r$.

4. If we know the geometric arrangement of the cameras, then we can turn the disparity map into distances by *triangulation*. This step is called *reprojection*, and the output is a depth map.

We start with the last step to motivate the first three.

Triangulation

Assume that we have a perfectly undistorted, aligned, and measured stereo rig as shown in Figure 12-4: two cameras whose image planes are exactly coplanar with each other, with exactly parallel optical axes (the optical axis is the ray from the center of projection O through the principal point c and is also known as the *principal ray*[§]) that are a known distance apart, and with equal focal lengths $f_l = f_r$. Also, assume for now that the *principal points* c_x^{left} and c_x^{right} have been calibrated to have the same pixel coordinates in their respective left and right images. Please don't confuse these principal points with the center of the image. A principal point is where the principal ray intersects the imaging

---

* Here we give just a high-level understanding. For details, we recommend the following texts: Trucco and Verri [Trucco98], Hartley and Zisserman [Hartley06], Forsyth and Ponce [Forsyth03], and Shapiro and Stockman [Shapiro02]. The stereo rectification sections of these books will give you the background to tackle the original papers cited in this chapter.

† By "row-aligned" we mean that the two image planes are coplanar and that the image rows are exactly aligned (in the same direction and having the same y-coordinates).

‡ Every time we refer to left and right cameras you can also use vertically oriented up and down cameras, where disparities are in the y-direction rather than the x-direction.

§ Two parallel principal rays are said to intersect at infinity.

plane. This intersection depends on the optical axis of the lens. As we saw in Chapter 11, the image plane is rarely aligned exactly with the lens and so the center of the imager is almost never exactly aligned with the principal point.

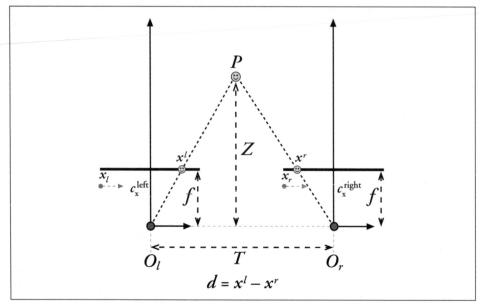

Figure 12-4. With a perfectly undistorted, aligned stereo rig and known correspondence, the depth Z can be found by similar triangles; the principal rays of the imagers begin at the centers of projection O_l and O_r, and extend through the principal points of the two image planes at c_l and c_r

Moving on, let's further assume the images are row-aligned and that every pixel row of one camera aligns exactly with the corresponding row in the other camera.* We will call such a camera arrangement *frontal parallel*. We will also assume that we can find a point P in the physical world in the left and the right image views at p_l and p_r, which will have the respective horizontal coordinates x^l and x^r.

In this simplified case, taking x^l and x^r to be the horizontal positions of the points in the left and right imager (respectively) allows us to show that the depth is inversely proportional to the disparity between these views, where the disparity is defined simply by $d = x^l - x^r$. This situation is shown in Figure 12-4, where we can easily derive the depth Z by using similar triangles. Referring to the figure, we have:[†]

---

* This makes for quite a few assumptions, but we are just looking at the basics right now. Remember that the process of rectification (to which we will return shortly) is how we get things done mathematically when these assumptions are not physically true. Similarly, in the next sentence we will temporarily "assume away" the correspondence problem.

† This formula is predicated on the principal rays intersecting at infinity. However, as you will see in "Calibrated Stereo Rectification" (later in this chapter), we derive stereo rectification relative to the principal points c_x^{left} and c_x^{right}. In our derivation, if the principal rays intersect at infinity then the principal points have the same coordinates and so the formula for depth holds as is. However, if the principal rays intersect at a finite distance then the principal points will not be equal and so the equation for depth becomes $Z = fT / (d - (c_x^{left} - c_x^{right}))$.

$$\frac{T-(x^l-x^r)}{Z-f}=\frac{T}{Z} \quad \Rightarrow \quad Z=\frac{fT}{x^l-x^r}$$

Since depth is inversely proportional to disparity, there is obviously a nonlinear relationship between these two terms. When disparity is near 0, small disparity differences make for large depth differences. When disparity is large, small disparity differences do not change the depth by much. The consequence is that stereo vision systems have high depth resolution only for objects relatively near the camera, as Figure 12-5 makes clear.

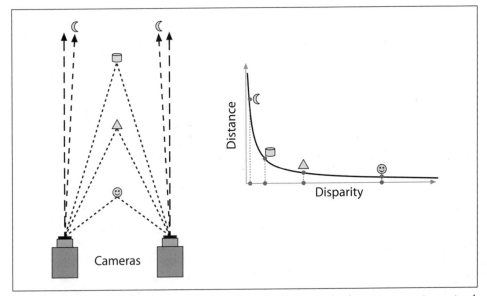

Figure 12-5. Depth and disparity are inversely related, so fine-grain depth measurement is restricted to nearby objects

We have already seen many coordinate systems in the discussion of calibration in Chapter 11. Figure 12-6 shows the 2D and 3D coordinate systems used in OpenCV for stereo vision. Note that it is a right-handed coordinate system: if you point your right index finger in the direction of X and bend your right middle finger in the direction of Y, then your thumb will point in the direction of the principal ray. The left and right imager pixels have image origins at upper left in the image, and pixels are denoted by coordinates (x_l, y_l) and (x_r, y_r), respectively. The center of projection are at O_l and O_r with principal rays intersecting the image plane at the principal point (not the center) (c_x, c_y). After mathematical rectification, the cameras are row-aligned (coplanar and horizontally aligned), displaced from one another by T, and of the same focal length f.

With this arrangement it is relatively easily to solve for distance. Now we must spend some energy on understanding how we can map a real-world camera setup into a geometry that resembles this ideal arrangement. In the real world, cameras will almost never be exactly aligned in the frontal parallel configuration depicted in Figure 12-4. Instead,

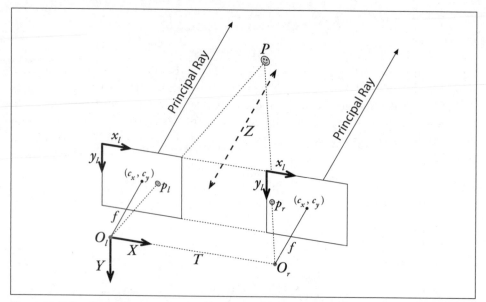

Figure 12-6. Stereo coordinate system used by OpenCV for undistorted rectified cameras: the pixel coordinates are relative to the upper left corner of the image, and the two planes are row-aligned; the camera coordinates are relative to the left camera's center of projection

we will mathematically find image projections and distortion maps that will rectify the left and right images into a frontal parallel arrangement. When designing your stereo rig, it is best to arrange the cameras approximately frontal parallel and as close to horizontally aligned as possible. This physical alignment will make the mathematical tranformations more tractable. If you don't align the cameras at least approximately, then the resulting mathematical alignment can produce extreme image distortions and so reduce or eliminate the stereo overlap area of the resulting images.* For good results, you'll also need synchronized cameras. If they don't capture their images at the exact same time, then you will have problems if anything is moving in the scene (including the cameras themselves). If you do not have synchronized cameras, you will be limited to using stereo with stationary cameras viewing static scenes.

Figure 12-7 depicts the real situation between two cameras and the mathematical alignment we want to achieve. To perform this mathematical alignment, we need to learn more about the geometry of two cameras viewing a scene. Once we have that geometry defined and some terminology and notation to describe it, we can return to the problem of alignment.

---

* The exception to this advice is that for applications where we want more resolution at close range; in this case, we tilt the cameras slightly in toward each other so that their principal rays intersect at a finite distance. After mathematical alignment, the effect of such inward verging cameras is to introduce an x-offset that is subtracted from the disparity. This may result in negative disparities, but we can thus gain finer depth resolution at the nearby depths of interest.

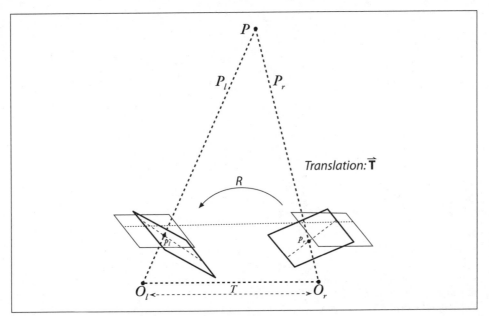

Figure 12-7. Our goal will be to mathematically (rather than physically) align the two cameras into one viewing plane so that pixel rows between the cameras are exactly aligned with each other

Epipolar Geometry

The basic geometry of a stereo imaging system is referred to as *epipolar geometry*. In essence, this geometry combines two pinhole models (one for each camera[*]) and some interesting new points called the *epipoles* (see Figure 12-8). Before explaining what these epipoles are good for, we will start by taking a moment to define them clearly and to add some new related terminology. When we are done, we will have a concise understanding of this overall geometry and will also find that we can narrow down considerably the possible locations of corresponding points on the two stereo cameras. This added discovery will be important to practical stereo implementations.

For each camera there is now a separate center of projection, O_l and O_r, and a pair of corresponding projective planes, Π_l and Π_r. The point P in the physical world has a projection onto each of the projective planes that we can label p_l and p_r. The new points of interest are the epipoles. An epipole e_l (resp. e_r) on image plane Π_l (resp. Π_r) is defined as the image of the center of projection of the other camera O_r (resp. O_l). The plane in space formed by the actual viewed point P and the two epipoles e_l and e_r (or, equivalently, through the two centers of projection O_r and O_l) is called the *epipolar plane*, and the lines $p_l e_l$ and $p_r e_r$ (from the points of projection to the corresponding epipolar points) are called the *epipolar lines*.[†]

---

[*] Since we are actually dealing with real lenses and not pinhole cameras, it is important that the two images be undistorted; see Chapter 11.

[†] You can see why the epipoles did not come up before: as the planes approach being perfectly parallel, the epipoles head out toward infinity!

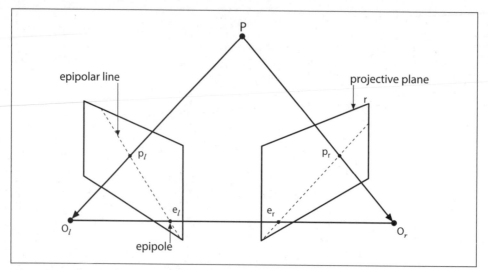

Figure 12-8. The epipolar plane is defined by the observed point P and the two centers of projection, O_l and O_r; the epipoles are located at the point of intersection of the line joining the centers of projection and the two projective planes

To understand the utility of the epipoles we first recall that, when we see a point in the physical world projected onto our right (or left) image plane, that point could actually be located anywhere along a entire line of points formed by the ray going from O_r out through p_r (or O_l through p_l) because, with just that single camera, we do not know the distance to the point we are viewing. More specifically, take for example the point P as seen by the camera on the right. Because that camera sees only p_r (the projection of P onto Π_r), the actual point P could be located anywhere on the line defined by p_r and O_r. This line obviously contains P, but it contains a lot of other points, too. What is interesting, though, is to ask what that line looks like projected onto the left image plane Π_l; in fact, it is the epipolar line defined by p_l and e_l. To put that into English, the image of all of the possible locations of a *point* seen in one imager is the *line* that goes through the corresponding point and the epipolar point on the other imager.

We'll now summarize some facts about stereo camera epipolar geometry (and why we care).

- Every 3D point in view of the cameras is contained in an epipolar plane that intersects each image in an epipolar line.

- Given a feature in one image, its matching view in the other image *must* lie along the corresponding epipolar line. This is known as the *epipolar constraint*.

- The epipolar constraint means that the possible two-dimensional search for matching features across two imagers becomes a one-dimensional search along the epipolar lines once we know the epipolar geometry of the stereo rig. This is not only a vast computational savings, it also allows us to reject a lot of points that could otherwise lead to spurious correspondences.

- Order is preserved. If points *A* and *B* are visible in both images and occur horizontally in that order in one imager, then they occur horizontally in that order in the other imager.*

The Essential and Fundamental Matrices

You might think that the next step would be to introduce some OpenCV function that computes these epipolar lines for us, but we actually need two more ingredients before we can arrive at that point. These ingredients are the *essential matrix E* and the *fundamental matrix F*.[†] The matrix *E* contains information about the translation and rotation that relate the two cameras in physical space (see Figure 12-9), and *F* contains the same information as *E* in addition to information about the intrinsics of both cameras.[‡] Because *F* embeds information about the intrinsic parameters, it relates the two cameras in pixel coordinates.

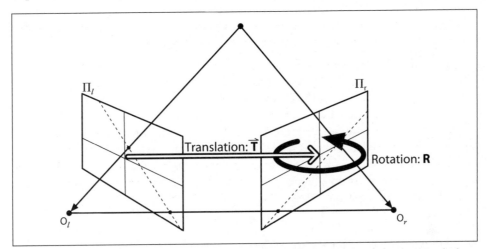

Figure 12-9. The essential geometry of stereo imaging is captured by the essential matrix E, which contains all of the information about the translation T and the rotation R, which describe the location of the second camera relative to the first in global coordinates

---

* Because of occlusions and areas of overlapping view, it is certainly possible that both cameras do not see the same points. Nevertheless, order is maintained. If points *A*, *B*, and *C* are arranged left to right on the left imager and if *B* is not seen on the right imager owing to occlusion, then the right imager will still see points *A* and *C* left to right.

† The next subsections are a bit mathy. If you do not like math then just skim over them; at least you'll have confidence that somewhere, someone understands all of this stuff. For simple applications, you can just use the machinery that OpenCV provides without the need for all of the details in these next few pages.

‡ The astute reader will recognize that *E* was described in almost the exact same words as the homography matrix *H* in the previous section. Although both are constructed from similar information, they are not the same matrix and should not be confused. An essential part of the definition of *H* is that we were considering a plane viewed by a camera and thus could relate one point in that plane to the point on the camera plane. The matrix *E* makes no such assumption and so will only be able to relate a point in one image to a line in the other.

Let's reinforce the differences between E and F. The essential matrix E is purely geometrical and knows nothing about imagers. It relates the location, in physical coordinates, of the point P as seen by the left camera to the location of the same point as seen by the right camera (i.e., it relates p_l to p_r). The fundamental matrix F relates the points on the image plane of one camera in image coordinates (pixels) to the points on the image plane of the other camera in image coordinates (for which we will use the notation q_l and q_r).

Essential matrix math

We will now submerge into some math so we can better understand the OpenCV function calls that do the hard work for our stereo geometry problems.

Given a point P, we would like to derive a relation which connects the observed locations p_l and p_r of P on the two imagers. This relationship will turn out to serve as the definition of the essential matrix. We begin by considering the relationship between p_l and p_r, the physical locations of the point we are viewing in the coordinates of the two cameras. These can be related by using epipolar geometry, as we have already seen.[*]

Let's pick one set of coordinates, left or right, to work in and do our calculations there. Either one is just as good, but we'll choose the coordinates centered on O_l of the left camera. In these coordinates, the location of the observed point is P_l and the origin of the other camera is located at T. The point P as seen by the right camera is P_r in that camera's coordinates, where $P_r = R(P_l - T)$. The key step is the introduction of the epipolar plane, which we already know relates all of these things. We could, of course, represent a plane any number of ways, but for our purpose it is most helpful to recall that the equation for all points $\mathbf{x}$ on a plane with normal vector $\mathbf{n}$ and passing through point $\mathbf{a}$ obey the following constraint:

$$(\mathbf{x} - \mathbf{a}) \cdot \mathbf{n} = 0$$

Recall that the epipolar plane contains the vectors P_l and T; thus, if we had a vector (e.g., $P_l \times T$) perpendicular to both,[†] then we could use that for $\mathbf{n}$ in our plane equation. Thus an equation for all possible points P_l through the point T and containing both vectors would be:[‡]

$$(P_l - T)^{\mathrm{T}} (T \times P_l) = 0$$

Remember that our goal was to relate q_l and q_r by first relating P_l and P_r. We draw P_r into the picture via our equality $P_r = R(P_l - T)$, which we can conveniently rewrite as $(P_l - T) = R^{-1} P_r$. Making this substitution and using that $R^{\mathrm{T}} = R^{-1}$ yields:

---

[*] Please do not confuse p_l and p_r, which are points on the projective image planes, with p_l and p_r, which are the locations of the point P in the coordinate frames of the two cameras.

[†] The cross product of vectors produces a third vector orthogonal to the first two. The direction is defined by the "right hand rule": if you point in the direction a and bend your middle finger in the direction b, then the cross product $a \times b$ points perpendicular to a and b in the direction of your thumb.

[‡] Here we have replaced the dot product with matrix multiplication by the transpose of the normal vector.

$$(R^{\mathrm{T}}P_r)^{\mathrm{T}}(T \times P_l) = 0$$

It is always possible to rewrite a cross product as a (somewhat bulky) matrix multiplication. We thus define the matrix S such that:

$$T \times P_l = SP_l \quad \Rightarrow \quad S = \begin{bmatrix} 0 & -T_z & T_y \\ T_z & 0 & -T_x \\ -T_y & T_x & 0 \end{bmatrix}$$

This leads to our first result. Making this substitution for the cross product gives:

$$(P_r)^{\mathrm{T}} RSP_l = 0$$

This product RS is what we define to be the essential matrix E, which leads to the compact equation:

$$(P_r)^{\mathrm{T}} EP_l = 0$$

Of course, what we really wanted was a relation between the points as we observe them on the imagers, but this is just a step away. We can simply substitute using the projection equations $p_l = f_l P_l / Z_l$ and $p_r = f_r P_r / Z_r$, and then divide the whole thing by $Z_l Z_r / f_l f_r$ to obtain our final result:

$$p_r^{\mathrm{T}} Ep_l = 0$$

This might look at first like it completely specifies one of the p-terms if the other is given, but E turns out to be a rank-deficient matrix* (the 3-by-3 essential matrix has rank 2) and so this actually ends up being an equation for a line. There are five parameters in the essential matrix—three for rotation and two for the direction of translation (scale is not set)—along with two other constraints. The two additional constraints on the essential matrix are: (1) the determinant is 0 because it is rank-deficient (a 3-by-3 matrix of rank 2); and (2) its two nonzero singular values are equal because the matrix S is skew-symmetric and R is a rotation matrix. This yields a total of seven constraints. Note again that E contains nothing intrinsic to the cameras in E; thus, it relates points to each other in physical or camera coordinates, not pixel coordinates.

Fundamental matrix math

The matrix E contains all of the information about the geometry of the two cameras relative to one another but no information about the cameras themselves. In practice, we are usually interested in pixel coordinates. In order to find a relationship between a pixel in one image and the corresponding epipolar line in the other image, we will have

---

* For a square n-by-n matrix like E, *rank deficient* essentially means that there are fewer than n nonzero eigenvalues. As a result, a system of linear equations specified by a rank-deficient matrix does not have a unique solution. If the rank (number of nonzero eigenvalues) is $n - 1$ then there will be a line formed by a set of points all of which satisfy the system of equations. A system specified by a matrix of rank $n - 2$ will form a plane, and so forth.

to introduce intrinsic information about the two cameras. To do this, for p (the pixel coordinate) we substitute q and the camera intrinsics matrix that relates them. Recall that $q = Mp$ (where M is the camera intrinsics matrix) or, equivalently, $p = M^{-1} q$. Hence our equation for E becomes:

$$q_r^T (M_r^{-1})^T E M_l^{-1} q_l = 0$$

Though this looks like a bit of a mess, we clean it up by defining the fundamental matrix F as:

$$F = (M_r^{-1})^T E M_l^{-1}$$

so that

$$q_r^T F q_l = 0$$

In a nutshell: the fundamental matrix F is just like the essential matrix E, except that F operates in image pixel coordinates whereas E operates in physical coordinates.* Just like E, the fundamental matrix F is of rank 2. The fundamental matrix has seven parameters, two for each epipole and three for the homography that relates the two image planes (the scale aspect is missing from the usual four parameters).

How OpenCV handles all of this

We can compute F, in a manner analogous to computing the image homography in the previous section, by providing a number of known correspondences. In this case, we don't even have to calibrate the cameras separately because we can solve directly for F, which contains implicitly the fundamental matrices for both cameras. The routine that does all of this for us is called cvFindFundamentalMat().

```
int cvFindFundamentalMat(
    const CvMat*    points1,
    const CvMat*    points2,
    CvMat*          fundamental_matrix,
    int             method        = CV_FM_RANSAC,
    double          param1        = 1.0,
    double          param2        = 0.99,
    CvMat*          status        = NULL
);
```

The first two arguments are N-by-2 or N-by-3† floating-point (single- or double-precision) matrices containing the corresponding N points that you have collected (they can also be N-by-1 multichannel matrices with two or three channels). The result is

---

* Note the equation that relates the fundamental matrix to the essential matrix. If we have rectified images and we normalize the points by dividing by the focal lengths, then the intrinsic matrix M becomes the identity matrix and $F = E$.

† You might be wondering what the N-by-3 or three-channel matrix is for. The algorithm will deal just fine with actual 3D points (x, y, z) measured on the calibration object. Three-dimensional points will end up being scaled to $(x/z, y/z)$, or you could enter 2D points in homogeneous coordinates $(x, y, 1)$, which will be treated in the same way. If you enter $(x, y, 0)$ then the algorithm will just ignore the 0. Using actual 3D points would be rare because usually you have only the 2D points detected on the calibration object.

fundamental_matrix, which should be a 3-by-3 matrix of the same precision as the points (in a special case the dimensions may be 9-by-3; see below).

The next argument determines the method to be used in computing the fundamental matrix from the corresponding points, and it can take one of four values. For each value there are particular restrictions on the number of points required (or allowed) in points1 and points2, as shown in Table 12-2.

Table 12-2. Restrictions on argument for method in cvFindFundamentalMat()

Value of method	Number of points	Algorithm
CV_FM_7POINT	$N = 7$	7-point algorithm
CV_FM_8POINT	$N \geq 8$	8-point algorithm
CV_FM_RANSAC	$N \geq 8$	RANSAC algorithm
CV_FM_LMEDS	$N \geq 8$	LMedS algorithm

The 7-point algorithm uses exactly seven points, and it uses the fact that the matrix F must be of rank 2 to fully constrain the matrix. The advantage of this constraint is that F is then always exactly of rank 2 and so cannot have one very small eigenvalue that is not quite 0. The disadvantage is that this constraint is not absolutely unique and so three different matrices might be returned (this is the case in which you should make fundamental_matrix a 9-by-3 matrix, so that all three returns can be accommodated). The 8-point algorithm just solves F as a linear system of equations. If more than eight points are provided then a linear least-squares error is minimized across all points. The problem with both the 7-point and 8-point algorithms is that they are extremely sensitive to outliers (even if you have many more than eight points in the 8-point algorithm). This is addressed by the RANSAC and LMedS algorithms, which are generally classified as robust methods because they have some capacity to recognize and remove outliers.*
For both methods, it is desirable to have many more than the minimal eight points.

The next two arguments are parameters used only by RANSAC and LMedS. The first, param1, is the maximum distance from a point to the epipolar line (in pixels) beyond which the point is considered an outlier. The second parameter, param2, is the desired confidence (between 0 and 1), which essentially tells the algorithm how many times to iterate.

The final argument, status, is optional; if used, it should be an N-by-1 matrix of type CV_8UC1, where N is the same as the length of points1 and points2. If this matrix is non-NULL, then RANSAC and LMedS will use it to store information about which points were ultimately considered outliers and which points were not. In particular, the appropriate

---

* The inner workings of RANSAC and LMedS are beyond the scope of this book, but the basic idea of RANSAC is to solve the problem many times using a random subset of the points and then take the particular solution closest to the average or the median solution. LMedS takes a subset of points, estimates a solution, then adds from the remaining points only those points that are "consistent" with that solution. You do this many times, take the set of points that fits the best, and throw away the others as "outliers". For more information, consult the original papers: Fischler and Bolles [Fischler81] for RANSAC; Rousseeuw [Rousseeuw84] for least median squares; and Inui, Kaneko, and Igarashi [Inui03] for line fitting using LMedS.

entry will be set to 0 if the point was decided to be an outlier and to 1 otherwise. For the other two methods, if this array is present then all values will be set to 1.

The return value of cvFindFundamentalMat() is an integer indicating the number of matrices found. It will be either 1 or 0 for all methods other than the 7-point algorithm, where it can also be 3. If the value is 0 then no matrix could be computed. The sample code from the OpenCV manual, shown in Example 12-2, makes this clear.

Example 12-2. Computing the fundamental matrix using RANSAC

```
int point_count = 100;
CvMat* points1;
CvMat* points2;
CvMat* status;
CvMat* fundamental_matrix;

points1 = cvCreateMat(1,point_count,CV_32FC2);
points2 = cvCreateMat(1,point_count,CV_32FC2);
status = cvCreateMat(1,point_count,CV_8UC1);

/* Fill the points here ... */
for( int i = 0; i < point_count; i++ )
{
    points1->data.fl[i*2]   = <x1,i>;  //These are points such as found
    points1->data.fl[i*2+1] = <y1,i>;  // on the chessboard calibration
    points2->data.fl[i*2]   = <x2,i>;  // pattern.
    points2->data.fl[i*2+1] = <y2,i>;
}

fundamental_matrix = cvCreateMat(3,3,CV_32FC1);
int fm_count = cvFindFundamentalMat( points1, points2,
                         fundamental_matrix,
                         CV_FM_RANSAC,1.0,0.99,status );
```

One word of warning—related to the possibility of returning 0—is that these algorithms can fail if the points supplied form *degenerate configurations*. These degenerate configurations arise when the points supplied provide less than the required amount of information, such as when one point appears more than once or when multiple points are collinear or coplanar with too many other points. It is important to always check the return value of cvFindFundamentalMat().

Computing Epipolar Lines

Now that we have the fundamental matrix, we want to be able to compute epipolar lines. The OpenCV function cvComputeCorrespondEpilines() computes, for a list of points in one image, the epipolar lines in the other image. Recall that, for any given point in one image, there is a different corresponding epipolar line in the other image. Each computed line is encoded in the form of a vector of three points (a, b, c) such that the epipolar line is defined by the equation:

$$ax + by + c = 0$$

To compute these epipolar lines, the function requires the fundamental matrix that we computed with cvFindFundamentalMat().

```
void cvComputeCorrespondEpilines(
  const CvMat* points,
  int          which_image,
  const CvMat* fundamental_matrix,
  CvMat*       correspondent_lines
);
```

Here the first argument, points, is the usual N-by-2 or N-by-3* array of points (which may be an N-by-1 multichannel array with two or three channels). The argument which_image must be either 1 or 2, and indicates which image the points are defined on (relative to the points1 and points2 arrays in cvFindFundamentalMat()), Of course, fundamental_matrix is the 3-by-3 matrix returned by cvFindFundamentalMat(). Finally, correspondent_lines is an N-by-3 array of floating-point numbers to which the result lines will be written. It is easy to see that the line equation $ax + by = c = 0$ is independent of the overall normalization of the parameters a, b, and c. By default they are normalized so that $a^2 + b^2 = 1$.

Stereo Calibration

We've built up a lot of theory and machinery behind cameras and 3D points that we can now put to use. This section will cover stereo calibration, and the next section will cover stereo rectification. *Stereo calibration* is the process of computing the geometrical relationship between the two cameras in space. In contrast, *stereo rectification* is the process of "correcting" the individual images so that they appear as if they had been taken by two cameras with row-aligned image planes (review Figures 12-4 and 12-7). With such a rectification, the optical axes (or principal rays) of the two cameras are parallel and so we say that they intersect at infinity. We could, of course, calibrate the two camera images to be in many other configurations, but here (and in OpenCV) we focus on the more common and simpler case of setting the principal rays to intersect at infinity.

Stereo calibration depends on finding the rotation matrix R and translation vector T between the two cameras, as depicted in Figure 12-9. Both R and T are calculated by the function cvStereoCalibrate(), which is similar to cvCalibrateCamera2() that we saw in Chapter 11 except that we now have two cameras and our new function can compute (or make use of any prior computation of) the camera, distortion, essential, or fundamental matrices. The other main difference between stereo and single-camera calibration is that, in cvCalibrateCamera2(), we ended up with a list of rotation and translation vectors between the camera and the chessboard views. In cvStereoCalibrate(), we seek a single rotation matrix and translation vector that relate the right camera to the left camera.

We've already shown how to compute the essential and fundamental matrices. But how do we compute R and T between the left and right cameras? For any given 3D point P in object coordinates, we can separately use single-camera calibration for the two cameras

* See the footnote on page 424.

to put P in the camera coordinates $P_l = R_l P + T_l$ and $P_r = R_r P + T_r$ for the left and right cameras, respectively. It is also evident from Figure 12-9 that the two views of P (from the two cameras) are related by $P_l = R^T (P_r - T)$,* where R and T are, respectively, the rotation matrix and translation vector between the cameras. Taking these three equations and solving for the rotation and translation separately yields the following simple relations:[†]

$$R = R_r (R_l)^T$$

$$T = T_r - RT_l$$

Given many joint views of chessboard corners, cvStereoCalibrate() uses cvCalibrate Camera2() to solve for rotation and translation parameters of the chessboard views for each camera separately (see the discussion in the "What's under the hood?" subsection of Chapter 11 to recall how this is done). It then plugs these left and right rotation and translation solutions into the equations just displayed to solve for the rotation and translation parameters between the two cameras. Because of image noise and rounding errors, each chessboard pair results in slightly different values for R and T. The cvStereoCalibrate() routine then takes the median values for the R and T parameters as the initial approximation of the true solution and then runs a robust Levenberg-Marquardt iterative algorithm to find the (local) minimum of the reprojection error of the chessboard corners for both camera views, and the solution for R and T is returned. To be clear on what stereo calibration gives you: the rotation matrix will put the right camera in the same plane as the left camera; this makes the two image planes coplanar but not row-aligned (we'll see how row-alignment is accomplished in the Stereo Rectification section below).

The function cvStereoCalibrate() has a lot of parameters, but they are all fairly straightforward and many are the same as for cvCalibrateCamera2() in Chapter 11.

```
bool cvStereoCalibrate(
    const CvMat*    objectPoints,
    const CvMat*    imagePoints1,
    const CvMat*    imagePoints2,
    const CvMat*    npoints,
    CvMat*          cameraMatrix1,
    CvMat*          distCoeffs1,
    CvMat*          cameraMatrix2,
    CvMat*          distCoeffs2,
    CvSize          imageSize,
    CvMat*          R,
    CvMat*          T,
    CvMat*          E,
    CvMat*          F,
```

---

* Let's be careful about what these terms mean: P_l and P_r denote the locations of the 3D point P from the coordinate system of the left and right cameras respectively; R_l and T_l (resp., R_r and T_r) denote the rotation and translation vectors from the camera to the 3D point for the left (resp. right) camera; and R and T are the rotation and translation that bring the right-camera coordinate system into the left.

† The left and right cameras can be reversed in these equations either by reversing the subscripts in both equations or by reversing the subscripts and dropping the transpose of R in the translation equation only.

```
        CvTermCriteria termCrit,
        int             flags=CV_CALIB_FIX_INTRINSIC
);
```

The first parameter, objectPoints, is an *N*-by-3 matrix containing the physical coordinates of each of the *K* points on each of the *M* images of the 3D object such that *N* = *K* × *M*. When using chessboards as the 3D object, these points are located in the coordinate frame attached to the object—setting, say, the upper left corner of the chessboard as the origin (and usually choosing the *Z*-coordinate of the points on the chessboard plane to be 0), but any known 3D points may be used as discussed with cvCalibrateCamera2().

We now have two cameras, denoted by "1" and "2" appended to the appropriate parameter names.[*] Thus we have imagePoints1 and imagePoints2, which are *N*-by-2 matrices containing the left and right pixel coordinates (respectively) of all of the object reference points supplied in objectPoints. If you performed calibration using a chessboard for the two cameras, then imagePoints1 and imagePoints2 are just the respective returned values for the *M* calls to cvFindChessboardCorners() for the left and right camera views.

The argument npoints contains the number of points in each image supplied as an *M*-by-1 matrix.

The parameters cameraMatrix1 and cameraMatrix2 are the 3-by-3 camera matrices, and distCoeffs1 and distCoeffs2 are the 5-by-1 distortion matrices for cameras 1 and 2, respectively. Remember that, in these matrices, the first two radial parameters come first; these are followed by the two tangential parameters and finally the third radial parameter (see the discussion in Chapter 11 on distortion coefficients). The third radial distortion parameter is last because it was added later in OpenCV's development; it is mainly used for wide-angle (fish-eye) camera lenses. The use of these camera intrinsics is controlled by the flags parameter. If flags is set to CV_CALIB_FIX_INTRINSIC, then these matrices are used as is in the calibration process. If flags is set to CV_CALIB_USE_INTRINSIC_GUESS, then these matrices are used as a starting point to optimize further the intrinsic and distortion parameters for each camera and will be set to the refined values on return from cvStereoCalibrate(). You may additively combine other settings of flags that have possible values that are exactly the same as for cvCalibrateCamera2(), in which case these parameters will be computed from scratch in cvStereoCalibrate(). That is, you can compute the intrinsic, extrinsic, and stereo parameters in a single pass using cvStereoCalibrate().[†]

The parameter imageSize is the image size in pixels. It is used only if you are refining or computing intrinsic parameters, as when flags is not equal to CV_CALIB_FIX_INTRINSIC.

---

[*] For simplicity, think of "1" as denoting the left camera and "2" as denoting the right camera. You can interchange these as long as you consistently treat the resulting rotation and translation solutions in the opposite fashion to the text discussion. The most important thing is to physically align the cameras so that their scan lines approximately match in order to achieve good calibration results.

[†] Be careful: Trying to solve for too many parameters at once will sometimes cause the solution to diverge to nonsense values. Solving systems of equations is something of an art, and you must verify your results. You can see some of these considerations in the calibration and rectification code example, where we check our calibration results by using the epipolar constraint.

The terms R and T are output parameters that are filled on function return with the rotation matrix and translation vector (relating the right camera to the left camera) that we seek. The parameters E and F are optional. If they are not set to NULL, then cvStereo Calibrate() will calculate and fill these 3-by-3 essential and fundamental matrices. We have seen termCrit many times before. It sets the internal optimization either to terminate after a certain number of iterations or to stop when the computed parameters change by less than the threshold indicated in the termCrit structure. A typical argument for this function is cvTermCriteria(CV_TERMCRIT_ITER + CV_TERMCRIT_EPS, 100, 1e-5).

Finally, we've already discussed the flags parameter somewhat. If you've calibrated both cameras and are sure of the result, then you can "hard set" the previous single-camera calibration results by using CV_CALIB_FIX_INTRINSIC. If you think the two cameras' initial calibrations were OK but not great, you can use it to refine the intrinsic and distortion parameters by setting flags to CV_CALIB_USE_INTRINSIC_GUESS. If the cameras have not been individually calibrated, you can use the same settings as we used for the flags parameter in cvCalibrateCamera2() in Chapter 11.

Once we have either the rotation and translation values (R, T) or the fundamental matrix F, we may use these results to rectify the two stereo images so that the epipolar lines are arranged along image rows and the scan lines are the same across both images. Although R and T don't define a unique stereo rectification, we'll see how to use these terms (together with other constraints) in the next section.

Stereo Rectification

It is easiest to compute the stereo disparity when the two image planes align exactly (as shown in Figure 12-4). Unfortunately, as discussed previously, a perfectly aligned configuration is rare with a real stereo system, since the two cameras almost never have exactly coplanar, row-aligned imaging planes. Figure 12-7 shows the goal of stereo rectification: We want to reproject the image planes of our two cameras so that they reside in the exact same plane, with image rows perfectly aligned into a frontal parallel configuration. How we choose the specific plane in which to mathematically align the cameras depends on the algorithm being used. In what follows we discuss two cases addressed by OpenCV.

We want the image rows between the two cameras to be aligned after rectification so that stereo correspondence (finding the same point in the two different camera views) will be more reliable and computationally tractable. Note that reliability and computational efficiency are both enhanced by having to search only one row for a match with a point in the other image. The result of aligning horizontal rows within a common image plane containing each image is that the epipoles themselves are then located at infinity. That is, the image of the center of projection in one image is parallel to the other image plane. But because there are an infinite number of possible frontal parallel planes to choose from, we will need to add more constraints. These include maximizing view overlap and/or minimizing distortion, choices that are made by the algorithms discussed in what follows.

The result of the process of aligning the two image planes will be eight terms, four each for the left and the right cameras. For each camera we'll get a distortion vector distCoeffs, a rotation matrix R_{rect} (to apply to the image), and the rectified and unrectified camera matrices (M_{rect} and M, respectively). From these terms, we can make a map, using cvInitUndistortRectifyMap() (to be discussed shortly), of where to interpolate pixels from the original image in order to create a new rectified image.*

There are many ways to compute our rectification terms, of which OpenCV implements two: (1) Hartley's algorithm [Hartley98], which can yield uncalibrated stereo using just the fundamental matrix; and (2) Bouguet's algorithm,[†] which uses the rotation and translation parameters from two calibrated cameras. Hartley's algorithm can be used to derive structure from motion recorded by a single camera but may (when stereo rectified) produce more distorted images than Bouguet's calibrated algorithm. In situations where you can employ calibration patterns—such as on a robot arm or for security camera installations—Bouguet's algorithm is the natural one to use.

Uncalibrated stereo rectification: Hartley's algorithm

Hartley's algorithm attempts to find homographies that map the epipoles to infinity while minimizing the computed disparities between the two stereo images; it does this simply by matching points between two image pairs. Thus, we bypass having to compute the camera intrinsics for the two cameras because such intrinsic information is implicitly contained in the point matches. Hence we need only compute the fundamental matrix, which can be obtained from any matched set of seven or more points between the two views of the scene via cvFindFundamentalMat() as already described. Alternatively, the fundamental matrix can be computed from cvStereoCalibrate().

The advantage of Hartley's algorithm is that online stereo calibration can be performed simply by observing points in the scene. The disadvantage is that we have no sense of image scale. For example, if we used a chessboard for generating point matches then we would not be able to tell if the chessboard were 100 meters on each side and far away or 100 centimeters on each side and nearby. Neither do we explicitly learn the intrinsic camera matrix, without which the cameras might have different focal lengths, skewed pixels, different centers of projection, and/or different principal points. As a result, we can determine 3D object reconstruction only up to a projective transform. What this means is that different scales or projections of an object can appear the same to us (i.e., the feature points have the same 2D coordinates even though the 3D objects differ). Both of these issues are illustrated in Figure 12-10.

---

* Stereo rectification of an image in OpenCV is possible only when the epipole is outside of the image rectangle. Hence this rectification algorithm may not work with stereo configurations that are characterized by either a very wide baseline or when the cameras point towards each other too much.

† The Bouguet algorithm is a completion and simplification of the method first presented by Tsai [Tsai87] and Zhang [Zhang99; Zhang00]. Jean-Yves Bouguet never published this algorithm beyond its well-known implementation in his Camera Calibration Toolbox Matlab.

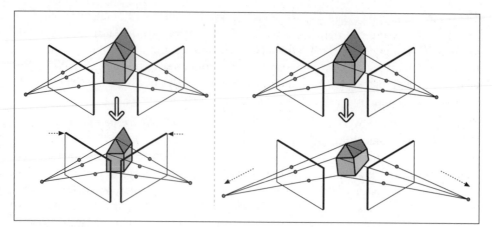

Figure 12-10. Stereo reconstruction ambiguity: if we do not know object size, then different size objects can appear the same depending on their distance from the camera (left); if we don't know the camera instrinsics, then different projections can appear the same—for example, by having different focal lengths and principal points

Assuming we have the fundamental matrix F, which required seven or more points to compute, Hartley's algorithm proceeds as follows (see Hartley's original paper [Hartley98] for more details).

1. We use the fundamental matrix to compute the two epipoles via the relations $Fe_l = 0$ and $(e_r)^T F = 0$ for the left and right epipoles, respectively.

2. We seek a first homography H_r, which will map the right epipole to the 2D homogeneous point at infinity $(1, 0, 0)^T$. Since a homography has seven constraints (scale is missing), and we use three to do the mapping to infinity, we have 4 degrees of freedom left in which to choose our H_r. These 4 degrees of freedom are mostly freedom to make a mess since most choices of H_r will result in highly distorted images. To find a good H_r, we choose a point in the image where we want minimal distortion to happen, allowing only rigid rotation and translation not shearing there. A reasonable choice for such a point is the image origin and we'll further assume that the epipole $(e_r)^T = (f, 0, 1)$ lies on the x-axis (a rotation matrix will accomplish this below). Given these coordinates, the matrix

$$G = \begin{pmatrix} 1 & 0 & 0 \\ 0 & 1 & 0 \\ -1/k & 0 & 1 \end{pmatrix}$$

will take such an epipole to infinity.

3. For a selected point of interest in the right image (we chose the origin), we compute the translation T that will take that point to the image origin (0 in our case) and the rotation R that will take the epipole to $(e_r)^T = (f, 0, 1)$. The homography we want will then be $H_r = GRT$.

4. We next search for a matching homography H_l that will send the left epipole to infinity and align the rows of the two images. Sending the left epipole to infinity is easily done by using up three constraints as in step 2. To align the rows, we just use the fact that aligning the rows minimizes the total distance between all matching points between the two images. That is, we find the H_l that minimizes the total disparity in left-right matching points $\sum_i d(H_l\, p_i^l, H_r p_i^r)$. These two homographies define the stereo rectification.

Although the details of this algorithm are a bit tricky, `cvStereoRectify Uncalibrated()` does all the hard work for us. The function is a bit misnamed because it does not rectify uncalibrated stereo images; rather, it computes homographies that may be used for rectification. The algorithm call is

```
int cvStereoRectifyUncalibrated(
    const CvMat* points1,
    const CvMat* points2,
    const CvMat* F,
        CvSize imageSize,
        CvMat* Hl,
        CvMat* Hr,
        double threshold
);
```

In `cvStereoRectifyUncalibrated()`, the algorithm takes as input an array of 2-by-K corresponding points between the left and right images in the arrays `points1` and `points2`. The fundamental matrix we calculated above is passed as the array F. We are familiar with `imageSize`, which just describes the width and height of the images that were used during calibration. Our return rectifying homographies are returned in the function variables `Hl` and `Hr`. Finally, if the distance from points to their corresponding epilines exceeds a set `threshold`, the corresponding point is eliminated by the algorithm.*

If our cameras have roughly the same parameters and are set up in an approximately horizontally aligned frontal parallel configuration, then our eventual rectified outputs from Hartley's algorithm will look very much like the calibrated case described next. If we know the size or the 3D geometry of objects in the scene, we can obtain the same results as the calibrated case.

Calibrated stereo rectification: Bouguet's algorithm

Given the rotation matrix and translation (R, T) between the stereo images, Bouguet's algorithm for stereo rectification simply attempts to minimize the amount of change reprojection produces for each of the two images (and thereby minimize the resulting reprojection distortions) while maximizing common viewing area.

To minimize image reprojection distortion, the rotation matrix R that rotates the right camera's image plane into the left camera's image plane is split in half between the two

---

* Hartley's algorithm works best for images that have been rectified previously by single-camera calibration. It won't work at all for images with high distortion. It is rather ironic that our "calibration-free" routine works only for undistorted image inputs whose parameters are typically derived from prior calibration. For another uncalibrated 3D approach, see Pollefeys [Pollefeys99a].

cameras; we call the two resulting rotation matrixes r_l and r_r for the left and right camera, respectively. Each camera rotates half a rotation, so their principal rays each end up parallel to the vector sum of where their original principal rays had been pointing. As we have noted, such a rotation puts the cameras into coplanar alignment but not into row alignment. To compute the R_{rect} that will take the left camera's epipole to infinity and align the epipolar lines horizontally, we create a rotation matrix by starting with the direction of the epipole e_1 itself. Taking the principal point (c_x, c_y) as the left image's origin, the (unit normalized) direction of the epipole is directly along the translation vector between the two cameras' centers of projection:

$$e_1 = \frac{T}{\|T\|}$$

The next vector, e_2, must be orthogonal to e_1 but is otherwise unconstrained. For e_2, choosing a direction orthogonal to the principal ray (which will tend to be along the image plane) is a good choice. This is accomplished by using the cross product of e_1 with the direction of the principal ray and then normalizing so that we've got another unit vector:

$$e_2 = \frac{[-T_y\ T_x\ 0]^T}{\sqrt{T_x^2 + T_y^2}}$$

The third vector is just orthogonal to e_1 and e_2; it can be found using the cross product:

$$e_3 = e_1 \times e_2$$

Our matrix that takes the epipole in the left camera to infinity is then:

$$R_{rect} = \begin{bmatrix} (e_1)^T \\ (e_2)^T \\ (e_3)^T \end{bmatrix}$$

This matrix rotates the left camera about the center of projection so that the epipolar lines become horizontal and the epipoles are at infinity. The row alignment of the two cameras is then achieved by setting:

$$R_l = R_{rect}r_l$$

$$R_r = R_{rect}r_r$$

We will also compute the rectified left and right camera matrices M_{rect_l} and M_{rect_r} but return them combined with projection matrices P_l and P_r:

$$P_l = M_{rect\_l}P_l' = \begin{bmatrix} f_{x\_l} & \alpha_l & c_{x\_l} \\ 0 & f_{y\_l} & c_{y\_l} \\ 0 & 0 & 1 \end{bmatrix}\begin{bmatrix} 1 & 0 & 0 & 0 \\ 0 & 1 & 0 & 0 \\ 0 & 0 & 1 & 0 \end{bmatrix}$$

and

$$P_r = M_{rect\_r} P'_r = \begin{bmatrix} f_{x\_r} & \alpha_r & c_{x\_r} \\ 0 & f_{y\_r} & c_{y\_r} \\ 0 & 0 & 1 \end{bmatrix} \begin{bmatrix} 1 & 0 & 0 & T_x \\ 0 & 1 & 0 & 0 \\ 0 & 0 & 1 & 0 \end{bmatrix}$$

(here α_l and α_r allow for a pixel skew factor that in modern cameras is almost always 0). The projection matrices take a 3D point in homogeneous coordinates to a 2D point in homogeneous coordinates as follows:

$$P \begin{bmatrix} X \\ Y \\ Z \\ 1 \end{bmatrix} = \begin{bmatrix} x \\ y \\ w \end{bmatrix}$$

where the screen coordinates can be calculated as $(x/w, y/w)$. Points in two dimensions can also then be reprojected into three dimensions given their screen coordinates and the camera intrinsics matrix. The reprojection matrix is:

$$Q = \begin{bmatrix} 1 & 0 & 0 & -c_x \\ 0 & 1 & 0 & -c_y \\ 0 & 0 & 0 & f \\ 0 & 0 & -1/T_x & (c_x - c'_x)/T_x \end{bmatrix}$$

Here the parameters are from the left image except for c'_x, which is the principal point x coordinate in the right image. If the principal rays intersect at infinity, then $c_x = c'_x$ and the term in the lower right corner is 0. Given a two-dimensional homogeneous point and its associated disparity d, we can project the point into three dimensions using:

$$Q \begin{bmatrix} x \\ y \\ d \\ 1 \end{bmatrix} = \begin{bmatrix} X \\ Y \\ Z \\ W \end{bmatrix}$$

The 3D coordinates are then $(X/W, Y/W, Z/W)$.

Applying the Bouguet rectification method just described yields our ideal stereo configuration as per Figure 12-4. New image centers and new image bounds are then chosen for the rotated images so as to maximize the overlapping viewing area. Mainly this just sets a uniform camera center and a common maximal height and width of the two image areas as the new stereo viewing planes.

```
void cvStereoRectify(
    const CvMat* cameraMatrix1,
    const CvMat* cameraMatrix2,
```

```
        const CvMat* distCoeffs1,
        const CvMat* distCoeffs2,
              CvSize imageSize,
        const CvMat* R,
        const CvMat* T,
              CvMat* Rl,
              CvMat* Rr,
              CvMat* Pl,
              CvMat* Pr,
              CvMat* Q=0,
              int    flags=CV_CALIB_ZERO_DISPARITY
);
```

For cvStereoRectify(),* we input the familiar original camera matrices and distortion vectors returned by cvStereoCalibrate(). These are followed by imageSize, the size of the chessboard images used to perform the calibration. We also pass in the rotation matrix R and translation vector T between the right and left cameras that was also returned by cvStereoCalibrate().

Return parameters are Rl and Rr, the 3-by-3 row-aligned rectification rotations for the left and right image planes as derived in the preceding equations. Similarly, we get back the 3-by-4 left and right projection equations Pl and Pr. An optional return parameter is Q, the 4-by-4 reprojection matrix described previously.

The flags parameter is defaulted to set disparity at infinity, the normal case as per Figure 12-4. Unsetting flags means that we want the cameras verging toward each other (i.e., slightly "cross-eyed") so that zero disparity occurs at a finite distance (this might be necessary for greater depth resolution in the proximity of that particular distance).

If the flags parameter was not set to CV_CALIB_ZERO_DISPARITY, then we must be more careful about how we achieve our rectified system. Recall that we rectified our system relative to the principal points (c_x, c_y) in the left and right cameras. Thus, our measurements in Figure 12-4 must also be relative to these positions. Basically, we have to modify the distances so that $\tilde{x}^r = x^r - c_x^{\text{right}}$ and $\tilde{x}^l = x^l - c_x^{\text{left}}$. When disparity has been set to infinity, we have $c_x^{\text{left}} = c_x^{\text{right}}$ (i.e., when CV_CALIB_ZERO_DISPARITY is passed to cvStereoRectify()), and we can pass plain pixel coordinates (or disparity) to the formula for depth. But if cvStereoRectify() is called without CV_CALIB_ZERO_DISPARITY then $c_x^{\text{left}} \neq c_x^{\text{right}}$ in general. Therefore, even though the formula $Z = fT/(x^l - x^r)$ remains the same, one should keep in mind that x^l and x^r are not counted from the image center but rather from the respective principal points c_x^{left} and c_x^{right}, which could differ from x^l and x^r. Hence, if you computed disparity $d = x^l - x^r$ then it should be adjusted before computing Z: $Z fT/(d - (c_x^{\text{left}} - c_x^{\text{right}}))$.

Rectification map

Once we have our stereo calibration terms, we can pre-compute left and right rectification lookup maps for the left and right camera views using separate calls to cvInitUndistort

---

* Again, cvStereoRectify() is a bit of a misnomer because the function computes the terms that we can use for rectification but doesn't actually rectify the stereo images.

RectifyMap(). As with any image-to-image mapping function, a forward mapping (in which we just compute where pixels go from the source image to the destination image) will not, owing to floating-point destination locations, hit all the pixel locations in the destination image, which thus will look like Swiss cheese. So instead we work backward: for each integer pixel location in the destination image, we look up what floating-point coordinate it came from in the source image and then interpolate from its surrounding source pixels a value to use in that integer destination location. This source lookup typically uses bilinear interpolation, which we encountered with cvRemap() in Chapter 6.

The process of rectification is illustrated in Figure 12-11. As shown by the equation flow in that figure, the actual rectification process proceeds backward from (c) to (a) in a process known as reverse mapping. For each integer pixel in the rectified image (c), we find its coordinates in the undistorted image (b) and use those to look up the actual (floating-point) coordinates in the raw image (a). The floating-point coordinate pixel value is then interpolated from the nearby integer pixel locations in the original source image, and that value is used to fill in the rectified integer pixel location in the destination image (c). After the rectified image is filled in, it is typically cropped to emphasize the overlapping areas between the left and right images.

The function that implements the math depicted in Figure 12-11 is called cvInitUndistort RectifyMap(). We call this function twice, once for the left and once for the right image of stereo pair.

```
void cvInitUndistortRectifyMap(
    const CvMat* M,
    const CvMat* distCoeffs,
    const CvMat* Rrect,
    const CvMat* Mrect,
    CvArr*       mapx,
    CvArr*       mapy
);
```

The cvInitUndistortRectifyMap() function takes as input the 3-by-3 camera matrix M, the rectified 3-by-3 camera matrix Mrect, the 3-by-3 rotation matrix Rrect, and the 5-by-1 camera distortion parameters in distCoeffs.

If we calibrated our stereo cameras using cvStereoRectify(), then we can read our input to cvInitUndistortRectifyMap() straight out of cvStereoRectify() using first the left parameters to rectify the left camera and then the right parameters to rectify the right camera. For Rrect, use Rl or Rr from cvStereoRectify(); for M, use cameraMatrix1 or cameraMatrix2. For Mrect we could use the first three columns of the 3-by-4 Pl or Pr from cvStereoRectify(), but as a convenience the function allows us to pass Pl or Pr directly and it will read Mrect from them.

If, on the other hand, we used cvStereoRectifyUncalibrated() to calibrate our stereo cameras, then we must preprocess the homography a bit. Although we could—in principle and in practice—rectify stereo without using the camera intrinsics, OpenCV does not have a function for doing this directly. If we do not have Mrect from some prior calibration, the proper procedure is to set Mrect equal to M. Then, for Rrect in

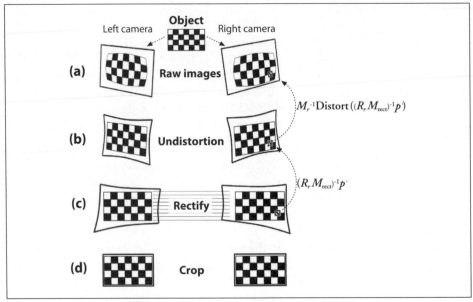

Figure 12-11. Stereo rectification: for the left and right camera, the raw image (a) is undistorted (b) and rectified (c) and finally cropped (d) to focus on overlapping areas between the two cameras; the rectification computation actually works backward from (c) to (a)

cvInitUndistortRectifyMap(), we need to compute $R_{\text{rect\_}l} = M_{\text{rect\_}l}^{-1} H_l M_l$ (or just $R_{\text{rect\_}l} = M_l^{-1} H_l M_l$ if $M_{\text{rect\_}l}^{-1}$ is unavailable) and $R_{\text{rect\_}r} = M_{\text{rect\_}r}^{-1} H_r M_r$ (or just $R_{\text{rect\_}r} = M_r^{-1} H_r M_r$ if $M_{\text{rect\_}r}^{-1}$ is unavailable) for the left and the right rectification, respectively. Finally, we will also need the distortion coefficients for each camera to fill in the 5-by-1 distCoeffs parameters.

The function cvInitUndistortRectifyMap() returns lookup maps mapx and mapy as output. These maps indicate from where we should interpolate source pixels for each pixel of the destination image; the maps can then be plugged directly into cvRemap(), a function we first saw in Chapter 6. As we mentioned, the function cvInitUndistortRectifyMap() is called separately for the left and the right cameras so that we can obtain their distinct mapx and mapy remapping parameters. The function cvRemap() may then be called, using the left and then the right maps each time we have new left and right stereo images to rectify. Figure 12-12 shows the results of stereo undistortion and rectification of a stereo pair of images. Note how feature points become horizontally aligned in the undistorted rectified images.

Stereo Correspondence

Stereo correspondence—matching a 3D point in the two different camera views—can be computed only over the visual areas in which the views of the two cameras overlap. Once again, this is one reason why you will tend to get better results if you arrange your cameras to be as nearly frontal parallel as possible (at least until you become expert at stereo vision). Then, once we know the physical coordinates of the cameras or the sizes

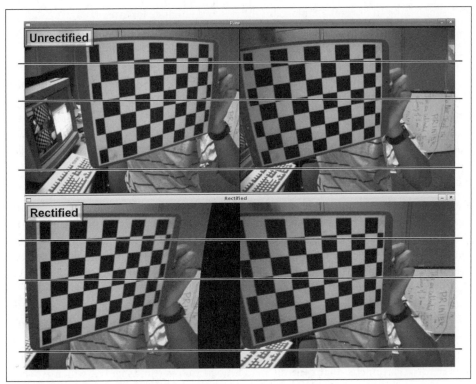

Figure 12-12. Stereo rectification: original left and right image pair (upper panels) and the stereo rectified left and right image pair (lower panels); note that the barrel distortion (in top of chessboard patterns) has been corrected and the scan lines are aligned in the rectified images

of objects in the scene, we can derive depth measurements from the triangulated disparity measures $d = x^l - x^r$ (or $d = x^l - x^r - (c_x^{\text{left}} - c_x^{\text{right}})$ if the principal rays intersect at a finite distance) between the corresponding points in the two different camera views. Without such physical information, we can compute depth only up to a scale factor. If we don't have the camera instrinsics, as when using Hartley's algorithm, we can compute point locations only up to a projective transform (review Figure 12-10).

OpenCV implements a fast and effective block-matching stereo algorithm, cvFindStereo CorrespondenceBM(), that is similar to the one developed by Kurt Konolige [Konolige97]; it works by using small "sum of absolute difference" (SAD) windows to find matching points between the left and right stereo rectified images.* This algorithm finds only strongly matching (high-texture) points between the two images. Thus, in a highly textured scene such as might occur outdoors in a forest, every pixel might have computed depth. In a very low-textured scene, such as an indoor hallway, very few points might register depth. There are three stages to the block-matching stereo correspondence algorithm, which works on undistorted, rectified stereo image pairs:

---

* This algorithm is available in an FPGA stereo hardware system from Videre (see [Videre]).

1. Prefiltering to normalize image brightness and enhance texture.

2. Correspondence search along horizontal epipolar lines using an SAD window.

3. Postfiltering to eliminate bad correspondence matches.

In the prefiltering step, the input images are normalized to reduce lighting differences and to enhance image texture. This is done by running a window—of size 5-by-5, 7-by-7 (the default), . . ., 21-by-21 (the maximum)—over the image. The center pixel I_c under the window is replaced by $\min[\max(I_c - \overline{I}, -I_{cap}), I_{cap}]$, where $\overline{I}$ is the average value in the window and I_{cap} is a positive numeric limit whose default value is 30. This method is invoked by a CV_NORMALIZED_RESPONSE flag. The other possible flag is CV_LAPLACIAN_OF_GAUSSIAN, which runs a peak detector over a smoothed version of the image.

Correspondence is computed by a sliding SAD window. For each feature in the left image, we search the corresponding row in the right image for a best match. After rectification, each row is an epipolar line, so the matching location in the right image must be along the same row (same y-coordinate) as in the left image; this matching location can be found if the feature has enough texture to be detectable and if it is not occluded in the right camera's view (see Figure 12-16). If the left feature pixel coordinate is at (x_0, y_0) then, for a horizontal frontal parallel camera arrangement, the match (if any) must be found on the same row and at, or to the left of, x_0; see Figure 12-13. For frontal parallel cameras, x_0 is at zero disparity and larger disparities are to the left. For cameras that are angled toward each other, the match may occur at negative disparities (to the right of x_0). The first parameter that controls matching search is minDisparity, which is where the matching search should start. The default for minDisparity is 0. The disparity search is then carried out over numberOfDisparities counted in pixels (the default is 64 pixels). Disparities have discrete, subpixel resolution that is set by the parameter subPixelDisparities (the default is 16 subdisparities per pixel). Reducing the number of disparities to be searched can help cut down computation time by limiting the length of a search for a matching point along an epipolar line. Remember that large disparities represent closer distances.

Setting the minimum disparity and the number of disparities to be searched establishes the *horopter*, the 3D volume that is covered by the search range of the stereo algorithm. Figure 12-14 shows disparity search limits of five pixels starting at three different disparity limits: 20, 17, and 16. Each disparity limit defines a plane at a fixed depth from the cameras (see Figure 12-15). As shown in Figure 12-14, each disparity limit—together with the number of disparities—sets a different horopter at which depth can be detected. Outside of this range, depth will not be found and will represent a "hole" in the depth map where depth is not known. Horopters can be made larger by decreasing the baseline distance T between the cameras, by making the focal length smaller, by increasing the stereo disparity search range, or by increasing the pixel width.

Correspondence within the horopter has one in-built constraint, called the *order constraint*, which simply states that the order of the features cannot change from the left view to the right. There may be *missing* features—where, owing to occlusion and noise,

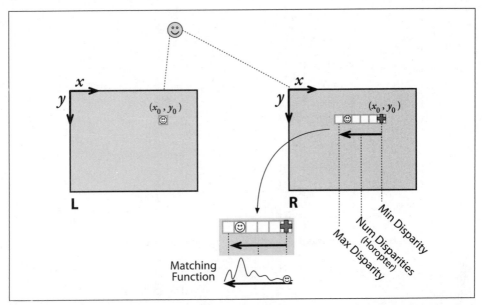

Figure 12-13. Any right-image match of a left-image feature must occur on the same row and at (or to the left of) the same coordinate point, where the match search starts at the minDisparity point (here, 0) and moves to the left for the set number of disparities; the characteristic matching function of window-based feature matching is shown in the lower part of the figure

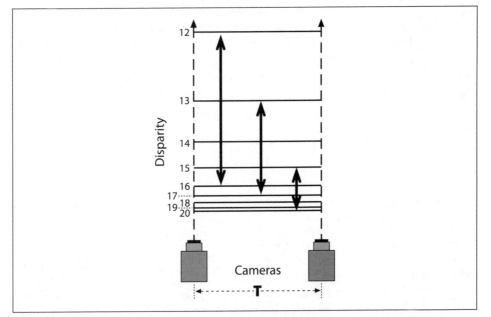

Figure 12-14. Each line represents a plane of constant disparity in integer pixels from 20 to 12; a disparity search range of five pixels will cover different horopter ranges, as shown by the vertical arrows, and different maximal disparity limits establish different horopters

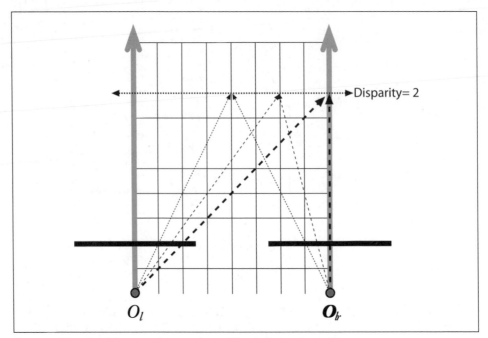

Figure 12-15. A fixed disparity forms a plane of fixed distance from the cameras

some features found on the left cannot be found on the right—but the ordering of those features that are found remains the same. Similarly, there may be many features on the right that were not identified on the left (these are called *insertions*), but insertions do not change the *order* of features although they may spread those features out. The procedure illustrated in Figure 12-16 reflects the ordering constraint when matching features on a horizontal scan line.

Given the smallest allowed disparity increment Δd, we can determine smallest achievable depth range resolution ΔZ by using the formula:

$$\Delta Z = \frac{Z^2}{fT} \Delta d$$

It is useful to keep this formula in mind so that you know what kind of depth resolution to expect from your stereo rig.

After correspondence, we turn to postfiltering. The lower part of Figure 12-13 shows a typical matching function response as a feature is "swept" from the minimum disparity out to maximum disparity. Note that matches often have the characteristic of a strong central peak surrounded by side lobes. Once we have candidate feature correspondences between the two views, postfiltering is used to prevent false matches. OpenCV makes use of the matching function pattern via a uniquenessRatio parameter (whose default value is 12) that filters out matches, where uniquenessRatio > (match_val–min_match)/ min_match.

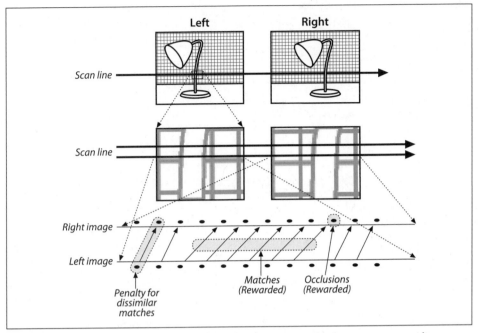

Figure 12-16. Stereo correspondence starts by assigning point matches between corresponding rows in the left and right images: left and right images of a lamp (upper panel); an enlargement of a single scan line (middle panel); visualization of the correspondences assigned (lower panel).

To make sure that there is enough texture to overcome random noise during matching, OpenCV also employs a textureThreshold. This is just a limit on the SAD window response such that no match is considered whose response is below the textureThreshold (the default value is 12). Finally, block-based matching has problems near the boundaries of objects because the matching window catches the foreground on one side and the background on the other side. This results in a local region of large and small disparities that we call *speckle*. To prevent these borderline matches, we can set a speckle detector over a speckle window (ranging in size from 5-by-5 up to 21-by-21) by setting speckleWindowSize, which has a default setting of 9 for a 9-by-9 window. Within the speckle window, as long as the minimum and maximum detected disparities are within speckleRange, the match is allowed (the default range is set to 4).

Stereo vision is becoming crucial to surveillance systems, navigation, and robotics, and such systems can have demanding real-time performance requirements. Thus, the stereo correspondence routines are designed to run fast. Therefore, we can't keep allocating all the internal scratch buffers that the correspondence routine needs each time we call cvFindStereoCorrespondenceBM().

The block-matching parameters and the internal scratch buffers are kept in a data structure named CvStereoBMState:

```
typedef struct CvStereoBMState {
  //pre filters (normalize input images):
```

```
    int        preFilterType;
    int        preFilterSize;//for 5x5 up to 21x21
    int        preFilterCap;
  //correspondence using Sum of Absolute Difference (SAD):
    int        SADWindowSize; // Could be 5x5,7x7, ..., 21x21
    int        minDisparity;
    int        numberOfDisparities;//Number of pixels to search
  //post filters (knock out bad matches):
    int        textureThreshold; //minimum allowed
    float      uniquenessRatio;// Filter out if:
                               // [ match_val - min_match <
                               // uniqRatio*min_match ]
                               // over the corr window area
    int        speckleWindowSize;//Disparity variation window
    int        speckleRange;//Acceptable range of variation in window
  // temporary buffers
    CvMat* preFilteredImg0;
    CvMat* preFilteredImg1;
    CvMat* slidingSumBuf;
} CvStereoBMState;
```

The state structure is allocated and returned by the function cvCreateStereoBMState().
This function takes the parameter preset, which can be set to any one of the following.

CV_STEREO_BM_BASIC
    Sets all parameters to their default values

CV_STEREO_BM_FISH_EYE
    Sets parameters for dealing with wide-angle lenses

CV_STEREO_BM_NARROW
    Sets parameters for stereo cameras with narrow field of view

This function also takes the optional parameter numberOfDisparities; if nonzero, it
overrides the default value from the preset. Here is the specification:

```
CvStereoBMState* cvCreateStereoBMState(
    int presetFlag=CV_STEREO_BM_BASIC,
    int numberOfDisparities=0
);
```

The state structure, CvStereoBMState{}, is released by calling

```
void cvReleaseBMState(
    CvStereoBMState **BMState
);
```

Any stereo correspondence parameters can be adjusted at any time between cvFindStereo
CorrespondenceBM calls by directly assigning new values of the state structure fields. The
correspondence function will take care of allocating/reallocating the internal buffers as
needed.

Finally, cvFindStereoCorrespondenceBM() takes in rectified image pairs and outputs a
disparity map given its state structure:

```
void cvFindStereoCorrespondenceBM(
    const CvArr      *leftImage,
```

```
    const CvArr      *rightImage,
    CvArr            *disparityResult,
    CvStereoBMState *BMState
);
```

Stereo Calibration, Rectification, and Correspondence Code

Let's put this all together with code in an example program that will read in a number of chessboard patterns from a file called *list.txt*. This file contains a list of alternating left and right stereo (chessboard) image pairs, which are used to calibrate the cameras and then rectify the images. Note once again that we're assuming you've arranged the cameras so that their image scan lines are roughly physically aligned and such that each camera has essentially the same field of view. This will help avoid the problem of the epipole being within the image* and will also tend to maximize the area of stereo overlap while minimizing the distortion from reprojection.

In the code (Example 12-3), we first read in the left and right image pairs, find the chessboard corners to subpixel accuracy, and set object and image points for the images where all the chessboards could be found. This process may optionally be displayed. Given this list of found points on the found good chessboard images, the code calls cvStereoCalibrate() to calibrate the camera. This calibration gives us the camera matrix _M and the distortion vector _D for the two cameras; it also yields the rotation matrix _R, the translation vector _T, the essential matrix _E, and the fundamental matrix _F.

Next comes a little interlude where the accuracy of calibration is assessed by checking how nearly the points in one image lie on the epipolar lines of the other image. To do this, we undistort the original points using cvUndistortPoints() (see Chapter 11), compute the epilines using cvComputeCorrespondEpilines(), and then compute the dot product of the points with the lines (in the ideal case, these dot products would all be 0). The accumulated absolute distance forms the error.

The code then optionally moves on to computing the rectification maps using the uncalibrated (Hartley) method cvStereoRectifyUncalibrated() or the calibrated (Bouguet) method cvStereoRectify(). If uncalibrated rectification is used, the code further allows for either computing the needed fundamental matrix from scratch or for just using the fundamental matrix from the stereo calibration. The rectified images are then computed using cvRemap(). In our example, lines are drawn across the image pairs to aid in seeing how well the rectified images are aligned. An example result is shown in Figure 12-12, where we can see that the barrel distortion in the original images is largely corrected from top to bottom and that the images are aligned by horizontal scan lines.

Finally, if we rectified the images then we initialize the block-matching state (internal allocations and parameters) using cvCreateBMState(). We can then compute the disparity maps by using cvFindStereoCorrespondenceBM(). Our code example allows you to use either horizontally aligned (left-right) or vertically aligned (top-bottom) cameras; note,

---

* OpenCV does not (yet) deal with the case of rectifying stereo images when the epipole is within the image frame. See, for example, Pollefeys, Koch, and Gool [Pollefeys99b] for a discussion of this case.

however, that for the vertically aligned case the function cvFindStereoCorrespondenceBM()
can compute disparity only for the case of uncalibrated rectification unless you add
code to transpose the images yourself. For horizontal camera arrangements, cvFind
StereoCorrespondenceBM() can find disparity for calibrated or for uncalibrated rectified
stereo image pairs. (See Figure 12-17 in the next section for example disparity results.)

Example 12-3. Stereo calibration, rectification, and correspondence

```
#include "cv.h"
#include "cxmisc.h"
#include "highgui.h"
#include "cvaux.h"
#include <vector>
#include <string>
#include <algorithm>
#include <stdio.h>
#include <ctype.h>

using namespace std;

//
// Given a list of chessboard images, the number of corners (nx, ny)
// on the chessboards, and a flag called useCalibrated (0 for Hartley
// or 1 for Bouguet stereo methods). Calibrate the cameras and display the
// rectified results along with the computed disparity images.
//
static void
StereoCalib(const char* imageList, int nx, int ny, int useUncalibrated)
{
    int displayCorners = 0;
    int showUndistorted = 1;
    bool isVerticalStereo = false;//OpenCV can handle left-right
                                  //or up-down camera arrangements
    const int maxScale = 1;
    const float squareSize = 1.f; //Set this to your actual square size
    FILE* f = fopen(imageList, "rt");
    int i, j, lr, nframes, n = nx*ny, N = 0;
    vector<string> imageNames[2];
    vector<CvPoint3D32f> objectPoints;
    vector<CvPoint2D32f> points[2];
    vector<int> npoints;
    vector<uchar> active[2];
    vector<CvPoint2D32f> temp(n);
    CvSize imageSize = {0,0};
    // ARRAY AND VECTOR STORAGE:
    double M1[3][3], M2[3][3], D1[5], D2[5];
    double R[3][3], T[3], E[3][3], F[3][3];
    CvMat _M1 = cvMat(3, 3, CV_64F, M1 );
    CvMat _M2 = cvMat(3, 3, CV_64F, M2 );
    CvMat _D1 = cvMat(1, 5, CV_64F, D1 );
    CvMat _D2 = cvMat(1, 5, CV_64F, D2 );
    CvMat _R = cvMat(3, 3, CV_64F, R );
    CvMat _T = cvMat(3, 1, CV_64F, T );
    CvMat _E = cvMat(3, 3, CV_64F, E );
```

Example 12-3. Stereo calibration, rectification, and correspondence (continued)

```
    CvMat _F = cvMat(3, 3, CV_64F, F );
    if( displayCorners )
        cvNamedWindow( "corners", 1 );
// READ IN THE LIST OF CHESSBOARDS:
    if( !f )
    {
        fprintf(stderr, "can not open file %s\n", imageList );
        return;
    }
    for(i=0;;i++)
    {
        char buf[1024];
        int count = 0, result=0;
        lr = i % 2;
        vector<CvPoint2D32f>& pts = points[lr];
        if( !fgets( buf, sizeof(buf)-3, f ))
            break;
        size_t len = strlen(buf);
        while( len > 0 && isspace(buf[len-1]))
            buf[--len] = '\0';
        if( buf[0] == '#')
            continue;
        IplImage* img = cvLoadImage( buf, 0 );
        if( !img )
            break;
        imageSize = cvGetSize(img);
        imageNames[lr].push_back(buf);
//FIND CHESSBOARDS AND CORNERS THEREIN:
        for( int s = 1; s <= maxScale; s++ )
        {
            IplImage* timg = img;
            if( s > 1 )
            {
                timg = cvCreateImage(cvSize(img->width*s,img->height*s),
                    img->depth, img->nChannels );
                cvResize( img, timg, CV_INTER_CUBIC );
            }
            result = cvFindChessboardCorners( timg, cvSize(nx, ny),
                &temp[0], &count,
                CV_CALIB_CB_ADAPTIVE_THRESH |
                CV_CALIB_CB_NORMALIZE_IMAGE);
            if( timg != img )
                cvReleaseImage( &timg );
            if( result || s == maxScale )
                for( j = 0; j < count; j++ )
                {
                    temp[j].x /= s;
                    temp[j].y /= s;
                }
            if( result )
                break;
        }
        if( displayCorners )
```

Example 12-3. Stereo calibration, rectification, and correspondence (continued)

```
        {
            printf("%s\n", buf);
            IplImage* cimg = cvCreateImage( imageSize, 8, 3 );
            cvCvtColor( img, cimg, CV_GRAY2BGR );
            cvDrawChessboardCorners( cimg, cvSize(nx, ny), &temp[0],
                count, result );
            cvShowImage( "corners", cimg );
            cvReleaseImage( &cimg );
            if( cvWaitKey(0) == 27 ) //Allow ESC to quit
                exit(-1);
        }
        else
            putchar('.');
        N = pts.size();
        pts.resize(N + n, cvPoint2D32f(0,0));
        active[lr].push_back((uchar)result);
    //assert( result != 0 );
        if( result )
        {
         //Calibration will suffer without subpixel interpolation
            cvFindCornerSubPix( img, &temp[0], count,
                cvSize(11, 11), cvSize(-1,-1),
                cvTermCriteria(CV_TERMCRIT_ITER+CV_TERMCRIT_EPS,
                30, 0.01) );
            copy( temp.begin(), temp.end(), pts.begin() + N );
        }
        cvReleaseImage( &img );
    }
    fclose(f);
    printf("\n");
// HARVEST CHESSBOARD 3D OBJECT POINT LIST:
    nframes = active[0].size();//Number of good chessboads found
    objectPoints.resize(nframes*n);
    for( i = 0; i < ny; i++ )
        for( j = 0; j < nx; j++ )
        objectPoints[i*nx + j] =
        cvPoint3D32f(i*squareSize, j*squareSize, 0);
    for( i = 1; i < nframes; i++ )
        copy( objectPoints.begin(), objectPoints.begin() + n,
        objectPoints.begin() + i*n );
    npoints.resize(nframes,n);
    N = nframes*n;
    CvMat _objectPoints = cvMat(1, N, CV_32FC3, &objectPoints[0] );
    CvMat _imagePoints1 = cvMat(1, N, CV_32FC2, &points[0][0] );
    CvMat _imagePoints2 = cvMat(1, N, CV_32FC2, &points[1][0] );
    CvMat _npoints = cvMat(1, npoints.size(), CV_32S, &npoints[0] );
    cvSetIdentity(&_M1);
    cvSetIdentity(&_M2);
    cvZero(&_D1);
    cvZero(&_D2);

// CALIBRATE THE STEREO CAMERAS
    printf("Running stereo calibration ...");
```

Example 12-3. Stereo calibration, rectification, and correspondence (continued)

```
    fflush(stdout);
    cvStereoCalibrate( &_objectPoints, &_imagePoints1,
        &_imagePoints2, &_npoints,
        &_M1, &_D1, &_M2, &_D2,
        imageSize, &_R, &_T, &_E, &_F,
        cvTermCriteria(CV_TERMCRIT_ITER+
        CV_TERMCRIT_EPS, 100, 1e-5),
        CV_CALIB_FIX_ASPECT_RATIO +
        CV_CALIB_ZERO_TANGENT_DIST +
        CV_CALIB_SAME_FOCAL_LENGTH );
    printf(" done\n");
// CALIBRATION QUALITY CHECK
// because the output fundamental matrix implicitly
// includes all the output information,
// we can check the quality of calibration using the
// epipolar geometry constraint: m2^t*F*m1=0
    vector<CvPoint3D32f> lines[2];
    points[0].resize(N);
    points[1].resize(N);
    _imagePoints1 = cvMat(1, N, CV_32FC2, &points[0][0] );
    _imagePoints2 = cvMat(1, N, CV_32FC2, &points[1][0] );
    lines[0].resize(N);
    lines[1].resize(N);
    CvMat _L1 = cvMat(1, N, CV_32FC3, &lines[0][0]);
    CvMat _L2 = cvMat(1, N, CV_32FC3, &lines[1][0]);
//Always work in undistorted space
    cvUndistortPoints( &_imagePoints1, &_imagePoints1,
        &_M1, &_D1, 0, &_M1 );
    cvUndistortPoints( &_imagePoints2, &_imagePoints2,
        &_M2, &_D2, 0, &_M2 );
    cvComputeCorrespondEpilines( &_imagePoints1, 1, &_F, &_L1 );
    cvComputeCorrespondEpilines( &_imagePoints2, 2, &_F, &_L2 );
    double avgErr = 0;
    for( i = 0; i < N; i++ )
    {
        double err = fabs(points[0][i].x*lines[1][i].x +
            points[0][i].y*lines[1][i].y + lines[1][i].z)
            + fabs(points[1][i].x*lines[0][i].x +
            points[1][i].y*lines[0][i].y + lines[0][i].z);
        avgErr += err;
    }
    printf( "avg err = %g\n", avgErr/(nframes*n) );
//COMPUTE AND DISPLAY RECTIFICATION
    if( showUndistorted )
    {
        CvMat* mx1 = cvCreateMat( imageSize.height,
            imageSize.width, CV_32F );
        CvMat* my1 = cvCreateMat( imageSize.height,
            imageSize.width, CV_32F );
        CvMat* mx2 = cvCreateMat( imageSize.height,
            imageSize.width, CV_32F );
        CvMat* my2 = cvCreateMat( imageSize.height,
```

Example 12-3. Stereo calibration, rectification, and correspondence (continued)

```
        imageSize.width, CV_32F );
    CvMat* img1r = cvCreateMat( imageSize.height,
        imageSize.width, CV_8U );
    CvMat* img2r = cvCreateMat( imageSize.height,
        imageSize.width, CV_8U );
    CvMat* disp = cvCreateMat( imageSize.height,
        imageSize.width, CV_16S );
    CvMat* vdisp = cvCreateMat( imageSize.height,
        imageSize.width, CV_8U );
    CvMat* pair;
    double R1[3][3], R2[3][3], P1[3][4], P2[3][4];
    CvMat _R1 = cvMat(3, 3, CV_64F, R1);
    CvMat _R2 = cvMat(3, 3, CV_64F, R2);
// IF BY CALIBRATED (BOUGUET'S METHOD)
    if( useUncalibrated == 0 )
    {
        CvMat _P1 = cvMat(3, 4, CV_64F, P1);
        CvMat _P2 = cvMat(3, 4, CV_64F, P2);
        cvStereoRectify( &_M1, &_M2, &_D1, &_D2, imageSize,
            &_R, &_T,
            &_R1, &_R2, &_P1, &_P2, 0,
            0/*CV_CALIB_ZERO_DISPARITY*/ );
        isVerticalStereo = fabs(P2[1][3]) > fabs(P2[0][3]);
//Precompute maps for cvRemap()
        cvInitUndistortRectifyMap(&_M1,&_D1,&_R1,&_P1,mx1,my1);
        cvInitUndistortRectifyMap(&_M2,&_D2,&_R2,&_P2,mx2,my2);
    }
//OR ELSE HARTLEY'S METHOD
    else if( useUncalibrated == 1 || useUncalibrated == 2 )
    // use intrinsic parameters of each camera, but
    // compute the rectification transformation directly
    // from the fundamental matrix
    {
        double H1[3][3], H2[3][3], iM[3][3];
        CvMat _H1 = cvMat(3, 3, CV_64F, H1);
        CvMat _H2 = cvMat(3, 3, CV_64F, H2);
        CvMat _iM = cvMat(3, 3, CV_64F, iM);
//Just to show you could have independently used F
        if( useUncalibrated == 2 )
            cvFindFundamentalMat( &_imagePoints1,
            &_imagePoints2, &_F);
        cvStereoRectifyUncalibrated( &_imagePoints1,
            &_imagePoints2, &_F,
            imageSize,
            &_H1, &_H2, 3);
        cvInvert(&_M1, &_iM);
        cvMatMul(&_H1, &_M1, &_R1);
        cvMatMul(&_iM, &_R1, &_R1);
        cvInvert(&_M2, &_iM);
        cvMatMul(&_H2, &_M2, &_R2);
        cvMatMul(&_iM, &_R2, &_R2);
//Precompute map for cvRemap()
```

```
                cvInitUndistortRectifyMap(&_M1,&_D1,&_R1,&_M1,mx1,my1);

                cvInitUndistortRectifyMap(&_M2,&_D1,&_R2,&_M2,mx2,my2);
        }
        else
            assert(0);
        cvNamedWindow( "rectified", 1 );
// RECTIFY THE IMAGES AND FIND DISPARITY MAPS
        if( !isVerticalStereo )
            pair = cvCreateMat( imageSize.height, imageSize.width*2,
            CV_8UC3 );
        else
            pair = cvCreateMat( imageSize.height*2, imageSize.width,
            CV_8UC3 );
//Setup for finding stereo correspondences
        CvStereoBMState *BMState = cvCreateStereoBMState();
        assert(BMState != 0);
        BMState->preFilterSize=41;
        BMState->preFilterCap=31;
        BMState->SADWindowSize=41;
        BMState->minDisparity=-64;
        BMState->numberOfDisparities=128;
        BMState->textureThreshold=10;
        BMState->uniquenessRatio=15;
        for( i = 0; i < nframes; i++ )
        {
            IplImage* img1=cvLoadImage(imageNames[0][i].c_str(),0);
            IplImage* img2=cvLoadImage(imageNames[1][i].c_str(),0);
            if( img1 && img2 )
            {
                CvMat part;
                cvRemap( img1, img1r, mx1, my1 );
                cvRemap( img2, img2r, mx2, my2 );
                if( !isVerticalStereo || useUncalibrated != 0 )
                {
                // When the stereo camera is oriented vertically,
                // useUncalibrated==0 does not transpose the
                // image, so the epipolar lines in the rectified
                // images are vertical. Stereo correspondence
                // function does not support such a case.
                    cvFindStereoCorrespondenceBM( img1r, img2r, disp,
                        BMState);
                    cvNormalize( disp, vdisp, 0, 256, CV_MINMAX );
                    cvNamedWindow( "disparity" );
                    cvShowImage( "disparity", vdisp );
                }
                if( !isVerticalStereo )
                {
                    cvGetCols( pair, &part, 0, imageSize.width );
                    cvCvtColor( img1r, &part, CV_GRAY2BGR );
                    cvGetCols( pair, &part, imageSize.width,
                        imageSize.width*2 );
```

```
                    cvCvtColor( img2r, &part, CV_GRAY2BGR );
                    for( j = 0; j < imageSize.height; j += 16 )
                        cvLine( pair, cvPoint(0,j),
                            cvPoint(imageSize.width*2,j),
                            CV_RGB(0,255,0));
                }
                else
                {
                    cvGetRows( pair, &part, 0, imageSize.height );
                    cvCvtColor( img1r, &part, CV_GRAY2BGR );
                    cvGetRows( pair, &part, imageSize.height,
                        imageSize.height*2 );
                    cvCvtColor( img2r, &part, CV_GRAY2BGR );
                    for( j = 0; j < imageSize.width; j += 16 )
                        cvLine( pair, cvPoint(j,0),
                            cvPoint(j,imageSize.height*2),
                            CV_RGB(0,255,0));
                }
                cvShowImage( "rectified", pair );
                if( cvWaitKey() == 27 )
                    break;
            }
            cvReleaseImage( &img1 );
            cvReleaseImage( &img2 );
        }
        cvReleaseStereoBMState(&BMState);
        cvReleaseMat( &mx1 );
        cvReleaseMat( &my1 );
        cvReleaseMat( &mx2 );
        cvReleaseMat( &my2 );
        cvReleaseMat( &img1r );
        cvReleaseMat( &img2r );
        cvReleaseMat( &disp );
    }
}
int main(void)
{
    StereoCalib("list.txt", 9, 6, 1);
    return 0;
}
```

Depth Maps from 3D Reprojection

Many algorithms will just use the disparity map directly—for example, to detect
whether or not objects are on (stick out from) a table. But for 3D shape matching, 3D
model learning, robot grasping, and so on, we need the actual 3D reconstruction or
depth map. Fortunately, all the stereo machinery we've built up so far makes this easy.
Recall the 4-by-4 reprojection matrix Q introduced in the section on calibrated stereo
rectification. Also recall that, given the disparity d and a 2D point (x, y), we can derive
the 3D depth using

$$Q \begin{bmatrix} x \\ y \\ d \\ 1 \end{bmatrix} = \begin{bmatrix} X \\ Y \\ Z \\ W \end{bmatrix}$$

where the 3D coordinates are then (X/W, Y/W, Z/W). Remarkably, Q encodes whether or not the cameras' lines of sight were converging (cross eyed) as well as the camera baseline and the principal points in both images. As a result, we need not explicitly account for converging or frontal parallel cameras and may instead simply extract depth by matrix multiplication. OpenCV has two functions that do this for us. The first, which you are already familiar with, operates on an array of points and their associated disparities. It's called cvPerspectiveTransform:

```
void cvPerspectiveTransform(
    const CvArr  *pointsXYD,
    CvArr* result3DPoints,
    const CvMat  *Q
);
```

The second (and new) function cvReprojectImageTo3D() operates on whole images:

```
void cvReprojectImageTo3D(
    CvArr *disparityImage,
    CvArr *result3DImage,
    CvArr *Q
);
```

This routine takes a single-channel disparityImage and transforms each pixel's (x, y) coordinates along with that pixel's disparity (i.e., a vector $[x\ y\ d]^T$) to the corresponding 3D point (X/W, Y/W, Z/W) by using the 4-by-4 reprojection matrix Q. The output is a three-channel floating-point (or a 16-bit integer) image of the same size as the input.

Of course, both functions let you pass an arbitrary perspective transformation (e.g., the canonical one) computed by cvStereoRectify or a superposition of that and the arbitrary 3D rotation, translation, et cetera.

The results of cvReprojectImageTo3D() on an image of a mug and chair are shown in Figure 12-17.

Structure from Motion

Structure from motion is an important topic in mobile robotics as well as in the analysis of more general video imagery such as might come from a handheld camcorder. The topic of structure from motion is a broad one, and a great deal of research has been done in this field. However, much can be accomplished by making one simple observation: In a static scene, an image taken by a camera that has moved is no different than an image taken by a second camera. Thus all of our intuition, as well as our mathematical and algorithmic machinery, is immediately portable to this situation. Of course, the descriptor

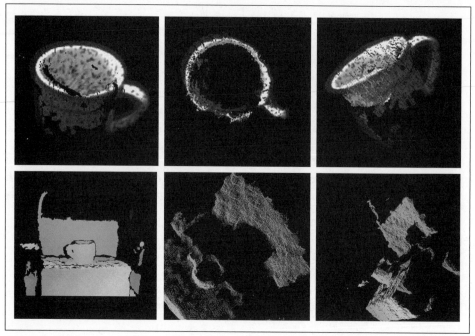

Figure 12-17. Example output of depth maps (for a mug and a chair) computed using cvFindStereo-CorrespondenceBM() and cvReprojectImageTo3D() (image courtesy of Willow Garage)

"static" is crucial, but in many practical situations the scene is either static or sufficiently static that the few moved points can be treated as outliers by robust fitting methods.

Consider the case of a camera moving through a building. If the environment is relatively rich in recognizable features, as might be found with optical flow techniques such as cvCalcOpticalFlowPyrLK(), then we should be able to compute correspondences between enough points—from frame to frame—to reconstruct not only the trajectory of the camera (this information is encoded in the essential matrix E, which can be computed from the fundamental matrix F and the camera intrinsics matrix M) but also, indirectly, the overall three-dimensional structure of the building and the locations of all the aforementioned features in that building. The cvStereoRectifyUncalibrated() routine requires only the fundamental matrix in order to compute the basic structure of a scene up to a scale factor.

Fitting Lines in Two and Three Dimensions

A final topic of interest in this chapter is that of general line fitting. This can arise for many reasons and in a many contexts. We have chosen to discuss it here because one especially frequent context in which line fitting arises is that of analyzing points in three dimensions (although the function described here can also fit lines in two dimensions). Line-fitting algorithms generally use statistically robust techniques [Inui03, Meer91,

Rousseeuw87]. The OpenCV line-fitting algorithm `cvFitLine()` can be used whenever line fitting is needed.

```
void cvFitLine(
  const CvArr* points,
  int          dist_type,
  double       param,
  double       reps,
  double       aeps,
  float*       line
);
```

The array points can be an N-by-2 or N-by-3 matrix of floating-point values (accommodating points in two or three dimensions), or it can be a sequence of cvPointXXX structures.* The argument dist_type indicates the distance metric that is to be minimized across all of the points (see Table 12-3).

Table 12-3. Metrics used for computing dist_type values

Value of dist_type	Metric
CV_DIST_L2	$\rho(r)=\dfrac{r^2}{2}$
CV_DIST_L1	$\rho(r)=r$
CV_DIST_L12	$\rho(r)=\left[\sqrt{1+\dfrac{r^2}{2}}-1\right]$
CV_DIST_FAIR	$\rho(r)=C^2\left[\dfrac{r}{C}-\log\left(1+\dfrac{r}{C}\right)\right],\ C=1.3998$
CV_DIST_WELSCH	$\rho(r)=\dfrac{C^2}{2}\left[1-\exp\left(\dfrac{r}{C}\right)^2\right],\ C=2.9846$
CV_DIST_HUBER	$\rho(r)=\begin{cases} r^2/2 & r<C \\ C(r-C/2) & r\geq C \end{cases}\ C=1.345$

The parameter param is used to set the parameter C listed in Table 12-3. This can be left set to 0, in which case the listed value from the table will be selected. We'll get back to reps and aeps after describing line.

The argument line is the location at which the result is stored. If points is an N-by-2 array, then line should be a pointer to an array of four floating-point numbers (e.g., float array[4]). If points is an N-by-3 array, then line should be a pointer to an array of six floating-point numbers (e.g., float array[6]). In the former case, the return values will be (v_x, v_y, x_0, y_0), where (v_x, v_y) is a normalized vector parallel to the fitted line and (x_0, y_0)

---

* Here XXX is used as a placeholder for anything like 2D32f or 3D64f.

is a point on that line. Similarly, in the latter (three-dimensional) case, the return values will be $(v_x, v_y, v_z, x_0, y_0, z_0)$, where (v_x, v_y, v_z) is a normalized vector parallel to the fitted line and (x_0, y_0, z_0) is a point on that line. Given this line representation, the estimation accuracy parameters reps and aeps are as follows: reps is the requested accuracy of x0, y0[, z0] estimates and aeps is the requested angular accuracy for vx, vy[, vz]. The OpenCV documentation recommends values of 0.01 for both accuracy values.

cvFitLine() can fit lines in two or three dimensions. Since line fitting in two dimensions is commonly needed and since three-dimensional techniques are of growing importance in OpenCV (see Chapter 14), we will end with a program for line fitting, shown in Example 12-4.* In this code we first synthesize some 2D points noisily around a line, then add some random points that have nothing to do with the line (called *outlier* points), and finally fit a line to the points and display it. The cvFitLine() routine is good at ignoring the outlier points; this is important in real applications, where some measurements might be corrupted by high noise, sensor failure, and so on.

Example 12-4. Two-dimensional line fitting

```
#include "cv.h"
#include "highgui.h"
#include <math.h>

int main( int argc, char** argv )
{
  IplImage* img = cvCreateImage( cvSize( 500, 500 ), 8, 3 );
  CvRNG rng = cvRNG(-1);

  cvNamedWindow( "fitline", 1 );

  for(;;) {

    char key;
    int i;
    int count    = cvRandInt(&rng)%100 + 1;
    int outliers = count/5;
    float a      = cvRandReal(&rng)*200;
    float b      = cvRandReal(&rng)*40;
    float angle  = cvRandReal(&rng)*CV_PI;
    float cos_a  = cos(angle);
    float sin_a  = sin(angle);
    CvPoint pt1, pt2;
    CvPoint* points = (CvPoint*)malloc( count * sizeof(points[0]));
    CvMat pointMat = cvMat( 1, count, CV_32SC2, points );
    float line[4];
    float d, t;

    b = MIN(a*0.3, b);

    // generate some points that are close to the line
    //
```

* Thanks to Vadim Pisarevsky for generating this example.

Example 12-4. Two-dimensional line fitting (continued)

```
    for( i = 0; i < count - outliers; i++ ) {
        float x = (cvRandReal(&rng)*2-1)*a;
        float y = (cvRandReal(&rng)*2-1)*b;
        points[i].x = cvRound(x*cos_a - y*sin_a + img->width/2);
        points[i].y = cvRound(x*sin_a + y*cos_a + img->height/2);
    }

    // generate "completely off" points
    //
    for( ; i < count; i++ ) {
        points[i].x = cvRandInt(&rng) % img->width;
        points[i].y = cvRandInt(&rng) % img->height;
    }

    // find the optimal line
    //
    cvFitLine( &pointMat, CV_DIST_L1, 1, 0.001, 0.001, line );
    cvZero( img );

    // draw the points
    //
    for( i = 0; i < count; i++ )
      cvCircle(
        img,
        points[i],
        2,
        (i < count - outliers) ? CV_RGB(255, 0, 0) : CV_RGB(255,255,0),
        CV_FILLED, CV_AA,
        0
      );

    // ... and the line long enough to cross the whole image
    d = sqrt((double)line[0]*line[0] + (double)line[1]*line[1]);
    line[0] /= d;
    line[1] /= d;
    t = (float)(img->width + img->height);
    pt1.x = cvRound(line[2] - line[0]*t);
    pt1.y = cvRound(line[3] - line[1]*t);
    pt2.x = cvRound(line[2] + line[0]*t);
    pt2.y = cvRound(line[3] + line[1]*t);
    cvLine( img, pt1, pt2, CV_RGB(0,255,0), 3, CV_AA, 0 );

    cvShowImage( "fitline", img );

    key = (char) cvWaitKey(0);
    if( key == 27 || key == 'q' || key == 'Q' ) // 'ESC'
        break;
    free( points );
  }

  cvDestroyWindow( "fitline" );
  return 0;
}
```

Exercises

1. Calibrate a camera using `cvCalibrateCamera2()` and at least 15 images of chessboards. Then use `cvProjectPoints2()` to project an arrow orthogonal to the chessboards (the surface normal) into each of the chessboard images using the rotation and translation vectors from the camera calibration.

2. *Three-dimensional joystick.* Use a simple known object with at least four measured, non-coplanar, trackable feature points as input into the POSIT algorithm. Use the object as a 3D joystick to move a little stick figure in the image.

3. In the text's bird's-eye view example, with a camera above the plane looking out horizontally along the plane, we saw that the homography of the ground plane had a horizon line beyond which the homography wasn't valid. How can an infinite plane have a horizon? Why doesn't it just appear to go on forever?

   > Hint: Draw lines to an equally spaced series of points on the plane going out away from the camera. How does the angle from the camera to each next point on the plane change from the angle to the point before?

4. Implement a bird's-eye view in a video camera looking at the ground plane. Run it in real time and explore what happens as you move objects around in the normal image versus the bird's-eye view image.

5. Set up two cameras or a single camera that you move between taking two images.

   a. Compute, store, and examine the fundamental matrix.

   b. Repeat the calculation of the fundamental matrix several times. How stable is the computation?

6. If you had a calibrated stereo camera and were tracking moving points in both cameras, can you think of a way of using the fundamental matrix to find tracking errors?

7. Compute and draw epipolar lines on two cameras set up to do stereo.

8. Set up two video cameras, implement stereo rectification and experiment with depth accuracy.

   a. What happens when you bring a mirror into the scene?

   b. Vary the amount of texture in the scene and report the results.

   c. Try different disparity methods and report on the results.

9. Set up stereo cameras and wear something that is textured over one of your arms. Fit a line to your arm using all the `dist_type` methods. Compare the accuracy and reliability of the different methods.

CHAPTER 13

Machine Learning

What Is Machine Learning

The goal of *machine learning* (ML)* is to turn data into information. After learning from a collection of data, we want a machine to be able to answer questions about the data: What other data is most similar to this data? Is there a car in the image? What ad will the user respond to? There is often a cost component, so this question could become: "Of the products that we make the most money from, which one will the user most likely buy if we show them an ad for it?" Machine learning turns data into information by extracting rules or patterns from that data.

Training and Test Set

Machine learning works on data such as temperature values, stock prices, color intensities, and so on. The data is often preprocessed into *features*. We might, for example, take a database of 10,000 face images, run an edge detector on the faces, and then collect features such as edge direction, edge strength, and offset from face center for each face. We might obtain 500 such values per face or a feature vector of 500 entries. We could then use machine learning techniques to construct some kind of model from this collected data. If we only want to see how faces fall into different groups (wide, narrow, etc.), then a *clustering* algorithm would be the appropriate choice. If we want to learn to predict the age of a person from (say) the pattern of edges detected on his or her face, then a *classifier* algorithm would be appropriate. To meet our goals, machine learning algorithms analyze our collected features and adjust weights, thresholds, and other parameters to maximize performance according to those goals. This process of parameter adjustment to meet a goal is what we mean by the term *learning*.

---

* Machine learning is a vast topic. OpenCV deals mostly with statistical machine learning rather than things that go under the name "Bayesian networks", "Markov random fields", or "graphical models". Some good texts in machine learning are by Hastie, Tibshirani, and Friedman [Hastie01], Duda and Hart [Duda73], Duda, Hart, and Stork [Duda00], and Bishop [Bishop07]. For discussions on how to parallelize machine learning, see Ranger et al. [Ranger07] and Chu et al. [Chu07].

It is always important to know how well machine learning methods are working, and this can be a subtle task. Traditionally, one breaks up the original data set into a large training set (perhaps 9,000 faces, in our example) and a smaller test set (the remaining 1,000 faces). We can then run our classifier over the training set to learn our age prediction model given the data feature vectors. When we are done, we can test the age prediction classifier on the remaining images in the test set.

The test set is not used in training, and we do not let the classifier "see" the test set age labels. We run the classifier over each of the 1,000 faces in the test set of data and record how well the ages it predicts from the feature vector match the actual ages. If the classifier does poorly, we might try adding new features to our data or consider a different type of classifier. We'll see in this chapter that there are many kinds of classifiers and many algorithms for training them.

If the classifier does well, we now have a potentially valuable model that we can deploy on data in the real world. Perhaps this system will be used to set the behavior of a video game based on age. As the person prepares to play, his or her face will be processed into 500 (edge direction, edge strength, offset from face center) features. This data will be passed to the classifier; the age it returns will set the game play behavior accordingly. After it has been deployed, the classifier sees faces that it never saw before and makes decisions according to what it learned on the training set.

Finally, when developing a classification system, we often use a validation data set. Sometimes, testing the whole system at the end is too big a step to take. We often want to tweak parameters along the way before submitting our classifier to final testing. We can do this by breaking the original 10,000-face data set into three parts: a training set of 8,000 faces, a validation set of 1,000 faces, and a test set of 1,000 faces. Now, while we're running through the training data set, we can "sneak" pretests on the validation data to see how we are doing. Only when we are satisfied with our performance on the validation set do we run the classifier on the test set for final judgment.

Supervised and Unsupervised Data

Data sometimes has no labels; we might just want to see what kinds of groups the faces settle into based on edge information. Sometimes the data has labels, such as age. What this means is that machine learning data may be *supervised* (i.e., may utilize a teaching "signal" or "label" that goes with the data feature vectors). If the data vectors are unlabeled then the machine learning is *unsupervised*.

Supervised learning can be *categorical*, such as learning to associate a name to a face, or the data can have *numeric* or *ordered* labels, such as age. When the data has names (categories) as labels, we say we are doing *classification*. When the data is numeric, we say we are doing *regression*: trying to fit a numeric output given some categorical or numeric input data.

Supervised learning also comes in shades of gray: It can involve one-to-one pairing of labels with data vectors or it may consist of *deferred learning* (sometimes called *reinforcement learning*). In reinforcement learning, the data label (also called the *reward* or *punishment*) can come long after the individual data vectors were observed. When a mouse is running down a maze to find food, the mouse may experience a series of turns before it finally finds the food, its reward. That reward must somehow cast its influence back on all the sights and actions that the mouse took before finding the food. Reinforcement learning works the same way: the system receives a delayed signal (a reward or a punishment) and tries to infer a policy for future runs (a way of making decisions; e.g., which way to go at each step through the maze). Supervised learning can also have partial labeling, where some labels are missing (this is also called *semisupervised learning*), or noisy labels, where some labels are just wrong. Most ML algorithms handle only one or two of the situations just described. For example, the ML algorithms might handle classification but not regression; the algorithm might be able to do semisupervised learning but not reinforcement learning; the algorithm might be able to deal with numeric but not categorical data; and so on.

In contrast, often we don't have labels for our data and are interested in seeing whether the data falls naturally into groups. The algorithms for such unsupervised learning are called *clustering algorithms*. In this situation, the goal is to group unlabeled data vectors that are "close" (in some predetermined or possibly even some learned sense). We might just want to see how faces are distributed: Do they form clumps of thin, wide, long, or short faces? If we're looking at cancer data, do some cancers cluster into groups having different chemical signals? Unsupervised clustered data is also often used to form a feature vector for a higher-level supervised classifier. We might first cluster faces into face types (wide, narrow, long, short) and then use that as an input, perhaps with other data such as average vocal frequency, to predict the gender of a person.

These two common machine learning tasks, classification and clustering, overlap with two of the most common tasks in computer vision: recognition and segmentation. This is sometimes referred to as "the what" and "the where". That is, we often want our computer to name the object in an image (recognition, or "what") and also to say where the object appears (segmentation, or "where"). Because computer vision makes such heavy use of machine learning, OpenCV includes many powerful machine learning algorithms in the ML library, located in the .../ *opencv/ml* directory.

 The OpenCV machine learning code is general. That is, although it is highly useful for vision tasks, the code itself is not specific to vision. One could learn, say, genomic sequences using the appropriate routines. Of course, our concern here is mostly with object recognition given feature vectors derived from images.

Generative and Discriminative Models

Many algorithms have been devised to perform learning and clustering. OpenCV supports some of the most useful currently available statistical approaches to machine learning. Probabilistic approaches to machine learning, such as Bayesian networks or graphical models, are less well supported in OpenCV, partly because they are newer and still under active development. OpenCV tends to support *discriminative algorithms*, which give us the probability of the label given the data ($P(L \mid D)$), rather than *generative algorithms*, which give the distribution of the data given the label ($P(D \mid L)$). Although the distinction is not always clear, discriminative models are good for yielding predictions given the data while generative models are good for giving you more powerful representations of the data or for conditionally synthesizing new data (think of "imagining" an elephant; you'd be generating data given a condition "elephant").

It is often easier to interpret a generative model because it models (correctly or incorrectly) the cause of the data. Discriminative learning often comes down to making a decision based on some threshold that may seem arbitrary. For example, suppose a patch of road is identified in a scene partly because its color "red" is less than 125. But does this mean that red = 126 is definitely not road? Such issues can be hard to interpret. With generative models you are usually dealing with conditional distributions of data given the categories, so you can develop a feel for what it means to be "close" to the resulting distribution.

OpenCV ML Algorithms

The machine learning algorithms included in OpenCV are given in Table 13-1. All algorithms are in the *ML* library with the exception of Mahalanobis and K-means, which are in *CVCORE*, and face detection, which is in *CV*.

Table 13-1. Machine learning algorithms supported in OpenCV, original references to the algorithms are provided after the descriptions

Algorithm	Comment
Mahalanobis	A distance measure that accounts for the "stretchiness" of the data space by dividing out the covariance of the data. If the covariance is the identity matrix (identical variance), then this measure is identical to the Euclidean distance measure [Mahalanobis36].
K-means	An unsupervised clustering algorithm that represents a distribution of data using K centers, where K is chosen by the user. The difference between this algorithm and expectation maximization is that here the centers are not Gaussian and the resulting clusters look more like soap bubbles, since centers (in effect) compete to "own" the closest data points. These cluster regions are often used as sparse histogram bins to represent the data. Invented by Steinhaus [Steinhaus56], as used by Lloyd [Lloyd57].
Normal/Naïve Bayes classifier	A generative classifier in which features are assumed to be Gaussian distributed and statistically independent from each other, a strong assumption that is generally not true. For this reason, it's often called a "naïve Bayes" classifier. However, this method often works surprisingly well. Original mention [Maron61; Minsky61].
Decision trees	A discriminative classifier. The tree finds one data feature and a threshold at the current node that best divides the data into separate classes. The data is split and we recursively repeat the procedure down the left and right branches of the tree. Though not often the top performer, it's often the first thing you should try because it is fast and has high functionality [Breiman84].

Table 13-1. Machine learning algorithms supported in OpenCV, original references to the algorithms are provided after the descriptions (continued)

Algorithm	Comment
Boosting	A discriminative group of classifiers. The overall classification decision is made from the combined weighted classification decisions of the group of classifiers. In training, we learn the group of classifiers one at a time. Each classifier in the group is a "weak" classifier (only just above chance performance). These weak classifiers are typically composed of single-variable decision trees called "stumps". In training, the decision stump learns its classification decisions from the data and also learns a weight for its "vote" from its accuracy on the data. Between training each classifier one by one, the data points are re-weighted so that more attention is paid to data points where errors were made. This process continues until the total error over the data set, arising from the combined weighted vote of the decision trees, falls below a set threshold. This algorithm is often effective when a large amount of training data is available [Freund97].
Random trees	A discriminative forest of many decision trees, each built down to a large or maximal splitting depth. During learning, each node of each tree is allowed to choose splitting variables only from a random subset of the data features. This helps ensure that each tree becomes a statistically independent decision maker. In run mode, each tree gets an unweighted vote. This algorithm is often very effective and can also perform regression by averaging the output numbers from each tree [Ho95]; implemented: [Breiman01].
Face detector / Haar classifier	An object detection application based on a clever use of boosting. The OpenCV distribution comes with a trained frontal face detector that works remarkably well. You may train the algorithm on other objects with the software provided. It works well for rigid objects and characteristic views [Viola04].
Expectation maximization (EM)	A generative unsupervised algorithm that is used for clustering. It will fit N multidimensional Gaussians to the data, where N is chosen by the user. This can be an effective way to represent a more complex distribution with only a few parameters (means and variances). Often used in segmentation. Compare with K-means listed previously [Dempster77].
K-nearest neighbors	The simplest possible discriminative classifier. Training data are simply stored with labels. Thereafter, a test data point is classified according to the majority vote of its K nearest other data points (in a Euclidean sense of nearness). This is probably the simplest thing you can do. It is often effective but it is slow and requires lots of memory [Fix51].
Neural networks / Multilayer perceptron (MLP)	A discriminative algorithm that (almost always) has "hidden units" between output and input nodes to better represent the input signal. It can be slow to train but is very fast to run. Still the top performer for things like letter recognition [Werbos74; Rumelhart88].
Support vector machine (SVM)	A discriminative classifier that can also do regression. A distance function between any two data points in a higher-dimensional space is defined. (Projecting data into higher dimensions makes the data more likely to be linearly separable.) The algorithm learns separating hyperplanes that maximally separate the classes in the higher dimension. It tends to be among the best with limited data, losing out to boosting or random trees only when large data sets are available [Vapnik95].

Using Machine Learning in Vision

In general, all the algorithms in Table 13-1 take as input a data vector made up of many features, where the number of features might well number in the thousands. Suppose your task is to recognize a certain type of object—for example, a person. The first problem that you will encounter is how to collect and label training data that falls into positive (there is a person in the scene) and negative (no person) cases. You will soon realize that people appear at different scales: their image may consist of just a few pixels, or you may be looking at an ear that fills the whole screen. Even worse, people will often be occluded: a man inside a car; a woman's face; one leg showing behind a tree. You need to define what you actually mean by saying a person is in the scene.

Next, you have the problem of collecting data. Do you collect it from a security camera, go to *http://www.flicker.com* and attempt to find "person" labels, or both (and more)? Do you collect movement information? Do you collect other information, such as whether a gate in the scene is open, the time, the season, the temperature? An algorithm that finds people on a beach might fail on a ski slope. You need to capture the variations in the data: different views of people, different lightings, weather conditions, shadows, and so on.

After you have collected lots of data, how will you label it? You must first decide on what you mean by "label". Do you want to know where the person is in the scene? Are actions (running, walking, crawling, following) important? You might end up with a million images or more. How will you label all that? There are many tricks, such as doing background subtraction in a controlled setting and collecting the segmented foreground humans who come into the scene. You can use data services to help in classification; for example, you can pay people to label your images through Amazon's "mechanical turk" (*http://www.mturk.com/mturk/welcome*). If you arrange things to be simple, you can get the cost down to somewhere around a penny per label.

After labeling the data, you must decide which features to extract from the objects. Again, you must know what you are after. If people always appear right side up, there's no reason to use rotation-invariant features and no reason to try to rotate the objects beforehand. In general, you must find features that express some invariance in the objects, such as scale-tolerant histograms of gradients or colors or the popular SIFT features.[*] If you have background scene information, you might want to first remove it to make other objects stand out. You then perform your image processing, which may consist of normalizing the image (rescaling, rotation, histogram equalization, etc.) and computing many different feature types. The resulting data vectors are each given the label associated with that object, action, or scene.

Once the data is collected and turned into feature vectors, you often want to break up the data into training, validation, and test sets. It is a "best practice" to do your learning, validation, and testing within a cross-validation framework. That is, the data is divided into K subsets and you run many training (possibly validation) and test sessions, where each session consists of different sets of data taking on the roles of training (validation) and test.[†] The test results from these separate sessions are then averaged to get the final performance result. Cross-validation gives a more accurate picture of how the classifier

---

[*] See Lowe's SIFT feature demo (*http://www.cs.ubc.ca/~lowe/keypoints/*).

[†] One typically does the train (possibly validation) and test cycle five to ten times.

will perform when deployed in operation on novel data. (We'll have more to say about this in what follows.)

Now that the data is prepared, you must choose your classifier. Often the choice of classifier is dictated by computational, data, or memory considerations. For some applications, such as online user preference modeling, you must train the classifier rapidly. In this case, nearest neighbors, normal Bayes, or decision trees would be a good choice. If memory is a consideration, decision trees or neural networks are space efficient. If you have time to train your classifier but it must run quickly, neural networks are a good choice, as are normal Bayes classifiers and support vector machines. If you have time to train but need high accuracy, then boosting and random trees are likely to fit your needs. If you just want an easy, understandable sanity check that your features are chosen well, then decision trees or nearest neighbors are good bets. For best "out of the box" classification performance, try boosting or random trees first.

 There is no "best" classifier (see *http://en.wikipedia.org/wiki/No_free_lunch_theorem*). Averaged over all possible types of data distributions, all classifiers perform the same. Thus, we cannot say which algorithm in Table 13-1 is the "best". Over any given data distribution or set of data distributions, however, there is usually a best classifier. Thus, when faced with real data it's a good idea to try many classifiers. Consider your purpose: Is it just to get the right score, or is it to interpret the data? Do you seek fast computation, small memory requirements, or confidence bounds on the decisions? Different classifiers have different properties along these dimensions.

Variable Importance

Two of the algorithms in Table 13-1 allow you to assess a variable's importance.[*] Given a vector of features, how do you determine the importance of those features for classification accuracy? Binary decision trees do this directly: they are trained by selecting which variable best splits the data at each node. The top node's variable is the most important variable; the next-level variables are the second most important, and so on. Random trees can measure variable importance using a technique developed by Leo Breiman;[†] this technique can be used with any classifier, but so far it is implemented only for decision and random trees in OpenCV.

One use of variable importance is to reduce the number of features your classifier must consider. Starting with many features, you train the classifier and then find the importance of each feature relative to the other features. You can then discard unimportant features. Eliminating unimportant features improves speed performance (since it eliminates the processing it took to compute those features) and makes training and testing quicker. Also, if you don't have enough data, which is often the case, then eliminating

---

[*] This is known as "variable importance" even though it refers to the importance of a variable (noun) and not the fluctuating importance (adjective) of a variable.

[†] Breiman's variable importance technique is described in "Looking Inside the Black Box" (*www.stat.berkeley.edu/~breiman/wald2002-2.pdf*).

unimportant variables can increase classification accuracy; this yields faster processing with better results.

Breiman's variable importance algorithm runs as follows.

1. Train a classifier on the training set.

2. Use a validation or test set to determine the accuracy of the classifier.

3. For every data point and a chosen feature, randomly choose a new value for that feature from among the values the feature has in the rest of the data set (called "sampling with replacement"). This ensures that the distribution of that feature will remain the same as in the original data set, but now the actual structure or meaning of that feature is erased (because its value is chosen at random from the rest of the data).

4. Train the classifier on the altered set of training data and then measure the accuracy of classification on the altered test or validation data set. If randomizing a feature hurts accuracy a lot, then that feature is very important. If randomizing a feature does not hurt accuracy much, then that feature is of little importance and is a candidate for removal.

5. Restore the original test or validation data set and try the next feature until we are done. The result is an ordering of each feature by its importance.

This procedure is built into random trees and decision trees. Thus, you can use random trees or decision trees to decide which variables you will actually use as features; then you can use the slimmed-down feature vectors to train the same (or another) classifier.

Diagnosing Machine Learning Problems

Getting machine learning to work well can be more of an art than a science. Algorithms often "sort of" work but not quite as well as you need them to. That's where the art comes in; you must figure out what's going wrong in order to fix it. Although we can't go into all the details here, we'll give an overview of some of the more common problems you might encounter.* First, some rules of thumb: More data beats less data, and better features beat better algorithms. If you design your features well—maximizing their independence from one another and minimizing how they vary under different conditions—then almost any algorithm will work well. Beyond that, there are two common problems:

Bias

    Your model assumptions are too strong for the data, so the model won't fit well.

Variance

    Your algorithm has memorized the data *including* the noise, so it can't generalize.

Figure 13-1 shows the basic setup for statistical machine learning. Our job is to model the true function *f* that transforms the underlying inputs to some output. This function may

---

* Professor Andrew Ng at Stanford University gives the details in a web lecture entitled "Advice for Applying Machine Learning" (*http://www.stanford.edu/class/cs229/materials/ML-advice.pdf*).

be a regression problem (e.g., predicting a person's age from their face) or a category prediction problem (e.g., identifying a person given their facial features). For problems in the real world, noise and unconsidered effects can cause the observed outputs to differ from the theoretical outputs. For example, in face recognition we might learn a model of the measured distance between eyes, mouth, and nose to identify a face. But lighting variations from a nearby flickering bulb might cause noise in the measurements, or a poorly manufactured camera lens might cause a systematic distortion in the measurements that wasn't considered as part of the model. These affects will cause accuracy to suffer.

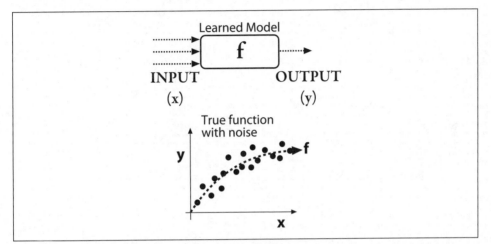

Figure 13-1. Setup for statistical machine learning: we train a classifier to fit a data set; the true model f is almost always corrupted by noise or unknown influences

Figure 13-2 shows under- and overfitting of data in the upper two panels and the consequences in terms of error with training set size in the lower two panels. On the left side of Figure 13-2 we attempt to train a classifier to predict the data in the lower panel of Figure 13-1. If we use a model that's too restrictive—indicated here by the heavy, straight dashed line—then we can never fit the underlying true parabola *f* indicated by the thinner dashed line. Thus, the fit to both the training data and the test data will be poor, even with a lot of data. In this case we have bias because both training and test data are predicted poorly. On the right side of Figure 13-2 we fit the training data exactly, but this produces a nonsense function that fits every bit of noise. Thus, it memorizes the training data as well as the noise in that data. Once again, the resulting fit to the test data is poor. Low training error combined with high test error indicates a variance (overfit) problem.

Sometimes you have to be careful that you are solving the correct problem. If your training and test set error are low but the algorithm does not perform well in the real world, the data set may have been chosen from unrealistic conditions—perhaps because these conditions made collecting or simulating the data easier. If the algorithm just cannot reproduce the test or training set data, then perhaps the algorithm is the wrong one to use or the features that were extracted from the data are ineffective or the "signal" just isn't in the data you collected. Table 13-2 lays out some possible fixes to the problems

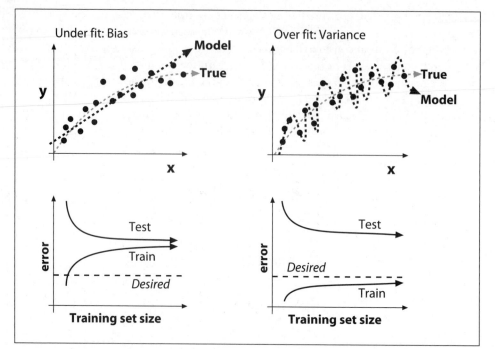

Figure 13-2. Poor model fitting in machine learning and its effect on training and test prediction performance, where the true function is graphed by the lighter dashed line at top: an underfit model for the data (upper left) yields high error in predicting the training and the test set (lower left), whereas an overfit model for the data (upper right) yields low error in the training data but high error in the test data (lower right)

we've described here. Of course, this is not a complete list of the possible problems or solutions. It takes careful thought and design of what data to collect and what features to compute in order for machine learning to work well. It can also take some systematic thinking to diagnose machine learning problems.

Table 13-2. Problems encountered in machine learning and possible solutions to try; coming up with better features will help any problem

Problem	Possible Solutions
Bias	• More features can help make a better fit.
	• Use a more powerful algorithm.
Variance	• More training data can help smooth the model.
	• Fewer features can reduce overfitting.
	• Use a less powerful algorithm.
Good test/train, bad real world	• Collect a more realistic set of data.
Model can't learn test or train	• Redesign features to better capture invariance in the data.
	• Collect new, more relevant data.
	• Use a more powerful algorithm.

Cross-validation, bootstrapping, ROC curves, and confusion matrices

Finally, there are some basic tools that are used in machine learning to measure results. In supervised learning, one of the most basic problems is simply knowing how well your algorithm has performed: How accurate is it at classifying or fitting the data? You might think: "Easy, I'll just run it on my test or validation data and get the result." But for real problems, we must account for noise, sampling fluctuations, and sampling errors. Simply put, your test or validation set of data might not accurately reflect the actual distribution of data. To get closer to "guessing" the true performance of the classifier, we employ the technique of *cross-validation* and/or the closely related technique of *bootstrapping.**

In its most basic form, cross-validation involves dividing the data into K different subsets of data. You train on $K - 1$ of the subsets and test on the final subset of data (the "validation set") that wasn't trained on. You do this K times, where each of the K subsets gets a "turn" at being the validation set, and then average the results.

Bootstrapping is similar to cross-validation, but the validation set is selected at random from the training data. Selected points for that round are used only in test, not training. Then the process starts again from scratch. You do this N times, where each time you randomly select a new set of validation data and average the results in the end. Note that this means some and/or many of the data points are reused in different validation sets, but the results are often superior compared to cross-validation.

Using either one of these techniques can yield more accurate measures of actual performance. This increased accuracy can in turn be used to tune parameters of the learning system as you repeatedly change, train, and measure.

Two other immensely useful ways of assessing, characterizing, and tuning classifiers are plotting the *receiver operating characteristic* (ROC) and filling in a confusion matrix; see Figure 13-3. The ROC curve measures the response over the performance parameter of the classifier over the full range of settings of that parameter. Let's say the parameter is a threshold. Just to make this more concrete, suppose we are trying to recognize yellow flowers in an image and that we have a threshold on the color yellow as our detector. Setting the yellow threshold extremely high would mean that the classifier would fail to recognize any yellow flowers, yielding a false positive rate of 0 but at the cost of a true positive rate also at 0 (lower left part of the curve in Figure 13-3). On the other hand, if the yellow threshold is set to 0 then any signal at all counts as a recognition. This means that all of the true positives (the yellow flowers) are recognized as well as all the false positives (orange and red flowers); thus we have a false positive rate of 100% (upper right part of the curve in Figure 13-3). The best possible ROC curve would be one that follows the *y*-axis up to 100% and then cuts horizontally over to the upper right corner. Failing that, the closer the curve comes to the upper left corner, the better. One can compute the fraction of area under the ROC curve versus the total area of the ROC plot as a summary statistic of merit: The closer that ratio is to 1 the better is the classifier.

* For more information on these techniques, see "What Are Cross-Validation and Bootstrapping?" (*http://www.faqs.org/faqs/ai-faq/neural-nets/part3/section-12.html*).

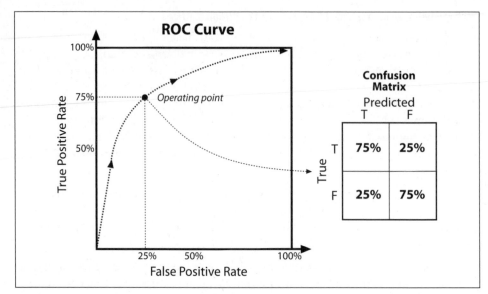

Figure 13-3. Receiver operating curve (ROC) and associated confusion matrix: the former shows the response of correct classifications to false positives along the full range of varying a performance parameter of the classifier; the latter shows the false positives (false recognitions) and false negatives (missed recognitions)

Figure 13-3 also shows a *confusion matrix*. This is just a chart of true and false positives along with true and false negatives. It is another quick way to assess the performance of a classifier: ideally we'd see 100% along the NW-SE diagonal and 0% elsewhere. If we have a classifier that can learn more than one class (e.g., a multilayer perceptron or random forest classifier can learn many different class labels at once), then the confusion matrix generalizes to many classes and you just keep track of the class to which each labeled data point was assigned.

Cost of misclassification. One thing we haven't discussed much here is the cost of misclassification. That is, if our classifier is built to detect poisonous mushrooms (we'll see an example that uses such a data set shortly) then we are willing to have more false negatives (edible mushrooms mistaken as poisonous) as long as we minimize false positives (poisonous mushrooms mistaken as edible). The ROC curve can help with this; we can set our ROC parameter to choose an operation point lower on the curve—toward the lower left of the graph in Figure 13-3. The other way of doing this is to weight false positive errors more than false negatives when generating the ROC curve. For example, you can set each false positive error to count as much as ten false negatives.* Some OpenCV machine learning algorithms, such as decision trees and SVM, can regulate this balance of "hit rate versus false alarm" by specifying prior probabilities of the classes themselves

---

* This is useful if you have some specific a priori notion of the relative cost of the two error types. For example, the cost of misclassifying one product as another in a supermarket checkout would be easy to quantify exactly beforehand.

(which classes are expected to be more likely and which less) or by specifying weights of the individual training samples.

Mismatched feature variance. Another common problem with training some classifiers arises when the feature vector comprises features of widely different variances. For instance, if one feature is represented by lowercase ASCII characters then it ranges over only 26 different values. In contrast, a feature that is represented by the count of biological cells on a microscope slide might vary over several billion values. An algorithm such as K-nearest neighbors might then see the first feature as relatively constant (nothing to learn from) compared to the cell-count feature. The way to correct this problem is to pre-process each feature variable by normalizing for its variance. This practice is acceptable provided the features are not correlated with each other; when features are correlated, you can normalize by their average variance or by their covariance. Some algorithms, such as decision trees,* are not adversely affected by widely differing variance and so this precaution need not be taken. A rule of thumb is that if the algorithm depends in some way on a distance measure (e.g., weighted values) then you should normalize for variance. One may normalize all features at once and account for their covariance by using the Mahalanobis distance, which is discussed later in this chapter.†

We now turn to discussing some of the machine learning algorithms supported in OpenCV, most of which are found in the *.../opencv/ml* directory. We start with some of the class methods that are universal across the ML sublibrary.

Common Routines in the ML Library

This chapter is written to get you up and running with the machine learning algorithms. As you try out and become comfortable with different methods, you'll also want to reference the *.../opencv/docs/ref/opencvref_ml.htm* manual that installs with OpenCV and/or the online OpenCV Wiki documentation (*http://opencvlibrary.sourceforge.net/*). *Because* this portion of the library is under active development, you will want to know about the latest and greatest available tools.

All the routines in the ML library‡ are written as C++ classes and all derived from the CvStatModel class, which holds the methods that are universal to all the algorithms. These methods are listed in Table 13-3. Note that in the CvStatModel there are two ways of storing and recalling the model from disk: save() versus write() and load() versus read(). For machine learning models, you should use the much simpler save()

---

* Decision trees are not affected by variance differences in feature variables because each variable is searched only for effective separating thresholds. In other words, it doesn't matter how large the variable's range is as long as a clear separating value can be found.

† Readers familiar with machine learning or signal processing might recognize this as a technique for "whitening" the data.

‡ Note that the Haar classifier, Mahalanobis, and K-means algorithms were written before the ML library was created and so are in *cv* and *cvcore* libraries instead.

and load(), which essentially wrap the more complex write() and read() functions into an interface that writes and reads XML and YAML to and from disk. Beyond that, for learning from data the two most important functions, predict() and train(), vary by algorithm and will be discussed next.

Table 13-3. Base class methods for the machine learning (ML) library

CvStatModel:: Methods	Description

save(
    const char* filename,
    const char* name   = 0
)
	Saves learned model in XML or YMAL. Use this method for storage.

load(
    const char* filename,
    const char* name=0
 );
	Calls clear() and then loads XML or YMAL model. Use this method for recall.

clear()	De-allocates all memory. Ready for reuse.

bool train(
  –data points–,
  [flags]
  –responses–,
  [flags etc]
) ;
	The training function to learn a model of the dataset. Training is specific to the algorithm and so the input parameters will vary.

float predict(
    const CvMat* sample
    [,<prediction_params>]
) const;
	After training, use this function to predict the label or value of a new training point or points.

Constructor, Destructor:	
CvStatModel(); CvStatModel(const CvMat* train_data ...);	Default constructor and constructor that allows creation and training of the model in one shot.
CvStatModel::~CvStatModel();	The destructor of the ML model.
Write/Read support (but use save/load above instead):	

write(
  CvFileStorage* storage,
  const char*   name
 );
	Generic CvFileStorage structured write to disk, located in the *cvcore* library (discussed in Chapter 3) and called by save().

read(
  CvFileStorage* storage,
  CvFileNode*   node
 );
	Generic file read to CvFileStorage structure, located in the *cvcore* library and called by load().

Training

The training prototype is as follows:

```
bool CvStatModel::train(
  const CvMat* train_data,
  [int tflag,]            ...,
  const CvMat* responses,  ...,
```

```
    [const CvMat* var_idx,]    ...,
    [const CvMat* sample_idx,] ...,
    [const CvMat* var_type,]   ...,
    [const CvMat* missing_mask,]
    <misc_training_alg_params> ...
);
```

The `train()` method for the machine learning algorithms can assume different forms according to what the algorithm can do. All algorithms take a CvMat matrix pointer as training data. This matrix must be of type 32FC1 (32-bit, floating-point, single-channel). CvMat does allow for multichannel images, but machine learning algorithms take only a single channel—that is, just a two-dimensional matrix of numbers. Typically this matrix is organized as rows of data points, where each "point" is represented as a vector of features. Hence the columns contain the individual features for each data point and the data points are stacked to yield the 2D single-channel training matrix. To belabor the topic: the typical data matrix is thus composed of (rows, columns) = (data points, features). However, some algorithms can handle transposed matrices directly. For such algorithms you may use the `tflag` parameter to tell the algorithm that the training points are organized in columns. This is just a convenience so that you won't have to transpose a large data matrix. When the algorithm can handle both row-order and column-order data, the following flags apply.

`tflag = CV_ROW_SAMPLE`
    Means that the feature vectors are stored as rows (default)

`tflag = CV_COL_SAMPLE`
    Means that the feature vectors are stored as columns

The reader may well ask: What if my training data is not floating-point numbers but instead is letters of the alphabet or integers representing musical notes or names of plants? The answer is: Fine, just turn them into unique 32-bit floating-point numbers when you fill the CvMat. If you have letters as features or labels, you can cast the ASCII character to floats when filling the data array. The same applies to integers. As long as the conversion is unique, things should work—but remember that some routines are sensitive to widely differing variances among features. It's generally best to normalize the variance of features as discussed previously. With the exception of the tree-based algorithms (decision trees, random trees, and boosting) that support both categorical and ordered input variables, all other OpenCV ML algorithms work only with ordered inputs. A popular technique for making ordered-input algorithms also work with categorical data is to represent them in 1-radix notation; for example, if the input variable color may have seven different values then it may be replaced by seven binary variables, where one and only one of the variables may be set to 1.

The parameter responses are either categorical labels such as "poisonous" or "nonpoisonous", as with mushroom identification, or are regression values (numbers) such as body temperatures taken with a thermometer. The response values or "labels" are usually a one-dimensional vector of one value per data point—except for neural networks, which can have a vector of responses for each data point. Response values are one of two types: For categorical responses, the type can be integer (32SC1); for regression values, the response is 32-bit floating-point (32FC1). Observe also that some algorithms can deal only with classification problems and others only with regression; but others can handle both. In this last case, the type of output variable is passed either as a separate parameter or as a last element of a var_type vector, which can be set as follows.

CV_VAR_CATEGORICAL
>   Means that the output values are discrete class labels

CV_VAR_ORDERED (= CV_VAR_NUMERICAL)
>   Means that the output values are ordered; that is, different values can be compared as numbers and so this is a regression problem

The types of input variables can also be specified using var_type. However, algorithms of the regression type can handle only ordered-input variables. Sometimes it is possible to make up an ordering for categorical variables as long as the order is kept consistent, but this can sometimes cause difficulties for regression because the pretend "ordered" values may jump around wildly when they have no physical basis for their imposed order.

Many models in the ML library may be trained on a selected feature subset and/or on a selected sample subset of the training set. To make this easier for the user, the method train() usually includes the vectors var_idx and sample_idx as parameters. These may be defaulted to "use all data" by passing NULL values for these parameters, but var_idx can be used to indentify variables (features) of interest and sample_idx can identify data points of interest. Using these, you may specify which features and which sample points on which to train. Both vectors are either single-channel integer (CV_32SC1) vectors— that is, lists of zero-based indices—or single-channel 8-bit (CV_8UC1) masks of active variables/samples, where a nonzero value signifies active. The parameter sample_idx is particularly helpful when you've read in a chunk of data and want to use some of it for training and some of it for test *without* breaking it into two different vectors.

Additionally, some algorithms can handle missing measurements. For example, when the authors were working with manufacturing data, some measurement features would end up missing during the time that workers took coffee breaks. Sometimes experimental data simply is forgotten, such as forgetting to take a patient's temperature one day during a medical experiment. For such situations, the parameter missing_mask, an 8-bit matrix of the same dimensions as train_data, is used to mark the missed values (nonzero elements of the mask). Some algorithms cannot handle missing values, so the missing points should be interpolated by the user before training or the corrupted records should be rejected in advance. Other algorithms, such as decision tree and naïve Bayes, handle missing values in different ways. Decision trees use alternative splits (called "surrogate splits" by Breiman); the naïve Bayes algorithm infers the values.

Usually, the previous model state is cleared by clear() before running the training procedure. However, some algorithms may optionally update the model learning with the new training data instead of starting from scratch.

Prediction

When using the method predict(), the var_idx parameter that specifies which features were used in the train() method is remembered and then used to extract only the necessary components from the input sample. The general form of the predict() method is as follows:

```
float CvStatMode::predict(
    const CvMat* sample
    [, <prediction_params>]
) const;
```

This method is used to predict the response for a new input data vector. When using a classifier, predict() returns a class label. For the case of regression, this method returns a numerical value. Note that the input sample must have as many components as the train_data that was used for training. Additional prediction_params are algorithm-specific and allow for such things as missing feature values in tree-based methods. The function suffix const tells us that prediction does not affect the internal state of the model, so this method is thread-safe and can be run in parallel, which is useful for web servers performing image retrieval for multiple clients and for robots that need to accelerate the scanning of a scene.

Controlling Training Iterations

Although the iteration control structure CvTermCriteria has been discussed in other chapters, it is used by several machine learning routines. So, just to remind you of what the function is, we repeat it here.

```
typedef struct CvTermCriteria {
    int     type;       /* CV_TERMCRIT_ITER and/or CV_TERMCRIT_EPS */
    int     max_iter;   /* maximum number of iterations */
    double  epsilon;    /* stop when error is below this value  */
}
```

The integer parameter max_iter sets the total number of iterations that the algorithm will perform. The epsilon parameter sets an error threshold stopping criteria; when the error drops below this level, the routine stops. Finally, the type tells which of these two criteria to use, though you may add the criteria together and so use both (CV_TERMCRIT_ITER | CV_TERMCRIT_EPS). The defined values for term_crit.type are:

```
#define CV_TERMCRIT_ITER    1
#define CV_TERMCRIT_NUMBER  CV_TERMCRIT_ITER
#define CV_TERMCRIT_EPS     2
```

Let's now move on to describing specific algorithms that are implemented in OpenCV. We will start with the frequently used Mahalanobis distance metric and then go into some detail on one unsupervised algorithm (K-means); both of these may be found in the *cxcore* library. We then move into the machine learning library proper with the normal Bayes classifier, after which we discuss decision-tree algorithms (decision trees, boosting, random trees, and Haar cascade). For the other algorithms we'll provide short descriptions and usage examples.

Mahalanobis Distance

The *Mahalanobis distance* is a distance measure that accounts for the covariance or "stretch" of the space in which the data lies. If you know what a *Z-score* is then you can think of the Mahalanobis distance as a multidimensional analogue of the *Z-score*. Figure 13-4(a) shows an initial distribution between three sets of data that make the vertical sets look closer together. When we normalize the space by the covariance in the data, we see in Figure 13-4(b) that that horizontal data sets are actually closer together. This sort of thing occurs frequently; for instance, if we are comparing people's height in meters with their age in days, we'd see very little variance in height to relate to the large variance in age. By normalizing for the variance we can obtain a more realistic comparison of variables. Some classifiers such as K-nearest neighbors deal poorly with large differences in variance, whereas other algorithms (such as decision trees) don't mind it.

We can already get a hint for what the Mahalanobis distance must be by looking at Figure 13-4;* we must somehow divide out the covariance of the data while measuring distance. First, let us review what covariance is. Given a list X of N data points, where each data point may be of dimension (vector length) K with mean vector μ_x (consisting of individual means $\mu_{1,...,K}$), the covariance is a K-by-K matrix given by:

$$\Sigma = E[(X - \mu_x)(X - \mu_x)^T]$$

where $E[\cdot]$ is the expectation operator. OpenCV makes computing the covariance matrix easy, using

```
void cvCalcCovarMatrix(
    const CvArr** vects,
    int           count,
    CvArr*        cov_mat,
    CvArr*        avg,
    int           flags
);
```

This function is a little bit tricky. Note that vects is a pointer to a pointer of CvArr. This implies that we have vects[0] through vects[count-1], but it actually depends on the flags settings as described in what follows. Basically, there are two cases.

1. Vects is a 1D vector of pointers to 1D vectors or 2D matrices (the two dimensions are to accommodate images). That is, each vects[i] can point to a 1D or a 2D vector, which occurs if neither CV_COV_ROWS nor CV_COV_COLS is set. The accumulating covariance computation is scaled or divided by the number of data points given by count if CV_COVAR_SCALE is set.

2. Often there is only one input vector, so use only vects[0] if either CV_COVAR_ROWS or CV_COVAR_COLS is set. If this is set, then scaling by the value given by count is ignored

---

* Note that Figure 13-4 has a diagonal covariance matrix, which entails independent X and Y variance rather than actual covariance. This was done to make the explanation simple. In reality, data is often "stretched" in much more interesting ways.

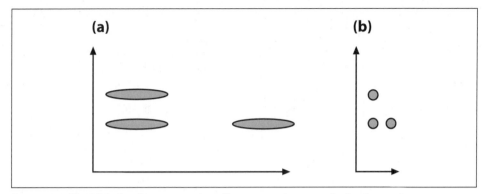

Figure 13-4. The Mahalanobis computation allows us to reinterpret the data's covariance as a "stretch" of the space: (a) the vertical distance between raw data sets is less than the horizontal distance; (b) after the space is normalized for variance, the horizontal distance between data sets is less than the vertical distance

in favor of the number of actual data vectors contained in vects[0]. All the data points are then in:

a. the rows of vects[0] if CV_COVAR_ROWS is set; or

b. the columns of vects[0] if instead CV_COVAR_COLS is set. You cannot set both row and column flags simultaneously (see flag descriptions for more details).

Vects can be of types 8UC1, 16UC1, 32FC1, or 64FC1. In any case, vects contains a list of K-dimensional data points. To reiterate: count is how many vectors there are in vects[] for case 1 (CV_COVAR_ROWS and CV_COVAR_COLS not set); for case 2a and 2b (CV_COVAR_ROWS or CV_COVAR_COLS is set), count is ignored and the actual number of vectors in vects[0] is used instead. The resulting K-by-K covariance matrix will be returned in cov_mat, and it can be of type CV_32FC1 or CV_64FC1. Whether or not the vector avg is used depends on the settings of flags (see listing that follows). If avg is used then it has the same type as vects and contains the K-feature averages across vects. The parameter flags can have many combinations of settings formed by adding values together (for more complicated applications, refer to the *.../opencv/docs/ref/opencvref_cxcore.htm* documentation). In general, you will set flags to one of the following.

CV_COVAR_NORMAL
   Do the regular type of covariance calculation as in the previously displayed equation. Average the results by the number in count if CV_COVAR_SCALE is not set; otherwise, average by the number of data points in vects[0].

CV_COVAR_SCALE
   Normalize the computed covariance matrix.

CV_COVAR_USE_AVG
   Use the avg matrix instead of automatically calculating the average of each feature. Setting this saves on computation time if you already have the averages (e.g., by

having called cvAvg() yourself); otherwise, the routine will compute these averages for you.*

Most often you will combine your data into one big matrix, let's say by rows of data points; then flags would be set as flags = CV_COVAR_NORMAL | CV_COVAR_SCALE | CV_COVAR_ROWS.

We now have the covariance matrix. For Mahalanobis distance, however, we'll need to divide out the variance of the space and so will need the *inverse* covariance matrix. This is easily done by using:

```
double cvInvert(
  const CvArr* src,
  CvArr*       dst,
  int          method = CV_LU
);
```

In cvInvert(), the src matrix should be the covariance matrix calculated before and dst should be a same sized matrix, which will be filled with the inverse on return. You could leave the method at its default value, CV_LU, but it is better to set the method to CV_SVD_SYM.[†]

With the inverse covariance matrix Σ^{-1} finally in hand, we can move on to the Mahalanobis distance measure. This measure is much like the Euclidean distance measure, which is the square root of the sum of squared differences between two vectors $\mathbf{x}$ and $\mathbf{y}$, but it divides out the covariance of the space:

$$D_{\text{Mahalanobis}}(\mathbf{x},\mathbf{y}) = \sqrt{(\mathbf{x}-\mathbf{y})^{\mathrm{T}}\,\Sigma^{-1}\,(\mathbf{x}-\mathbf{y})}$$

This distance is just a number. Note that if the covariance matrix is the identity matrix then the Mahalanobis distance is equal to the Euclidean distance. We finally arrive at the actual function that computes the Mahalanobis distance. It takes two input vectors (vec1 and vec2) and the inverse covariance in mat, and it returns the distance as a double:

```
double cvMahalanobis(
  const CvArr* vec1,
  const CvArr* vec2,
  CvArr*       mat
);
```

The Mahalanobis distance is an important measure of similarity between two different data points in a multidimensional space, but is not a clustering algorithm or classifier itself. Let us now move on, starting with the most frequently used clustering algorithm: K-means.

---

* A precomputed average data vector should be passed if the user has a more statistically justified value of the average or if the covariance matrix is computed by blocks.

† CV_SVD could also be used in this case, but it is somewhat slower and less accurate than CV_SVD_SYM. CV_SVD_SYM, even if it is slower than CV_LU, still should be used if the dimensionality of the space is much smaller than the number of data points. In such a case the overall computing time will be dominated by cvCalcCovarMatrix() anyway. So it may be wise to spend a little bit more time on computing inverse covariance matrix more accurately (much more accurately, if the set of points is concentrated in a subspace of a smaller dimensionality). Thus, CV_SVD_SYM is usually the best choice for this task.

K-Means

K-means is a clustering algorithm implemented in the *cxcore* because it was written long before the ML library. K-means attempts to find the natural clusters or "clumps" in the data. The user sets the desired number of clusters and then K-means rapidly finds a good placement for those cluster centers, where "good" means that the cluster centers tend to end up located in the middle of the natural clumps of data. It is one of the most used clustering techniques and has strong similarities to the expectation maximization algorithm for Gaussian mixture (implemented as CvEM() in the ML library) as well as some similarities to the mean-shift algorithm discussed in Chapter 9 (implemented as cvMeanShift() in the CV library). K-means is an iterative algorithm and, as implemented in OpenCV, is also known as Lloyd's algorithm* or (equivalently) "Voronoi iteration". The algorithm runs as follows.

1. Take as input (a) a data set and (b) desired number of clusters K (chosen by the user).

2. Randomly assign cluster center locations.

3. Associate each data point with its nearest cluster center.

4. Move cluster centers to the centroid of their data points.

5. Return to step 3 until convergence (centroid does not move).

Figure 13-5 diagrams K-means in action; in this case, it takes just two iterations to converge. In real cases the algorithm often converges rapidly, but it can sometimes require a large number of iterations.

Problems and Solutions

K-means is an extremely effective clustering algorithm, but it does have three problems.

1. K-means isn't guaranteed to find the best possible solution to locating the cluster centers. However, it is guaranteed to converge to some solution (i.e., the iterations won't continue indefinitely).

2. K-means doesn't tell you how many cluster centers you should use. If we had chosen two or four clusters for the example of Figure 13-5, then the results would be different and perhaps nonintuitive.

3. K-means presumes that the covariance in the space either doesn't matter or has already been normalized (cf. our discussion of the Mahalanobis distance).

Each one of these problems has a "solution", or at least an approach that helps. The first two of these solutions depend on "explaining the variance of the data". In K-means, each cluster center "owns" its data points and we compute the variance of those points.

---

* S. P. Lloyd, "Least Squares Quantization in PCM," *IEEE Transactions on Information Theory 28* (1982), 129–137.

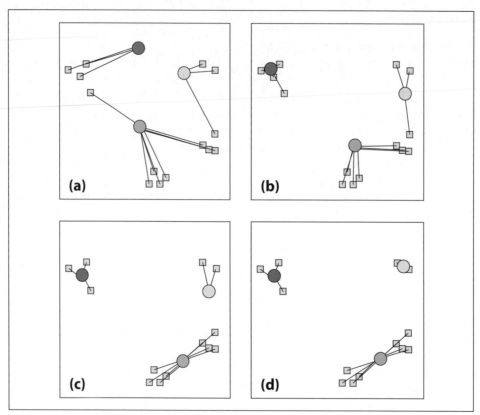

Figure 13-5. K-means in action for two iterations: (a) cluster centers are placed randomly and each data point is then assigned to its nearest cluster center; (b) cluster centers are moved to the centroid of their points; (c) data points are again assigned to their nearest cluster centers; (d) cluster centers are again moved to the centroid of their points

The best clustering minimizes the variance without causing too much complexity (too many clusters). With that in mind, the listed problems can be ameliorated as follows.

1. Run K-means several times, each with different placement of the cluster centers (easy to do, since OpenCV places the centers at random); then choose the run whose results exhibit the least variance.

2. Start with one cluster and try an increasing number of clusters (up to some limit), each time employing the method of #1 as well. Usually the total variance will shrink quite rapidly, after which an "elbow" will appear in the variance curve; this indicates that a new cluster center does not significantly reduce the total variance. Stop at the elbow and keep that many cluster centers.

3. Multiply the data by the inverse covariance matrix (as described in the "Mahalanobis Distance" section). For example, if the input data vectors D are organized as rows with one data point per row, then normalize the "stretch" in the space by computing a new data vector D^*, where $D^* = D\Sigma^{-1/2}$.

K-Means Code

The call for K-means is simple:

```
void cvKMeans2(
    const CvArr*    samples,
    int             cluster_count,
    CvArr*          labels,
    CvTermCriteria  termcrit
);
```

The `samples` array is a matrix of multidimensional data points, one per row. There is a little subtlety here in that each element of the data point may be either a regular floating-point vector of CV_32FC1 numbers or a multidimensional point of type CV_32FC2 or CV_32FC3 or even CV_32FC(K).* The parameter `cluster_count` is simply how many clusters you want, and the return vector `labels` contains the final cluster index for each data point. We encountered `termcrit` in the section "Common Routines in the ML Library" and in the "Controlling Training Iterations" subsection.

It's instructive to see a complete example of K-means in code (Example 13-1), because the data generation sections can be used to test other machine learning routines.

Example 13-1. Using K-means

```
#include "cxcore.h"
#include "highgui.h"

void main( int argc, char** argv )
{
    #define MAX_CLUSTERS 5
    CvScalar color_tab[MAX_CLUSTERS];
    IplImage* img = cvCreateImage( cvSize( 500, 500 ), 8, 3 );
    CvRNG rng = cvRNG(0xffffffff);

    color_tab[0] = CV_RGB(255,0,0);
    color_tab[1] = CV_RGB(0,255,0);
    color_tab[2] = CV_RGB(100,100,255);
    color_tab[3] = CV_RGB(255,0,255);
    color_tab[4] = CV_RGB(255,255,0);

    cvNamedWindow( "clusters", 1 );

    for(;;)
    {
        int k, cluster_count = cvRandInt(&rng)%MAX_CLUSTERS + 1;
        int i, sample_count = cvRandInt(&rng)%1000 + 1;
        CvMat* points = cvCreateMat( sample_count, 1, CV_32FC2 );
        CvMat* clusters = cvCreateMat( sample_count, 1, CV_32SC1 );

        /* generate random sample from multivariate
```

---

* This is exactly equivalent to an N-by-K matrix in which the N rows are the data points, the K columns are the individual components of each point's location, and the underlying data type is 32FC1. Recall that, owing to the memory layout used for arrays, there is no distinction between these representations.

Example 13-1. Using K-means (continued)

```
          Gaussian distribution */
      for( k = 0; k < cluster_count; k++ )
      {
          CvPoint center;
          CvMat point_chunk;
          center.x = cvRandInt(&rng)%img->width;
          center.y = cvRandInt(&rng)%img->height;
          cvGetRows( points, &point_chunk,
                     k*sample_count/cluster_count,
                     k == cluster_count - 1 ? sample_count :
                     (k+1)*sample_count/cluster_count );
          cvRandArr( &rng, &point_chunk, CV_RAND_NORMAL,
                     cvScalar(center.x,center.y,0,0),
                     cvScalar(img->width/6, img->height/6,0,0) );
      }

      /* shuffle samples */
      for( i = 0; i < sample_count/2; i++ )
      {
          CvPoint2D32f* pt1 = (CvPoint2D32f*)points->data.fl +
                              cvRandInt(&rng)%sample_count;
          CvPoint2D32f* pt2 = (CvPoint2D32f*)points->data.fl +
                              cvRandInt(&rng)%sample_count;
          CvPoint2D32f temp;
          CV_SWAP( *pt1, *pt2, temp );
      }

      cvKMeans2( points, cluster_count, clusters,
                 cvTermCriteria( CV_TERMCRIT_EPS+CV_TERMCRIT_ITER,
                                 10, 1.0 ));
      cvZero( img );
      for( i = 0; i < sample_count; i++ )
      {
          CvPoint2D32f pt = ((CvPoint2D32f*)points->data.fl)[i];
          int cluster_idx = clusters->data.i[i];
          cvCircle( img, cvPointFrom32f(pt), 2,
                    color_tab[cluster_idx], CV_FILLED );
      }

      cvReleaseMat( &points );
      cvReleaseMat( &clusters );

      cvShowImage( "clusters", img );

      int key = cvWaitKey(0);
      if( key == 27 ) // 'ESC'
          break;
  }
}
```

In this code we included *highgui.h* to use a window output interface and *cxcore.h* because it contains Kmeans2(). In main(), we set up the coloring of returned clusters for display, set the upper limit to how many cluster centers can be chosen at random to MAX_

CLUSTERS (here 5) in cluster_count, and allow up to 1,000 data points, where the random value for this is kept in sample_count. In the outer for{} loop, which repeats until the Esc key is hit, we allocate a floating point matrix points to contain sample_count data points (in this case, a single column of 2D data points CV_32FC2) and allocate an integer matrix clusters to contain their resulting cluster labels, 0 through cluster_count - 1.

We next enter a data generation for{} loop that can be reused for testing other algorithms. For each cluster, we fill in the points array in successive chunks of size sample_count/cluster_count. Each chunk is filled with a normal distribution, CV_RAND_NORMAL, of 2D (CV_32FC2) data points centered on a randomly chosen 2D center.

The next for{} loop merely shuffles the resulting total "pack" of points. We then call cvKMeans2(), which runs until the largest movement of a cluster center is less than 1 (but allowing no more than ten iterations).

The final for{} loop just draws the results. This is followed by de-allocating the allocated arrays and displaying the results in the "clusters" image. Finally, we wait indefinitely (cvWaitKey(0)) to allow the user another run or to quit via the Esc key.

Naïve/Normal Bayes Classifier

The preceding routines are from *cxcore*. We'll now start discussing the machine learning (ML) library section of OpenCV. We'll begin with OpenCV's simplest supervised classifier, CvNormalBayesClassifier, which is called both a *normal Bayes* classifier and a *naïve Bayes* classifier. It's "naïve" because it assumes that all the features are independent from one another even though this is seldom the case (e.g., finding one eye usually implies that another eye is lurking nearby). Zhang discusses possible reasons for the sometimes surprisingly good performance of this classifier [Zhang04]. Naïve Bayes is not used for regression, but it's an effective classifier that can handle multiple classes, not just two. This classifier is the simplest possible case of what is now a large and growing field known as Bayesian networks, or "probabilistic graphical models". Bayesian networks are causal models; in Figure 13-6, for example, the face features in an image are caused by the existence of a face. In use, the face variable is considered a *hidden variable* and the face features—via image processing operations on the input image—constitute the observed evidence for the existence of a face. We call this a *generative* model because the face causally generates the face features. Conversely, we might start by assuming the face node is active and then randomly sample what features are probabilistically generated given that face is active.* This top-down generation of data with the same statistics as the learned causal model (here, the face) is a useful ability that a purely discriminative model does not possess. For example, one might generate faces for computer graphics display, or a robot might literally "imagine" what it should do next by generating scenes, objects, and interactions. In contrast to Figure 13-6, a discriminative model would have the direction of the arrows reversed.

---

* Generating a face would be silly with the naïve Bayes algorithm because it assumes independence of features. But a more general Bayesian network can easily build in feature dependence as needed.

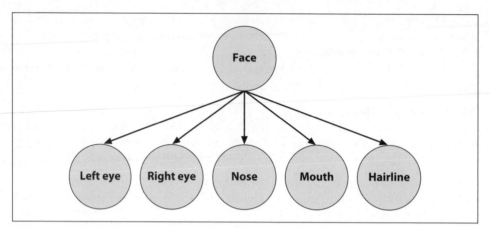

Figure 13-6. A (naïve) Bayesian network, where the lower-level features are caused by the presence of an object (the face)

Bayesian networks are a deep and initially difficult field to understand, but the naïve Bayes algorithm derives from a simple application of Bayes' law. In this case, the probability (denoted p) of a face given the features (denoted, left to right in Figure 13-6, as LE, RE, N, M, H) is:

$$p(\text{face} \mid \text{LE}, \text{RE}, \text{N}, \text{M}, \text{H}) = \frac{p(\text{LE}, \text{RE}, \text{N}, \text{M}, \text{H} \mid \text{face})\, p(\text{face})}{p(\text{LE}, \text{RE}, \text{N}, \text{M}, \text{H})}$$

Just so you'll know, in English this equation means:

$$\text{posterior probability} = \frac{\text{likelihood} \times \text{prior probability}}{\text{evidence}}$$

In practice, we compute some evidence and then decide what object caused it. Since the computed evidence stays the same for the objects, we can drop that term. If we have many models then we need only find the one with the maximum numerator. The numerator is exactly the joint probability of the model with the data: $p(\text{face, LE, RE, N, M, H})$. We can then use the definition of conditional probability to derive the joint probability:

$$p(\text{face}, \text{LE}, \text{RE}, \text{N}, \text{M}, \text{H})$$
$$= p(\text{face})\, p(\text{LE} \mid \text{face})\, p(\text{RE} \mid \text{face}, \text{LE})\, p(\text{N} \mid \text{face}, \text{LE}, \text{RE})$$
$$\times p(\text{M} \mid \text{face}, \text{LE}, \text{RE}, \text{N})\, p(\text{H} \mid \text{face}, \text{LE}, \text{RE}, \text{N}, \text{M})$$

Applying our assumption of independence of features, the conditional features drop out. So, generalizing face to "object" and particular features to "all features", we obtain the reduced equation:

$$p(\text{object, all features}) = p(\text{object}) \prod_{i=1}^{\text{all features}} p(\text{feature}_i \mid \text{object})$$

To use this as an overall classifier, we learn models for the objects that we want. In run mode we compute the features and find the object that maximizes this equation. We typically then test to see if the probability for that "winning" object is over a given threshold. If it is, then we declare the object to be found; if not, we declare that no object was recognized.

 If (as frequently occurs) there is only one object of interest, then you might ask: "The probability I'm computing is the probability relative to what?" In such cases, there is always an implicit second object—namely, the background—which is everything that is *not* the object of interest that we're trying to learn and recognize.

Learning the models is easy. We take many images of the objects; we then compute features over those objects and compute the fraction of how many times a feature occurred over the training set for each object. In practice, we don't allow zero probabilities because that would eliminate the chance of an object existing; hence zero probabilities are typically set to some very low number. In general, if you don't have much data then simple models such as naïve Bayes will tend to outperform more complex models, which will "assume" too much about the data (bias).

Naïve/Normal Bayes Code

The training method for the normal Bayes classifier is:

```
bool CvNormalBayesClassifier::train(
    const CvMat* _train_data,
    const CvMat* _responses,
    const CvMat* _var_idx    = 0,
    const CvMat* _sample_idx = 0,
    bool         update      = false
);
```

This follows the generic method for training described previously, but it allows only data for which each row is a training point (i.e., as if tflag=CV_ROW_SAMPLE). Also, the input _train_data is a single-column CV_32FC1 vector that can only be of type ordered, CV_VAR_ORDERED (numbers). The output label _responses is a vector column that can only be of categorical type CV_VAR_CATEGORICAL (integers, even if contained in a float vector). The parameters _var_idx and _sample_idx are optional; they allow you to mark (respectively) features and data points that you want to use. Mostly you'll use all features and data and simply pass NULL for these vectors, but _sample_idx can be used to divide the training and test sets, for example. Both vectors are either single-channel integer (CV_32SC1) zero-based indexes or 8-bit (CV_8UC1) mask values, where 0 means to skip. Finally, update can be set to merely update the normal Bayes learning rather than to learn a new model from scratch.

The prediction for method for CvNormalBayesClassifier computes the most probable class for its input vectors. One or more input data vectors are stored as rows of the samples matrix. The predictions are returned in corresponding rows of the results vector. If there is only a single input in samples, then the resulting prediction is returned as a float value by the `predict` method and the `results` array may be set to NULL (the default). The format for the prediction method is:

```
float CvNormalBayesClassifier::predict(
    const CvMat* samples,
    CvMat*       results = 0
) const;
```

We move next to a discussion of tree-based classifiers.

Binary Decision Trees

We will go through decision trees in detail, since they are highly useful and use most of the functionality in the machine learning library (and thus serve well as an instructional example). Binary decision trees were invented by Leo Breiman and colleagues,[*] who named them *classification and regression tree* (CART) algorithms. This is the decision tree algorithm that OpenCV implements. The gist of the algorithm is to define an impurity metric relative to the data in every node of the tree. For example, when using regression to fit a function, we might use the sum of squared differences between the true value and the predicted value. We want to minimize the sum of differences (the "impurity") in each node of the tree. For categorical labels, we define a measure that is minimal when most values in a node are of the same class. Three common measures to use are *entropy, Gini index,* and *misclassification* (all are described in this section). Once we have such a metric, a binary decision tree searches through the feature vector to find which feature combined with which threshold most purifies the data. By convention, we say that features above the threshold are "true" and that the data thus classified will branch to the left; the other data points branch right. This procedure is then used recursively down each branch of the tree until the data is of sufficient purity or until the number of data points in a node reaches a set minimum.

The equations for node impurity $i(N)$ are given next. We must deal with two cases, regression and classification.

Regression Impurity

For regression or function fitting, the equation for node impurity is simply the square of the difference in value between the node value y and the data value x. We want to minimize:

$$i(N) = \sum_j (y_j - x_j)^2$$

Classification Impurity

For classification, decision trees often use one of three methods: *entropy impurity, Gini impurity,* or *misclassification impurity*. For these methods, we use the notation $P(\omega_j)$ to

---

[*] L. Breiman, J. Friedman, R. Olshen, and C. Stone, *Classification and Regression Trees* (1984), Wadsworth.

denote the fraction of patterns at node N that are in class ω_j. Each of these impurities has slightly different effects on the splitting decision. Gini is the most commonly used, but all the algorithms attempt to minimize the impurity at a node. Figure 13-7 graphs the impurity measures that we want to minimize.

Entropy impurity

$$i(N) = -\sum_j P(\omega_j) \log P(\omega_j)$$

Gini impurity

$$i(N) = \sum_{j \neq i} P(\omega_i) P(\omega_j)$$

Misclassification impurity

$$i(N) = 1 - \max P(\omega_j)$$

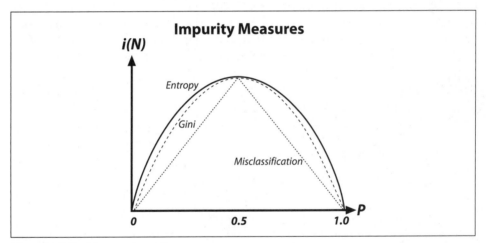

Figure 13-7. Decision tree impurity measures

Decision trees are perhaps the most widely used classification technology. This is due to their simplicity of implementation, ease of interpretation of results, flexibility with different data types (categorical, numerical, unnormalized and mixes thereof), ability to handle missing data through surrogate splits, and natural way of assigning importance to the data features by order of splitting. Decision trees form the basis of other algorithms such as boosting and random trees, which we will discuss shortly.

Decision Tree Usage

In what follows we describe perhaps more than enough for you to get decision trees working well. However, there are many more methods for accessing nodes, modifying splits, and so forth. For that level of detail (which few readers are likely ever to need) you should consult the user manual *.../opencv/docs/ref/opencvref_ml.htm*, particularly with regard to the classes CvDTree{}, the training class CvDTreeTrainData{}, and the nodes CvDTreeNode{} and splits CvDTreeSplit{}.

For a pragmatic introduction, we start by dissecting a specific example. In the *.../opencv/samples/c* directory, there is a *mushroom.cpp* file that runs decision trees on the *agaricus-lepiota.data* data file. This data file consists of a label "p" or "e" (denoting poisonous or edible, respectively) followed by 22 categorical attributes, each represented by a single letter. Observe that the data file is given in "comma separated value" (CSV) format, where the features' values are separated from each other by commas. In *mushroom.cpp* there is a rather messy function mushroom_read_database() for reading in this particular data file. This function is rather overspecific and brittle but mainly it's just filling three arrays as follows. (1) A floating-point matrix data[][], which has dimensions rows = number of data points by columns = number of features (22 in this case) and where all the features are converted from their categorical letter values to floating-point numbers. (2) A character matrix missing[][], where a "true" or "1" indicates a missing value that is indicated in the raw data file by a question mark and where all other values are set to 0. (3) A floating-point vector responses[], which contains the poison "p" or edible "e" response cast in floating-point values. In most cases you would write a more general data input program. We'll now discuss the main working points of *mushroom.cpp*, all of which are called directly or indirectly from main() in the program.

Training the tree

For training the tree, we fill out the tree parameter structure CvDTreeParams{}:

```
struct CvDTreeParams {

    int    max_categories;        //Until pre-clustering
    int    max_depth;             //Maximum levels in a tree
    int    min_sample_count;      //Don't split a node if less
    int    cv_folds;              //Prune tree with K fold cross-validation
    bool   use_surrogates;        //Alternate splits for missing data
    bool   use_1se_rule;          //Harsher pruning
    bool   truncate_pruned_tree;  //Don't "remember" pruned branches
    float  regression_accuracy;   //One of the "stop splitting" criteria
    const float* priors;          //Weight of each prediction category

    CvDTreeParams() : max_categories(10), max_depth(INT_MAX),
      min_sample_count(10), cv_folds(10), use_surrogates(true),
      use_1se_rule(true), truncate_pruned_tree(true),
      regression_accuracy(0.01f), priors(NULL) { ; }

    CvDTreeParams(
        int         _max_depth,
        int         _min_sample_count,
        float       _regression_accuracy,
        bool        _use_surrogates,
        int         _max_categories,
        int         _cv_folds,
        bool        _use_1se_rule,
```

```
    bool            _truncate_pruned_tree,
    const float*    _priors
  );
}
```

In the structure, max_categories has a default value of 10. This limits the number of categorical values before which the decision tree will precluster those categories so that it will have to test no more than $2^{\text{max\_categories}}-2$ possible value subsets.* This isn't a problem for ordered or numerical features, where the algorithm just has to find a threshold at which to split left or right. Those variables that have more categories than max_categories will have their category values clustered down to max_categories possible values. In this way, decision trees will have to test no more than max_categories levels at a time. This parameter, when set to a low value, reduces computation at the cost of accuracy.

The other parameters are fairly self-explanatory. The last parameter, priors, can be crucial. It sets the relative weight that you give to misclassification. That is, if the weight of the first category is 1 and the weight of the second category is 10, then each mistake in predicting the second category is equivalent to making 10 mistakes in predicting the first category. In the code we have edible and poisonous mushrooms, so we "punish" mistaking a poisonous mushroom for an edible one 10 times more than mistaking an edible mushroom for a poisonous one.

The template of the methods for training a decision tree is shown below. There are two methods: the first is used for working directly with decision trees; the second is for ensembles (as used in boosting) or forests (as used in random trees).

```
// Work directly with decision trees:
  bool CvDTree::train(
  const CvMat*    _train_data,
  int             _tflag,
  const CvMat*    _responses,
  const CvMat*    _var_idx      = 0,
  const CvMat*    _sample_idx   = 0,
  const CvMat*    _var_type     = 0,
  const CvMat*    _missing_mask = 0,
  CvDTreeParams params          = CvDTreeParams()
);

// Method that ensembles of decision trees use to call individual
```

---

* More detail on categorical vs. ordered splits: Whereas a split on an ordered variable has the form "if $x <$ a then go left, else go right", a split on a categorical variable has the form "if $x \in \{v_1, v_2, v_3, \dots, v_k\}$ then go left, else go right", where the v_i are some possible values of the variable. Thus, if a categorical variable has N possible values then, in order to find a best split on that variable, one needs to try $2^N - 2$ subsets (empty and full subset are excluded). Thus, an approximate algorithm is used whereby all N values are grouped into $K \leq$ max_categories clusters (via the K-mean algorithm) based on the statistics of the samples in the currently analyzed node. Thereafter, the algorithm tries different combinations of the clusters and chooses the best split, which often gives quite a good result. Note that for the two most common tasks, two-class classification and regression, the optimal categorical split (i.e., the best subset of values) can be found efficiently without any clustering. Hence the clustering is applied only in $n > 2$-class classification problems for categorical variables with $N >$ max_categories possible values. Therefore, you should think twice before setting max_categories to anything greater than 20, which would imply more than a million operations for each split!

```
// training for each tree in the ensemble
bool CvDTree::train(
  CvDTreeTrainData* _train_data,
  const CvMat*       _subsample_idx
);
```

In the train() method, we have the floating-point _train_data[][] matrix. In that matrix, if _tflag is set to CV_ROW_SAMPLE then each row is a data point consisting of a vector of features that make up the columns of the matrix. If tflag is set to CV_COL_SAMPLE, the row and column meanings are reversed. The _responses[] argument is a floating-point vector of values to be predicted given the data features. The other parameters are optional. The vector _var_idx indicates features to include, and the vector _sample_idx indicates data points to include; both of these vectors are either zero-based integer lists of values to skip or 8-bit masks of active (1) or skip (0) values (see our general discussion of the train() method earlier in the chapter). The byte (CV_8UC1) vector _var_type is a zero-based mask for each feature type (CV_VAR_CATEGORICAL or CV_VAR_ORDERED*); its size is equal to the number of features plus 1. That last entry is for the response type to be learned. The byte-valued _missing_mask[][] matrix is used to indicate missing values with a 1 (else 0 is used). Example 13-2 details the creation and training of a decision tree.

Example 13-2. Creating and training a decision tree

```
float priors[] = { 1.0, 10.0}; // Edible vs poisonous weights

CvMat* var_type;

var_type = cvCreateMat( data->cols + 1, 1, CV_8U );

cvSet( var_type, cvScalarAll(CV_VAR_CATEGORICAL) ); // all these vars
                                                    // are categorical
CvDTree* dtree;
dtree = new CvDTree;
dtree->train(
  data,
  CV_ROW_SAMPLE,
  responses,
  0,
  0,
  var_type,
  missing,
  CvDTreeParams(
    8,      // max depth
    10,     // min sample count
    0,      // regression accuracy: N/A here
    true,   // compute surrogate split,
            //   since we have missing data
    15,     // max number of categories
            //   (use suboptimal algorithm for
            //   larger numbers)
    10,     // cross-validations
```

* CV_VAR_ORDERED is the same thing as CV_VAR_NUMERICAL.

Example 13-2. Creating and training a decision tree (continued)

```
        true,   // use 1SE rule => smaller tree
        true,   // throw away the pruned tree branches
        priors  // the array of priors, the bigger
                //   p_weight, the more attention
                //   to the poisonous mushrooms
   )
);
```

In this code the decision tree dtree is declared and allocated. The dtree->train() method is then called. In this case, the vector of responses[] (poisonous or edible) was set to the ASCII value of "p" or "e" (respectively) for each data point. After the train() method terminates, dtree is ready to be used for predicting new data. The decision tree may also be saved to disk via save() and loaded via load() (each method is shown below).* Between the saving and the loading, we reset and zero out the tree by calling the clear() method.

```
    dtree->save("tree.xml","MyTree");

    dtree->clear();

    dtree->load("tree.xml","MyTree");
```

This saves and loads a tree file called *tree.xml*. (Using the *.xml* extension stores an XML data file; if we used a *.yml* or *.yaml* extension, it would store a YAML data file.) The optional "MyTree" is a tag that labels the tree within the *tree.xml* file. As with other statistical models in the machine learning module, multiple objects cannot be stored in a single *.xml* or *.yml* file when using save(); for multiple storage one needs to use cvOpenFileStorage() and write(). However, load() is a different story: this function can load an object by its name even if there is some other data stored in the file.

The function for prediction with a decision tree is:

```
    CvDTreeNode* CvDTree::predict(
      const CvMat* _sample,
      const CvMat* _missing_data_mask = 0,
      bool         raw_mode           = false
    ) const;
```

Here _sample is a floating-point vector of features used to predict; _missing_data_mask is a byte vector of the same length and orientation[†] as the _sample vector, in which non-zero values indicate a missing feature value. Finally, raw_mode indicates unnormalized data with "false" (the default) or "true" for normalized input categorical data values. This is mainly used in ensembles of trees to speed up prediction. Normalizing data to fit within the (0, 1) interval is simply a computational speedup because the algorithm then knows the bounds in which data may fluctuate. Such normalization has no effect on accuracy. This method returns a node of the decision tree, and you may access the

---

[*] As mentioned previously, save() and load() are convenience wrappers for the more complex functions write() and read().

[†] By "same . . . orientation" we mean that if the sample is a 1-by-N vector the mask must be 1-by-N, and if the sample is N-by-1 then the mask must be N-by-1.

predicted value using (CvDTreeNode *)->value which is returned by the dtree->predict() method (see CvDTree::predict() described previously):

```
double r = dtree->predict( &sample, &mask )->value;
```

Finally, we can call the useful var_importance() method to learn about the importance of the individual features. This function will return an *N*-by-1 vector of type double (CV_64FC1) containing each feature's relative importance for prediction, where the value 1 indicates the highest importance and 0 indicates absolutely not important or useful for prediction. Unimportant features may be eliminated on a second-pass training. (See Figure 13-12 for a display of variable importance.) The call is as follows:

```
const CvMat* var_importance = dtree->get_var_importance();
```

As demonstrated in the *.../opencv/samples/c/mushroom.cpp* file, individual elements of the importance vector may be accessed directly via

```
double val = var_importance->data.db[i];
```

Most users will only train and use the decision trees, but advanced or research users may sometimes wish to examine and/or modify the tree nodes or the splitting criteria. As stated in the beginning of this section, the information for how to do this is in the ML documentation that ships with OpenCV at *.../opencv/docs/ref/opencvref_ ml.htm#ch_dtree*, which can also be accessed via the OpenCV Wiki (*http://opencvlibrary .sourceforge.net/*). The sections of interest for such advanced analysis are the class structure CvDTree{}, the training structure CvDTreeTrainData{}, the node structure CvDTree-Node{}, and its contained split structure CvDTreeSplit{}.

Decision Tree Results

Using the code just described, we can learn several things about edible or poisonous mushrooms from the *agaricus-lepiota.data* file. If we just train a decision tree without pruning, so that it learns the data perfectly, we get the tree shown in Figure 13-8. Although the full decision tree learns the training set of data perfectly, remember the lesson of Figure 13-2 (overfitting). What we've done in Figure 13-8 is to memorize the data together with its mistakes and noise. Thus, it is unlikely to perform well on real data. That is why OpenCV decision trees and CART type trees typically include an additional step of penalizing complex trees and pruning them back until complexity is in balance with performance. There are other decision tree implementations that grow the tree only until complexity is balanced with performance and so combine the pruning phase with the learning phase. However, during development of the ML library it was found that trees that are fully grown first and then pruned (as implemented in OpenCV) performed better than those that combine training with pruning in their generation phase.

Figure 13-9 shows a pruned tree that still does quite well (but not perfectly) on the training set but will probably perform better on real data because it has a better balance between bias and variance. Yet this classifier has an serious shortcoming: Although it performs well on the data, it still labels poisonous mushrooms as edible 1.23% of the time. Perhaps we'd be happier with a worse classifier that labeled many edible mushrooms as poisonous provided it never invited us to eat a poisonous mushroom! Such

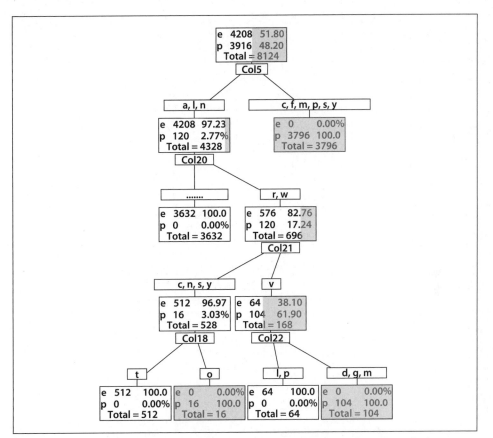

Figure 13-8. Full decision tree for poisonous (p) or edible (e) mushrooms: this tree was built out to full complexity for 0% error on the training set and so would probably suffer from variance problems on test or real data (the dark portion of a rectangle represents the poisonous portion of mushrooms at that phase of categorization)

a classifier can be created by intentionally biasing the classifier and/or the data. This is sometimes referred to as *adding a cost* to the classifier. In our case, we want to add a higher cost for misclassifying poisonous mushrooms than for misclassifying edible mushrooms. Cost can be imposed "inside" a classifier by changing the weighting of how much a "bad" data point counts versus a "good" one. OpenCV allows you to do this by adjusting the priors vector in the CvDTreeParams{} structure passed to the train() method, as we have discussed previously. Even without going inside the classifier code, we can impose a prior cost by duplicating (or resampling from) "bad" data. Duplicating "bad" data points implicitly gives a higher weight to the "bad" data, a technique that can work with any classifier.

Figure 13-10 shows a tree where a 10 × bias was imposed against poisonous mushrooms. This tree makes no mistakes on poisonous mushrooms at a cost of many more mistakes on edible mushrooms—a case of "better safe than sorry". Confusion matrices for the (pruned) unbiased and biased trees are shown in Figure 13-11.

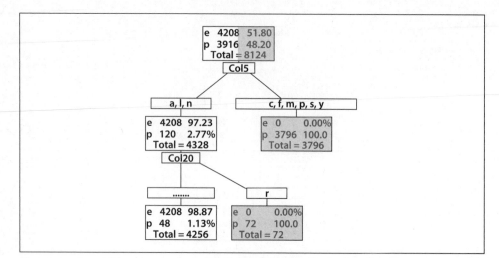

Figure 13-9. Pruned decision tree for poisonous (p) and edible (e) mushrooms: despite being pruned, this tree shows low error on the training set and would likely work well on real data

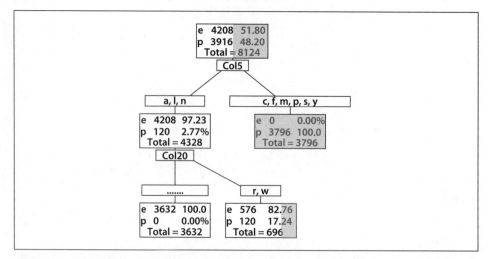

Figure 13-10. An edible mushroom decision tree with 10 × bias against misidentification of poisonous mushrooms as edible; note that the lower right rectangle, though containing a vast majority of edible mushrooms, does not contain a 10 × majority and so would be classified as inedible

Finally, we can learn something more from the data by using the *variable importance* machinery that comes with the tree-based classifiers in OpenCV.* Variable importance measurement techniques were discussed in a previous subsection, and they involve successively perturbing each feature and then measuring the effect on classifier performance. Features that cause larger drops in performance when perturbed are more important. Also, decision trees directly show importance via the splits they found in the

---

* Variable importance techniques may be used with any classifier, but at this time OpenCV implements them only with tree-based methods.

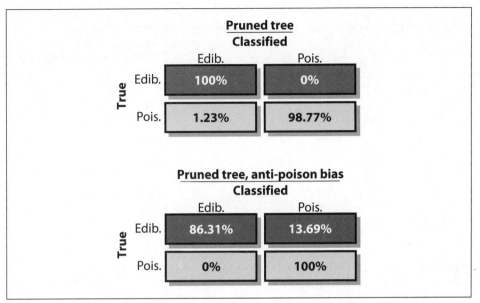

Figure 13-11. Confusion matrices for (pruned) edible mushroom decision trees: the unbiased tree yields better overall performance (top panel) but sometimes misclassifies poisonous mushrooms as edible; the biased tree does not perform as well overall (lower panel) but never misclassifies poisonous mushrooms

data: the first splits are presumably more important than later splits. Splits can be a useful indicator of importance, but they are done in a "greedy" fashion—finding which split most purifies the data *now*. It is often the case that doing a worse split first leads to better splits later, but these trees won't find this out.* The variable importance for poisonous mushrooms is shown in Figure 13-12 for both the unbiased and the biased trees. Note that the order of important variables changes depending on the bias of the trees.

Boosting

Decision trees are extremely useful, but they are often not the best-performing classifiers. In this and the next section we present two techniques, *boosting* and *random trees*, that use trees in their inner loop and so inherit many of the useful properties of trees (e.g., being able to deal with mixed and unnormalized data types and missing features). These techniques typically perform at or near the state of the art; thus they are often the best "out of the box" supervised classification techniques[†] available in the library.

Within in the field of supervised learning there is a *meta-learning* algorithm (first described by Michael Kerns in 1988) called *statistical boosting*. Kerns wondered whether

---

* OpenCV (following Breiman's technique) computes variable importance across all the splits, including surrogate ones, which decreases the possible negative effect that CART's greedy splitting algorithm would have on variable importance ratings.

† Recall that the "no free lunch" theorem informs us that there is no a priori "best" classifier. But on many data sets of interest in vision, boosting and random trees perform quite well.

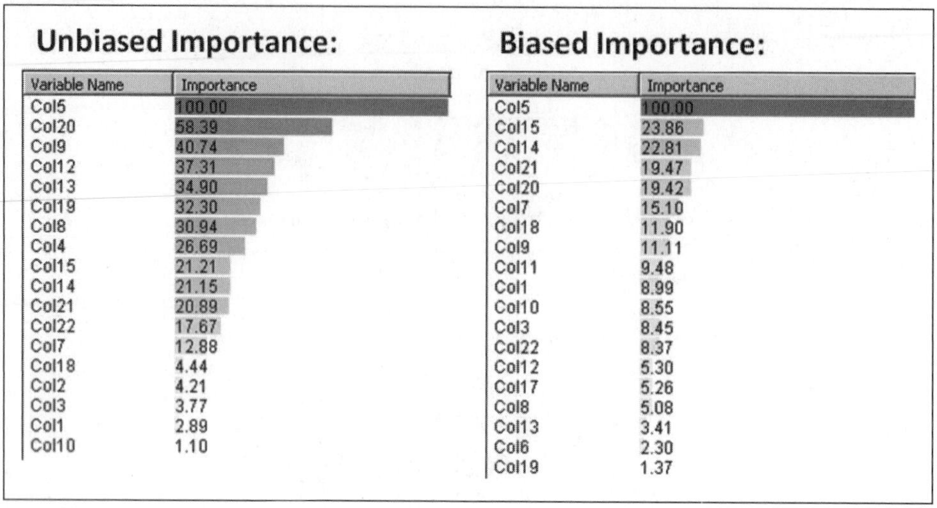

Unbiased Importance:

Variable Name	Importance
Col5	100.00
Col20	58.39
Col9	40.74
Col12	37.31
Col13	34.90
Col19	32.30
Col8	30.94
Col4	26.69
Col15	21.21
Col14	21.15
Col21	20.89
Col22	17.67
Col7	12.88
Col18	4.44
Col2	4.21
Col3	3.77
Col1	2.89
Col10	1.10

Biased Importance:

Variable Name	Importance
Col5	100.00
Col15	23.86
Col14	22.81
Col21	19.47
Col20	19.42
Col7	15.10
Col18	11.90
Col9	11.11
Col11	9.48
Col1	8.99
Col10	8.55
Col3	8.45
Col22	8.37
Col12	5.30
Col17	5.26
Col8	5.08
Col13	3.41
Col6	2.30
Col19	1.37

Figure 13-12. Variable importance for edible mushroom as measured by an unbiased tree (left panel) and a tree biased against poison (right panel)

it is possible to learn a strong classifier out of many weak classifiers.* The first boosting algorithm, known as *AdaBoost*, was formulated shortly thereafter by Freund and Schapire.[†] OpenCV ships with four types of boosting:

- CvBoost :: DISCRETE (discrete AdaBoost)
- CvBoost :: REAL (real AdaBoost)
- CvBoost :: LOGIT (LogitBoost)
- CvBoost :: GENTLE (gentle AdaBoost)

Each of these are variants of the original AdaBoost, and often we find that the "real" and "gentle" forms of AdaBoost work best. *Real AdaBoost* is a technique that utilizes confidence-rated predictions and works well with categorical data. *Gentle AdaBoost* puts less weight on outlier data points and for that reason is often good with regression data. *LogitBoost* can also produce good regression fits. Because you need only set a flag, there's no reason not to try all types on a data set and then select the boosting method that works best.[‡] Here we'll describe the original AdaBoost. For classification it should

---

* The output of a "weak classifier" is only weakly correlated with the true classifications, whereas that of a "strong classifier" is strongly correlated with true classifications. Thus, weak and strong are defined in a statistical sense.

† Y. Freund and R. E. Schapire, "Experiments with a New Boosting Algorithm", in *Machine Learning: Proceedings of the Thirteenth International Conference* (Morgan Kauman, San Francisco, 1996), 148–156.

‡ This procedure is an example of the machine learning metatechnique known as *voodoo learning* or *voodoo programming*. Although unprincipled, it is often an effective method of achieving the best possible performance. Sometimes, after careful thought, one can figure out why the best-performing method was the best, and this can lead to a deeper understanding of the data. Sometimes not.

be noted that, as implemented in OpenCV, boosting is a two-class (yes-or-no) classifier*
(unlike the decision tree or random tree classifiers, which can handle multiple classes at
once). Of the different OpenCV boosting methods, LogitBoost and GentleBoost (referenced in the "Boosting Code" subsection to follow) can be used to perform regression in
addition to binary classification.

AdaBoost

Boosting algorithms are used to train T weak classifiers h_t, $t \in \{1,...,T\}$. These classifiers
are generally very simple individually. In most cases these classifiers are decision trees
with only one split (called *decision stumps*) or at most a few levels of splits (perhaps up to
three). Each of the classifiers is assigned a weighted vote α_t in the final decision-making
process. We use a labeled data set of input feature vectors x_i, each with scalar label y_i
(where $i = 1,...,M$ data points). For AdaBoost the label is binary, $y_i \in \{-1,+1\}$, though it
can be any floating-point number in other algorithms. We initialize a data point weighting distribution $D_t(i)$ that tells the algorithm how much misclassifying a data point will
"cost". The key feature of boosting is that, as the algorithm progresses, this cost will
evolve so that weak classifiers trained later will focus on the data points that the earlier
trained weak classifiers tended to do poorly on. The algorithm is as follows.

1. $D_1(i) = 1/m$, $i = 1,...,m$.

2. For $t = 1,...,T$:

    a. Find the classifier h_t that minimizes the $D_t(i)$ weighted error:

    b. $h_t = \arg\min_{h_j \in H} \varepsilon_j$, where $\varepsilon_j = \sum_{i=1}^{m} D_t(i)$ (for $y_i \neq h_j(x_i)$) as long as $\varepsilon_j < 0.5$;
    else quit.

    c. Set the h_t voting weight $\alpha_t = \frac{1}{2}\log[(1-\varepsilon_t)/\varepsilon_t]$, where ε_t is the arg min error from
    step 2b.

    d. Update the data point weights: $D_{t+1}(i) = [D_t(i)\exp(-\alpha_t y_i h_t(x_i))]/Z_t$, where Z_t
    normalizes the equation over all data points i.

Note that, in step 2b, if we can't find a classifier with less than a 50% error rate then we
quit; we probably need better features.

When the training algorithm just described is finished, the final strong classifier takes
a new input vector x and classifies it using a weighted sum over the learned weak classifiers h_t:

$$H(x) = \text{sign}\left(\sum_{t=1}^{T} \alpha_t h_t(x)\right)$$

---

* There is a trick called *unrolling* that can be used to adapt any binary classifier (including boosting) for
  N-class classification problems, but this makes both training and prediction significantly more expensive.
  See .../opencv/samples/c/letter_recog.cpp.

Here, the sign function converts anything positive into a 1 and anything negative into a –1 (zero remains 0).

Boosting Code

There is example code in *…/opencv/samples/c/letter_recog.cpp* that shows how to use boosting, random trees and back-propagation (aka multilayer perception, MLP). The code for boosting is similar to the code for decision trees but with its own control parameters:

```
struct CvBoostParams : public CvDTreeParams {
    int    boost_type;        // CvBoost:: DISCRETE, REAL, LOGIT, GENTLE
    int    weak_count;        // How many classifiers
    int    split_criteria;    // CvBoost:: DEFAULT, GINI, MISCLASS, SQERR
    double weight_trim_rate;
    CvBoostParams();
    CvBoostParams(
        int          boost_type,
        int          weak_count,
        double       weight_trim_rate,
        int          max_depth,
        bool         use_surrogates,
        const float* priors
    );
};
```

In CvDTreeParams, boost_type selects one of the four boosting algorithms listed previously. The split_criteria is one of the following.

- CvBoost :: DEFAULT (use the default for the particular boosting method)

- CvBoost :: GINI (default option for real AdaBoost)

- CvBoost :: MISCLASS (default option for discrete AdaBoost)

- CvBoost :: SQERR (least-square error; only option available for LogitBoost and gentle AdaBoost)

The last parameter, weight_trim_rate, is for computational savings and is used as described next. As training goes on, many data points become unimportant. That is, the weight $D_t(i)$ for the ith data point becomes very small. The weight_trim_rate is a threshold between 0 and 1 (inclusive) that is implicitly used to throw away some training samples in a given boosting iteration. For example, suppose weight_trim_rate is set to 0.95. This means that samples with summary weight $\leq 1.0-0.95 = 0.05$ (5%) do not participate in the next iteration of training. Note the words "next iteration". The samples are not discarded forever. When the next weak classifier is trained, the weights are computed for all samples and so some previously insignificant samples may be returned back to the next training set. To turn this functionality off, set the weight_trim_rate value to 0.

Observe that CvBoostParams{} inherits from CvDTreeParams{}, so we may set other parameters that are related to decision trees. In particular, if we are dealing with features that may be missing* then we can set use_surrogates to CvDTreeParams::use_surrogates, which will ensure that alternate features on which the splitting is based are stored at each node. An important option is that of using priors to set the "cost" of false positives. Again, if we are learning edible or poisonous mushrooms then we might set the priors to be float priors[] = {1.0, 10.0}; then each error of labeling a poisonous mushroom edible would cost ten times as much as labeling an edible mushroom poisonous.

The CvBoost class contains the member weak, which is a CvSeq* pointer to the weak classifiers that inherits from CvDTree decision trees.[†] For LogitBoost and GentleBoost, the trees are regression trees (trees that predict floating-point values); decision trees for the other methods return only votes for class 0 (if positive) or class 1 (if negative). This contained class sequence has the following prototype:

```
class CvBoostTree: public CvDTree {

public:
    CvBoostTree();
    virtual ~CvBoostTree();
    virtual bool train(
      CvDTreeTrainData* _train_data,
      const CvMat*      subsample_idx,
      CvBoost*          ensemble
    );
    virtual void scale( double s );
    virtual void read(
      CvFileStorage*    fs,
      CvFileNode*       node,
      CvBoost*          ensemble,
      CvDTreeTrainData* _data
    );
    virtual void clear();

protected:
    ...
    CvBoost* ensemble;

};
```

Training is almost the same as for decision trees, but there is an extra parameter called update that is set to false (0) by default. With this setting, we train a whole new ensemble of weak classifiers from scratch. If update is set to true (1) then we just add new weak classifiers onto the existing group. The function prototype for training a boosted classifier is:

---

* Note that, for computer vision, features are computed from an image and then fed to the classifier; hence they are almost never "missing". Missing features arise often in data collected by humans—for example, forgetting to take the patient's temperature one day.

† The naming of these objects is somewhat nonintuitive. The object of type CvBoost is the boosted tree classifier. The objects of type CvBoostTree are the weak classifiers that constitute the overall boosted strong classifier. Presumably, the weak classifiers are typed as CvBoostTree because they derive from CvDTree (i.e., they are little trees in themselves, albeit possibly so little that they are just stumps). The member variable weak of CvBoost points to a sequence enumerating the weak classifiers of type CvBoostTree.

```
bool CvBoost::train(
    const CvMat*   _train_data,
    int            _tflag,
    const CvMat*   _responses,
    const CvMat*   _var_idx      = 0,
    const CvMat*   _sample_idx   = 0,
    const CvMat*   _var_type     = 0,
    const CvMat*   _missing_mask = 0,
    CvBoostParams  params        = CvBoostParams(),
    bool           update        = false
);
```

An example of training a boosted classifier may be found in .../*opencv/samples/c/ letter_recog.cpp*. The training code snippet is shown in Example 13-3.

Example 13-3. Training snippet for boosted classifiers

```
var_type = cvCreateMat( var_count + 2, 1, CV_8U );

cvSet( var_type, cvScalarAll(CV_VAR_ORDERED) );

// the last indicator variable, as well
// as the new (binary) response are categorical
//
cvSetReal1D( var_type, var_count, CV_VAR_CATEGORICAL );
cvSetReal1D( var_type, var_count+1, CV_VAR_CATEGORICAL );

// Train the classifier
//
boost.train(
  new_data,
  CV_ROW_SAMPLE,
  responses,
  0,
  0,
  var_type,
  0,
  CvBoostParams( CvBoost::REAL, 100, 0.95, 5, false, 0 )
);

cvReleaseMat( &new_data );
cvReleaseMat( &new_responses );
```

The prediction function for boosting is also similar to that for decision trees:

```
float CvBoost::predict(
    const CvMat* sample,
    const CvMat* missing        = 0,
    CvMat*       weak_responses = 0,
    CvSlice      slice          = CV_WHOLE_SEQ,
    bool         raw_mode       = false
) const;
```

To perform a simple prediction, we pass in the feature vector sample and then predict() returns the predicted value. Of course, there are a variety of optional parameters. The first of these is the missing feature mask, which is the same as it was for decision trees;

it consists of a byte vector of the same dimension as the sample vector, where nonzero values indicate a missing feature. (Note that this mask cannot be used unless you have trained the classifier with the use_surrogates parameter set to CvDTreeParams::use_surrogates.)

If we want to get back the responses of each of the weak classifiers, we can pass in a floating-point CvMat vector, weak_responses, with length equal to the number of weak classifiers. If weak_responses is passed, CvBoost::predict will fill the vector with the response of each individual classifier:

```
CvMat* weak_responses = cvCreateMat(
    1,
    boostedClassifier.get_weak_predictors()->total,
    CV_32F
);
```

The next prediction parameter, slice, indicates which contiguous subset of the weak classifiers to use; it can be set by

```
inline CvSlice cvSlice( int start, int end );
```

However, we usually just accept the default and leave slice set to "every weak classifier" (CvSlice slice=CV_WHOLE_SEQ). Finally, we have the raw_mode, which is off by default but can be turned on by setting it to true. This parameter is exactly the same as for decision trees and indicates that the data is prenormalized to save computation time. Normally you won't need to use this. An example call for boosted prediction is

```
boost.predict( temp_sample, 0, weak_responses );
```

Finally, some auxiliary functions may be of use from time to time. We can remove a weak classifier from the learned model via

```
void CvBoost::prune( CvSlice slice );
```

We can also return all the weak classifiers for examination:

```
CvSeq* CvBoost::get_weak_predictors();
```

This function returns a CvSeq of pointers to CvBoostTree.

Random Trees

OpenCV contains a *random trees* class, which is implemented following Leo Breiman's theory of *random forests*.[*] Random trees can learn more than one class at a time simply by collecting the class "votes" at the leaves of each of many trees and selecting the class receiving the maximum votes as the winner. Regression is done by averaging the values across the leaves of the "forest". Random trees consist of randomly perturbed decision trees and are among the best-performing classifiers on data sets studied while the ML library was being assembled. Random trees also have the potential for parallel implementation, even on nonshared memory systems, a feature that lends itself to increased use in the future. The basic subsystem on which random trees are built is once again a decision tree. This decision tree is built all the way down until it's *pure*. Thus (cf. the upper right

---

[*] Most of Breiman's work on random forests is conveniently collected on a single website (*http://www.stat .berkeley.edu/users/breiman/RandomForests/cc_home.htm*).

panel of Figure 13-2), each tree is a high-variance classifier that nearly perfectly learns its training data. To counterbalance the high variance, we average together many such trees (hence the name random trees).

Of course, averaging trees will do us no good if the trees are all very similar to each other. To overcome this, random trees cause each tree to be different by randomly selecting a different feature subset of the total features from which the tree may learn at each node. For example, an object-recognition tree might have a long list of potential features: color, texture, gradient magnitude, gradient direction, variance, ratios of values, and so on. Each node of the tree is allowed to choose from a random subset of these features when determining how best to split the data, and each subsequent node of the tree gets a new, randomly chosen subset of features on which to split. The size of these random subsets is often chosen as the square root of the number of features. Thus, if we had 100 potential features then each node would randomly choose 10 of the features and find a best split of the data from among those 10 features. To increase robustness, random trees use an *out of bag* measure to verify splits. That is, at any given node, training occurs on a new subset of the data that is randomly selected *with replacement,** and the rest of the data—those values not randomly selected, called "out of bag" (or OOB) data—are used to estimate the performance of the split. The OOB data is usually set to have about one third of all the data points.

Like all tree-based methods, random trees inherit many of the good properties of trees: surrogate splits for missing values, handling of categorical and numerical values, no need to normalize values, and easy methods for finding variables that are important for prediction. Random trees also used the OOB error results to estimate how well it will do on unseen data. If the training data has a similar distribution to the test data, this OOB performance prediction can be quite accurate.

Finally, random trees can be used to determine, for any two data points, their *proximity* (which in this context means "how alike" they are, not "how near" they are). The algorithm does this by (1) "dropping" the data points into the trees, (2) counting how many times they end up in the same leaf, and (3) dividing this "same leaf" count by the total number of trees. A proximity result of 1 is exactly similar and 0 means very dissimilar. This proximity measure can be used to identify outliers (those points very unlike any other) and also to cluster points (group close points together).

Random Tree Code

We are by now familiar with how the ML library works, and random trees are no exception. It starts with a parameter structure, CvRTParams, which it inherits from decision trees:

```
struct CvRTParams : public CvDTreeParams {

    bool            calc_var_importance;
    int             nactive_vars;
```

* This means that some data points might be randomly repeated.

```
CvTermCriteria term_crit;

CvRTParams() : CvDTreeParams(
  5, 10, 0, false,
  10, 0, false, false,
  0
), calc_var_importance(false), nactive_vars(0) {

  term_crit = cvTermCriteria(
    CV_TERMCRIT_ITER | CV_TERMCRIT_EPS,
    50,
    0.1
  );
}

CvRTParams(
  int          _max_depth,
  int          _min_sample_count,
  float        _regression_accuracy,
  bool         _use_surrogates,
  int          _max_categories,
  const float* _priors,
  bool         _calc_var_importance,
  int          _nactive_vars,
  int          max_tree_count,
  float        forest_accuracy,
  int          termcrit_type,
);

};
```

The key new parameters in CvRTParams are calc_var_importance, which is just a switch to calculate the variable importance of each feature during training (at a slight cost in additional computation time). Figure 13-13 shows the variable importance computed on a subset of the mushroom data set that ships with OpenCV in the *…/opencv/samples/c/ agaricus-lepiota.data* file. The nactive_vars parameter sets the size of the randomly selected subset of features to be tested at any given node and is typically set to the square root of the total number of features; term_crit (a structure discussed elsewhere in this chapter) is the control on the maximum number of trees. For learning random trees, in term_crit the max_iter parameter sets the total number of trees; epsilon sets the "stop learning" criteria to cease adding new trees when the error drops below the OOB error; and the type tells which of the two stopping criteria to use (usually it's both: CV_TERMCRIT_ ITER | CV_TERMCRIT_EPS).

Random trees training has the same form as decision trees training (see the deconstruction of CvDTree::train() in the subsection on "Training the Tree") except that is uses the CvRTParam structure:

```
bool CvRTrees::train(
  const CvMat* train_data,
  int          tflag,
  const CvMat* responses,
  const CvMat* comp_idx    = 0,
```

```
const CvMat* sample_idx   = 0,
const CvMat* var_type     = 0,
const CvMat* missing_mask = 0,
CvRTParams   params       = CvRTParams()
);
```

Variable Name	RandomTrees	Boosting	DecisionTree
Col5	100.00	100.00	100.00
Col20	35.20	58.89	57.37
Col21	16.47	6.11	34.51
Col19	13.35	4.57	26.11
Col9	13.01	43.15	45.96
Col13	10.02	24.47	26.85
Col8	9.52	37.51	42.28
Col12	9.09	27.66	28.90
Col22	8.29	0.28	20.00
Col7	6.08	0.10	21.33
Col15	4.06	1.84	21.41
Col11	3.52	0.44	16.29
Col4	3.12		14.67
Col14	2.98	0.25	20.81
Col18	2.68		0.70
Col3	2.56	0.11	9.15
Col2	2.22	0.39	12.14
Col10	1.79		2.67
Col1	0.41	0.24	7.26
Col17	0.18	0.32	0.54
Col0			
Col6			
Col16			

Figure 13-13. Variable importance over the mushroom data set for random trees, boosting, and decision trees: random trees used fewer significant variables and achieved the best prediction (100% correct on a randomly selected test set covering 20% of data)

An example of calling the train function for a multiclass learning problem is provided in the samples directory that ships with OpenCV; see the *.../opencv/samples/c/letter_recog.cpp* file, where the random trees classifier is named forest.

```
forest.train(
  data,
  CV_ROW_SAMPLE,
  responses,
  0,
  sample_idx,
  var_type,
  0,
  CvRTParams(10,10,0,false,15,0,true,4,100,0.01f,CV_TERMCRIT_ITER)
);
```

Random trees prediction has a form similar to that of the decision trees prediction function CvDTree::predict, but rather than return a CvDTreeNode* pointer it returns the average return value over all the trees in the forest. The `missing` mask is an optional parameter of the same dimension as the `sample` vector, where nonzero values indicate a missing feature value in `sample`.

```
double CvRTrees::predict(
  const CvMat* sample,
  const CvMat* missing = 0
) const;
```

An example prediction call from the *letter_recog.cpp* file is

```
double r;
CvMat  sample;

cvGetRow( data, &sample, i );

r = forest.predict( &sample );
r = fabs((double)r - responses->data.fl[i]) <= FLT_EPSILON ? 1 : 0;
```

In this code, the return variable r is converted into a count of correct predictions.

Finally, there are random tree analysis and utility functions. Assuming that `CvRTParams::calc_var_importance` is set in training, we can obtain the relative importance of each variable by

```
const CvMat* CvRTrees::get_var_importance() const;
```

See Figure 13-13 for an example of variable importance for the mushroom data set from random trees. We can also obtain a measure of the learned random trees model proximity of one data point to another by using the call

```
float CvRTrees::get_proximity(
  const CvMat* sample_1,
  const CvMat* sample_2
) const;
```

As mentioned previously, the returned proximity is 1 if the data points are identical and 0 if the points are completely different. This value is usually between 0 and 1 for two data points drawn from a distribution similar to that of the training set data.

Two other useful functions give the total number of trees or the data structure containing a given decision tree:

```
int          get_tree_count() const; // How many trees are in the forest
CvForestTree* get_tree(int i) const;  // Get an individual decision tree
```

Using Random Trees

We've remarked that the random trees algorithm often performs the best (or among the best) on the data sets we tested, but the best policy is still to try many classifiers once you have your training data defined. We ran random trees, boosting, and decision trees on the mushroom data set. From the 8,124 data points we randomly extracted 1,624 test points, leaving the remainder as the training set. After training these three tree-based classifiers with their default parameters, we obtained the results shown in Table 13-4 on the test set. The mushroom data set is fairly easy and so—although random trees did the best—it wasn't such an overwhelming favorite that we can definitively say which of the three classifiers works better on this particular data set.

Table 13-4. Results of tree-based methods on the OpenCV mushroom data set (1,624 randomly chosen test points with no extra penalties for misclassifying poisonous mushrooms)

Classifier	Performance Results
Random trees	100%
AdaBoost	99%
Decision trees	98%

What is more interesting is the variable importance (which we also measured from the classifiers), shown in Figure 13-13. The figure shows that random trees and boosting each used significantly fewer important variables than required by decision trees. Above 15% significance, random trees used only three variables and boosting used six whereas decision trees needed thirteen. We could thus shrink the feature set size to save computation and memory and still obtain good results. Of course, for the decision trees algorithm you have just a single tree while for random trees and AdaBoost you must evaluate multiple trees; thus, which method has the least computational cost depends on the nature of the data being used.

Face Detection or Haar Classifier

We now turn to the final tree-based technique in OpenCV: the *Haar classifier*, which builds a *boosted rejection cascade*. It has a different format from the rest of the ML library in OpenCV because it was developed earlier as a full-fledged face-recognition application. Thus, we cover it in detail and show how it can be trained to recognize faces and other rigid objects.

Computer vision is a broad and fast-changing field, so the parts of OpenCV that implement a specific technique—rather than a component algorithmic piece—are more at risk of becoming out of date. The face detector that comes with OpenCV is in this "risk" category. However, face detection is such a common need that it is worth having a baseline technique that works fairly well; also, the technique is built on the well-known and often used field of statistical boosting and thus is of more general use as well. In fact, several companies have engineered the "face" detector in OpenCV to detect "mostly rigid" objects (faces, cars, bikes, human body) by training new detectors on many thousands of selected training images for each view of the object. This technique has been used to create state-of-the-art detectors, although with a different detector trained for each view or pose of the object. Thus, the Haar classifier is a valuable tool to keep in mind for such recognition tasks.

OpenCV implements a version of the face-detection technique first developed by Paul Viola and Michael Jones—commonly known as the *Viola-Jones detector*[*]—and later

---

[*] P. Viola and M. J. Jones, "Rapid Object Detection Using a Boosted Cascade of Simple Features," *IEEE CVPR* (2001).

extended by Rainer Lienhart and Jochen Maydt[*] to use *diagonal features* (more on this distinction to follow). OpenCV refers to this detector as the "Haar classifier" because it uses Haar features[†] or, more precisely, Haar-like wavelets that consist of adding and subtracting rectangular image regions before thresholding the result. OpenCV ships with a set of pretrained object-recognition files, but the code also allows you to train and store new object models for the detector. We note once again that the training (createsamples(), haartraining()) and detecting (cvHaarDetectObjects()) code works well on any objects (not just faces) that are consistently textured and mostly rigid.

The pretrained objects that come with OpenCV for this detector are in .../opencv/data/ haarcascades, where the model that works best for frontal face detection is *haarcascade_ frontalface_alt2.xml*. Side face views are harder to detect accurately with this technique (as we shall describe shortly), and those shipped models work less well. If you end up training good object models, perhaps you will consider contributing them as open source back to the community.

Supervised Learning and Boosting Theory

The Haar classifier that is included in OpenCV is a supervised classifier (these were discussed at the beginning of the chapter). In this case we typically present histogram- and size-equalized image patches to the classifier, which are then labeled as containing (or not containing) the object of interest, which for this classifier is most commonly a face.

The Viola-Jones detector uses a form of AdaBoost but organizes it as a *rejection cascade* of nodes, where each node is a multitree AdaBoosted classifier designed to have high (say, 99.9%) detection rate (low false negatives, or missed faces) at the cost of a low (near 50%) rejection rate (high false positives, or "nonfaces" wrongly classified). For each node, a "not in class" result at any stage of the cascade terminates the computation, and the algorithm then declares that no face exists at that location. Thus, true class detection is declared only if the computation makes it through the entire cascade. For instances where the true class is rare (e.g., a face in a picture), rejection cascades can greatly reduce total computation because most of the regions being searched for a face terminate quickly in a nonclass decision.

Boosting in the Haar cascade

Boosted classifiers were discussed earlier in this chapter. For the Viola-Jones rejection cascade, the weak classifiers that it boosts in each node are decision trees that often are only one level deep (i.e., "decision stumps"). A decision stump is allowed just one decision of the following form: "Is the value v of a particular feature f above or below some threshold t"; then, for example, a "yes" indicates face and a "no" indicates no face:

---

[*] R. Lienhart and J. Maydt, "An Extended Set of Haar-like Features for Rapid Object Detection," *IEEE ICIP* (2002), 900–903.

[†] This is technically not correct. The classifier uses the threshold of the sums and differences of rectangular regions of data produced by any feature detector, which may include the Haar case of rectangles of raw (grayscale) image values. Henceforth we will use the term "Haar-like" in deference to this distinction.

$$f_i = \begin{cases} +1 & v_i \geq t_i \\ -1 & v_i < t_i \end{cases}$$

The number of Haar-like features that the Viola-Jones classifier uses in each weak classifier can be set in training, but mostly we use a single feature (i.e., a tree with a single split) or at most about three features. Boosting then iteratively builds up a classifier as a weighted sum of these kinds of weak classifiers. The Viola-Jones classifier uses the classification function:

$$F = \text{sign}(w_1 f_1 + w_2 f_2 + \cdots + w_n f_n)$$

Here, the sign function returns –1 if the number is less than 0, 0 if the number equals 0, and +1 if the number is positive. On the first pass through the data set, we learn the threshold t_1 of f_1 that best classifies the input. Boosting then uses the resulting errors to calculate the weighted vote w_1. As in traditional AdaBoost, each feature vector (data point) is also reweighted low or high according to whether it was classified correctly or not* in that iteration of the classifier. Once a node is learned this way, the surviving data from higher up in the cascade is used to train the next node and so on.

Viola-Jones Classifier Theory

The Viola-Jones classifier employs AdaBoost at each node in the cascade to learn a high detection rate at the cost of low rejection rate multitree (mostly multistump) classifier at each node of the cascade. This algorithm incorporates several innovative features.

1. It uses Haar-like input features: a threshold applied to sums and differences of rectangular image regions.

2. Its *integral image* technique enables rapid computation of the value of rectangular regions or such regions rotated 45 degrees (see Chapter 6). This data structure is used to accelerate computation of the Haar-like input features.

3. It uses statistical boosting to create binary (face–not face) classification nodes characterized by high detection and weak rejection.

4. It organizes the weak classifier nodes of a rejection cascade. In other words: the first group of classifiers is selected that best detects image regions containing an object while allowing many mistaken detections; the next classifier group[†] is the second-best at detection with weak rejection; and so forth. In test mode, an object is detected only if it makes it through the entire cascade.[‡]

---

* There is sometimes confusion about boosting lowering the classification weight on points it classifies correctly in training and raising the weight on points it classified wrongly. The reason is that boosting attempts to focus on correcting the points that it has "trouble" on and to ignore points that it already "knows" how to classify. One of the technical terms for this is that boosting is a margin maximize.

† Remember that each "node" in a rejection cascade is an AdaBoosted group of classifiers.

‡ This allows the cascade to run quickly, because it almost immediately rejects image regions that don't contain the object (and hence need not process through the rest of the cascade).

The Haar-like features used by the classifier are shown in Figure 13-14. At all scales, these features form the "raw material" that will be used by the boosted classifiers. They are rapidly computed from the integral image (see Chapter 6) representing the original grayscale image.

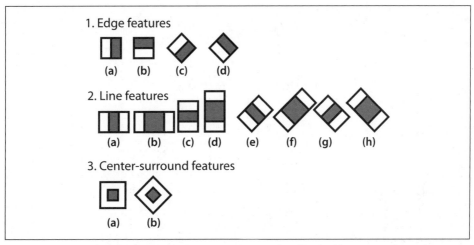

Figure 13-14. Haar-like features from the OpenCV source distribution (the rectangular and rotated regions are easily calculated from the integral image): in this diagrammatic representation of the wavelets, the light region is interpreted as "add that area" and the dark region as "subtract that area"

Viola and Jones organized each boosted classifier group into nodes of a rejection cascade, as shown in Figure 13-15. In the figure, each of the nodes F_j contains an entire boosted cascade of groups of decision stumps (or trees) trained on the Haar-like features from faces and nonfaces (or other objects the user has chosen to train on). Typically, the nodes are ordered from least to most complex so that computations are minimized (simple nodes are tried first) when rejecting easy regions of the image. Typically, the boosting in each node is tuned to have a very high detection rate (at the usual cost of many false positives). When training on faces, for example, almost all (99.9%) of the faces are found but many (about 50%) of the nonfaces are erroneously "detected" at each node. But this is OK because using (say) 20 nodes will still yield a face detection rate (through the whole cascade) of $0.999^{20} \approx 98\%$ with a false positive rate of only $0.5^{20} \approx 0.0001\%$!

During the run mode, a search region of different sizes is swept over the original image. In practice, 70–80% of nonfaces are rejected in the first two nodes of the rejection cascade, where each node uses about ten decision stumps. This quick and early "attentional reject" vastly speeds up face detection.

Works well on . . .

This technique implements face detection but is not limited to faces; it also works fairly well on other (mostly rigid) objects that have distinguishing views. That is, front views

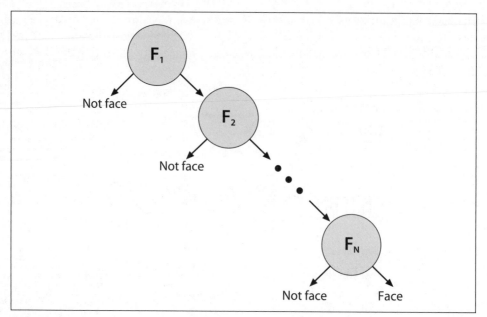

Figure 13-15. Rejection cascade used in the Viola-Jones classifier: each node represents a multitree boosted classifier ensemble tuned to rarely miss a true face while rejecting a possibly small fraction of nonfaces; however, almost all nonfaces have been rejected by the last node, leaving only true faces

of faces work well; backs, sides, or fronts of cars work well; but side views of faces or "corner" views of cars work less well—mainly because these views introduce variations in the template that the "blocky" features (see next paragraph) used in this detector cannot handle well. For example, a side view of a face must catch part of the changing background in its learned model in order to include the profile curve. To detect side views of faces, you may try *haarcascade_profileface.xml*, but to do a better job you should really collect much more data than this model was trained with and perhaps expand the data with different backgrounds behind the face profiles. Again, profile views are hard for this classifier because it uses block features and so is forced to attempt to learn the background variability that "peaks" through the informative profile edge of the side view of faces. In training, it's more efficient to learn only (say) right profile views. Then the test procedure would be to (1) run the right-profile detector and then (2) flip the image on its vertical axis and run the right-profile detector again to detect left-facing profiles.

As we have discussed, detectors based on these Haar-like features work well with "blocky" features—such as eyes, mouth, and hairline—but work less well with tree branches, for example, or when the object's outline shape is its most distinguishing characteristic (as with a coffee mug).

All that being said, if you are willing to gather lots of good, well-segmented data on fairly rigid objects, then this classifier can still compete with the best, and its construction as a rejection cascade makes it very fast to run (though not to train, however). Here "lots of data" means thousands of object examples and tens of thousands of nonobject examples.

By "good" data we mean that one shouldn't mix, for instance, tilted faces with upright faces; instead, keep the data divided and use two classifiers, one for tilted and one for upright. "Well-segmented" data means data that is consistently boxed. Sloppiness in box boundaries of the training data will often lead the classifier to correct for fictitious variability in the data. For example, different placement of the eye locations in the face data location boxes can lead the classifier to assume that eye locations are not a geometrically fixed feature of the face and so can move around. Performance is almost always worse when a classifier attempts to adjust to things that aren't actually in the real data.

Code for Detecting Faces

The detect_and_draw() code shown in Example 13-4 will detect faces and draw their found locations in different-colored rectangles on the image. As shown in the fourth through seventh (comment) lines, this code presumes that a previously trained classifier cascade has been loaded and that memory for detected faces has been created.

Example 13-4. Code for detecting and drawing faces

```
// Detect and draw detected object boxes on image
// Presumes 2 Globals:
// Cascade is loaded by:
//     cascade = (CvHaarClassifierCascade*)cvLoad( cascade_name,
//   0, 0, 0 );
// AND that storage is allocated:
// CvMemStorage* storage = cvCreateMemStorage(0);
//
void detect_and_draw(
  IplImage* img,
  Double    scale = 1.3
){
    static CvScalar colors[] = {
      {{0,0,255}},   {{0,128,255}},{{0,255,255}},{{0,255,0}},
      {{255,128,0}},{{255,255,0}},{{255,0,0}},   {{255,0,255}}
    }; //Just some pretty colors to draw with

    // IMAGE PREPARATION:
    //
    IplImage* gray = cvCreateImage( cvSize(img->width,img->height), 8, 1 );
    IplImage* small_img = cvCreateImage(
      cvSize( cvRound(img->width/scale), cvRound(img->height/scale)), 8, 1
    );
    cvCvtColor( img, gray, CV_BGR2GRAY );
    cvResize( gray, small_img, CV_INTER_LINEAR );
    cvEqualizeHist( small_img, small_img );

    // DETECT OBJECTS IF ANY
    //
    cvClearMemStorage( storage );
    CvSeq* objects = cvHaarDetectObjects(
      small_img,
      cascade,
      storage,
```

Example 13-4. Code for detecting and drawing faces (continued)

```
   1.1,
   2,
   0 /*CV_HAAR_DO_CANNY_PRUNING*/,
   cvSize(30, 30)
);

// LOOP THROUGH FOUND OBJECTS AND DRAW BOXES AROUND THEM
//
for(int i = 0; i < (objects ? objects->total : 0); i++ ) {
  CvRect* r = (CvRect*)cvGetSeqElem( objects, i );
  cvRectangle(
    img,
    cvPoint(r.x,r.y),
    cvPoint(r.x+r.width,r.y+r.height),
    colors[i%8]
  )
}
cvReleaseImage( &graygray );
cvReleaseImage( &small_img );
}
```

For convenience, in this code the detect_and_draw() function has a static array of color vectors colors[] that can be indexed to draw found faces in different colors. The classifier works on grayscale images, so the color BGR image img passed into the function is converted to grayscale using cvCvtColor() and then optionally resized in cvResize(). This is followed by histogram equalization via cvEqualizeHist(), which spreads out the brightness values—necessary because the integral image features are based on differences of rectangle regions and, if the histogram is not balanced, these differences might be skewed by overall lighting or exposure of the test images. Since the classifier returns found object rectangles as a sequence object CvSeq, we need to clear the global storage that we're using for these returns by calling cvClearMemStorage(). The actual detection takes place just above the for{} loop, whose parameters are discussed in more detail below. This loop steps through the found face rectangle regions and draws them in different colors using cvRectangle(). Let us take a closer look at detection function call:

```
CvSeq* cvHaarDetectObjects(
    const CvArr*              image,
    CvHaarClassifierCascade*  cascade,
    CvMemStorage*             storage,
    double                    scale_factor   = 1.1,
    int                       min_neighbors  = 3,
    int                       flags          = 0,
    CvSize                    min_size       = cvSize(0,0)
);
```

CvArr image is a grayscale image. If region of interest (ROI) is set, then the function will respect that region. Thus, one way of speeding up face detection is to trim down the image boundaries using ROI. The classifier cascade is just the Haar feature cascade that we loaded with cvLoad() in the face detect code. The storage argument is an OpenCV "work buffer" for the algorithm; it is allocated with cvCreateMemStorage(0) in the face detection code and cleared for reuse with cvClearMemStorage(storage). The cvHaarDetectObjects() function scans the input image for faces at all scales. Setting the scale_factor parameter determines how big of a jump there is between each scale; setting this to a higher value means faster computation time at the cost of possible missed detections if the scaling misses faces of certain sizes. The min_neighbors parameter is a control for preventing false detection. Actual face locations in an image tend to get multiple "hits" in the same area because the surrounding pixels and scales often indicate a face. Setting this to the default (3) in the face detection code indicates that we will only decide a face is present in a location if there are at least three overlapping detections. The flags parameter has four valid settings, which (as usual) may be combined with the Boolean OR operator. The first is CV_HAAR_DO_CANNY_PRUNING. Setting flags to this value causes flat regions (no lines) to be skipped by the classifier. The second possible flag is CV_HAAR_SCALE_IMAGE, which tells the algorithm to scale the image rather than the detector (this can yield some performance advantages in terms of how memory and cache are used). The next flag option, CV_HAAR_FIND_BIGGEST_OBJECT, tells OpenCV to return only the largest object found (hence the number of objects returned will be either one or none).* The final flag is CV_HAAR_DO_ROUGH_SEARCH, which is used only with CV_HAAR_FIND_BIGGEST_OBJECT. This flag is used to terminate the search at whatever scale the first candidate is found (with enough neighbors to be considered a "hit"). The final parameter, min_size, is the smallest region in which to search for a face. Setting this to a larger value will reduce computation at the cost of missing small faces. Figure 13-16 shows results for using the face-detection code on a scene with faces.

Learning New Objects

We've seen how to load and run a previously trained classifier cascade stored in an XML file. We used the cvLoad() function to load it and then used cvHaarDetectObjects() to find objects similar to the ones it was trained on. We now turn to the question of how to train our own classifiers to detect other objects such as eyes, walking people, cars, et cetera. We do this with the OpenCV *haartraining* application, which creates a classifier given a training set of positive and negative samples. The four steps of training a classifier are described next. (For more details, see the *haartraining* reference manual supplied with OpenCV in the *opencv/apps/HaarTraining/doc* directory.)

1. Gather a data set consisting of examples of the object you want to learn (e.g., front views of faces, side views of cars). These may be stored in one or more directories indexed by a text file in the following format:

   ```
   <path>/img_name_1 count_1 x11 y11 w11 h11 x12 y12 . . .
   <path>/img_name_2 count_2 x21 y21 w21 h21 x22 y22 . . .
        . . .
   ```

   Each of these lines contains the path (if any) and file name of the image containing the object(s). This is followed by the count of how many objects are in that image and then

---

* It is best not to use CV_HAAR_DO_CANNY_PRUNING with CV_HAAR_FIND_BIGGEST_OBJECT. Using both will seldom yield a performance gain; in fact, the net effect will often be a performance loss.

Figure 13-16. Face detection on a park scene: some tilted faces are not detected, and there is also a false positive (shirt near the center); for the 1054-by-851 image shown, more than a million sites and scales were searched to achieve this result in about 1.5 seconds on a 2 GHz machine

a list of rectangles containing the objects. The format of the rectangles is the *x*- and *y*-coordinates of the upper left corner followed by the width and height in pixels.

To be more specific, if we had a data set of faces located in directory *data/faces/*, then the index file *faces.idx* might look like this:

```
data/faces/face_000.jpg 2 73 100 25 37 133 123 30 45
data/faces/face_001.jpg 1 155 200 55 78
. . .
```

If you want your classifier to work well, you will need to gather a lot of high-quality data (1,000–10,000 positive examples). "High quality" means that you've removed all unnecessary variance from the data. For example, if you are learning faces, you should align the eyes (and preferably the nose and mouth) as much as possible. The intuition here is that otherwise you are teaching the classifier that eyes need not appear at fixed locations in the face but instead could be anywhere within some region. Since this is not true of real data, your classifier will not perform as well. One strategy is to first train a cascade on a subpart, say "eyes", which are easier to align. Then use eye detection to find the eyes and rotate/resize the face until the eyes are aligned. For asymmetric data, the "trick" of flipping an image on its vertical axis was described previously in the subsection "Works well on . . .".

2. Use the utility application createsamples to build a vector output file of the positive samples. Using this file, you can repeat the training procedure below on many runs, trying different parameters while using the same vector output file. For example:

```
createsamples -vec faces.vec -info faces.idx -w 30 -h 40
```

This reads in the *faces.idx* file described in step 1 and outputs a formatted training file, *faces.vec*. Then createsamples extracts the positive samples from the images before normalizing and resizing them to the specified width and height (here, 30-by-40). Note that createsamples can also be used to synthesize data by applying geometric transformations, adding noise, altering colors, and so on. This procedure could be used (say) to learn a corporate logo, where you take just one image and put it through various distortions that might appear in real imagery. More details can be found in the OpenCV reference manual *haartraining* located in */apps/HaarTraining/doc/*.

3. The Viola-Jones cascade is a binary classifier: It simply decides whether or not ("yes" or "no") the object in an image is similar to the training set. We've described how to collect and process the "yes" samples that contained the object of choice. Now we turn to describing how to collect and process the "no" samples so that the classifier can learn what does *not* look like our object. Any image that doesn't contain the object of interest can be turned into a negative sample. It is best to take the "no" images from the same type of data we will test on. That is, if we want to learn faces in online videos, for best results we should take our negative samples from comparable frames (i.e., other frames from the same video). However, respectable results can still be achieved using negative samples taken from just about anywhere (e.g., CD or Internet image collections). Again we put the images into one or more directories and then make an index file consisting of a list of image filenames, one per line. For example, an image index file called *backgrounds.idx* might contain the following path and filenames of image collections:

```
data/vacations/beach.jpg
data/nonfaces/img_043.bmp
data/nonfaces/257-5799_IMG.JPG
   . . .
```

4. *Training*. Here's an example training call that you could type on a command line or create using a batch file:

```
Haartraining /
  -data face_classifier_take_3 /
  -vec faces.vec -w 30 -h 40 /
  -bg backgrounds.idx /
  -nstages 20 /
  -nsplits 1 /
  [-nonsym] /
  -minhitrate 0.998 /
  -maxfalsealarm 0.5
```

In this call the resulting classifier will be stored in *face_classifier_take_3.xml*. Here faces.vec is the set of positive samples (sized to width-by-height = 30-by-40), and random images extracted from backgrounds.idx will be used as negative samples. The cascade is set to have 20 (-nstages) stages, where every stage is trained to have a detection rate (-minhitrate) of 0.998 or higher. The false hit rate (-maxfalsealarm) has been set at 50% (or lower) each stage to allow for the overall hit rate of 0.998. The weak classifiers are specified in this case as "stumps", which means they can have only one split (-nsplits); we could ask for more, and this might improve the results in some cases. For more complicated objects one might use as many as six splits, but mostly you want to keep this smaller and use no more than three splits.

Even on a fast machine, training may take several hours to a day, depending on the size of the data set. The training procedure must test approximately 100,000 features within the training window over all positive and negative samples. This search is parallelizable and can take advantage of multicore machines (using OpenMP via the Intel Compiler). This parallel version is the one shipped with OpenCV.

Other Machine Learning Algorithms

We now have a good feel for how the ML library in OpenCV works. It is designed so that new algorithms and techniques can be implemented and embedded into the library easily. In time, it is expected that more new algorithms will appear. This section looks briefly at four machine learning routines that have recently been added to OpenCV. Each implements a well-known learning technique, by which we mean that a substantial body of literature exists on each of these methods in books, published papers, and on the Internet. For more detailed information you should consult the literature and also refer to the *.../opencv/docs/ref/opencvref_ml.htm* manual.

Expectation Maximization

Expectation maximization (EM) is another popular clustering technique. OpenCV supports EM only with Gaussian mixtures, but the technique itself is much more general. It involves multiple iterations of taking the most likely (average or "expected") guess given your current model and then adjusting that model to maximize its chances of being right. In OpenCV, the EM algorithm is implemented in the CvEM{} class and simply involves fitting a mixture of Gaussians to the data. Because the user provides the number of Gaussians to fit, the algorithm is similar to K-means.

K-Nearest Neighbors

One of the simplest classification techniques is *K-nearest neighbors* (KNN), which merely stores all the training data points. When you want to classify a new point, look up its K nearest points (for K an integer number) and then label the new point according to which set contains the majority of its K neighbors. This algorithm is implemented in the CvKNearest{} class in OpenCV. The KNN classification technique can be very effective, but it requires that you store the entire training set; hence it can use a lot of memory and become quite slow. People often cluster the training set to reduce its size before using this method. Readers interested in how dynamically adaptive nearest neighbor type techniques might be used in the brain (and in machine learning) can see Grossberg [Grossberg87] or a more recent summary of advances in Carpenter and Grossberg [Carpenter03].

Multilayer Perceptron

The *multilayer perceptron* (MLP; also known as *back-propagation*) is a neural network that still ranks among the top-performing classifiers, especially for text recognition. It can be rather slow in training because it uses gradient descent to minimize error by adjusting weighted connections between the numerical classification nodes within the layers. In test mode, however, it is quite fast: just a series of dot products followed by a squashing function. In OpenCV it is implemented in the CvANN_MLP{} class, and its use is documented in the *.../opencv/samples/c/letter_recog.cpp* file. Interested readers will find details on using MLP effectively for text and object recognition in LeCun, Bottou, Bengio, and Haffner [LeCun98a]. Implementation and tuning details are given in LeCun, Bottou, and Muller [LeCun98b]. New work on brainlike hierarchical networks that propagate probabilities can be found in Hinton, Osindero, and Teh [Hinton06].

Support Vector Machine

With lots of data, boosting or random trees are usually the best-performing classifiers. But when your data set is limited, the *support vector machine* (SVM) often works best. This *N*-class algorithm works by projecting the data into a higher-dimensional space (creating new dimensions out of combinations of the features) and then finding the optimal linear separator between the classes. In the original space of the raw input data, this high-dimensional linear classifier can become quite nonlinear. Hence we can use linear classification techniques based on *maximal between-class separation* to produce nonlinear classifiers that in some sense optimally separate classes in the data. With enough additional dimensions, you can almost always perfectly separate data classes. This technique is implemented in the CvSVM{} class in OpenCV's ML library.

These tools are closely tied to many computer vision algorithms that range from finding feature points via trained classification to tracking to segmenting scenes and also include the more straightforward tasks of classifying objects and clustering image data.

Exercises

1. Consider trying to learn the next stock price from several past stock prices. Suppose you have 20 years of daily stock data. Discuss the effects of various ways of turning your data into training and testing data sets. What are the advantages and disadvantages of the following approaches?

   a. Take the even-numbered points as your training set and the odd-numbered points as your test set.

b. Randomly select points into training and test sets.

c. Divide the data in two, where the first half is for training and the second half for testing.

d. Divide the data into many small windows of several past points and one prediction point.

2. Figure 13-17 depicts a distribution of "false" and "true" classes. The figure also shows several potential places (a, b, c, d, e, f, g) where a threshold could be set.

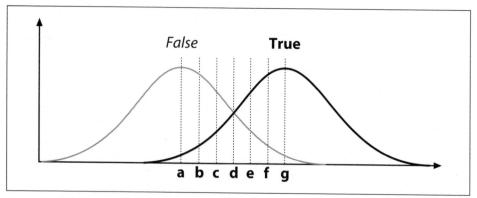

Figure 13-17. A Gaussian distribution of two classes, "false" and "true"

a. Draw the points a–g on an ROC curve.

b. If the "true" class is poisonous mushrooms, at which letter would you set the threshold?

c. How would a decision tree split this data?

3. Refer to Figure 13-1.

a. Draw how a decision tree would approximate the true curve (the dashed line) with three splits (here we seek a regression, not a classification model).

 The "best" split for a regression takes the average value of the data values contained in the leaves that result from the split. The output values of a regression-tree fit thus look like a staircase.

b. Draw how a decision tree would fit the true data in seven splits.

c. Draw how a decision tree would fit the noisy data in seven splits.

d. Discuss the difference between (b) and (c) in terms of overfitting.

4. Why do the splitting measures (e.g., Gini) still work when we want to learn multiple classes in a single decision tree?

5. Review Figure 13-4, which depicts a two-dimensional space with unequal variance at left and equalized variance at right. Let's say that these are feature values related to a classification problem. That is, data near one "blob" belongs to one of two

classes while data near another blob belongs to the same or another of two classes. Would the variable importance be different between the left or the right space for:

a. decision trees?

b. K-nearest neighbors?

c. naïve Bayes?

6. Modify the sample code for data generation in Example 13-1—near the top of the outer for{} loop in the K-means section—to produce a randomly generated labeled data set. We'll use a single normal distribution of 10,000 points centered at pixel (63, 63) in a 128-by-128 image with standard deviation (img->width/6, img->height/6). To label these data, we divide the space into four quadrants centered at pixel (63, 63). To derive the labeling probabilities, we use the following scheme. If $x < 64$ we use a 20% probability for *class A*; else if $x \geq 64$ we use a 90% factor for *class A*. If $y < 64$ we use a 40% probability for *class A*; else if $y \geq 64$ we use a 60% factor for *class A*. Multiplying the x and y probabilities together yields the total probability for *class A* by quadrant with values listed in the 2-by-2 matrix shown. If a point isn't labeled A, then it is labeled B by default. For example, if $x < 64$ and $y < 64$, we would have an 8% chance of a point being labeled *class A* and a 92% chance of that point being labeled *class B*. The four-quadrant matrix for the probability of a point being labeled *class A* (and if not, it's *class B*) is:

$0.2 \times 0.6 = 0.12$	$0.9 \times 0.6 = 0.54$
$0.2 \times 0.4 = 0.08$	$0.9 \times 0.4 = 0.36$

Use these quadrant odds to label the data points. For each data point, determine its quadrant. Then generate a random number from 0 to 1. If this is less than or equal to the quadrant odds, label that data point as *class A*; else label it *class B*. We will then have a list of labeled data points together with x and y as the features. The reader will note that the x-axis is more informative than the y-axis as to which class the data might be. Train random forests on this data and calculate the variable importance to show x is indeed more important than y.

7. Using the same data set as in exercise 6, use discrete AdaBoost to learn two models: one with weak_count set to 20 trees and one set to 500 trees. Randomly select a training and a test set from the 10,000 data points. Train the algorithm and report test results when the training set contains:

a. 150 data points;

b. 500 data points;

c. 1,200 data points;

d. 5,000 data points.

e. Explain your results. What is happening?

8. Repeat exercise 7 but use the random trees classifier with 50 and 500 trees.

9. Repeat exercise 7, but this time use 60 trees and compare random trees versus SVM.

10. In what ways is the random tree algorithm more robust against overfitting than decision trees?

11. Refer to Figure 13-2. Can you imagine conditions under which the test set error would be lower than the training set error?

12. Figure 13-2 was drawn for a regression problem. Label the first point on the graph *A*, the second point *B*, the third point *A*, the forth point *B* and so on. Draw a separation line for these two classes (*A* and *B)* that shows:

    a. bias;

    b. variance.

13. Refer to Figure 13-3.

    a. Draw the generic best-possible ROC curve.

    b. Draw the generic worst-possible ROC curve.

    c. Draw a curve for a classifier that performs randomly on its test data.

14. The "no free lunch" theorem states that no classifier is optimal over all distributions of labeled data. Describe a labeled data distribution over which no classifier described in this chapter would work well.

    a. What distribution would be hard for naïve Bayes to learn?

    b. What distribution would be hard for decision trees to learn?

    c. How would you preprocess the distributions in parts a and b so that the classifiers could learn from the data more easily?

15. Set up and run the Haar classifier to detect your face in a web camera.

    a. How much scale change can it work with?

    b. How much blur?

    c. Through what angles of head tilt will it work?

    d. Through what angles of chin down and up will it work?

    e. Through what angles of head yaw (motion left and right) will it work?

    f. Explore how tolerant it is of 3D head poses. Report on your findings.

16. Use blue or green screening to collect a flat hand gesture (static pose). Collect examples of other hand poses and of random backgrounds. Collect several hundred images and then train the Haar classifier to detect this gesture. Test the classifier in real time and estimate its detection rate.

17. Using your knowledge and what you've learned from exercise 16, improve the results you obtained in that exercise.

CHAPTER 14

OpenCV's Future

Past and Future

In Chapter 1 we saw something of OpenCV's past. This was followed by Chapters 2–13, in which OpenCV's present state was explored in detail. We now turn to OpenCV's future. Computer vision applications are growing rapidly, from product inspection to image and video indexing on the Web to medical applications and even to local navigation on Mars. OpenCV is also growing to accommodate these developments.

OpenCV has long received support from Intel Corporation and has more recently received support from Willow Garage (*www.willowgarage.com*), a privately funded new robotics research institute and technology incubator. Willow Garage's intent is to jumpstart civilian robotics by developing open and supported hardware and software infrastructure that now includes but goes beyond OpenCV. This has given OpenCV new resources for more rapid update and support, with several of the original developers of OpenCV now recontracted to help maintain and advance the library. These renewed resources are also intended to support and enable greater community contribution to OpenCV by allowing for faster code assessment and integration cycles.

One of the key new development areas for OpenCV is robotic perception. This effort focuses on 3D perception as well as 2D plus 3D object recognition since the combination of data types makes for better features for use in object detection, segmentation and recognition. Robotic perception relies heavily on 3D sensing, so efforts are under way to extend camera calibration, rectification and correspondence to multiple cameras and to camera + laser rangefinder combinations (see Figure 14-1).*

Should commercially available hardware warrant it, the "laser + camera calibration" effort will be generalized to include devices such as flash LIDAR and infrared wavefront devices. Additional efforts are aimed at developing triangulation with structured or laser light for extremely accurate depth sensing. The raw output of most depth-sensing

---

* At the time of this writing, these methods remain under development and are not yet in OpenCV.

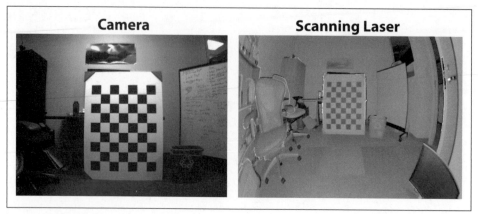

Figure 14-1. New 3D imager combinations: calibrating a camera (left) with the brightness return from a laser depth scanner (right). (Images courtesy of Hai Nguyen and Willow Garage)

methods is in the form of a 3D point cloud. Complementary efforts are thus planned to support turning the raw point clouds resulting from 3D depth perception into 3D meshes. 3D meshes will allow for 3D model capture of objects in the environment, segmenting objects in 3D and hence the ability for robots to grasp and manipulate such objects. Three-dimensional mesh generation can also be used to allow robots to move seamlessly from external 3D perception to internal 3D graphics representation for planning and then back out again for object registration, manipulation, and movement.

Along with sensing 3D objects, robots will need to recognize 3D objects and their 3D poses. To support this, several scalable methods of 2D plus 3D object recognition are being pursued. Creating capable robots subsumes most fields of computer vision and artificial intelligence, from accurate 3D reconstruction to tracking, identifying humans, object recognition, and image stitching and on to learning, control, planning, and decision making. Any higher-level task, such as planning, is made much easier by rapid and accurate depth perception and recognition. It is in these areas especially that OpenCV hopes to enable rapid advance by encouraging many groups to contribute and use ever better methods to solve the difficult problems of real-world perception, recognition, and learning.

OpenCV will, of course, support many other areas as well, from image and movie indexing on the web to security systems and medical analysis. The wishes of the general community will heavily influence OpenCV's direction and growth.

Directions

Although OpenCV does not have an absolute focus on real-time algorithms, it will continue to favor real-time techniques. No one can state future plans with certainty, but the following high-priority areas are likely to be addressed.

Applications

There are more "consumers" for full working applications than there are for low-level functionality. For example, more people will make use of a fully automatic stereo solution than a better subpixel corner detector. There will be several more full applications, such as extensible single-to-many camera calibration and rectification as well as 3D depth display GUI.

3D

As already mentioned, you can expect to see better support for 3D depth sensors and combinations of 2D cameras with 3D measurement devices. Also expect better stereo algorithms. Support for structured light is also likely.

Dense Optical Flow

Because we want to know how whole objects move (and partially to support 3D), OpenCV is long overdue for an efficient implementation of Black's [Black96] dense optical flow techniques.

Features

In support of better object recognition, you can expect a full-function tool kit that will have a framework for interchangeable interest-point detection and interchangeable keys for interest-point identification. This will include popular features such as SURF, HoG, Shape Context, MSER, Geometric Blur, PHOG, PHOW, and others. Support for 2D and 3D features is planned.

Infrastructure

This includes things like a wrapper class,* a good Python interface, GUI improvements, documentation improvements, better error handling, improved Linux support, and so on.

Camera Interface

More seamless handling of cameras is planned along with eventual support for cameras with higher dynamic range. Currently, most cameras support only 8 bits per color channel (if that), but newer cameras can supply 10 or 12 bits per channel.[†] The higher dynamic range of such cameras allows for better recognition and stereo registration because it enables them to detect the subtle textures and colors to which older, more narrow-range cameras are blind.

---

* Daniel Filip and Google have donated the fast, lightweight image class wrapper, WImage, which they developed for internal use, to OpenCV. It will be incorporated by the time this book is published, but too late for documentation in this version.

† Many expensive cameras claim up to 16 bits, but the authors have yet to see more than 10 actual bits of resolution, the rest being noise.

Specific Items

Many object recognition techniques in computer vision detect salient regions that change little between views. These *salient regions** can be tagged with some kind of *key*—for example, a histogram of image gradient directions around the salient point. Although all the techniques described in this section can be built with existing OpenCV primitives, OpenCV currently lacks direct implementation of the most popular interest-region detectors and feature keys.

OpenCV does include an efficient implementation of the Harris corner interest-point detectors, but it lacks direct support for the popular "maximal Laplacian over scale" detector developed by David Lowe [Lowe04] and for maximally stable extremal region (MSER) [Matas02] detectors and others.

Similarly, OpenCV lacks many of the popular keys, such as SURF gradient histogram grids [Bay06], that identify the salient regions. Also, we hope to include features such as histogram of oriented gradients (HoG) [Dalai05], Geometric Blur [Berg01], offset image patches [Torralba07], dense rapidly computed Gaussian scale variant gradients (DAISY) [Tola08], gradient location and orientation histogram (GLOH) [Mikolajczyk04], and, though patented, we want to add for reference the scale invariant feature transform (SIFT) descriptor [Lowe04] that started it all. Other learned feature descriptors that show promise are learned patches with orientation [Hinterstoisser08] and learned ratio points [Ozuysal07]. We'd also like to see contextual or meta-features such as pyramid match kernels [Grauman05], pyramid histogram embedding of other features, PHOW [Bosch07], Shape Context [Belongie00; Mori05], or other approaches that locate features by their probabilistic spatial distribution [Fei-Fei98]. Finally, some global features give the gist of an entire scene, which can be used to boost recognition by context [Oliva06]. All this is a tall order, and the OpenCV community is encouraged to develop and donate code for these and other features.

Other groups have demonstrated encouraging results using frameworks that employ efficient nearest neighbor matching to recognize objects using huge learned databases of objects [Nister06; Philbin07; Torralba08]. Putting in an efficient nearest neighbor framework is therefore suggested.

For robotics, we need object recognition (what) and object location (where). This suggests adding segmentation approaches building on Shi and Malik's work [Shi00] perhaps with faster implementations [Sharon06]. Recent approaches, however, use learning to provide recognition and segmentation together [Oppelt08; Schroff08; Sivic08]. Direction of lighting [Sun98] and shape cues may be important [Zhang99; Prados05].

Along with better support for features and for 3D sensing should come support for visual odometry and visual SLAM (simultaneous localization and mapping). As we acquire more accurate depth perception and feature identification, we'll want to enable better navigation and 3D object manipulation. There is also discussion about creating

---

* These are also known as *interest points*.

a specialized vision interface to a ray-tracing package (e.g., perhaps the Manta open source ray-tracing software [Manta]) in order to generate better 3D object training sets.

Robots, security systems, and Web image and video search all need the ability to recognize objects; thus, OpenCV must refine the pattern-matching techniques in its machine learning library. In particular, OpenCV should first simplify its interface to the learning algorithms and then to give them good defaults so that they work "out of the box". Several new learning techniques may arise, some of which will work with two or more object classes at a time (as random forest does now in OpenCV). There is a need for scalable recognition techniques so that the user can avoid having to learn a completely new model for each object class. More allowances should also be made to enable ML classifiers to work with depth information and 3D features.

Markov random fields (MRFs) and *conditional random fields* (CRFs) are becoming quite popular in computer vision. These methods are often highly problem-specific, yet we would like to figure how they might be supported in a flexible way.

We'll also want methods of learning web-sized or automatically collected via moving robot databases, perhaps by incorporating Zisserman's suggestion for "approximate nearest neighbor" techniques as mentioned previously when dealing with millions or billions of data points. Similarly, we need much-accelerated boosting and Haar feature training support to allow scaling to larger object databases. Several of the ML library routines currently require that all the data reside in memory, severely limiting their use on large datasets. OpenCV will need to break free of such restrictions.

OpenCV also requires better documentation than is now available. This book helps of course, but the OpenCV manual needs an overhaul together with improved search capability. A high priority is incorporating better Linux support and a better external language interface—especially to allow easy vision programming with Python and Numpy. We'll also want to make sure that the machine learning library can be directly called from Python and its SciPy and Numpy packages.

For better developer community interaction, developer workshops may be held at major vision conferences. There are also efforts underway that propose vision "grand challenge" competitions with commensurate prize money.

OpenCV for Artists

There is a worldwide community of interactive artists who use OpenCV so that viewers can interact with their art in dynamic ways. The most commonly used routines for this application are face detection, optical flow, and tracking. We hope this book will enable artists to better understand and use OpenCV for their work, and we believe that the addition of better depth sensing will make interaction richer and more reliable. The focused effort on improving object recognition will allow different modes of interacting with art, because objects can then be used as modal controls. With the ability to capture 3D meshes, it may also be possible to "import" the viewer into the art and so allow the artist to gain a better feel for recognizing user action; this, in turn, could be used to enhance dynamic interaction. The needs and desires of the artistic community for using computer vision will receive enhanced priority in OpenCV's future.

Afterword

We've covered a lot of theory and practice in this book, and we've described some of the plans for what comes next. Of course, as we're developing the software, the hardware is also changing. Cameras are now cheaper and have proliferated from cell phones to traffic lights. A group of manufacturers are aiming to develop cell-phone projectors— perfect for robots, because most cell phones are lightweight, low-energy devices whose circuits already include an embedded camera. This opens the way for close-range portable structured light and thereby accurate depth maps, which are just what we need for robot manipulation and 3D object scanning.

Both authors participated in creating the vision system for Stanley, Stanford's robot racer that won the 2005 DARPA Grand Challenge. In that effort, a vision system coupled with a laser range scanner worked flawlessly for the seven-hour desert road race [Dahlkamp06]. For us, this drove home the power of combining vision with other perception systems: the previously unsolved problem of reliable road perception was converted into a solvable engineering challenge by merging vision with other forms of perception. It is our hope that—by making vision easier to use and more accessible through this book— others can add vision to their own problem-solving tool kits and thus find new ways to solve important problems. That is, with commodity camera hardware and OpenCV, people can start solving real problems such as using stereo vision as an automobile backup safety system, new game controls, and new security systems. Get hacking!

Computer vision has a rich future ahead, and it seems likely to be one of the key enabling technologies for the 21st century. Likewise, OpenCV seems likely to be (at least in part) one of the key enabling technologies for computer vision. Endless opportunities for creativity and profound contribution lie ahead. We hope that this book encourages, excites, and enables all who are interested in joining the vibrant computer vision community.

Bibliography

[Acharya05] T. Acharya and A. Ray, *Image Processing: Principles and Applications*, New York: Wiley, 2005.

[Adelson84] E. H. Adelson, C. H. Anderson, J. R. Bergen, P. J. Burt, and J. M. Ogden, "Pyramid methods in image processing," *RCA Engineer* 29 (1984): 33–41.

[Ahmed74] N. Ahmed, T. Natarajan, and K. R. Rao, "Discrete cosine transform," *IEEE Transactions on Computers* 23 (1974): 90–93.

[Al-Haytham1038] I. al-Haytham, *Book of Optics*, circa 1038.

[AMI] Applied Minds, *http://www.appliedminds.com*.

[Antonisse82] H. J. Antonisse, "Image segmentation in pyramids," *Computer Graphics and Image Processing* 19 (1982): 367–383.

[Arfken85] G. Arfken, "Convolution theorem," in *Mathematical Methods for Physicists*, 3rd ed. (pp. 810–814), Orlando, FL: Academic Press, 1985.

[Bajaj97] C. L. Bajaj, V. Pascucci, and D. R. Schikore, "The contour spectrum," *Proceedings of IEEE Visualization 1997* (pp. 167–173), 1997.

[Ballard81] D. H. Ballard, "Generalizing the Hough transform to detect arbitrary shapes," *Pattern Recognition* 13 (1981): 111–122.

[Ballard82] D. Ballard and C. Brown, *Computer Vision*, Englewood Cliffs, NJ: Prentice-Hall, 1982.

[Bardyn84] J. J. Bardyn et al., "Une architecture VLSI pour un operateur de filtrage median," *Congres reconnaissance des formes et intelligence artificielle* (vol. 1, pp. 557–566), Paris, 25–27 January 1984.

[Bay06] H. Bay, T. Tuytelaars, and L. V. Gool, "SURF: Speeded up robust features," *Proceedings of the Ninth European Conference on Computer Vision* (pp. 404–417), May 2006.

[Bayes1763] T. Bayes, "An essay towards solving a problem in the doctrine of chances. By the late Rev. Mr. Bayes, F.R.S. communicated by Mr. Price, in a letter to John

Canton, A.M.F.R.S.," *Philosophical Transactions, Giving Some Account of the Present Undertakings, Studies and Labours of the Ingenious in Many Considerable Parts of the World* 53 (1763): 370–418.

[Beauchemin95] S. S. Beauchemin and J. L. Barron, "The computation of optical flow," *ACM Computing Surveys* 27 (1995): 433–466.

[Belongie00] S. Belongie, J. Malik, and J. Puzicha, "Shape context: A new descriptor for shape matching and object recognition," NIPS 2000, Computer Vision Group, University of California, Berkeley, 2000.

[Berg01] A. C. Berg and J. Malik, "Geometric blur for template matching," *IEEE Conference on Computer Vision and Pattern Recognition* (vol. 1, pp. 607–614), Kauai, Hawaii, 2001.

[Bhattacharyya43] A. Bhattacharyya, "On a measure of divergence between two statistical populations defined by probability distributions," *Bulletin of the Calcutta Mathematical Society* 35 (1943): 99–109.

[Bishop07] C. M. Bishop, *Pattern Recognition and Machine Learning*, New York: Springer-Verlag, 2007.

[Black92] M. J. Black, "Robust incremental optical flow" (YALEU-DCS-RR-923), Ph.D. thesis, Department of Computer Science, Yale University, New Haven, CT, 1992.

[Black93] M. J. Black and P. Anandan, "A framework for the robust estimation of optical flow," *Fourth International Conference on Computer Vision* (pp. 231–236), May 1993.

[Black96] M. J. Black and P. Anandan, "The robust estimation of multiple motions: Parametric and piecewise-smooth flow fields," *Computer Vision and Image Understanding* 63 (1996): 75–104.

[Bobick96] A. Bobick and J. Davis, "Real-time recognition of activity using temporal templates," *IEEE Workshop on Applications of Computer Vision* (pp. 39–42), December 1996.

[Borgefors86] G. Borgefors, "Distance transformations in digital images," *Computer Vision, Graphics and Image Processing* 34 (1986): 344–371.

[Bosch07] A. Bosch, A. Zisserman, and X. Muñoz, "Image classification using random forests and ferns," *IEEE International Conference on Computer Vision*, Rio de Janeiro, October 2007.

[Bouguet] J.-Y. Bouguet, "Camera calibration toolbox for Matlab," retrieved June 2, 2008, from *http://www.vision.caltech.edu/bouguetj/calib_doc/index.html*.

[BouguetAlg] J.-Y. Bouguet, "The calibration toolbox for Matlab, example 5: Stereo rectification algorithm" (code and instructions only), *http://www.vision.caltech.edu/bouguetj/calib_doc/htmls/example5.html*.

[Bouguet04] J.-Y. Bouguet, "Pyramidal implementation of the Lucas Kanade feature tracker description of the algorithm," *http://robots.stanford.edu/cs223b04/algo_tracking.pdf*.

[Bracewell65] R. Bracewell, "Convolution" and "Two-dimensional convolution," in *The Fourier Transform and Its Applications* (pp. 25–50 and 243–244), New York: McGraw-Hill, 1965.

[Bradski00] G. Bradski and J. Davis, "Motion segmentation and pose recognition with motion history gradients," *IEEE Workshop on Applications of Computer Vision*, 2000.

[Bradski98a] G. R. Bradski, "Real time face and object tracking as a component of a perceptual user interface," *Proceedings of the 4th IEEE Workshop on Applications of Computer Vision*, October 1998.

[Bradski98b] G. R. Bradski, "Computer video face tracking for use in a perceptual user interface," *Intel Technology Journal* Q2 (1998): 705–740.

[Breiman01] L. Breiman, "Random forests," *Machine Learning* 45 (2001): 5–32.

[Breiman84] L. Breiman, J. H. Friedman, R. A. Olshen, and C. J. Stone, *Classification and Regression Trees*, Monteray, CA: Wadsworth, 1984.

[Bresenham65] J. E. Bresenham, "Algorithm for computer control of a digital plotter," *IBM Systems Journal* 4 (1965): 25–30.

[Bronshtein97] I. N. Bronshtein, and K. A. Semendyayev, *Handbook of Mathematics*, 3rd ed., New York: Springer-Verlag, 1997.

[Brown66] D. C. Brown, "Decentering distortion of lenses," *Photogrammetric Engineering* 32 (1966): 444–462.

[Brown71] D. C. Brown, "Close-range camera calibration," *Photogrammetric Engineering* 37 (1971): 855–866.

[Burt81] P. J. Burt, T. H. Hong, and A. Rosenfeld, "Segmentation and estimation of image region properties through cooperative hierarchical computation," *IEEE Transactions on Systems, Man, and Cybernetics* 11 (1981): 802–809.

[Burt83] P. J. Burt and E. H. Adelson, "The Laplacian pyramid as a compact image code," *IEEE Transactions on Communications* 31 (1983): 532–540.

[Canny86] J. Canny, "A computational approach to edge detection," *IEEE Transactions on Pattern Analysis and Machine Intelligence* 8 (1986): 679–714.

[Carpenter03] G. A. Carpenter and S. Grossberg, "Adaptive resonance theory," in M. A. Arbib (Ed.), *The Handbook of Brain Theory and Neural Networks*, 2nd ed. (pp. 87–90), Cambridge, MA: MIT Press, 2003.

[Carr04] H. Carr, J. Snoeyink, and M. van de Panne, "Progressive topological simplification using contour trees and local spatial measures," *15th Western Computer Graphics Symposium*, Big White, British Columbia, March 2004.

[Chen05] D. Chen and G. Zhang, "A new sub-pixel detector for x-corners in camera calibration targets," *WSCG Short Papers* (2005): 97–100.

[Chetverikov99] D. Chetverikov and Zs. Szabo, "A simple and efficient algorithm for detection of high curvature points in planar curves," *Proceedings of the 23rd Workshop of the Austrian Pattern Recognition Group* (pp. 175–184), 1999.

[Chu07] C.-T. Chu, S. K. Kim, Y.-A. Lin, Y. Y. Yu, G. Bradski, A. Y. Ng, and K. Olukotun, "Map-reduce for machine learning on multicore," *Proceedings of the Neural Information Processing Systems Conference* (vol. 19, pp. 304–310), 2007.

[Clarke98] T. A. Clarke and J. G. Fryer, "The Development of Camera Calibration Methods and Models," *Photogrammetric Record* 16 (1998): 51–66.

[Colombari07] A. Colombari, A. Fusiello, and V. Murino, "Video objects segmentation by robust background modeling," *International Conference on Image Analysis and Processing* (pp. 155–164), September 2007.

[Comaniciu99] D. Comaniciu and P. Meer, "Mean shift analysis and applications," *IEEE International Conference on Computer Vision* (vol. 2, p. 1197), 1999.

[Comaniciu03] D. Comaniciu, "Nonparametric information fusion for motion estimation," *IEEE Conference on Computer Vision and Pattern Recognition* (vol. 1, pp. 59–66), 2003.

[Conrady1919] A. Conrady, "Decentering lens systems," *Monthly Notices of the Royal Astronomical Society* 79 (1919): 384–390.

[Cooley65] J. W. Cooley and O. W. Tukey, "An algorithm for the machine calculation of complex Fourier series," *Mathematics of Computation* 19 (1965): 297–301.

[Dahlkamp06] H. Dahlkamp, A. Kaehler, D. Stavens, S. Thrun, and G. Bradski, "Self-supervised monocular road detection in desert terrain," *Robotics: Science and Systems*, Philadelphia, 2006.

[Dalai05] N. Dalai, and B. Triggs, "Histograms of oriented gradients for human detection," *Computer Vision and Pattern Recognition* (vol. 1, pp. 886–893), June 2005.

[Davis97] J. Davis and A. Bobick, "The representation and recognition of action using temporal templates" (Technical Report 402), MIT Media Lab, Cambridge, MA, 1997.

[Davis99] J. Davis and G. Bradski, "Real-time motion template gradients using Intel CVLib," *ICCV Workshop on Framerate Vision*, 1999.

[Delaunay34] B. Delaunay, "Sur la sphère vide," *Izvestia Akademii Nauk SSSR, Otdelenie Matematicheskikh i Estestvennykh Nauk* 7 (1934): 793–800.

[DeMenthon92] D. F. DeMenthon and L. S. Davis, "Model-based object pose in 25 lines of code," *Proceedings of the European Conference on Computer Vision* (pp. 335–343), 1992.

[Dempster77] A. Dempster, N. Laird, and D. Rubin, "Maximum likelihood from incomplete data via the EM algorithm," *Journal of the Royal Statistical Society, Series B* 39 (1977): 1–38.

[Det] "History of matrices and determinants," *http://www-history.mcs.st-and.ac.uk/history/HistTopics/Matrices_and_determinants.html.*

[Douglas73] D. Douglas and T. Peucker, "Algorithms for the reduction of the number of points required for represent a digitized line or its caricature," *Canadian Cartographer* 10(1973): 112–122.

[Duda72] R. O. Duda and P. E. Hart, "Use of the Hough transformation to detect lines and curves in pictures," *Communications of the Association for Computing Machinery* 15 (1972): 11–15.

[Duda73] R. O. Duda and P. E. Hart, *Pattern Recognition and Scene Analysis*, New York: Wiley, 1973.

[Duda00] R. O. Duda, P. E. Hart, and D. G. Stork, *Pattern Classification*, New York: Wiley, 2001.

[Farin04] D. Farin, P. H. N. de With, and W. Effelsberg, "Video-object segmentation using multi-sprite background subtraction," *Proceedings of the IEEE International Conference on Multimedia and Expo*, 2004.

[Faugeras93] O. Faugeras, *Three-Dimensional Computer Vision: A Geometric Viewpoint*, Cambridge, MA: MIT Press, 1993.

[Fei-Fei98] L. Fei-Fei, R. Fergus, and P. Perona, "A Bayesian approach to unsupervised one-shot learning of object categories," *Proceedings of the Ninth International Conference on Computer Vision* (vol. 2, pp. 1134–1141), October 2003.

[Felzenszwalb63] P. F. Felzenszwalb and D. P. Huttenlocher, "Distance transforms of sampled functions" (Technical Report TR2004-1963), Department of Computing and Information Science, Cornell University, Ithaca, NY, 1963.

[FFmpeg] "Ffmpeg summary," *http://en.wikipedia.org/wiki/Ffmpeg.*

[Fischler81] M. A. Fischler and R. C. Bolles, "Random sample concensus: A paradigm for model fitting with applications to image analysis and automated cartography," *Communications of the Association for Computing Machinery* 24 (1981): 381–395.

[Fitzgibbon95] A. W. Fitzgibbon and R. B. Fisher, "A buyer's guide to conic fitting," *Proceedings of the 5th British Machine Vision Conference* (pp. 513–522), Birmingham, 1995.

[Fix51] E. Fix, and J. L. Hodges, "Discriminatory analysis, nonparametric discrimination: Consistency properties" (Technical Report 4), USAF School of Aviation Medicine, Randolph Field, Texas, 1951.

[Forsyth03] D. Forsyth and J. Ponce, *Computer Vision: A Modern Approach*, Englewood Cliffs, NJ: Prentice-Hall, 2003.

[FourCC] "FourCC summary," *http://en.wikipedia.org/wiki/Fourcc.*

[FourCC85] J. Morrison, "EA IFF 85 standard for interchange format files," *http://www.szonye.com/bradd/iff.html.*

[Fourier] "Joseph Fourier," *http://en.wikipedia.org/wiki/Joseph_Fourier.*

[Freeman67] H. Freeman, "On the classification of line-drawing data," *Models for the Perception of Speech and Visual Form* (pp. 408–412), 1967.

[Freeman95] W. T. Freeman and M. Roth, "Orientation histograms for hand gesture recognition," *International Workshop on Automatic Face and Gesture Recognition* (pp. 296–301), June 1995.

[Freund97] Y. Freund and R. E. Schapire, "A decision-theoretic generalization of on-line learning and an application to boosting," *Journal of Computer and System Sciences* 55 (1997): 119–139.

[Fryer86] J. G. Fryer and D. C. Brown, "Lens distortion for close-range photogrammetry," *Photogrammetric Engineering and Remote Sensing* 52 (1986): 51–58.

[Fukunaga90] K. Fukunaga, *Introduction to Statistical Pattern Recognition*, Boston: Academic Press, 1990.

[Galton] "Francis Galton," *http://en.wikipedia.org/wiki/Francis_Galton.*

[GEMM] "Generalized matrix multiplication summary," *http://notvincenz.blogspot.com/2007/06/generalized-matrix-multiplication.html.*

[Göktürk01] S. B. Göktürk, J.-Y. Bouguet, and R. Grzeszczuk, "A data-driven model for monocular face tracking," *Proceedings of the IEEE International Conference on Computer Vision* (vol. 2, pp. 701–708), 2001.

[Göktürk02] S. B. Göktürk, J.-Y. Bouguet, C. Tomasi, and B. Girod, "Model-based face tracking for view-independent facial expression recognition," *Proceedings of the Fifth IEEE International Conference on Automatic Face and Gesture Recognition* (pp. 287–293), May 2002.

[Grauman05] K. Grauman and T. Darrell, "The pyramid match kernel: Discriminative classification with sets of image features," *Proceedings of the IEEE International Conference on Computer Vision*, October 2005.

[Grossberg87] S. Grossberg, "Competitive learning: From interactive activation to adaptive resonance," *Cognitive Science* 11 (1987): 23–63.

[Harris88] C. Harris and M. Stephens, "A combined corner and edge detector," *Proceedings of the 4th Alvey Vision Conference* (pp. 147–151), 1988.

[Hartley98] R. I. Hartley, "Theory and practice of projective rectification," *International Journal of Computer Vision* 35 (1998): 115–127.

[Hartley06] R. Hartley and A. Zisserman, *Multiple View Geometry in Computer Vision*, Cambridge, UK: Cambridge University Press, 2006.

[Hastie01] T. Hastie, R. Tibshirani, and J. Friedman, *The Elements of Statistical Learning: Data Mining, Inference and Prediction*, New York: Springer-Verlag, 2001.

[Heckbert90] P. Heckbert, *A Seed Fill Algorithm* (Graphics Gems I), New York: Academic Press, 1990.

[Heikkila97] J. Heikkila and O. Silven, "A four-step camera calibration procedure with implicit image correction," *Proceedings of the 1997 Conference on Computer Vision and Pattern Recognition* (p. 1106), 1997.

[Hinterstoisser08] S. Hinterstoisser, S. Benhimane, V. Lepetit, P. Fua, and N. Navab, "Simultaneous recognition and homography extraction of local patches with a simple linear classifier," *British Machine Vision Conference*, Leeds, September 2008.

[Hinton06] G. E. Hinton, S. Osindero, and Y. Teh, "A fast learning algorithm for deep belief nets," *Neural Computation* 18 (2006): 1527–1554.

[Ho95] T. K. Ho, "Random decision forest," *Proceedings of the 3rd International Conference on Document Analysis and Recognition* (pp. 278–282), August 1995.

[Homma85] K. Homma and E.-I. Takenaka, "An image processing method for feature extraction of space-occupying lesions," *Journal of Nuclear Medicine* 26 (1985): 1472–1477.

[Horn81] B. K. P. Horn and B. G. Schunck, "Determining optical flow," *Artificial Intelligence* 17 (1981): 185–203.

[Hough59] P. V. C. Hough, "Machine analysis of bubble chamber pictures," *Proceedings of the International Conference on High Energy Accelerators and Instrumentation* (pp. 554–556), 1959.

[Hu62] M. Hu, "Visual pattern recognition by moment invariants," *IRE Transactions on Information Theory* 8 (1962): 179–187.

[Huang95] Y. Huang and X. H. Zhuang, "Motion-partitioned adaptive block matching for video compression," *International Conference on Image Processing* (vol. 1, p. 554), 1995.

[Iivarinen97] J. Iivarinen, M. Peura, J. Särelä, and A. Visa, "Comparison of combined shape descriptors for irregular objects," *8th British Machine Vision Conference*, 1997.

[Intel] Intel Corporation, *http://www.intel.com/*.

[Inui03] K. Inui, S. Kaneko, and S. Igarashi, "Robust line fitting using LmedS clustering," *Systems and Computers in Japan* 34 (2003): 92–100.

[IPL] Intel Image Processing Library (IPL), *www.cc.gatech.edu/dvfx/readings/iplman.pdf*.

[IPP] Intel Integrated Performance Primitives, *http://www.intel.com/cd/software/products/asmo-na/eng/219767.htm*.

[Isard98] M. Isard and A. Blake, "CONDENSATION: Conditional density propagation for visual tracking," *International Journal of Computer Vision* 29 (1998): 5–28.

[Jaehne95] B. Jaehne, *Digital Image Processing*, 3rd ed., Berlin: Springer-Verlag, 1995.

[Jaehne97] B. Jaehne, *Practical Handbook on Image Processing for Scientific Applications*, Boca Raton, FL: CRC Press, 1997.

[Jain77] A. Jain, "A fast Karhunen-Loeve transform for digital restoration of images degraded by white and colored noise," *IEEE Transactions on Computers* 26 (1997): 560–571.

[Jain86] A. Jain, *Fundamentals of Digital Image Processing*, Englewood Cliffs, NJ: Prentice-Hall, 1986.

[Johnson84] D. H. Johnson, "Gauss and the history of the fast Fourier transform," *IEEE Acoustics, Speech, and Signal Processing Magazine* 1 (1984): 14–21.

[Kalman60] R. E. Kalman, "A new approach to linear filtering and prediction problems," *Journal of Basic Engineering* 82 (1960): 35–45.

[Kim05] K. Kim, T. H. Chalidabhongse, D. Harwood, and L. Davis, "Real-time foreground-background segmentation using codebook model," *Real-Time Imaging* 11 (2005): 167–256.

[Kimme75] C. Kimme, D. H. Ballard, and J. Sklansky, "Finding circles by an array of accumulators," *Communications of the Association for Computing Machinery* 18 (1975): 120–122.

[Kiryati91] N. Kiryati, Y. Eldar, and A. M. Bruckshtein, "A probablistic Hough transform," *Pattern Recognition* 24 (1991): 303–316.

[Konolige97] K. Konolige, "Small vision system: Hardware and implementation," *Proceedings of the International Symposium on Robotics Research* (pp. 111–116), Hayama, Japan, 1997.

[Kreveld97] M. van Kreveld, R. van Oostrum, C. L. Bajaj, V. Pascucci, and D. R. Schikore, "Contour trees and small seed sets for isosurface traversal," *Proceedings of the 13th ACM Symposium on Computational Geometry* (pp. 212–220), 1997.

[Lagrange1773] J. L. Lagrange, "Solutions analytiques de quelques problèmes sur les pyramides triangulaires," in *Oeuvres* (vol. 3), 1773.

[Laughlin81] S. B. Laughlin, "A simple coding procedure enhances a neuron's information capacity," *Zeitschrift für Naturforschung* 9/10 (1981): 910–912.

[LeCun98a] Y. LeCun, L. Bottou, Y. Bengio, and P. Haffner, "Gradient-based learning applied to document recognition," *Proceedings of the IEEE* 86 (1998): 2278–2324.

[LeCun98b] Y. LeCun, L. Bottou, G. Orr, and K. Muller, "Efficient BackProp," in G. Orr and K. Muller (Eds.), *Neural Networks: Tricks of the Trade*, New York: Springer-Verlag, 1998.

[Lens] "Lens (optics)," *http://en.wikipedia.org/wiki/Lens_(optics)*.

[Liu07] Y. Z. Liu, H. X. Yao, W. Gao, X. L. Chen, and D. Zhao, "Nonparametric background generation," *Journal of Visual Communication and Image Representation* 18 (2007): 253–263.

[Lloyd57] S. Lloyd, "Least square quantization in PCM's" (Bell Telephone Laboratories Paper), 1957. ["Lloyd's algorithm" was later published in *IEEE Transactions on Information Theory* 28 (1982): 129–137.]

[Lowe04] D. G. Lowe, "Distinctive image features from scale-invariant keypoints," *International Journal of Computer Vision* 60 (2004): 91–110.

[LTI] LTI-Lib, Vision Library, *http://ltilib.sourceforge.net/doc/homepage/index.shtml.*

[Lucas81] B. D. Lucas and T. Kanade, "An iterative image registration technique with an application to stereo vision," *Proceedings of the 1981 DARPA Imaging Understanding Workshop* (pp. 121–130), 1981.

[Lucchese02] L. Lucchese and S. K. Mitra, "Using saddle points for subpixel feature detection in camera calibration targets," *Proceedings of the 2002 Asia Pacific Conference on Circuits and Systems* (pp. 191–195), December 2002.

[Mahal] "Mahalanobis summary," *http://en.wikipedia.org/wiki/Mahalanobis_distance.*

[Mahalanobis36] P. Mahalanobis, "On the generalized distance in statistics," *Proceedings of the National Institute of Science* 12 (1936): 49–55.

[Manta] Manta Open Source Interactive Ray Tracer, *http://code.sci.utah.edu/Manta/index.php/Main_Page.*

[Maron61] M. E. Maron, "Automatic indexing: An experimental inquiry," *Journal of the Association for Computing Machinery* 8 (1961): 404–417.

[Marr82] D. Marr, *Vision*, San Francisco: Freeman, 1982.

[Martins99] F. C. M. Martins, B. R. Nickerson, V. Bostrom, and R. Hazra, "Implementation of a real-time foreground/background segmentation system on the Intel architecture," *IEEE International Conference on Computer Vision Frame Rate Workshop*, 1999.

[Matas00] J. Matas, C. Galambos, and J. Kittler, "Robust detection of lines using the progressive probabilistic Hough transform," *Computer Vision Image Understanding* 78 (2000): 119–137.

[Matas02] J. Matas, O. Chum, M. Urba, and T. Pajdla, "Robust wide baseline stereo from maximally stable extremal regions," *Proceedings of the British Machine Vision Conference* (pp. 384–396), 2002.

[Meer91] P. Meer, D. Mintz, and A. Rosenfeld, "Robust regression methods for computer vision: A review," *International Journal of Computer Vision* 6 (1991): 59–70.

[Merwe00] R. van der Merwe, A. Doucet, N. de Freitas, and E. Wan, "The unscented particle filter," *Advances in Neural Information Processing Systems*, December 2000.

[Meyer78] F. Meyer, "Contrast feature extraction," in J.-L. Chermant (Ed.), *Quantitative Analysis of Microstructures in Material Sciences, Biology and Medicine* [Special issue of *Practical Metallography*], Stuttgart: Riederer, 1978.

[Meyer92] F. Meyer, "Color image segmentation," *Proceedings of the International Conference on Image Processing and Its Applications* (pp. 303–306), 1992.

[Mikolajczyk04] K. Mikolajczyk and C. Schmid, "A performance evaluation of local descriptors," *IEEE Transactions on Pattern Analysis and Machine Intelligence* 27 (2004): 1615–1630.

[Minsky61] M. Minsky, "Steps toward artificial intelligence," *Proceedings of the Institute of Radio Engineers* 49 (1961): 8–30.

[Mokhtarian86] F. Mokhtarian and A. K. Mackworth, "Scale based description and recognition of planar curves and two-dimensional shapes," *IEEE Transactions on Pattern Analysis and Machine Intelligence* 8 (1986): 34–43.

[Mokhtarian88] F. Mokhtarian, "Multi-scale description of space curves and three-dimensional objects," *IEEE Conference on Computer Vision and Pattern Recognition* (pp. 298–303), 1988.

[Mori05] G. Mori, S. Belongie, and J. Malik, "Efficient shape matching using shape contexts," *IEEE Transactions on Pattern Analysis and Machine Intelligence* 27 (2005): 1832–1837.

[Morse53] P. M. Morse and H. Feshbach, "Fourier transforms," in *Methods of Theoretical Physics* (Part I, pp. 453–471), New York: McGraw-Hill, 1953.

[Neveu86] C. F. Neveu, C. R. Dyer, and R. T. Chin, "Two-dimensional object recognition using multiresolution models," *Computer Vision Graphics and Image Processing* 34 (1986): 52–65.

[Ng] A. Ng, "Advice for applying machine learning," *http://www.stanford.edu/class/cs229/materials/ML-advice.pdf*.

[Nistér06] D. Nistér and H. Stewénius, "Scalable recognition with a vocabulary tree," *IEEE Conference on Computer Vision and Pattern Recognition*, 2006.

[O'Connor02] J. J. O'Connor and E. F. Robertson, "Light through the ages: Ancient Greece to Maxwell," *http://www-groups.dcs.st-and.ac.uk/~history/HistTopics/Light_1.html*.

[Oliva06] A. Oliva and A. Torralba, "Building the gist of a scene: The role of global image features in recognition visual perception," *Progress in Brain Research* 155 (2006): 23–36.

[Opelt08] A. Opelt, A. Pinz, and A. Zisserman, "Learning an alphabet of shape and appearance for multi-class object detection," *International Journal of Computer Vision* (2008).

[OpenCV] Open Source Computer Vision Library (OpenCV), *http://sourceforge.net/projects/opencvlibrary/*.

[OpenCV Wiki] Open Source Computer Vision Library Wiki, *http://opencvlibrary .sourceforge.net/*.

[OpenCV YahooGroups] OpenCV discussion group on Yahoo, *http://groups.yahoo .com/group/OpenCV*.

[Ozuysal07] M. Ozuysal, P. Fua, and V. Lepetit, "Fast keypoint recognition in ten lines of code," *Proceedings of the IEEE Conference on Computer Vision and Pattern Recognition*, 2007.

[Papoulis62] A. Papoulis, *The Fourier Integral and Its Applications*, New York: McGraw-Hill, 1962.

[Pascucci02] V. Pascucci and K. Cole-McLaughlin, "Efficient computation of the topology of level sets," *Proceedings of IEEE Visualization 2002* (pp. 187–194), 2002.

[Pearson] "Karl Pearson," *http://en.wikipedia.org/wiki/Karl_Pearson*.

[Philbin07] J. Philbin, O. Chum, M. Isard, J. Sivic, and A. Zisserman, "Object retrieval with large vocabularies and fast spatial matching," *Proceedings of the IEEE Conference on Computer Vision and Pattern Recognition*, 2007.

[Pollefeys99a] M. Pollefeys, "Self-calibration and metric 3D reconstruction from uncalibrated image sequences," Ph.D. thesis, Katholieke Universiteit, Leuven, 1999.

[Pollefeys99b] M. Pollefeys, R. Koch, and L. V. Gool, "A simple and efficient rectification method for general motion," *Proceedings of the 7th IEEE Conference on Computer Vision*, 1999.

[Porter84] T. Porter and T. Duff, "Compositing digital images," *Computer Graphics* 18 (1984): 253–259.

[Prados05] E. Prados and O. Faugeras, "Shape from shading: A well-posed problem?" *Proceedings of the IEEE Conference on Computer Vision and Pattern Recognition*, 2005.

[Ranger07] C. Ranger, R. Raghuraman, A. Penmetsa, G. Bradski, and C. Kozyrakis, "Evaluating mapreduce for multi-core and multiprocessor systems," *Proceedings of the 13th International Symposium on High-Performance Computer Architecture* (pp. 13–24), 2007.

[Reeb46] G. Reeb, "Sur les points singuliers d'une forme de Pfaff completement integrable ou d'une fonction numerique," *Comptes Rendus de l'Academie des Sciences de Paris* 222 (1946): 847–849.

[Rodgers88] J. L. Rodgers and W. A. Nicewander, "Thirteen ways to look at the correlation coefficient," *American Statistician* 42 (1988): 59–66.

[Rodrigues] "Olinde Rodrigues," *http://en.wikipedia.org/wiki/Benjamin_Olinde_ Rodrigues*.

[Rosenfeld73] A. Rosenfeld and E. Johnston, "Angle detection on digital curves," *IEEE Transactions on Computers* 22 (1973): 875–878.

[Rosenfeld80] A. Rosenfeld, "Some Uses of Pyramids in Image Processing and Segmentation," *Proceedings of the DARPA Imaging Understanding Workshop* (pp. 112–120), 1980.

[Rousseeuw84] P. J. Rousseeuw, "Least median of squares regression," *Journal of the American Statistical Association*, 79 (1984): 871–880.

[Rousseeuw87] P. J. Rousseeuw and A. M. Leroy, *Robust Regression and Outlier Detection*, New York: Wiley, 1987.

[Rubner98a] Y. Rubner, C. Tomasi, and L. J. Guibas, "Metrics for distributions with applications to image databases," *Proceedings of the 1998 IEEE International Conference on Computer Vision* (pp. 59–66), Bombay, January 1998.

[Rubner98b] Y. Rubner and C. Tomasi, "Texture metrics," *Proceeding of the IEEE International Conference on Systems, Man, and Cybernetics* (pp. 4601–4607), San Diego, October 1998.

[Rubner00] Y. Rubner, C. Tomasi, and L. J. Guibas, "The earth mover's distance as a metric for image retrieval," *International Journal of Computer Vision* 40 (2000): 99–121.

[Rumelhart88] D. E. Rumelhart, G. E. Hinton, and R. J. Williams, "Learning internal representations by error propagation," in D. E. Rumelhart, J. L. McClelland, and PDP Research Group (Eds.), *Parallel Distributed Processing. Explorations in the Microstructures of Cognition* (vol. 1, pp. 318–362), Cambridge, MA: MIT Press, 1988.

[Russ02] J. C. Russ, *The Image Processing Handbook*, 4th ed. Boca Raton, FL: CRC Press, 2002.

[Scharr00] H. Scharr, "Optimal operators in digital image processing," Ph.D. thesis, Interdisciplinary Center for Scientific Computing, Ruprecht-Karls-Universität, Heidelberg, *http://www.fz-juelich.de/icg/icg-3/index.php?index=195*.

[Schiele96] B. Schiele and J. L. Crowley, "Object recognition using multidimensional receptive field histograms," *European Conference on Computer Vision* (vol. I, pp. 610–619), April 1996.

[Schmidt66] S. Schmidt, "Applications of state-space methods to navigation problems," in C. Leondes (Ed.), *Advances in Control Systems* (vol. 3, pp. 293–340), New York: Academic Press, 1966.

[Schroff08] F. Schroff, A. Criminisi, and A. Zisserman, "Object class segmentation using random forests," *Proceedings of the British Machine Vision Conference*, 2008.

[Schwartz80] E. L. Schwartz, "Computational anatomy and functional architecture of the striate cortex: A spatial mapping approach to perceptual coding," *Vision Research* 20 (1980): 645–669.

[Schwarz78] A. A. Schwarz and J. M. Soha, "Multidimensional histogram normalization contrast enhancement," *Proceedings of the Canadian Symposium on Remote Sensing* (pp. 86–93), 1978.

[Semple79] J. Semple and G. Kneebone, *Algebraic Projective Geometry*, Oxford, UK: Oxford University Press, 1979.

[Serra83] J. Serra, *Image Analysis and Mathematical Morphology*, New York: Academic Press, 1983.

[Sezgin04] M. Sezgin and B. Sankur, "Survey over image thresholding techniques and quantitative performance evaluation," *Journal of Electronic Imaging* 13 (2004): 146–165.

[Shapiro02] L. G. Shapiro and G. C. Stockman, *Computer Vision*, Englewood Cliffs, NJ: Prentice-Hall, 2002.

[Sharon06] E. Sharon, M. Galun, D. Sharon, R. Basri, and A. Brandt, "Hierarchy and adaptivity in segmenting visual scenes," *Nature* 442 (2006): 810–813.

[Shaw04] J. R. Shaw, "QuickFill: An efficient flood fill algorithm," *http://www.codeproject.com/gdi/QuickFill.asp*.

[Shi00] J. Shi and J. Malik, "Normalized cuts and image segmentation," *IEEE Transactions on Pattern Analysis and Machine Intelligence* 22 (2000): 888–905.

[Shi94] J. Shi and C. Tomasi, "Good features to track," *9th IEEE Conference on Computer Vision and Pattern Recognition*, June 1994.

[Sivic08] J. Sivic, B. C. Russell, A. Zisserman, W. T. Freeman, and A. A. Efros, "Unsupervised discovery of visual object class hierarchies," *Proceedings of the IEEE Conference on Computer Vision and Pattern Recognition*, 2008.

[Smith78] A. R. Smith. "Color gamut transform pairs," *Computer Graphics* 12 (1978): 12–19.

[Smith79] A. R. Smith, "Painting tutorial notes," Computer Graphics Laboratory, New York Institute of Technology, Islip, NY, 1979.

[Sobel73] I. Sobel and G. Feldman, "A 3 × 3 Isotropic Gradient Operator for Image Processing," in R. Duda and P. Hart (Eds.), *Pattern Classification and Scene Analysis* (pp. 271–272), New York: Wiley, 1973.

[Steinhaus56] H. Steinhaus, "Sur la division des corp materiels en parties," *Bulletin of the Polish Academy of Sciences and Mathematics* 4 (1956): 801–804.

[Sturm99] P. F. Sturm and S. J. Maybank, "On plane-based camera calibration: A general algorithm, singularities, applications," *IEEE Conference on Computer Vision and Pattern Recognition*, 1999.

[Sun98] J. Sun and P. Perona, "Where is the sun?" *Nature Neuroscience* 1 (1998): 183–184.

[Suzuki85] S. Suzuki and K. Abe, "Topological structural analysis of digital binary images by border following," *Computer Vision, Graphics and Image Processing* 30 (1985): 32–46.

[SVD] "SVD summary," *http://en.wikipedia.org/wiki/Singular_value_decomposition*.

[Swain91] M. J. Swain and D. H. Ballard, "Color indexing," *International Journal of Computer Vision* 7 (1991): 11–32.

[Tanguay00] D. Tanguay, "Flying a Toy Plane," *IEEE Computer Society Conference on Computer Vision and Pattern Recognition* (p. 2231), 2000.

[Teh89] C. H. Teh, R. T. Chin, "On the detection of dominant points on digital curves," *IEEE Transactions on Pattern Analysis and Machine Intelligence* 11 (1989): 859–872.

[Telea04] A. Telea, "An image inpainting technique based on the fast marching method," *Journal of Graphics Tools* 9 (2004): 25–36.

[Thrun05] S. Thrun, W. Burgard, and D. Fox, *Probabilistic Robotics: Intelligent Robotics and Autonomus Agents*, Cambridge, MA: MIT Press, 2005.

[Thrun06] S. Thrun, M. Montemerlo, H. Dahlkamp, D. Stavens, A. Aron, J. Diebel, P. Fong, J. Gale, M. Halpenny, G. Hoffmann, K. Lau, C. Oakley, M. Palatucci, V. Pratt, P. Stang, S. Strohband, C. Dupont, L.-E. Jendrossek, C. Koelen, C. Markey, C. Rummel, J. van Niekerk, E. Jensen, P. Alessandrini, G. Bradski, B. Davies, S. Ettinger, A. Kaehler, A. Nefian, and P. Mahoney. "Stanley, the robot that won the DARPA Grand Challenge," *Journal of Robotic Systems* 23 (2006): 661–692.

[Titchmarsh26] E. C. Titchmarsh, "The zeros of certain integral functions," *Proceedings of the London Mathematical Society* 25 (1926): 283–302.

[Tola08] E. Tola, V. Lepetit, and P. Fua, "A fast local descriptor for dense matching," *Proceedings of the IEEE International Conference on Computer Vision and Pattern Recognition*, June 2008.

[Tomasi98] C. Tomasi and R. Manduchi, "Bilateral filtering for gray and color images," *Sixth International Conference on Computer Vision* (pp. 839–846), New Delhi, 1998.

[Torralba07] A. Torralba, K. P. Murphy, and W. T. Freeman, "Sharing visual features for multiclass and multiview object detection," *IEEE Transactions on Pattern Analysis and Machine Intelligence* 29 (2007): 854–869.

[Torralba08] A. Torralba, R. Fergus, and Y. Weiss, "Small codes and large databases for recognition," *Proceedings of the IEEE International Conference on Computer Vision and Pattern Recognition*, June 2008.

[Toyama99] K. Toyama, J. Krumm, B. Brumitt, and B. Meyers, "Wallflower: Principles and practice of background maintenance," *Proceedings of the 7th IEEE International Conference on Computer Vision* (pp. 255–261), 1999.

[Trace] "Matrix trace summary," *http://en.wikipedia.org/wiki/Trace_(linear_algebra)*.

[Trucco98] E. Trucco and A. Verri, *Introductory Techniques for 3-D Computer Vision*, Englewood Cliffs, NJ: Prentice-Hall, 1998.

[Tsai87] R. Y. Tsai, "A versatile camera calibration technique for high accuracy 3D machine vision metrology using off-the-shelf TV cameras and lenses," *IEEE Journal of Robotics and Automation* 3 (1987): 323–344.

[Vandevenne04] L. Vandevenne, "Lode's computer graphics tutorial, flood fill," *http://student.kuleuven.be/~m0216922/CG/floodfill.html*.

[Vapnik95] V. Vapnik, *The Nature of Statistical Learning Theory*, New York: Springer-Verlag, 1995.

[Videre] Videre Design, "Stereo on a chip (STOC)," *http://www.videredesign.com/templates/stoc.htm*.

[Viola04] P. Viola and M. J. Jones, "Robust real-time face detection," *International Journal of Computer Vision* 57 (2004): 137–154.

[VXL] VXL, Vision Library, *http://vxl.sourceforge.net/*.

[Welsh95] G. Welsh and G. Bishop, "An introduction to the Kalman filter" (Technical Report TR95-041), University of North Carolina, Chapel Hill, NC, 1995.

[Werbos74] P. Werbos, "Beyond regression: New tools for prediction and analysis in the behavioural sciences," Ph.D. thesis, Economics Department, Harvard University, Cambridge, MA, 1974.

[WG] Willow Garage, *http://www.willowgarage.com*.

[Wharton71] W. Wharton and D. Howorth, *Principles of Television Reception*, London: Pitman, 1971.

[Xu96] G. Xu and Z. Zhang, *Epipolar Geometry in Stereo, Motion and Object Recognition*, Dordrecht: Kluwer, 1996

[Zhang96] Z. Zhang, "Parameter estimation techniques: A tutorial with application to conic fitting," *Image and Vision Computing* 15 (1996): 59–76.

[Zhang99] R. Zhang, P.-S. Tsi, J. E. Cryer, and M. Shah, "Shape form shading: A survey," *IEEE Transactions on Pattern Analysis and Machine Intelligence* 21 (1999): 690 –706.

[Zhang99] Z. Zhang, "Flexible camera calibration by viewing a plane from unknown orientations," *Proceedings of the 7th International Conference on Computer Vision* (pp. 666–673), Corfu, September 1999.

[Zhang00] Z. Zhang, "A flexible new technique for camera calibration," *IEEE Transactions on Pattern Analysis and Machine Intelligence* 22 (2000): 1330–1334.

[Zhang04] H. Zhang, "The optimality of naive Bayes," *Proceedings of the 17th International FLAIRS Conference*, 2004.

Index

C

Intel Compiler, 516
Intel Corporation, 521
Intel Research, 6
Intel website for IPP, 9
intensity bumps/holes, finding, 115
intentional bias, 493–495, 496
interpolation, 130, 162, 163, 176
intersection method, histograms, 202
intrinsic parameters, defined, 371
intrinsics matrix, defined, 373, 392
inverse transforms, 179
IPAN algorithm, 246, 247
IPL (Image Processing Library), 42
IplImage data structure
    compared with RGB, 32
    element functions, 38, 39
    overview of, 42
    variables, 17, 42, 45–47
IPP (Integrated Performance Primitives),
    1, 7–10, 86, 179

J

Jacobi's method, 61, 406
Jaehne, B., 132
Jones, M. J., 506–511, 515

K

K-means algorithm, 462, 472, 479–483
K-nearest neighbor (KNN), 463, 471, 516
Kalman filter, 350–363
    blending factor (Kalman gain), 357
    extended, 363
    limitations of, 364
    mathematics of, 351–353, 355–358
    OpenCV and, 358–363
    overview of, 350
kernel density estimation, 338
kernels
    convolution, 144
    custom, 118–120
    defined, 115, 338
    shape values, 120
    support of, 144
Kerns, Michael, 495
key-frame, handling of, 21
Konolige, Kurt, 439
Kuriakin, Valery, 6

L

Lagrange multiplier, 336
Laplacian operator, 150–152

Laplacian pyramid, defined, 131, 132
learning, 459
Lee, Shinn, 6
lens distortion model, 371, 375–377, 378,
    391, 416
lenses, 370
Levenberg-Marquardt algorithm, 428
licensing terms, 2, 8
Lienhart, Rainer, 507
linear transformation, 56
lines
    drawing, 77–78
    epipolar, 454–457
    finding, 25, 153
    (see also Delaunay triangulation)
link strength, 298
Linux systems, 1, 8, 9, 15, 94, 523, 525
Lloyd algorithm, 479
LMedS algorithm, 425
log-polar transforms, 174–177
Lowe, David, 524
Lowe SIFT demo, 464
Lucas-Kanade (sparse) method, 316, 317,
    323–334, 335

M

Machine Learning Library (MLL), 1, 11–13,
    471–475
machine learning, overview of, 459–466
MacOS systems, 1, 10, 15, 92, 94
MacPowerPC, 15
Mahalonobis distance, 49, 66, 462–471,
    476–478
malloc() function, 223
Manhattan distance, 208
Manta open source ray-tracing, 524
Markov random fields (MRFs), 525
masks, 47, 120, 124, 135
matching methods
    Bhattacharyya, 202
    block, 322, 336, 439, 443–444
    contours, 251–259
    hierarchical, 256–259
    histogram, 201–206
    Hu moments, 253–256, 347, 348
    template, 214–219
Matlab interface, 1, 109, 431
matrix
    accessing data in, 34, 36–41
    array, comparison with, 40
    creating, 34, 35

ROC (receiver operating characteristic), 469, 470
Rodrigues, Olinde, 402
Rodrigues transform, 401–402, 406
ROI (region of interest), 43–46, 52
Rom, H., 207
Rosenfeld-Johnson algorithm, 245
rotation matrix, defined, 379–381
rotation vector, defined, 392
Ruby interface, 1
running average, 276

S

SAD (sum of absolute difference), 439, 443
salient regions, 523
scalable recognition techniques, 524
scalar tuples, 32
scale-invariant feature transform (SIFT), 321, 464, 524
scene modeling, 267
scene transitions, 193
Schapire, R. E., 496
Scharr filter, 150, 343
SciPy, 525
scrambled covariance matrix, 54–55
seed point, 124
segmentation, overview of, 265
self-cleaning procedure, 25
sequences, 134, 223–234
    accessing, 134, 226
    block size, 231
    converting to array, 233
    copying, 227–229
    creating, 224–226
    deleting, 226
    inserting and removing elements from, 231
    moving, 227–229
    partitioning, 229, 230
    readers, 231–233
    sorting, 228
    stack, using as, 229
    writers, 231–233
setup, OpenCV, 16
Shape Context, 523, 524
Shi and Tomasi corners, 318, 321
SHT (standard Hough transform), 156
SIFT (scale-invariant feature transform), 321, 464, 524
silhouettes, object, 342–346
simultaneous localization and mapping (SLAM), 524

singularity threshold, 76
singular value decomposition, 60, 61, 75, 391
SLAM (simultaneous localization and mapping), 524
slider trackbar, 20–22, 99–102, 105, 242
smoothing, 22–24, 109–115
Sobel derivatives, 145, 148–151, 158, 318, 343
software, additional needed, 8
Software Performance Libraries group, 6
SourceForge site, 8
spatial coherence, 324
speckle noise, 117, 443
spectrum multiplication, 179
square differences matching method, 215
stack, sequence as a, 229
standard Hough transform (SHT), 156
Stanford's "Stanley" robot, 2, 526
statistical machine learning, 467
stereo imaging
    calibration, 427–430, 445–452
    correspondence, 438–445
    overview of, 415
    rectification, 427, 433, 438, 439, 452
stereo reconstruction ambiguity, 432
strong classifiers, 496, 499
structured light, 523
subpixel corners, 319–321, 383, 523
summary characteristics, 247
sum of absolute difference (SAD), 439, 443
Sun systems, 15
superpixels, 265
supervised/unsupervised data, 460
support vector machine (SVM), 463, 470, 517
SURF gradient histogram grids, 523, 524
SVD (singular value decomposition), 60, 61, 75, 391
SVM (support vector machine), 463, 470, 517
switches, 101

T

tangential distortions, 375–377, 378
Taylor series, 375
Teh-Chin algorithm, 245
temporal persistence, 324
test sets, 460–464
text, drawing, 80–82
texture descriptors, 14
textured scene, high and low, 439
thresholds
    actions above/below, 135
    adaptive, 138–141

About the Authors

Dr. Gary Rost Bradski is a consulting professor in the CS department at the Stanford University AI Lab, where he mentors robotics, machine learning, and computer vision research. He is also senior scientist at Willow Garage (*http://www.willowgarage.com*), a recently founded robotics research institute/incubator. He holds a B.S. in EECS from UC Berkeley and a Ph.D. from Boston University. He has 20 years of industrial experience applying machine learning and computer vision, spanning option-trading operations at First Union National Bank, to computer vision at Intel Research, to machine learning in Intel Manufacturing, and several startup companies in between.

Gary started the Open Source Computer Vision Library (OpenCV, *http://sourceforge.net/projects/opencvlibrary*), which is used around the world in research, in government, and commercially; the statistical Machine Learning Library (which comes with OpenCV); and the Probabilistic Network Library (PNL). The vision libraries helped develop a notable part of the commercial Intel Performance Primitives Library (IPP, *http://tinyurl.com/36ua5s*). Gary also organized the vision team for Stanley, the Stanford robot that won the DARPA Grand Challenge autonomous race across the desert for a $2M team prize, and he helped found the Stanford AI Robotics project at Stanford (*http://www.cs.stanford.edu/group/stair*) working with Professor Andrew Ng. Gary has more than 50 publications and 13 issued patents with 18 pending. He lives in Palo Alto, CA, with his wife and three daughters and bikes road or mountain as much as he can.

Dr. Adrian Kaehler is a senior scientist at Applied Minds Corporation. His current research includes topics in machine learning, statistical modeling, computer vision, and robotics. Adrian received his Ph.D. in Theoretical Physics from Columbia University in 1998. He has since held positions at Intel Corporation and the Stanford University AI Lab and was a member of the winning Stanley race team in the DARPA Grand Challenge. He has a variety of published papers and patents in physics, electrical engineering, computer science, and robotics.

Colophon

The image on the cover of *Learning OpenCV* is a giant, or great, peacock moth (*Saturnia pyri*). Native to Europe, the moth's range includes southern France and Italy, the Iberian Peninsula, and parts of Siberia and northern Africa. It inhabits open landscapes with scattered trees and shrubs and can often be found in parklands, orchards, and vineyards, where it rests under shade trees during the day.

The largest of the European moths, giant peacock moths have a wingspan of up to six inches; their size and nocturnal nature can lead some observers to mistake them for bats. Their wings are gray and grayish-brown with accents of white and yellow. In the center of each wing, giant peacock moths have a large eyespot, a distinctive pattern most commonly associated with the birds they are named for.

The cover image is from Cassell's *Natural History*, Volume 5. The cover font is Adobe ITC Garamond. The text font is Linotype Birka; the heading font is Adobe Myriad Condensed; and the code font is LucasFont's TheSansMonoCondensed.